W9-BGR-050

INTELLIGENT DESIGN AND MANUFACTURING

INTELLIGENT DESIGN AND MANUFACTURING

Edited by

Andrew Kusiak
Intelligent Systems Laboratory,
Department of Industrial Engineering,
The University of Iowa

Copyright © 1992 by John Wiley & Sons, Inc.

Library of Congress Cataloging in Publication Data:
Kusiak, Andrew.
    Intelligent design and manufacturing / Andrew Kusiak.
       p.   cm.
    "A Wiley-interscience publication."
    Includes bibliographical references.
    1. Production planning--Automation. 2. CAD/CAM systems.
  3. Flexible manufacturing systems. I. Title.
    TS176.K8688 1992
    670.42--dc20   91-14404
    ISBN 0-471-53473-0   CIP

Printed in the United States of America

10 9 8 7 6 5 4 3 2 1

Printed and bound by Courier Companies, Inc.

In Memory of My Mother

CONTRIBUTORS

Jaekyoung Ahn
Intelligent Systems Laboratory,
Department of Industrial Engineering,
The University of Iowa,
Iowa City, Iowa

Bopaya Bidanda
Department of Industrial Engineering,
University of Pittsburgh,
Pittsburgh, Pennsylvania

John Cesarone
IIT Research Institute,
10 West 35th Street,
Chicago, Illinois

Wen-Chyuan Chiang
Department of Management,
College of Business Administration,
The University of Texas at Austin,
Austin, Texas

Suranjan De
Department of Decision and
   Information Systems,
Leavey School of Business
   Administration,
Santa Clara University,
Santa Clara, California

Gavin A. Finn
Stone & Webster Engineering
   Corporation,
245 Summer Street,
Boston, Massachusetts

Hironobu Gotoda
Department of Information Science,
Faculty of Science,
The University of Tokyo,
Tokyo, Japan

John R. Goulding
Department of Mechanical
   Engineering,
School of Engineering and Applied
   Science,
Portland State University,
Portland, Oregon

Genaro J. Gutierrez
Department of Management,
College of Business Administration,
The University of Texas at Austin,
Austin, Texas

Sanjay B. Joshi
Department of Industrial and
   Management Systems Engineering,
Pennsylvania State University,
University Park, Pennsylvania

Srikanth M. Kannapan
Xerox Design Research Institute,
Cornell University,
Ithaca, New York

Gerald M. Knapp
Department of Industrial Engineering,
The Unversity of Iowa,
Iowa City, Iowa

Panagiotis Kouvelis
Department of Management,
College of Business Administration,
The University of Texas at Austin,
Austin, Texas

Tosiyasu L. Kunii
Department of Information Science,
Faculty of Science,
The University of Tokyo
Tokyo, Japan

Andrew Kusiak
Intelligent Systems Laboratory,
Department of Industrial Engineering,
The University of Iowa,
Iowa City, Iowa

Shanling Li
Management Department,
College of Business Administration,
The University of Texas at Austin,
Austin, Texas

Debasish N. Mallick
Department of Management,
College of Business Administration,
The University of Texas at Austin,
Austin, Texas

Alex Markovsky
AT&T Bell Laboratories,
Crawfords Corner Road,
Holmdel, New Jersey

Kurt M. Marshek
Department of Mechanical
  Engineering,
The University of Texas at Austin,
Austin, Texas

C. K. Muralikrishnan
Department of Industrial Engineering,
University of Pittsburgh,
1042 Bendum Hall,
Pittsburgh, Pennsylvania

H. Muthsam
Fraunhofer Institute for
  Manufacturing Engineering and
  Automation (IPA),
12 Nobel Street,
Germany

S. Narasimhan
School of Computer Science,
Carnegie Mellon University,
Pittsburgh, Pennsylvania

D. Navin Chandra
School of Computer Science,
Carnegie Mellon University,
Pittsburgh, Pennsylvania

U. Negretto
Institute for Real-Time Computer
  Systems and Robotics,
University of Karlsruhe,
12 Kaiser Street,
Germany

Ioannis O. Pandelidis
Manufacturing Systems Research,
The Gillette Company,
Boston, Massachusetts

Kevin Paré
Tandy Instruments,
Fort Worth, Texas

Juan R. Pimentel
Universidad Poletecnica de Madrid,
DISAM,
Madrid, Spain

Michael Pinedo
Department of Industrial Engineering
  and Operations Research,
Columbia University,
New York, New York

John W. Priest
Department of Industrial Engineering,
The University of Texas at Arlington,
Arlington, Texas

Kenneth F. Reinschmidt
Stone & Webster Engineering
  Corporation,
245 Summer Street,
Boston, Massachusetts

Jose M. Sanchez
Instituto Technologica y de Estudios
  Superiors
de Monterey, Mexico

Jerry L. Sanders
Department of Industrial Engineering,
University of Wisconsin–Madison,
Madison, Wisconsin

Jami J. Shah
Department of Mechanical &
  Aerospace Engineering,
Arizona State University,
Tempe, Arizona

Yoshihisa Shinagawa
Department of Information Science,
Faculty of Science,
The University of Tokyo,
Tokyo, Japan

Subhash C. Singhal
AT&T Bell Laboratories,
Crawfords Corner Road,
Holmdel, New Jersey

Jeffrey S. Smith
Department of Industrial and
   Management Systems Engineering,
Pennsylvania State University,
207 Hammond Building,
University Park, Pennsylvania

K. P. Sycara
School of Computer Science,
Carnegie Mellon University,
Pittsburgh, Pennsylvania

Kwei Tang
Department of Quantitative Business
   Analysis,
Louisiana State University,
Baton Rouge, Louisiana

Horst Tempelmeier
Fachgebiet Productionswirtschaft,
Abteilung Betriebswirtschaftslehre,
Technical University of Braunschweig,
14 Pockel Street,
Germany

Ulrich A. W. Tetzlaff
Department of Decision Sciences and
   MIS,
School of Business Administration,
George Mason University,
Fairfax, Virginia

Devanath Tirupati
Management Department,
College of Business Administration,
The University of Texas at Austin,
Austin, Texas

Hsu-Pin (Ben) Wang
Department of Industrial Engineering,
The University of Iowa
Iowa City, Iowa

H. J. Warnecke
Fraunhofer Institute for
   Manufacturing Engineering and
   Automation (IPA),
12 Nobel Street,
Germany

Hormoz Zarefar
Department of Mechanical
   Engineering,
School of Engineering and Applied
   Science,
Portland State University,
Portland, Oregon

Maciej Zgorzelski
Department of Mechanical
   Engineering,
GMI Engineering & Management
   Institute,
Flint, Michigan

H.-J. Zimmermann
Institute of Operations Research,
RWTH Aachen,
Templergraben 64,
Germany

■ CONTENTS

The basic idea of concurrent design is to shorten the time horizon in which the design is performed. Concurrent engineering is a systematic approach to the integrated, simultaneous design of products and related processes, including manufacture and support. This definition emphasizes consideration of all elements of the product life cycle from concept through disposal, including quality, cost, schedule, and user requirements. The general features of concurrent engineering include a top-down design approach based on systems engineering principles, multifunction design teams, optimization of product and process characteristics, and execution in the engineering environment. Furthermore, the objective of concurrent engineering is that of simultaneously considering the life-cycle impacts during preliminary system design along with the paramount consideration of functionality.

In concurrent engineering, the product design is viewed as a strategic task that has a major effect on the subsequent production-related activities. Concurrent engineering's implementation takes a variety of forms and uses different methods and techniques; however, four key features have been identified as essential elements. These elements are a top-down systems engineering approach to provide a framework for explicit decision support and decomposition; use of multidiscipline teams to carry out integrated product and process design; application of quality engineering methods to achieve efficient and effective product and process optimization; and an integrated engineering environment to provide information for rapid and intelligent decision-making throughout the entire design process.

This book is intended to bridge the gap between design and manufacturing, two areas of great importance in concurrent engineering. Although the primary emphasis is on design and manufacturing, many issues important to successful applications of concurrent engineering are discussed. The problems arising in design and manufacturing are as diverse as the techniques and tools used for solving them. Some problems can be solved with simple heuristics, others may require more sophisticated optimization approaches, while others need to be tackled with artificial intelligence tools. The solution approaches discussed in each chapter represent the most appropriate tools as of today. As researchers and practitioners get new experiences, more appropriate tools will be developed and a better match between problems and solution approaches will be accomplished.

The material included in this book has been divided into six parts. The chapters included in Part I introduce the reader to the contemporary topics in design of products and components. The material grouped in Part II covers topics related to design of manufacturing systems. Part III emphasizes operational aspects of manufacturing systems. The chapters included in Part IV are concerned with quality engineering and management. Standardization issues in design and manufacturing are discussed in Part V. Some of the basic artificial intelligence tools and techniques used in design and manufacturing are presented in Part VI.

ANDREW KUSIAK

Iowa City, Iowa
December 1991

INTELLIGENT DESIGN AND MANUFACTURING

PART I

INTELLIGENT DESIGN AND
MANUFACTURING

DESIGN OF COMPONENTS AND PRODUCTS

■■■■■■ **CHAPTER ONE**

Engineering Design Methodologies: A New Perspective

SRIKANTH M. KANNAPAN*

Xerox Corporation, Design Research Institute, Theory Center, Cornell University

KURT M. MARSHEK

Department of Mechanical Engineering, The University of Texas at Austin

1.1. INTRODUCTION

The importance of design is widely recognized in delivering quality engineering products to highly competitive markets. As a result, substantial research effort in the area of engineering design has been directed at significantly improving the productivity of the design process.

Since design is an activity common to many fields of engineering, alternative models and methodologies of the design process have emerged that embody the popular paradigms and characteristics of their respective fields. This chapter is the result of a study of design methodologies across several fields of engineering (mechanical, software, electrical/electronics, and civil) with the objective of developing a common ground for exchange of ideas on approaches and tasks in design. Our emphasis is in categorizing the basic views of design (as a natural or artificial process) and categorizing the approaches developed by each view. Independent of the natural or artificial view, three basic tasks of design are identified and this framework is used to point to examples in the literature where the approaches have been applied to each task. Although the survey of the literature is extensive, no attempt is made to be uniform or exhaustive in referencing past and current research.

The organization of the chapter is as follows: Section 1.1 serves as an introduction; Section 1.2 distinguishes design and design methodology;

*This work was completed while the author was at the University of Texas at Austin.

Intelligent Design and Manufacturing, Edited by Andrew Kusiak.
ISBN 0-471-53473-0 © 1992 John Wiley & Sons, Inc.

Section 1.3 identifies the natural and artificial process views of design; Section 1.4 outlines the modes of design research and previews the approaches developed by the natural process view (Section 1.5) and the artificial process view (Section 1.6); Sections 1.7 and 1.8 deal with the basic task types of design selection, parametric redesign and design synthesis, and their interaction; Section 1.9 summarizes the status of research in design methodology and develops future goals and evaluation methods in engineering design research; Section 1.10 presents the conclusion of this study of engineering design methodologies.

1.2. DESIGN VERSUS DESIGN METHODOLOGY

First, it is important to distinguish between "design" and "design methodology"—a distinction that has sometimes been overlooked. *Design* by itself is primarily concerned with the question of "*what* to design" to satisfy some specified need. For example, the design of a new car is undertaken to satisfy some specified needs of the market, cost/performance ratio, aesthetics, and so on. A model of the context of design is given in Fig. 1.1

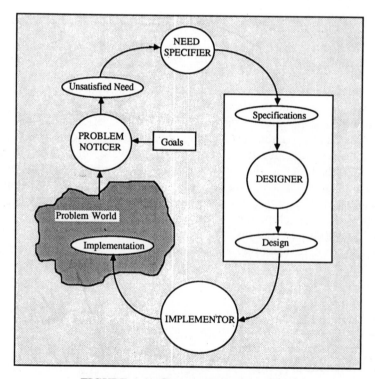

FIGURE 1.1. Context of design activity.

(possible iterations are not shown). In Fig. 1.1, design is viewed as an activity for converting specificational descriptions to design descriptions ready for implementation and, in general, includes within it the concerns of manufacturing (materials, processing, and equipment) and also concerns felt further down the life cycle of the product (e.g., maintenance and disposal). *Design methodology* is primarily concerned with the question of "how to design" so that in answering this question we may better model, teach, and aid/automate the finding of solutions to '*what* to design." Design methodology is thus a vehicle for the evolution of design activity from an art or skill to a science. Insofar as design activity (as in industrial product design) is the natural testing ground for design methodology, we propose that the term *engineering design research* or *research in engineering design* be reserved for research in design methodology where the object is not a product but the knowledge of how to design products. Thus even creative and innovative design of industrial products is not engineering design research unless the design activity itself is also under investigation.

1.3. NATURAL OR ARTIFICIAL SCIENCE

A fundamental issue that must be faced in engineering design research is whether design activity can be described by a natural or an artificial science. As we shall see, resolution of this issue leads to clarity in the modes of research appropriate for engineering design methodology.

There are arguments for the description of design activity by a natural as well as an artificial science. One may argue that design activity is essentially a natural process not only for the human species since the earliest of times but also for animals and insects in that all these natural organisms skillfully design artifacts and refashion their environment to satisfy some needs. On the other hand, one may view design activity as an artificial process that transforms symbolic specifications of need, and knowledge representations of design experience, to symbolic descriptions of products/processes. The distinguishing characteristic between the natural process view and the artificial process view is the following: the natural process view implies that design methodology is based on creativity and heuristic knowledge and hence in the realm of natural human behavior; the artificial process view implies that design methodology is formal and mathematical and hence in the realm of symbolic data representation/manipulation (as in a calculus or programming language). Of course, the two views are intertwined in that both arise from human intelligence and may be combined to mutual benefit.

It is evident that processes that are appropriate for computer execution (artificial processes) are not necessarily appropriate for natural human intelligence (natural process) and vice versa. Consider, for example, the case of stress analysis of mechanical and structural elements. Classical analytical methods (beam theory, shell theory, etc.) developed for solution of stress analysis problems that are traditionally found suitable for natural human intelligence are quite different in approach and methodology to finite element methods of stress analysis that have been found so successful for computer implementation. This characteristic also parallels the studies of intelligence as cognitive science (natural intelligence) and artificial intelligence. We should be convinced that, as in analysis fields and the study of intelligence, both views of design activity must be accommodated within engineering design research.

The value of accommodating both views of design activity within engineering design research are the following:

1. Investigation of design activity as a natural process will improve descriptive as well as prescriptive models of human design processes.
2. Investigation of design activity as an artificial process will lead to mathematical and computational tools for aiding/automating design.
3. Developments in 1 and 2 will lead to effective information and control flows between designers and computer tools.
4. While 1 leads to heuristics-based tools for design automation (expert systems), 2 leads to computational tools (calculi) for human designers.

Having defined the ambit of design methodology, in the following sections we describe a new perspective on approaches to design methodology and task types in design activity before returning to the examination of further goals and evaluation methods in engineering design research.

1.4. MODES OF ENGINEERING DESIGN RESEARCH

The modes of engineering design research fundamentally differ when design activity is viewed as a natural or artificial process.

Viewed as a natural process, the mode of research is *data abstraction*: the collection and measurement of behavioral data from (human) organisms executing design processes and the consequent abstraction of design theories from the data. Table 1.1 summarizes the three basic approaches to design methodology categorized by the data collection and measurement technique when design activity is viewed as a natural process.

Viewed as an artificial process, the mode of research is *model testing*: the testing of symbolic representation/manipulation models based on hypothesized design theories, in real world situations. A review of design methodologies across several fields of engineering reveals seven basic approaches to design methodology (see Table 1.2) categorized by the characteristics of the hypothesized artificial process model of design activity.

TABLE 1.1   Approaches to Design Methodology Viewed as a Natural Process

Approach	Technique for Data Collection and Measurement
1. Anecdotal	- Informal accounts of personal or group experiences of designers designing products/processes. - Measurement of behavioral data is highly qualitative for the most part.
2. Pedagogic	- Collective experiences of design teams over histories of design projects dividing design processes into time phases (needs analysis, preliminary design, detail design, etc.) and characterized by an iterative process over analysis, synthesis and evaluation. Also informally observes design teams in action and prescribes techniques (synectics, brainstorming, etc.) for managing human creativity. - Measurement of behavioral data is qualitative and quantitative.
3. Protocol Analysis	- Behavioral studies of designers in action in specific designs situations. Aspects oi behavior monitored include design data references, verbalization, writing and sketching times, etc. Collection of data can be scientific. - Measurement of behavioral data can be systematic and quantitative.

TABLE 1.2   Approaches to Design Methodology Viewed as an Artificial Process

Approach	Model characterization of Design Activity
1. System Science	black box theory, state theory, component integration theory, decision theory.
2. Problem Solving/ Planning	recursive problem decomposition, state-space search, constraint propagation.
3. Transformational	design description (program) transformation and language translation.
4. Database	centralized design library and independent design processes.
5. Algorithmic	finite deterministic processes including mathematical optimization techniques.
6. Axiomatic	axiomatization of general intuitively powerful design rules, proof of theorems.
7. Machine Learning	knowledge acquisition from instruction, examples, observation and discovery; execution of plans (e.g. by using analogies).

1.5. DESIGN METHODOLOGIES: NATURAL PROCESS

The approaches developed by the natural process view of design activity, namely, the anecdotal approach, the pedagogical approach, and the protocol studies approach, are described below.

1.5.1. Anecdotal Approach

The anecdotal approach relies on informal accounts of personal or group experiences of designers in designing products/processes in order to abstract general principles of "good" design. For example, Glegg (44) illustrates through examples described in conversational style how inventive ideas develop in embodied designs. French (42) beautifully draws the parallels between natural evolutionary design and engineering design through a host of examples. Love's (89) work is an example of the design management point of view where the combination of needs analysis, idea generation, and group management techniques required to develop successful designs is described. Anecdotal approaches usually concentrate on the subject of creative design and the measurement of behavioral data is highly qualitative for the most part.

1.5.2. Pedagogical Approach

The pedagogical approach refers to the way the design process itself is taught in engineering disciplines based on pedagogical models of the design process and the collective experience of engineers. The basic view of all pedagogical approaches is that design is an iterative process over the activities of analysis, synthesis, and evaluation, alternatively called divergence, transformation, and convergence, respectively. Each iteration is expected to make the design 'progressively less general and more detailed" (59, p. 64). Design activity is also split along the time scale of a project, such as feasibility study, preliminary design, detail design, planning, and execution. The stages are usually subdivided in to smaller steps that designers usually perform by experience, depending on the field of design (e.g., refs. 5, 9, and 56). Pahl and Beitz (113), building on the ideas of "systematic design" developed by Hansen, Rodenacker, Roth, Koller, and others in the German school, split the design process into the phases of clarification of the task, conceptual design, embodiment design, and detail design. They emphasize the development of structures of functional elements of a domain, selection of appropriate physical processes, and variation and combination of design solutions to each functional element.

Developments over the years have led to several techniques that can aid a design team: for example, brainstorming, synectics, morphological charts, boundary searching, and value analysis (see ref. 59 for details). These techniques combine the creative aspects of human designers with systematic methods of search, bookkeeping, and evaluation.

1.5.3. Protocol Analysis Approach

The protocol analysis approach focuses on the behavior of individual designers when presented by a specific design task that is carefully chosen to elicit the use of a design skill. Data on the behavior exhibited by the designer ("think aloud" verbalizations, sketching and drawing, book referencing, hand motions etc.), called a protocol (112), are recorded by voice and video tape. Measures of behavior are both quantitative (e.g., number of drawings produced, time taken to complete design) and qualitative (e.g., designer's motivation, design evaluation). The data are later analyzed to develop an information processing model of the subject designer. Akin (4) applied the protocol analysis approach to architectural design and categorized designer knowledge into design plans, transformation rules, and design symbols and identified forms of design plans used by designers. Stauffer et al. (129) describe the protocol analysis of mechanical designers presented with the tasks of (a) designing a positioning device ("flipper-dipper") and (b) designing the contacts and enclosure for batteries. They found that designers demonstrated behavior quite in contrast to those prescribed by the pedagogical approach (see Section 5.2). Waldron and others describe application of the protocol analysis approach to study the visual recall differences between expert and naive mechanical designers (142). A "depositional method" (where verbalization is stimulated when a decision is made) is used by Waldron and Waldron (143) to investigate how mechanical designers generate and use rigid and flexible constraints in the conceptual design of a robot manipulator elbow. Adelson (1) investigates analogical learning in software design using the protocol analysis approach and refines models of the mapping, evaluation, and debugging processes taking place in analogical learning.

1.6. DESIGN METHODOLOGIES: ARTIFICIAL PROCESS

The approaches developed by the artificial process view of design activity, namely, the system science approach, the problem solving/planning approach, the transformational approach, the database approach, the algorithmic approach, the axiomatic approach, and the machine learning approach, are described below.

1.6.1. System Science Approach

The system science approach distinguishes four theories, the first three of which essentially differ by the depth to which they model a system (see ref. 100):

1. *Black Box Theory*. This models a system only by the mapping of inputs to outputs with respect to the environment.

2. *State Theory.* This models a system by a vector of characteristic attributes representing the internal state of the system. The mapping of inputs to outputs of the system is correlated to the state of the system.

3. *Component Integration Theory.* This models a system by recognizing components whose input–output mappings are known. The input–output mapping of the system is derived from the component input–output mappings and the structure of the system.

4. *Decision Theory.* Here, all activities of modeling, design, and analysis are approached as decision-making activities. Each decision is made in a systematic way, taking into account interdependent actions: generate alternative actions, rank/order and choose actions after evaluating possible consequences according to specified criteria.

Techniques for structural modeling and design (refs. 84 and 111 provide recent reviews) based on these theories have been especially successful in large-scale engineering, economic, and social systems. The systems approach emphasizes the need for a close study of the problem environment before specifying design requirements. System science has also contributed substantially in developing theories for hierarchic structures and systems with mathematical bases (e.g., see ref. 101 on hierarchic systems and ref. 50 on structural models).

1.6.2. Problem Solving/Planning Approach

The problem solving/planning approach seeks to reduce a problem to subproblems recursively until subproblems solutions are directly known. Subproblem interdependencies are formulated as constraints to be satisfied. A plan of solution is expressed as a network, with solvable subproblems (executable subtasks) as nodes and subproblem interdependencies as arcs. Except for early discussions by Simon (125), problem solving and planning approaches to design are comparatively recent; for example, see refs. 15, 77, 97, and 98.

The search process involved in determining solution plans was emphasized in early research in problem solving and planning. The search is formulated in terms of problem states: Given an initial state, the attributes of a goal state, and a set of state change operators, problem solving involves determination of a sequence of operators that transform the initial state to a goal state. The path of search may be controlled and constrained in several ways [e.g., breadth first, depth first, heuristic ordered search. dependency directed backtracking (135)]. Depending on the problem, the primary objective is in finding goal states, or in finding plans, that is, sequences of operators leading to goal states. Here state change operators are treated generally; operators may have specific preconditions for applicability, in which case they can take the form of "if situation then action" (production) rules. Rule-based systems of this type are normally classified as production systems.

An important issue in search is that of combinatorial explosion; typically, search spaces are so large that blind searches and even a search directed by "shallow knowledge" can lead to impractical solution times. Current thinking to avoid this situation is to direct search by using "deep" structured knowledge and heuristics specific to the problem domain. However, two general mechanisms have resulted from research:

1. Means–ends analysis and its variations, where the "best" operator to be currently applied is determined by evaluating the difference it will make in the direction of the goal, the preconditions for applicability of this operator being recursively made current goals (originally in refs. 34, 39, and 125).
2. Propagation of constraints, where constraints applied to components of a system are propagated through the structure to solve for component attributes. In design analysis, this mechanism was originally proposed for network systems such as electrical circuits (135, 136). Stefik (130, 131) generalized the operators of constraint formulation, constraint propagation, and constraint satisfaction to hierarchical task networks. These methods are also used in qualitative reasoning (see ref. 13).

1.6.3. Transformational Approach

Here, the design process is viewed as a sequence of correctness preserving transformations (107). The initial design description is a specification of design requirements, which is transformed finally to a design description sufficient for implementation/manufacturing. Each transformation acts on a complete design description so as to change the level of description to a more concrete level (e.g., behavioral to structural) or to improve the design in some way but at the same level of description. The abstract-refinement theory (99, 148) is subsumed by this approach. In abstract refinement, a design description may be either exclusively refined (e.g., described in greater detail) or decomposed into parts that may be acted on independently. The couplings between decomposed parts are handled as constraints to be satisfied.

Transformations can be viewed as operations on programs that represent the design in some (formal) description language. Viewed in this manner, language translation [compiler theory (3)] and program transformation techniques may readily be applied as mechanisms for design description transformation. Translation involves recognition and replacement of program fragments of one language with those of another language that expresses greater design detail (so-called productions). Complex internal representations (parse trees, symbol tables, etc.) and mechanisms (recursive descent, etc.) are required for this purpose. Program transformations recognize and modify program fragments of a language with the goal of improving its efficiency while maintaining its correctness.

Program transformations require (a) transformation rules that express the equivalence of language constructs (program fragments and program schemas) and (b) a recognize-and-replace mechanism that operates on programs using the transformation rules. The language constructs are in terms of clauses of predicate logic, in terms of recursion equations in functional languages, and in terms of textual fragments in block structured languages. At a local level these constructs can be recognized by matching patterns directly in the program text, but global transformations are better achieved by operating on parse trees, dataflow graphs, and plan diagrams (119) that explicitly represent the roles and interactions of various parts of the program.

In general, program transformational methods involve search similar to problem solving/planning methods to determine the "best" sequence of transformation rule applications. Optimizing compilers normally substitute search with local program analysis. However, program transformation methods and heuristic search techniques offer such power and flexibility that concise and clear specifications may be transformed to computationally efficient implementations for several classes of problems. Wide spectrum languages based on logic or recursive functions can be used for expressing both the specifications and the implementations.

With specifications in predicate logic, transformations correspond to logical inferences from a database of facts and rules that encode knowledge about the problem domain, the implementation language (a subset of the specification language), and programming techniques. Program design using logic then corresponds to finding a constructive proof of the specifications. With specifications in functional language, the primary transformation mechanism is substitution; functional forms or recurrence equations are substituted with more efficient equivalents. Prominent work on program transformation has been conducted by Darlington (24), Burstall and Darlington (18), Manna and Waldinger (92), and Balzer et al. (6, 7). Partsch and Steinbruggen (114) provide a comprehensive review of existing methods and systems. Goldberg (45) surveys the use of logic and functional languages in program translation.

From the point of view of design, both translation and program transformations are required to act on design descriptions because, by the transformational approach, designs must traverse several levels of description whether these levels are represented implicitly or explicitly. This issue is touched on by Goldberg (45) in his discussion of the "granularity" of transformations.

1.6.4. Database Approach

The database approach is closely related to the idea of central storage of designs. Design activity is viewed as the execution of (usually independent) processes acting on a database that stores all aspects of the design from conception to manufacture. Two critical properties of the database are integrity and consistency. Integrity implies that functional relations between design data are maintained. Consistency implies that equivalence of redundant data is maintained. The design processes can act on the database at different abstraction levels, sequentially or in parallel, and may access and modify the design data. But the net effect of the processes, as design activity progresses, is to describe the design in more concrete terms, preserving integrity and consistency. The database approach, apart from providing consistent and integrity preserving schema, can also provide multiple views of design information to match individual design processes. This approach is described by Eastman (32) and Grabowski and Seiler (47).

A more complex data-oriented view results from the use of frame theory. Frames are data objects that provide an abstract template of "slots" that store stereotypical design information. This information includes data, defaults, and procedures for determining data. Designs can be represented as a hierarchical library of frames, manipulated by a frame manager, which handles the inheritance of information down the hierarchy, performs type checks on data, and controls data-triggered processes (demons). Frame theory, originated by Minsky (102), is well suited for design situations characterized by small "chunks" of highly specific design information, for example, as in detail design (90). Alternative representations such as networks and trees make data interrelationships more explicit. Fenves and Norabhoompipat (37) describe such a scheme for designing to standards (of structural engineering).

The database approach thus addresses the problem of design representation deemphasizing the rose of the mechanisms of design tasks. This approach is important when design representations are at less abstract structural levels and of even greater importance when libraries of previous designs are to be stored, accessed, and modified.

1.6.5. Algorithmic Approach

The algorithmic approach views design as a finite deterministic process. In particular, this approach expects that a finite sequence of known computational steps can convert a properly expressed specification or an initial design to a completed design. Cases where the entire design process is algorithmic are rare. However, parts of most design processes are algorithmic, especially where the emphasis is on numerical analysis and optimization. An important subclass of design problems reduce to sets of algebraic, differential, and integral equations that are solved by standard methods of symbolic algebra, calculus (differential/integral), and numerical procedures. Optimization techniques apply where design problems can be formulated in the standard mathematical form of objective functions and constraint equations. Many general techniques for optimization have been developed that are appropriate for problems with different mathematical characteristics.

The primary classes and their associated solution techniques are as follows (43, 52, 54, 88, 118):

1. *Linear Programming.* The objective function and the equality and inequality constraints are all linear; the Simplex method (and its extensions) is the classic method for finding optimum solutions. Other classes of problems such as goal programming and multicriteria programming deal with multiple competing objectives by ranking and weighting the objectives and determining a noninferior set of choices (pareto-optimality).

2. *Nonlinear Programming.* The objective function and constraint equalities and inequalities may be nonlinear; numerous techniques have been developed for solving nonlinear optimization problems falling mainly into the categories of linearization methods and penalty function methods. Quadratic programming deals with the special case where the objective function is quadratic and all the constraints are linear. Geometric programming deals with the class of problems where the design variables are restricted to be positive and the objective function and constraints are generalized polynomials.

3. *Integer Programming.* The variables are limited to integer domains while the objective function and constraints are typically linear; the primary techniques for solution are the branch-and-bound, implicit enumeration, and "cutting-plane" methods.

4. *Dynamic Programming.* The objective comprises multiple stages of decision-making with values (states) for variables being decided at each stage; "backward" and "forward" search techniques have been developed to determine the best sequence of decisions. The decision at each stage is reduced to one of the previous optimization problem types.

Algorithmic techniques in design that do not fit into the framework of equation solving or mathematical optimization are usually highly domain dependent and are not discussed here.

1.6.6. Axiomatic Approach

The idea here is to axiomatize a small set of general and intuitively powerful concepts and on this basis derive some theorems and corollaries that characterize the design process. Suh (134) proposes two axioms that may be used to guide search and evaluation in design. The axioms state that functional independence in the specifications must be maintained and information content in design objects must be minimized. Some design experiments are described by Suh where use of the axioms has led to reduced design search spaces. Yoshikawa (151) describes another axiomatic approach based on definitions of entity, attribute, and function. Three axioms are stated that serve to characterize the design process as an evolution of metamodels that add to design detail. Metamodels are models of models of design objects, which relate the concept of function to a topology of attributes (see refs. 151 and 152 for details). Four models of the design process are developed:

1. Catalogue model, when there is a one-to-one correspondence between specifications and stored solutions.
2. Calculation model, when the specifications can be partitioned so that a calculated combination of solutions can satisfy the specifications.
3. Production model, when design rules can be found that produce a solution in terms of an attribute, given a specification in terms of a function.
4. Paradigm model, when an initial solution that partially meets the specifications can be improved upon by identifying and changing attributes of the solution.

The applicability of these models are currently being investigated by Yoshikawa and others (73, 80, 152).

1.6.7. Machine Learning Approach

Machine learning is concerned with knowledge acquisition and skill refinement activities and is related to the problem solving approach. Current research has focused on knowledge acquisition that involves more complex symbolic structures and inference than skill refinement. It is now recognized that learning and designing are synergistic; designing provides experiences and goals for learning while learning guides designing (see refs. 25 and 107).

Learning strategies can be classified in ascending order of amount of inference as follows:

1. Rote learning: No inference; knowledge is directly implanted in the system (e.g., conventional programming).
2. Learning from instruction: System accepts instructions in a language and stores it in an internal, usable form (e.g., compilation).
3. Learning by analogy: Existing problem and solution knowledge is transformed and augmented to solve new problems similar to solved problems.
4. Learning from examples: Examples and counterexamples of a concept are used to induce a general, operational concept description.
5. Learning from observation and discovery: Unsupervised learning; the system observes or experiments with the problem and environment, focuses attention using criteria of "interestingness," and inductively learns concepts.

Here we consider 3, 4, and 5 of the above list. Models of analogical reasoning have been developed by Kling (74), Winston (147), and Carbonell (19, 20). Kling's work uses a theorem proving paradigm for expediting the search involved in solving a problem by identifying "useful" knowledge from the solution proof of a given analogical problem. Winston develops a theory for analogical learning and reasoning based on a semantic net representation with extensible relations. A situation is represented by situation parts as nodes and agent–action relations as arcs (extensible relations). Matching of situations represented in this way is shown to be the primary mechanism for analogical reasoning, with causal constraint relations playing an important role in determining similarities of importance between situations. Further details can be in ref. 147. Carbonell (20) integrates analogical reasoning with problem solving to formulate a search problem in plan–transformation space. Several operators are proposed for transformation such as deletion, splicing, and substitution. Search determines the sequence of such operators that transform the solution plan of the analogical problem to the solution plan of the current problem.

There are two models for learning from examples, inductive and deductive, with current research favoring the latter. Inductive methods search for similarities by directly examining examples for features and using a generalizer that is implicitly biased toward certain features that are considered important in the problem domain. Inductive methods are therefore data intensive. Examples of the use of such methods are STRIPS (40), MORRIS (103), LEX (104). Deductive methods rely on domain knowledge and models, and a declarative description of the concept under study to infer a proper operational description of the concept. These methods can generalize from a single training example by explaining how the training example is an example of the concept. Mitchell et al. (106) describe a Learning Apprentice (LEAP) design system based on such an approach.

The primary work in learning from observation and discovery in Lenat's (86, 87) EURISKO program, where a strong case is made for inductive inference from experimentation, guided by general heuristics. Use of EURISKO in invention and research (rather than design) of three-dimensional microelectronic devices is reported by Lenat et al. (85).

1.7. TASK TYPES IN DESIGN ACTIVITY

Whether approaches to design methodology are based on a natural process view or an artificial process view, the focus of an investigation must be selected beforehand. In order to provide a uniform framework for investigations of different kinds of design activity, we propose a categorization of three task types: *design selection*, *parametric redesign*, and *design synthesis*. This categorization agrees with Simon's interesting discussion on the logic of design (125, p. 140), where he distinguishes three cases:

1. "Where we are seeking a satisfactory alternative, once we have found a candidate we can ask: 'Does this alternative satisfy all the design criteria?'" This corresponds to the task of design selection.

2. "Of all possible worlds (those attainable for some admissible values of the action variables), which is the best (yields the highest value of the criterion function)?" Simon calls this optimization while we refer to a more general task of parametric redesign.

3. "Where the design alternatives are not given in any constructive sense but must be synthesized." This corresponds to the task of design synthesis.

It is important to note that a given design task need not fit neatly into one of these types; design tasks usually involve all three types in complex ways. The terms "field" (e.g., VLSI design), "approach" (e.g., transformational), and "task" (e.g., design selection) are disjoint concepts. Neither the sequence nor the type of object to which a task is applied is constrained by this view.

1.7.1. Design Selection

The design selection task simply involves selecting a design object from a set of alternatives. The term design object is used in a general sense; the task could be, for example, the selection of a material type for a component or the selection of even a partial or completed system design (see Fig. 1.2). The principal characteristic of the design selection task is that design objects that are complete at some level of description are already available as alternatives. Selection involves decision-making using specified criteria and knowledge of the attributes of the alternative designs and the design requirements. The basic executive mechanism needed for this task is that of attribute matching. When the attributes of a design match the required attributes optimally, according to the criteria specified, the design is select-

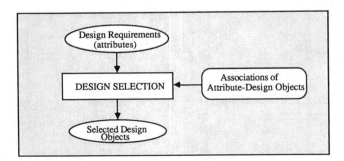

FIGURE 1.2. The task of design selection.

ed. Two approaches address this type of task directly: (a) the decision theoretic approach and (b) the problem solving approach.

In general, a decision problem involves the choice of an optimum plan or policy from a set of alternatives, possibly in the face of uncertain events. The basic theoretical model used in a decision problem is the decision tree, which "delineates all the alternatives and possible consequences" as described by Meredith et al. (100). Applied to design selection, decision models provide a systematic method for reducing specified criteria to a ranking of attributes and, consequently, a ranking of alternatives. Kuppuraju et al. (79) describe such a decision support system.

In the problem solving/planning approach a design selection task is posed as the problem of choosing a state-change operator or (sub)plan; that is (a part of) a known sequence of operators. Two strategies have emerged. By the first strategy, operators or (sub)plans are chosen by first testing whether preconditions for applicability are satisfied and then applying means–ends analysis. This corresponds to a "selection by simulation" strategy used by Fikes and Nilsson (39), Fikes et al. (40) in STRIPS, and other recent programs. The second strategy is to implement explicit choice rules that "RULE-IN" and "RULE-OUT" design plans, operators, and objects on the basis of their attributes. Using this strategy, McDermott (96, 97) describes the use of "choice protocols" for selection of electronic circuits and components. Stefik (130) describes the use of a similar strategy for eliminating alternatives in Molgen, a program that plans experiments in molecular genetics. McDermott (98) describes rule-based choice of computer system components by recognition of attributes in the R1 program. Kant (63) similarly uses "plausible implementation" rules in the LIBRA automatic programming system.

The well researched area of expert diagnostic systems addresses a similar form of problem as design selection, if the correspondence between symptom–disease and attribute–design is recognized. Qualitative representations of machine behavior together with abstraction operators can also be used as a means to classify designs for selection based on their behavior properties (61). Sandgren (123) develops a design tree representation for decisions taken during selection of mechanism types to satisfy multiple goals. Kannapan and Marshek (70) describe a framework for a library of mechanical transmission systems based on their function, behavior, and structure.

1.7.2. Parametric Redesign

This task essentially involves the determination of component attributes of a partial design: A prespecified design configuration is to be parametrically varied to achieve design requirements. The abstract formulation of a parametric redesign task is fairly algorithmic:

1. A structured symbolic model (e.g., graph, tree) of the design configuration is obtained where parameters of the design are clearly identified.
2. Structure laws (e.g., generalized Kirchhoff's laws) and the theories applicable to the entities (components) represented by the symbols are used to derive a mathematical model (e.g., set of differential equations).
3. The mathematical model is solved, perhaps iteratively, to determine values for the quantitative and qualitative parameters of the design which produce the behavior expressed by the design requirements.

This formulation, illustrated in Fig. 1.3, corresponds in general to the "design by analysis" paradigm familiar to all engineering disciplines (e.g., see refs. 62 and 94 for mechanical component design analysis methods). If the mathematical model can be cast into standard optimization problem form, it may be solved algorithmically to determine values of parameters (as in refs. 2, 43, 71, and 123). Qualitative reasoning or monotonicity analysis can also be combined with optimization techniques (2, 58). The algorithmic approach can, however, produce models of high complexity, avoidable in some cases by adopting the "propagation of constraints" strategy of the problem solving/planning approach. This allows the mathematical model to be built and solved incrementally. This strategy has been used effectively in parametric redesign, for example, in electrical circuit design by Sussman (136, 137) and in mechanical design by Brown (16). Similarly, the use of constraints for refinement of "indexed partial plans" for electrical circuit design is described by McDermott (96). Serrano and Gossard (124) develop a method for constraint representation and propagation while combining mathematical and geometric models of a parametric design.

Another interesting redesign strategy results from combining constraint propagation with causal reasoning as in the Steinberg–Mitchell (132) RE-DESIGN program for VLSI CAD. Causal reasoning is used to determine

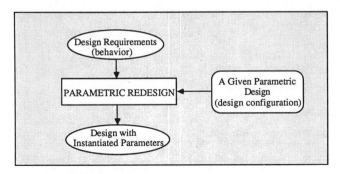

FIGURE 1.3. The task of parametric redesign.

input–output data relationships of circuit modules. In following the problem solving approach, explicit rules for redesign may also be used as shown by Dixon and Simmons (27) and Dixon et al. (29). Kannapan and Marshek (64) describe a graph-based representation and algorithm for managing parametric relationships. Wood and Antonsson (149, 150) apply an algorithmic approach to represent and propagate imprecision (uncertainty in choice of values) in parameters based on fuzzy set theory and a numerical method (fuzzy weighted average algorithm). Ward (144) applies an algorithmic approach to develop a labeled interval propagation calculus for catalog selection and parametric redesign.

Another powerful strategy for parametric variation is "generalization–specialization," which originates from applications of machine learning to problem solving/planning [originally in robot planning by Fikes et al. (40)]. This strategy can be described as follows: An instantiated model of a design/plan is used to create a generalization, that is, a parametric model; then the parametric model is specialized to a model with a new instantiation of parameters to solve a new problem. Both the generalization and specialization steps involve rule-directed search and theorem proving in some logic to determine proper parameters and instantiations (see refs. 104 and 105). Stefik (130, 131) infused ideas of planning-with-constraints into hierarchic specialization in the Molgen program to obtain a "least commitment" style of design. Three operators on constraints are distinguished: (a) constraint formulation, to guide hierarchic specialization of plans and design objects; (b) constraint propagation, to make subtask interaction explicit; and (c) constraint satisfaction, to determine parameter values.

The parametric redesign problem can be addressed by the machine learning approach at a higher level by storing an idealized development history, that is, the sequence of operators/transformations that resulted in a successful design starting from specifications. Parameters for variation are those aspects of the specification that do not affect the acceptability of a stored operator/transformation sequence. Parametric redesign with new instantiations of these parameters then simply corresponds to a replay of the development history (e.g., as in refs. 106, 108, 132, and 145).

Network structured models are appropriate for redesign in a large number of domains: for example, mechanical linkages, hydraulics and pneumatics, bond graphs of physical systems, electrical and electronic circuits, piping systems and chemical plants. Quantitative (122) and qualitative (78) simulations are useful evaluation tools for network models in guiding the process of parametric variation in design. During simulation the mathematical model is used to replicate the abstract operation of the designed system under specified environmental inputs (60, 81, 117). Outputs of simulation are normally the kinematic and dynamic behaviors of the system.

General architectures and techniques for parametric redesign are described by Sussman (135), Dixon et al. (28), Bowen and O'Grady (14), and

Kannapan and Marshek (67). Tools for the parametric redesign task have also been built using the database approach (32, 37, 47, 90) and the axiomatic approach (73, 80).

1.7.3. Design Synthesis

The task of design synthesis is to generate potentially successful design configurations; distinguished from parametric redesign wherein a pre-specified design configuration is to be parametrically varied, and from design selection wherein a design, complete at some level of description, is to be chosen. Synthesis, as it is commonly understood, involves putting "parts" together to form a "whole." The term "design synthesis" has been used to denote a range of concepts from redesign and detailing to innovation and discovery. Here, design synthesis refers precisely to the task of putting together primitives (components/events) of a domain to construct a system structure (design configuration) that satisfies some goals and constraints placed on the system. This definition corresponds correctly to the connotation of "design" and "synthesis" and presumes the knowledge of components and previous designs of a class. Figure 1.4 illustrates this task definition.

Models and tools for design synthesis that are generally applicable are yet to be developed. However, certain characteristics of the domains of Software Engineering (SE) and VLSI design have enabled the transformational approach to make a significant impact on synthesis tasks in these domains. Furthermore, researchers have sought to integrate the problem solving/planning approach within the transformational approach, attempting to make the goals–tasks–constraints structure more explicit (ref. 107 provides an excellent enumeration of these aspects of the design process). Recently, significant progress has also been made in the fields of civil engineering and mechanical engineering. The developments in these fields are briefly described next.

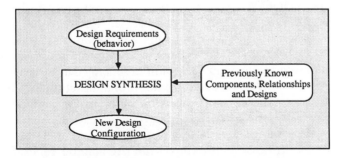

FIGURE 1.4. The task of design synthesis.

Software Engineering. In software engineering, the motivation for building synthesis tools is to enable programmers to build software systems from specifications that are concise and clear in their intent but that do not consider computational efficiency. The methodology is to write the specification programs preferably in a declarative language and use program transformations to interactively or automatically produce an implementation program that has a computationally efficient procedural interpretation.

The declarative languages preferred for specification are based on firm mathematical formalisms such as sets, functions, and relations, first-order logic (11, 126), and recursive functional equations (18, 24), although other mathematical notations have been used (92, 93). Clark and Darlington (22), Hogger (53), Clark and Sickel (21), Murray (109), and Ericksson (33) demonstrate methods for synthesis of programs from specifications using predicate logic. Software program development using stepwise refinement by logical transformations is developed by Wirth (148), Dijkstra (26), and Gries (48); Dromey (30) extends the approach toward constructive methods for program design. The need for procedural interpretation of the implemented program implies that either the implementation language (a) is procedural in itself (6, 8, 145) or (b) is a subset of a declarative language that can be interpreted procedurally by some abstract machine: functional languages as in refs. 18, 92, and 126 or a subset of predicate logic (e.g., Prolog—Horn clauses) as in ref. 11. Rich (119) describes an intermediate language for transforming LISP programs, called the plan calculus, based on dataflow and control-flow graph representations.

VLSI Design. The motivation for building synthesis tools in VLSI design is in providing electronic circuit and system designers with tools that hide the extreme design complexity of low-level structural design descriptions. This is achieved by accepting specification at an algorithmic behavior level that will nevertheless affect the implementation style and performance. (VLSI tools that require structural descriptions as specifications are not considered here.) The methodology is to write the specification program in a procedural language and use transformational and/or problem solving/planning approaches to produce a structural description that is realizable with the implementation technology (e.g., CMOS, TTL). The levels of design representation are, in order, behavior, register transfer, logic circuit, and geometric layout.

The specification program is parsed into an internal representation that makes the data, operations, and control interdependencies explicit. Typical internal representations are predicate logic, dataflow graph, finite state machine, and value trace (138). Program transformation methods are also applicable to these representations for improvement of efficiency in computation or for purposes of standardization. Formal methods of electrical/VLSI design and verification using logic are exemplified by Eveking (36), Gordon (46), and Hanna and Daeche (49). Two alternative methods are applicable beyond this stage. The first is a compilation method described as being subsumed by the transformational approach—silicon compilers, Mac-Pitts (127), Silc (12), and the EMUCS tool of the CMU-DA system (138). The second is a production system method corresponding to a problem solving/planning approach—the Design Automation Assistant (DAA) tool of the CMU-DA system (77).

Civil Engineering. Fenves (37) combines a problem solving/planning and database approach to outline a framework for civil and structural engineering design. Fenves et al. (38) and Maher and Fenves (91) use the framework to develop an expert system for preliminary structural design, including synthesis of the structural configuration, of high rise buildings. The expert system takes as input a three-dimensional grid for the structure. Rules that encode decisions on generic subsystems are searched depth first to generate alternative designs, which are later analyzed and evaluated. Powell and others (115) also investigate the data modeling requirements of structural engineering design.

Mathematical optimization methods (algorithmic approach) have been applied to a special form of the design synthesis task where discrete structural skeletons (typically trusses) are optimized for specified loads. The optimization process progressively removes members of the structure from a fully connected grid and relocates member joints to minimize structure weight (72, 128, 139). A novel boundary variational (algorithmic) approach is developed by Bendsoe and Kikuchi (10) for structural design, where the design synthesis problem is posed as the determination of the optimum distribution of holes in a structural material.

Mechanical Engineering. Early work in the combinatorial structural synthesis of mechanisms and machines is by Buchsbaum and Freudenstein (17) and Freudenstein and Maki (41). More recently, Dyer et al. (31) describe a design system called Edison; this system develops a machine learning approach to design synthesis. Based on an episodic memory-based model of device "functionality," Edison uses heuristics coded as rules to attempt the design of novel devices by memory generalization, analogy, and mutation (see also ref. 85). Murthy and Addanki (110) describe a design system called PROMPT, which introduces an intermediate level of reasoning for satisfying design constraints by using "modification operators." Modification operators are heuristics codes as rules that analyze and change the parametric model of a prototype (component) when the available prototype cannot be just parametrically varied to satisfy requirements of component behavior. Ulrich and Seering (140, 141) describe design systems that search for combinations of structural parts that satisfy a conjunction of "functional" attributes. A "design by debugging" strategy (135) is suggested by Ulrich (141) to attempt satisfaction of the specifications by generating bond graphs. Lai and Wilson (83) describe a specification language for description and improvement of synthesized designs. Mayer and Lu (95) describe an expert system for mechanical design.

Kannapan and Marshek (65, 66, 68, 69) develop a methodology for mechanical design called design synthetic reasoning based on an algebraic and predicate logic approach. Design synthetic reasoning comprises a design library and support for processes of design selection and synthesis, verification, derivation, and justification (66). Transformation of a specification of required behavior to a structure of systems and relationships is carried out using equivalence and implication rules in a predicate logic-based language. Steinberg et al. (133) use a problem solving/planning approach to encode design implementation heuristics, search control, and causal constraint propagation knowledge for mechanical transmission design. Hoover and Rinderle (55) describe a graph-based transformation system for design synthesis. Rinderle and Finger (121) describe a bond graph-based grammar and transformation methodology for dynamic system synthesis. Prabhu and Taylor (116) develop algebraic vector orientation and position transformation operators to synthesize a structure for the design of a mechanical device from specifications of behavior.

Kota (75) develops a problem decomposition method for hydraulic systems synthesis. Kota and Lee (76) also develop a matrix-based transformation method for design synthesis of machines. Cutkosky and Tenenbaum (23) demonstrate a methodology for synthesizing features on a mechanical part using "manufacturing mode" operators and rules. Kusiak and Szczerbicki (82) develop a system science approach to synthesis of structures from specifications of behavior. Hundal (57) describes a systematic approach to design synthesis of a mechanical system based on a database or catalog of available functional primitives. Williams (146) develops qualitative algebraic reasoning methods for synthesizing mechanical/hydraulic devices to achieve a required behavior. An algorithmic approach to synthesizing a dynamic system is proposed by Hauck and Taylor (51). A case-based reasoning and learning method (machine learning approach) is suggested for mechanism design synthesis by Esterline et al. (35).

1.7.4. Special Forms of Design Synthesis

Research efforts have also been directed at certain special forms of the task of design synthesis using a problem solving/planning approach. We distinguish three special forms of design synthesis:

1. Ordered subtasks extending partial design configurations. In this case, subtasks of design synthesis are known and can be temporally ordered, and also, components used to build the structure are known or can be determined from constraints. This situation is exemplified by the R1/XCON program of McDermott (98). In R1, the specifications are a list of components of a computer system and the task is to configure the computer system. The task can be accomplished by performing six subtasks in

sequence. Each subtask is executed by production rules that encode the constraints applicable in each subtask and thereby extending partial configurations of components without backtracking (except for the cabling subtask).

2. Designs as operator sequences. Here, the design to be synthesized is a linear sequence of operators, in which case it may be viewed as a planning problem, and so means–ends analysis and constraint propagation methods may be used. This class of problems is exemplified by Stefik's Molgen program (130, 131). Means–ends analysis, constraint formulation, constraint propagation, and constraint satisfaction are the primary mechanisms used within a hierarchical planning (and metaplanning) framework. This is a plan synthesis problem common to other areas as well, for example, in robot planning.

3. Designing from indexed partial solutions. Here, design requirements are matched with the specifications of indexed partial solutions. Indexed partial solutions are symbolic representations of design configurations, which may be parametrically varied. But when none of the partial solutions can directly satisfy the design requirements, some elementary mechanisms can be provided to compose partial solutions or decompose design requirements. McDermott's DESI/NASL problem solver (96, 97) is of this type. All redesign, compositions, and decompositions are uniformly treated as subtasks in a task network as described earlier under design approaches. Allowed compositions are defined by "RULE-TOGETHER" rules that are similar in form to the "RULE-IN" and "RULE-OUT" rules used for design selection.

1.8. DESIGN PROCESS TASK STRUCTURE

The tasks of design selection, parametric redesign, and design synthesis interact within the design process. Figure 1.5 shows the task structure

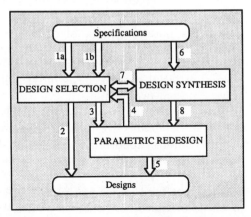

FIGURE 1.5. Task structure in the design process.

considering only the process aspects of design; knowledge requirements are not considered. Recalling the generality in the definition of the design selection task, several scenarios of design can be envisioned (arcs referred to below are shown in Fig. 1.5).

- A complete design (implementation description) is selected (arcs 1a, 2).
- A design configuration is selected and parametrically redesigned (arcs 1b, 3, 5); on conditions of failure in redesign, other design configurations are selected (arc 4).
- A design configuration is synthesized and parametrically redesigned (arcs 6, 8, 5) using design selection to choose from alternative components and configurations (arc 7).

1.9. RESEARCH STATUS AND FUTURE GOALS

The categorization of basic approaches to design methodology and tasks in design activity allows a concise assessment of the status of engineering design research. Table 1.3 summarizes the major areas of application of approaches (Table 1.2) to the design tasks, when viewing design activity as an artificial process.

The review of the literature indicates a lack of development in general approaches and tools for the task of design synthesis (reflected in increased research interest in "conceptual design"). Also, while general approaches have been developed for parametric redesign, software tools do not provide integrated support of the several approaches applicable (Table 1.3). Furthermore, while there is some clarity in identification and categorization of basic task types, the interactions between the task types in real-world design situations is largely unknown.

From the natural process viewpoint, in principle, each of the approaches listed in Table 1.1 is applicable to each of the design tasks. However, the real-world design activities investigated so far using a natural process view have not usually been chosen to fall precisely into a task type as described here (viz., design selection, parametric redesign, or design synthesis). Behavioral studies of design situations involving precisely one task type (and then of interacting task types) can provide a common frame of reference for comparison of natural and artificial process views of design in terms of the techniques used to solve a design problem.

Thus the research goals for engineering design research from both the natural and artificial process viewpoints are the development of models for the task types and task interactions supported by behavioral data abstraction (natural process) or model testing (artificial process) in real-world design situations.

The criteria of measurement for the achievement of engineering design research goals are in terms of the basic concerns of design methodology:

TABLE 1.3   Approaches Applied to Design Methodology Using Artificial Process View

Task		Approaches Applied
A.	Design Selection	(1) Systems Approach - decision theory
		(2) Problem Solving/Planning Approach - production systems
B.	Parametric Redesign	(1) Algorithmic Approach - mathematical optimization - iterative analysis-simulation
		(2) Problem Solving/Planning Approach - constraint propagation - recursive problem decomposition
		(3) Database Approach - designing to standards - frame systems
		(4) Machine Learning Approach - generalization/ specialization - replay of development history
		(5) Axiomatic Approach - "paradigm" model of CAD
C.	Design Synthesis	(1) Transformational Approach - logical expressions and inference rules - algebraic expressions, rules and grammars
		(2) Problem Solving/Planning Approach - refinement and constraint propagation - "design by debugging"
		(3) Algorithmic Approach - optimization by component/region elimination
		(4) Axiomatic Approach - axioms for function independence and information content - axioms for evolution of design metamodels
		(5) Machine Learning Approach - discovery heuristics

1. The modeling of natural design processes using descriptive design theories, the quality of the model being measured by its ability to predict and explain behavioral data in similar design situations.
2. The development of prescriptive design theories that lead to more productive human design activity, productivity being measured with respect to the design process cost versus design product value.
3. The development of computer software tools for computer-aided design activity, measured by the need for human interaction with computer tools and the productivity (as before) of design activity.

The research methodologies for the achievement of the above goals have already been outlined in Section 1.4 and depend on whether the natural process or artificial process view is held.

From the natural process view, the research methodology is:

1. Definition of the real-world design situations (initially of well established design situations with known solutions schemes).
2. Collection of behavioral data from designers in the defined design situation.
3. Abstraction of design theory from behavioral data.
4. Prediction of designers' behavior in new design situations and development of prescriptive design theories (for natural and the artificial use).

From the artificial process view, the research methodology is:

1. Hypothesis of a design theory for task types and/or task interactions.
2. Development of symbolic representation/manipulation models using the design theory.
3. Implementation of (computer software) tools based on the symbolic models.
4. Evaluation of degree of automation and productivity of (computer software) tools and development of prescriptive design theories (for natural and artificial use).

Of course, both research methodologies may be appropriately interfaced as an investigative procedure to develop models of the design process.

1.10. CONCLUSIONS

A new domain-independent perspective of engineering design presented in this chapter views design either as a natural or an artificial process and categorizes the design methodologies used in several fields of engineering (mechanical, software, electrical/electronics, and civil). Three approaches for the study of design are identified when design is viewed as a natural process. Viewed as an artificial process, seven basic approaches to design are distinguished. Independent of the natural or artificial process view, three basic tasks of design are identified and the framework of tasks and approaches is used to point to examples in the literature where the approaches have been applied to each task. The nature of interactions between basic task types in design situations is also described.

An important aspect of engineering design research emphasized in this chapter is the need for both the natural process view and the artificial process view of design activity to achieve the goals of modeling, teaching, and aiding/automating the process of solving design problems. The assessment of the status of engineering design research reveals that (a) generally applicable approaches, models, and tools for the task of design synthesis are at a preliminary stage of development; (b) the tasks of design selection and parametric redesign have several generally effective approaches but integrated tools for parametric redesign are yet to be developed; (c) behavioral studies of designers are leading to more detailed cognitive models of the design process but comparison with symbolic computer models require a commonality in the task type addressed; and (d) models of the interaction between basic design task types (design selection, parametric redesign, design synthesis) in real-world design situations are virtually nonexistent and require investigation.

ACKNOWLEDGMENTS

This work has been supported in part by the Texas Advanced Research Program (1989), the National Science Foundation, grant no. DMC8810503, and by a grant from the Challenge for Excellence endowment of the University of Texas at Austin. Any opinions, findings, conclusions, or recommendations expressed in this publication are those of the authors and do not necessarily reflect the views of the sponsors.

REFERENCES

1. Adelson, B. (1989). Cognitive Modeling: Explaining and Predicting How Designers Design, *NSF Engineering Design Research Conference*, University of Massachusetts, Amherst, June 11–14.
2. Agogino, A. M., and Almgren, A. (1987). Symbolic Computation in Computer Aided Optimal Design. In: *Expert Systems in Computer Aided Optimal Design*, edited by J. S. Gero, North-Holland, Amsterdam, pp. 267–284.
3. Aho, A. V., and Ullman, J.D. (1977). *Principles of Compiler Design*, Addison-Wesley, Reading, MA.
4. Akin, O. (1978). How Do Architects Design? In: *Artificial Intelligence and Pattern Recognition in Computer Aided Design*, edited by J.-C. Latombe, IFIP North-Holland, Amsterdam, pp. 65–103.
5. Asimow, M. (1962). *Introduction to Design*, Prentice-Hall, Englewood Cliffs, NJ.
6. Balzer, R., Goldman, N., and Wile, D. (1976). On the Transformational Implementation Approach to Programming, *Proceedings of the 2nd International Conference on Software Engineering*, pp. 337–343.
7. Balzer, R. (1981). Transformational Implementation: An Example, *IEEE Transactions on Software Engineering*, Vol. SE-7, No. 1, pp. 3–14.
8. Bauer, F.L., and Wossner, H. (1982). *Algorithmic Language and Program Development*, Springer-Verlag, New York.

9. Beitz, W. (1987). Systematic Approaches to Design of Technical Systems and Products, VDI 2221. In: *VDI Society for Product Development Design and Marketing*, Beuth Verlag, Berlin.

10. Bendsoe, M. P., and Kikuchi, N. (1988). Generating Optimal Topologies in Structural Design Using a Homogenization Method, *Computer Methods in Applied Mechanics and Engineering*, Vol. 71, pp. 197–224.

11. Bibel, W. (1980). Syntax Directed Semantics Supported Program Synthesis, *Artificial Intelligence*, Vol. 14, No. 3, October, pp. 243–261.

12. Blackman, T., Fox, J., and Rosebrugh, C. (1985). The Silc Silicon Compiler: Language and Features, *IEEE 22nd Design Automation Conference*, Paper 17.1, pp. 232–237.

13. Bobrow, D. (1985). Qualitative Reasoning About Physical Systems: An Introduction. In: *Qualitative Reasoning About Physical Systems*, edited by D. Bobrow, MIT Press, Cambridge, pp. 1–6.

14. Bowen, J., and O'Grady, P. (1990). A Technology for Building Life-Cycle Advisers, *Proceedings of Computers in Engineering 1990*, Boston, MA, August 5–9, Vol. 1, pp. 1–7.

15. Brown, D. C., and Chandrasekharan, B. (1983). An Approach to Expert Systems for Mechanical Design, *IEEE Computer Society Trends and Applications '83*, Gaithersburg, MD, May, pp. 173–180.

16. Brown, D. (1985). Capturing Mechanical Design Knowledge, *Proceedings of the 1985 ASME Computers in Engineering Conference*, Boston, MA, July, pp. 121–129.

17. Buchsbaum, F., and Freudenstein, F. (1970). Synthesis of Kinematic Structure of Geared Kinematic Chains and Other Mechanisms, *Journal of Mechanisms*, Vol. 5, pp. 357–392.

18. Burstall, R. M., and Darlington, J. (1977). A Transformation System for Developing Recursive Programs, *Journal of the ACM*, Vol. 24, No. 1, pp. 44–67.

19. Carbonell, J. G. (1981). A Computational Model of Analogical Problem Solving, *IJCAI*, Vol. 1, pp. 147–152.

20. Carbonell, J. G. (1983). Learning by Analogy: Formulating and Generalizing Plans from Past Experience. In: *Machine Learning: An Artificial Intelligence Approach*, edited by R. S. Michalski, J. G. Carbonell, and T. M. Mitchell, Tioga Press, Palo Alto, CA, pp. 137–160.

21. Clark, K., and Sickel, S. (1977). Predicate Logic: A Calculus for Deriving Programs, *Proceedings Fifth IJCAI*, Cambridge, MA, August 22–25, Vol. 1, pp. 419–420.

22. Clark, K., and Darlington, J. (1980). Algorithm Classification Through Synthesis, *Computer Journal*, Vol. 23, No. 1, pp. 61–65.

23. Cutkosky, M. R., and Tenenbaum, J. M. (1990). A Methodology and Computational Framework for Concurrent Product and Process Design, *Mechanism and Machine Theory*, Vol. 25, No. 3, pp. 365–381.

24. Darlington, J. (1981). An Experimental Program Transformation and Synthesis System, *Artificial Intelligence*, Vol. 16, pp. 1–46.

25. de Kleer, J., and Brown, J. S. (1981). Mental Models of Physical Mechanisms

and Their Acquisition. In: *Cognitive Skills and Their Acquisition*, edited by J. R. Anderson, Erlbaum, Hillsdale, NJ, pp. 285–309.

26. Dijkstra, E. W. (1976). *A Discipline of Programming*, Prentice-Hall, Englewood Cliffs, NJ.

27. Dixon, J. R., and Simmons, M. K. (1984). Expert Systems for Mechanical Design: V-Belt Drive Design as an Example of the Redesign Architecture, *Proceedings of the ASME CIME Conference*, Las Vegas, NV, August.

28. Dixon, J. R., Simmons, M. K., and Cohen, P. R. (1984). An Architecture for Applying Artificial Intelligence to Design, *Proceedings of the 21st ACM/IEEE Design Automation Conference*, Albuquerque, NM, June 25–27, pp. 634–640.

29. Dixon, J. R., Howe, A., Cohen, P. R., and Simmons, M. K. (1987). Dominic I: Progress Toward Domain Independence in Design by Iterative Redesign, *Engineering with Computers*, Vol. 2, pp. 137–145.

30. Dromey, R. G. (1988). Systematic Program Development, *IEEE Transactions on Software Engineering*, Vol. 14, No. 1, pp. 12–29.

31. Dyer, M. G., Flowers, M., and Hodges, J. (1986). Edison: An Engineering Design Invention System Operating Naively, *Proceedings of the 1st International Conference on Applications of Artificial Intelligence to Engineering Problems*, Southampton, UK, April, Vol. 1, pp. 327–341.

32. Eastman, C. M. (1978). The Representation of Design Problems and Maintenance of Their Structure. In: *Artificial Intelligence and Pattern Recognition in Computer Aided Design*, edited by J.-C. Latombe, IFIP/North-Holland, Amsterdam, pp. 335–371.

33. Eriksson, L.-H. (1984). Synthesis of a Unification Algorithm in a Logic Programming Calculus, *Journal of Logic Programming*, Vol. 1984, No. 1, pp. 3–18.

34. Ernst, G., and Newell, A. (1969). *GPS: A Case in Generality and Problem Solving*, Academic Press, Orlando.

35. Esterline, A., Bose, A., Shanmugavelu, I., Titus, J., Riley, D., and Erdman, A. (1991). Investigations into Early Stages of Mechanism Design. In: *Proceedings 1991 NSF Design and Manufacturing Systems Conference*, Society of Manufacturing Engineers, Dearborn, MI, 1009–1018.

36. Eveking, H. (1985). The Application of CHDL's to the Abstract Specification of Hardware. In: *Computer Hardware Description Languages and Their Applications*, edited by C.J. Koomen and T. Moto-oka, IFIP/North-Holland, Amsterdam, pp. 167–178.

37. Fenves, S., and Norabhoompipat, T. (1978). Potentials for Artificial Intelligence in Structural Engineering Design and Detailing. In: *Artificial Intelligence and Pattern Recognition in Computer Aided Design*, edited by J.-C. Latombe, IFIP/North-Holland, Amsterdam, pp. 105–122.

38. Fenves, S. J., Flemming, U., Hendrickson, C., Maher M. L., and Schmitt, G. (1988). An Integrated Software Environment for Building Design and Construction, *Computing in Civil Engineering: Microcomputers to Supercomputers, Proceedings of the 5th Conference*, Alexandria, VA, March 29–31, pp. 21–32.

39. Fikes, R. E., and Nilsson, N. J. (1971). STRIPS: A New Approach to the Applications of Theorem Proving to Problem Solving, *Artificial Intelligence*, Vol. 2, pp. 189–208.

40. Fikes, R. E., Hart, P. E., and Nilsson, N. J. (1972). Learning and Executing Generalized Robot Plans, *Artificial Intelligence*, Vol. 3, pp. 251–288.

41. Freudenstein, F., and Maki, E. R. (1979). The Creation of Mechanisms According to Kinematic Structure and Function, *Environment and Planning B*, Vol. 6, p. 375–391; also General Motors Research Lab. Publication GMR-3073.

42. French, M. J. (1988). *Invention and Evolution: Design in Nature and Engineering*, Cambridge University Press, Cambridge, UK.

43. Gero, J. S. (Editor) (1985). *Design Optimization*, Academic Press, Orlando.

44. Glegg, G. L. (1981). *The Development of Design*, Cambridge University Press, Cambridge, UK.

45. Goldberg, A. T. (1986). Knowledge-Based Programming: A Survey of Program Design and Construction Techniques, *IEEE Systems, Man and Cybernetics*, Vol. SE-12, No. 7, July, pp. 752–768.

46. Gordon, M. (1986). Why Higher-Order Logic Is a Good Formalism for Specifying and Verifying Hardware. In: *Formal Aspects of VLSI Design*, edited by G. J. Milne and P. A. Subrahmanyam, Elsevier Science Publishers, Amsterdam, pp. 153–177.

47. Grabowski, H., and Seiler, W. (1985). Techniques, Operations and Models for Functional and Preliminary Design Phases. In: *Design and Synthesis*, edited by H. Yoshikawa, Elsevier Science Publishers, Amsterdam, pp. 13–16.

48. Gries, D. (1981). *The Science of Programming*, Springer-Verlag, New York.

49. Hanna, F. K., and Daeche, N. (1986). Specification and Verification Using Higher Order Logic: A Case Study. In: *Formal Aspects of VLSI Design*, edited by G. J. Milne and P. A. Subrahmanyam, Elsevier Science Publishers, Amsterdam, pp. 179–213.

50. Harary, F., Norman, R., and Cartwright, D. (1965). *Structural Models: An Introduction to the Theory of Directed Graphs*, Wiley, New York.

51. Hauck, P. D., and Taylor, D. L. (1991). Design as a Series of Semantic Mappings. In: *Proceedings 1991 NSF Design and Manufacturing Systems Conference*, Society of Manufacturing Engineers, Dearborn, MI, pp. 1031–1040.

52. Himmelblau, D. (1972). *Applied Nonlinear Programming*, McGraw-Hill, New York.

53. Hogger, C. J. (1981). Derivation of Logic Programs, *Journal of the ACM*, Vol. 28, No. 2, pp. 372–392.

54. Holzman, A. G. (Editor) (1981). Mathematical Programming for Operations Researchers and Computer Scientists, Marcel Dekker, New York.

55. Hoover, S. P., and Rinderle, J. R. (1989). A Synthesis Strategy for Mechanical Devices, *Research in Engineering Design*, Vol. 1, No. 2, pp. 87–103.

56. Hubka, V. (1982). *Principles of Engineering Design* (translated and edited by E. E. Eder), Butterworths, London.

57. Hundal, M. S. (1990). A Systematic Method for Developing Function Structures, Solutions and Concept Variants, *Mechanism and Machine Theory*, Vol. 25, No. 3, pp. 243–256.

58. Jain, P., and Agogino, A. M. (1990). Theory of Design: An Optimization Perspective, *Mechanism and Machine Theory*, Vol. 25, No. 3, pp. 287–303.

59. Jones, J. C. (1980). *Design Methods: Seeds of Human Futures*, Wiley, New York.

60. Joskowicz, L. (1987). Shape and Function in Mechanical Devices, *Proceedings of the AAAI-87*, Seattle, WA, pp. 611–615.

61. Joskowicz, L. (1990). Mechanism Comparison and Classification for Design, *Research in Engineering Design*, Vol. 1, pp. 149–166.

62. Juvinall R. C., and Marshek, K. M. (1991). *Fundamentals of Machine Component Design*, Wiley, New York.

63. Kant, E. (1983). On the Efficient Synthesis of Efficient Programs, *Artificial Intelligence*, Vol. 20, pp. 253–305.

64. Kannapan, S., and Marshek, K. M. (1987). A Parametric Approach to Machine and Machine Element Design, *Mechanical Systems and Design Technical Report 198*, University of Texas at Austin, August. Also in *Journal of Mechanical Engineering Education*, Vol. 19, No. 3, 1991, pp. 197–211.

65. Kannapan, S. (1989). *Design Synthetic Reasoning: A Theoretical Basis and Methodology for Mechanical Design*, Ph.D. Dissertation, University of Texas at Austin, December.

66. Kannapan, S., and Marshek, K. M. (1990). An Algebraic and Predicate Logic Approach to Representation and Reasoning in Machine Design, *Mechanism and Machine Theory*, Vol. 25, No. 3, pp. 335–353.

67. Kannapan, S., and Marshek, K. M. (1993). An Approach to Parametric Machine Design and Negotiation in Concurrent Engineering. In: *Concurrent Engineering: Automation, Tools and Techniques*, edited by A. Kusiak, Wiley, New York.

68. Kannapan, S., and Marshek, K. M. (1992). Design Synthetic Reasoning, *Mechanism and Machine Theory* as Parts I, II, III papers, to be published.

69. Kannapan, S., and Marshek, K. M. (1991). Design Synthetic Reasoning: A Methodology for Mechanical Design, *Research in Engineering Design*, Vol. 2, No. 4, pp. 221–238.

70. Kannapan, S., Marshek, K. M., and Gerbert, G. (1991). A Framework for a Design Library for Mechanical Transmissions. In: *Proceedings of 1991 NSF Design and Manufacturing Systems Conference*, Society of Manufacturing Engineers, Dearborn, MI, pp. 1079–1088.

71. Karandikar, H., Srinivasan, R., Mistree, F., and Fuchs, W. J. (1989). Compromise: An Effective Approach for the Design of Pressure Vessels Using Composite Materials, *Computers and Structures*, Vol. 33, No. 6, pp. 1465–1477.

72. Kirsch, U. (1988). Applications of Mathematical Programming to the Design of Civil Engineering Structures. In: *Engineering Design: Better Results Through Operations Research Methods*, edited by R. R. Levary, Elsevier Science Publishing, Amsterdam, pp. 174–200.

73. Kitajima, K., and Yoshikawa, H. (1984). HIMADES-1: A Hierarchical Machine Design System Based on the Structure Model for a Machine, *CAD*, Vol. 16, No. 6, November, pp. 299–307.

74. Kling, R. E. (1971). A Paradigm for Reasoning by Analogy, *Artificial Intelligence*, Vol. 2, pp. 147–178.

75. Kota, S. (1990). Qualitative Motion Synthesis: Towards Automating Mechani-

cal Systems Configuration, *Proceedings of the NSF Design and Manufacturing Systems Conference*, Arizona State University, Tempe, January 8–12, pp. 77–91.

76. Kota, S., and Lee, C.-L. (1990). A Computational Model for Conceptual Design: Configuration of Hydraulic Systems, *Proceedings of NSF Design and Manufacturing Systems Conference*, Arizona State University, Tempe, January 8–12, pp. 93–104.

77. Kowalski, T. J., and Thomas, D. E. (1985). The VLSI Design Automation Assistant: What's in the Knowledge Base, *IEEE 22nd Design Automation Conference*, Paper 18-1, pp. 252–258.

78. Kuipers, B. (1986). Qualitative Simulation, *Artificial Intelligence*, Vol. 29, pp. 289–338.

79. Kuppuraju, N., Ittimakin, P., and Mistree, F. (1985). Design Through Selection: A Method that Works, *Design Studies*, Vol. 6, No. 2, April, pp. 91–105.

80. Kurumatani, K., and Yoshikawa, H. (1987). Representation of Design Knowledge Based on General Design Theory, *Proceedings of the International Conference on Engineering Design*, Boston, Vol. 2, pp. 723–730.

81. Kurumatani, K., Tomiyama, T., and Yoshikawa, H. (1990). Qualitative Representation of Machine Behaviors for Intelligent CAD Systems, *Mechanism and Machine Theory*, Vol. 25, No. 3, pp. 325–334.

82. Kusiak, A., and Szczerbicki, E. (1990). Conceptual Design System: A Modelling and Artificial Intelligence Approach, *Proceedings of the 2nd National Symposium on Concurrent Engineering*, Morgantown, WV, pp. 427–442.

83. Lai, K., and Wilson, W. R. D. (1987). FDL: A Language for Function Description and Rationalization in Mechanical Design, *Proceedings of Computers in Engineering Conference*, New York, pp. 87–94.

84. Lendaris, G. G. (1980). Structural Modeling: A Tutorial Guide, *IEEE Transactions on Systems, Man and Cybernetics*, Vol. SMC-10, No. 12, pp. 807–840.

85. Lenat, D. B., Sutherland, W. R., and Gibbons, J. (1982). Heuristic Search for New Microcircuit Structures: An Application of Artificial Intelligence, *The AI Magazine*, pp. 17–33, Summer.

86. Lenat, D. B. (1983). Theory Formation by Heuristic Search, *Artificial Intelligence*, Vol. 21, pp. 31–59.

87. Lenat, D. B. (1983). EURISKO: A Program that Learns New Heuristics and Domain Concepts, *Artificial Intelligence*, Vol. 21, pp. 61–98.

88. Levary, R. R. (Editor) (1988). *Engineering Design: Better Results Through Operations Research Methods*, Elsevier Science Publishing, Amsterdam.

89. Love, S. F. (1980). *Planning and Creating Successful Engineered Designs*, Van Nostrand Reinhold, New York.

90. Maeda, Y., Takeshige, A., Koguchi, T., Tomiyama, T., and Yoshikawa, H. (1985). Frame Operating System for CAD, *Design and Synthesis*, Elsevier Science Publishers, Amsterdam, pp. 13–16.

91. Maher, M., and Fenves, S. (1985). HI-RISE: An Expert System for the Preliminary Structural Design of High Rise Buildings. In: *Knowledge Engineering in Computer Aided Design*, edited by J. S. Gero, North-Holland, Amsterdam, pp. 125–135.

92. Manna, Z., and Waldinger, R. (1975). Knowledge and Reasoning in Program Synthesis, *Artificial Intelligence*, Vol. 6, No. 2, pp. 175–208.

93. Manna, Z., and Waldinger, R. (1980). A Deductive Approach to Program Synthesis, *ACM Transactions on Programming Languages and Systems*, Vol. 2, No. 1, pp. 90–121.

94. Marshek, K. (1987). *Design of Machine and Structural Parts*, Wiley, New York.

95. Mayer, A. K., and Lu, S. C.-Y. (1988). An AI-Based Approach for the Integration of Multiple Sources of Knowledge to Aid Engineering Design, *Journal of Mechanisms, Transmissions and Automation in Design*, Vol. 110, No. 3, pp. 316–323.

96. McDermott, D. (1978). Circuit Design as Problem Solving, *Proceedings IFIP Workshop on AI and Pattern Recognition in CAD*, North-Holland, Amsterdam.

97. McDermott, D. (1978). Planning and Acting, *Cognitive Science*, Vol. 2, pp. 71–109.

98. McDermott, J. (1982). R1: A Rule-Based Configurer of Computer Systems, *Artificial Intelligence*, Vol. 19, pp. 39–88.

99. Mead, C. and Conway, L. (1979). *An Introduction to VLSI Systems*, Addison-Wesley, London.

100. Meredith, D. D., Wong, K. W., Woodhead, R. W., and Wartman, R. H. (1985). *Design and Planning of Engineering Systems*, Prentice-Hall, Englewood Cliffs, NJ.

101. Mesarovic, M. D., Macko, D., and Takahara, Y. (1970). *Theory of Hierarchic Multilevel Systems*, Academic Press, Orlando.

102. Minsky, M. (1975). A Framework for Representing Knowledge. In: *The Psychology of Computer Vision*, edited by P. H. Winston, McGraw-Hill, New York.

103. Minton, S. (1985). Selectively Generalizing Plans for Problem-Solving, *IJCAI*, Vol. 1, pp. 596–599.

104. Mitchell, T. M., Utgoff, P. E., Nudel, B., and Banerji, R. (1981). Learning Problem-Solving Heuristics Through Practice, *IJCAI*, Vol. 1, pp. 127–134.

105. Mitchell, T. M. (1982). Generalization as Search, *Artificial Intelligence*, Vol. 18, pp. 203–226.

106. Mitchell, T. M., Mahadevan, S., and Steinberg, L. I. (1985). LEAP: A Learning Apprentice for VLSI Design, *IJCAI*, Los Angeles, CA, August.

107. Mostow, J. (1985). Towards Better Models of the Design Process, *The AI Magazine*, Spring, pp. 44–57.

108. Mostow, J., and Barley, M. (1987). Automated Reuse of Design Plans, *Proceedings of the International Conference on Engineering Design*, Boston, MA, August 17–20, pp. 632–647.

109. Murray, N. V. (1982). Completely Non-Clausal Theorem Proving, *Artificial Intelligence*, Vol. 18, pp. 67–85.

110. Murthy, S. S., and Addanki, S. (1987). PROMPT—An Innovative Design Tool, *Proceedings of AAAI-87*, pp. 637–642.

111. Nadler, G. (1985). Systems Methodology and Design, *IEEE Transactions on Systems, Man and Cybernetics*, Vol. SMC-15, No. 6, November/December, pp. 685–697.

112. Newell A., and Simon, H. A. (1972). *Human Problem Solving*, Prentice-Hall, Englewood Cliffs, NJ.

113. Pahl, G., and Beitz, W. (1988). *Engineering Design: A Systematic Approach*, edited by K. Wallace, The Design Council, London.

114. Partsch, H., and Steinbruggen, R. (1983). Program Transformation Systems, *Computing Surveys*, Vol. 15, No. 3, September, pp. 199–236.

115. Powell, G., Bhateja, R., Abdalla, G., An-Nashif, H., Martini, K., and Sause, R. (1988). A Database Concept for Computer Integrated Structural Engineering Design, *Computing in Civil Engineering: Microcomputers to Supercomputers, Proceedings of the 5th Conference*, Alexandria, VA, March 29–31, pp. 521–529.

116. Prabhu, D. R., and Taylor, D. L. (1989). Synthesis of Systems from Specifications Containing Orientations and Positions Associated with Flow Variables, *Advances in Design Automation, 1989 ASME 15th Design Automation Conference*, Montreal, Quebec, Canada, September 17–21, pp. 273–280.

117. Pu, P., and Badler, N. I. (1989). Design Knowledge Capture and Causal Simulation, *Intelligent CAD, I.* edited by H. Yoshikawa and D. Gossard, IFIP Elsevier Science Publishers, Amsterdam, pp. 201–212.

118. Reklaitis, G. V., Ravindran, A., and Ragsdell, K. M. (1983). *Engineering Optimization: Methods and Aplications*, Wiley, New York.

119. Rich, C. (1981). A Formal Representation of Plans in the Programmers Apprentice, *Proceedings of the 7th International Joint Conference on Artificial Intelligence*, Vancouver, Canada, August, pp. 1044–1052.

120. Rinderle, J. R. (1986). Function, Form, Fabrication Relations and Decomposition Strategies in Design, *Proceedings of the ASME Computers in Engineering Conference*, Chicago, IL, July.

121. Rinderle, J. R., and Finger, S. (1989). A Transformational Approach to Mechanical Design Synthesis, *Proceedings of the NSF Design and Manufacturing Systems Conference*, Arizona State University, Tempe, January 8–12, pp. 67–75.

122. Rosenberg, R. C., and Karnopp, D. C. (1983). *Introduction to Physical System Dynamics*, McGraw-Hill, New York.

123. Sandgren, E. (1990). A Multi-objective Design Tree Approach for the Optimization of Mechanisms, *Mechanism and Machine Theory*, Vol. 25, No. 3, pp. 257–272.

124. Serrano D., and Gossard, D. (1986). Combining Mathematical Models with Geometric Models in CAE Systems, ASME, *Proceedings of 1986 International Computers in Engineering Conference*, Chicago, IL, July.

125. Simon, H. A. (1969). *The Sciences of the Artificial*, MIT Press, Cambridge.

126. Smith, D., Kotik, G. B. and Westfold, S. J. (1985). Research on Knowledge Based Software Environments at Kestrel Institute, *IEEE Transactions on Software Engineering*, Vol. SE-11, November, pp. 1278–1295.

127. Southard, J. R. (1983). MacPitts: An Approach to Silicon Compilation, *IEEE Computer*, December, pp. 74–82.

128. Spillers, W. R. (1985). Shape Optimization of Structures. In: *Design Optimization*, edited by J. S. Gero, Academic Press, London, pp. 41–70.

129. Stauffer, L. A., Ullman, D. G., and Dietterich, T. G. (1987). Protocol

Analysis of Mechanical Engineering Design, *ICED 87*, Boston, MA, August, pp. 74–85.

130. Stefik, M. (1981). Planning with Constraints, (MOLGEN: Part 1), *Artificial Intelligence*, Vol. 16, pp. 111–140.

131. Stefik, M. (1981). Planning and Meta-Planning (MOLGEN: Part 2), *Artificial Intelligence*, Vol. 16, pp. 141–170.

132. Steinberg, L. I., and Mitchell, T. M. (1984). A Knowledge Based Approach to VLSI CAD: The REDESIGN System, *IEEE 21st Design Automation Conference*, June, Paper 26.2, pp. 412–418.

133. Steinberg, L., Langrana, N., and Fisher, G. (1989). MEET: Decomposition and Constraint Propagation in Mechanical Design, *NSF Engineering Design Research Conference*, University of Massachusetts, Amherst, June 11–14, pp. 363–375.

134. Suh, N. P. (1990). *The Principles of Design*, Oxford University Press, New York.

135. Sussman, G. J. (1977). Electrical Design: A Problem for Artificial Intelligence Research, *Proceeings 5th IJCAI*, MIT, Cambridge, August, Vol. 2, pp. 894–900.

136. Sussman, G. J. (1978). SLICES: At the Boundary Between Analysis and Synthesis, *Artificial Intelligence and Pattern Recognition in Computer Aided Design*, edited by J.-C. Latombe, IFIP/North-Holland, Amsterdam, pp. 261–298.

137. Sussman, G. J., and Steele, G. L. (1980). CONSTRAINTS—A Language for Expressing Almost-Hierarchical Descriptions, *Artificial Intelligence*, Vol. 14, pp. 1–39.

138. Thomas, D. E., Hitchcock, C. Y., Kowalski, T. J., Rajan, J. V., and Walker, R. A. (1983). Automatic Data Path Synthesis, *IEEE Computer*, December, pp. 59–70.

139. Topping, B. H. V. (1983). Shape Optimization of Skeletal Structures: A Review, *Journal of Structural Engineering*, Vol. 109, No. 8, pp. 1933–1951.

140. Ulrich, K. T., and Seering, W. P. (1987). Conceptual Design: Synthesis of Systems of Components. In: *ASME Winter Annual Meeting, Proceedings of the Symposium on Integrated and Intelligent Manufacturing: Analysis and Synthesis*, ASME, New York.

141. Ulrich, K. T. (1988). *Computational and Pre-Parametric Design*, Ph.D. Dissertation, Massachusetts Institute of Technology, Cambridge, September.

142. Waldron, M. B., Jelinek, W., Owen, D., and Waldron, K. J. (1987). A Study of Visual Recall Differences Between Expert and Naive Mechanical Designers, *ICED 87*, Boston, MA, August, pp. 86–90.

143. Waldron, M. B., and Waldron, K. J. (1989). Empirical Study on Generation of Constraints which Direct Design Decisions in Conceptual Mechanical Design, *NSF Engineering Design Research Conference*, University of Massachusetts, Amherst, June 11–14.

144. Ward, A. C. (1989). *A Theory of Quantitative Inference for Artifact Sets, Applied to a Mechanical Design Compiler*, Ph.D. Dissertation, Massachusetts Institute of Technology, Cambridge, January.

145. Wile, D. S. (1983). Program Developments: Formal Explanations of Im-

plementations, *Communications of the ACM*, Vol. 26, No. 11, November, pp. 902–911.

146. Williams, B. (1989). *Invention from First Principles via Topologies of Interaction*, Ph.D. Dissertation, Massachusetts Institute of Technology, Cambridge.

147. Winston, P. H. (1980). Learning and Reasoning by Analogy, *Communications of the ACM*, Vol. 23, No. 12, December, pp. 689–703.

148. Wirth, N. (1971). Program Development Through Stepwise Refinement, *Communications of the ACM*, Vol. 14, pp. 221–227.

149. Wood, K. L., and Antonsson, E. (1988). Computations with Imprecise Parameters in Engineering Design: Background and Theory, *Engineering Design Research Laboratory Report: 88-01*, California Institute of Technology, Pasadena, February.

150. Wood, K. L., and Antonsson, E. K. (1990). Modeling Imprecision and Uncertainty in Preliminary Engineering Design, *Mechanism and Machine Theory*, Vol. 25, No. 3, pp. 305–324.

151. Yoshikawa, H. (1981). General Design Theory and CAD Systems. In: *Man–Machine Communication in CAD/CAM*, edited by T. Sata and E. Warman, IFIP/North-Holland, Amsterdam, pp. 35–58.

152. Yoshikawa, H. (1989). General Design Theory as a Formal Theory of Design. In: *Intelligent CAD, Part I*, Proceedings of 1987 IFIP Workshop on Intelligent CAD, Boston, MA, Elsevier Science Publishers, Amsterdam, pp. 51–60.

CHAPTER TWO

Features in Design and Manufacturing

JAMI J. SHAH

Department of Mechanical and Aerospace Engineering, Arizona State University

2.1. INTRODUCTION

Features are generic shapes with which engineers associate certain attributes and knowledge useful in reasoning about the product. Features encapsulate the engineering significance of portions of the geometry and, as such, are important in product design, product definition, and reasoning for a variety of applications. Feature-based CAD/CAM systems have demonstrated some potential in creating attractive design environments and in automating geometric reasoning related to design function, performance evaluation, manufacturing and inspection process planning, NC programming, and other engineering tasks. Features technology is regarded by many as a key enabling technology for the implementation of concurrent engineering and CAD–CAM integration.

Traditional CAD systems are essentially geometric modelers that are not based on features; the database is either in terms of low-level geometric and topological entities organized hierarchically (Boundary Representation or BRep) or binary trees in terms of algebraic set operators applied to simple primitive objects (Constructive Solid Geometry or CSG). In traditional CAD, models are created from geometric primitive objects, such as cylinders, blocks, and spheres, or other geometric construction techniques like sweeping of profiles to get volumes, or fitting a skin to a series of cross sections arranged in space. Geometric modelers were designed to overcome the deficiencies of the earlier wireframe systems that could not distinguish between nonsense geometry and valid objects. Geometric modelers provide a mechanism to mathematically validate user-specified geometry. An important side benefit of this kind of representation is that algorithms are available for automatically calculating the mass and section properties of modeled objects.

Intelligent Design and Manufacturing, Edited by Andrew Kusiak.
ISBN 0-471-53473-0 © 1992 John Wiley & Sons, Inc.

Geometric modelers have proved deficient for most design tasks, as well as for many applications such as automated process planning, GT classification, and CMM path planning. They have instead found use in documentation of finalized designs, geometric arrangement planning, and as a front end to finite element mesh generation programs. There are two major reasons for the lack of CAD system use in the development of designs:

1. The geometric construction methods provided are too low level for mechanical design; designers cannot design directly in terms of features, such as countersunk blind holes and rectangular pockets.
2. Design changes are time consuming; the lack of associativity between defined entities means that changes do not propagate through the model. Each change has to be individually made by the designer.

The database produced by design is also lacking the information needed by downstream applications such as process planning. The reasoning used in process planning, GT classification, and automated part programming, to mention a few applications, is based on part features. In fact, much of the initial work on features seems to have come from a desire to devise methods to extract part geometry from geometric modelers from which process plans, GT codes, and NC programs could be generated. Features provide an additional layer of information in the new and upcoming CAD systems that is needed to make modelers more useful in design and to help integrate design with downstream applications.

2.2. FEATURE FUNDAMENTALS

2.2.1. Definition of a Feature

Although many different definitions appear in the literature, the essence of these definitions indicates that *features represent the engineering meaning of the geometry of a part or assembly*. Features can perhaps be thought of as building blocks for product definition or for geometric reasoning. In more explicit terms, a feature:

- is a physical constituent of a part
- is mappable to a generic shape (realizable or implicit)
- has engineering significance
- has predictable properties

Figure 2.1 shows an example part in terms of its shape features. The shape of a feature may be expressed in terms of dimension parameters, enumeration of geometric/topological entities and geometric/topological relations between composing entities, or in terms of a geometry construction proce-

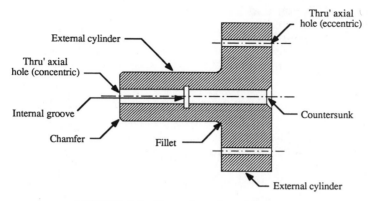

FIGURE 2.1. Examples of part features.

dure. The engineering meaning may involve the formalization of the function the feature serves, or how it can be produced, or what actions must be taken in the presence of this feature if one is performing some kind of evaluation, or how the feature "behaves" in various situations, and so on.

The components of a feature-based model, and the relations between them, are shown in Fig. 2.2. A mechanical assembly is defined in terms of its parts, part features, and attributes. Assembly attributes may include such information as (but not limited to) mating surfaces, fits/clearances, depth of insertion, and relative orientation vectors. Part attributes may include material specifications, part number, and administrative data. Feature attributes may be dimensions, shape, or size tolerances. Feature–feature relation attributes may have information about positioning, geometric constraints, and compatibility. Entity attributes could be surface finish or form tolerance. Entity–entity relation examples are adjacency and relative orientation (parallel/perpendicular).

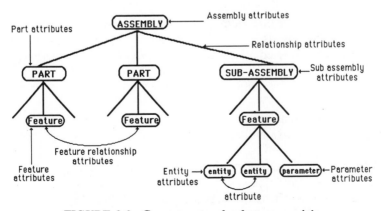

FIGURE 2.2. Components of a feature model.

2.2.2. Feature Types and Taxonomies

Since the term "feature" is used to include a wide variety of entities, it is necessary to distinguish between these types by using a subclassification. Some feature types are listed below:

Form features	Portions of nominal geometry; recurring shapes
Precision features	Deviations from nominal form/size/location (tolerances, finish)
Technological features	Nongeometric parameters related to function, performance, and so on
Material features	Material composition, treatment, condition, and so on
Assembly features	Part relative orientations, interaction surfaces, fits, kinematic relations

Feature-based CAD systems are currently only addressing form features; the discussion in this chapter is also restricted to form features.

The number of features is not finite but it may be possible to categorize features into groups or classes. Classification may be useful in the following ways: First, if features could be classified into families and their properties identified, then perhaps mechanisms could be designed to support each family instead of supporting special methods for each feature. Second, feature classification could perhaps lead to some common terminology. Third, a feature taxonomy could be useful in developing product data exchange standards.

Several schemes have been proposed for classification based entirely on shape, rather than the application. A scheme was developed by Pratt and Wilson (27) for CAM-I and adopted for the form features information model of Product Data Exchange Specification (PDES) (13). PDES classified features as follows:

Passages	Subtracted volumes that intersect the preexisting shape at both ends
Depressions	Subtracted volumes that intersect the preexisting shape at one end
Protrusions	Added volumes that intersect the preexisting shape at one end
Transitions	Regions involved in smoothing of intersection regions
Area features	Dimensionality 2 elements defined on faces of preexisting shape
Deformations	Shape changing operations such as bending and stretching

The PDES model is discussed further in Section 2.6.3. At the present time there are no universally accepted, or widely used, feature taxonomies.

2.2.3. Feature Properties

The shape, behavior, and engineering significance of a feature need to be encoded in its definition. One can separate these properties into two components: *generic* and *specific*. A hole, for example, has certain generic properties regardless of its size and specific location. The generic properties therefore need only be formalized and archived once for each feature. Specific instances then could simply refer to these generic properties. From a survey of features used in various applications, the following list of feature properties has been compiled:

Generic shape
Dimension parameters
Location method
Location parameters
Orientation method
Orientation parameters
Tolerances
Construction procedure for geometric model
Recognition algorithm, if applicable
Parameters inherited from other features
Inheritance rules and procedures (see later section for explanation)
Valid parent features and neighbors
Validation rules (see later section)
Nongeometric attributes (part number, function, etc.)

Figure 2.3 shows an example feature and its properties. Feature modelers usually provide a library of generic features, which have been formalized in terms of some of the properties listed above. In creating a model a user need only instance a feature from the library and it will automatically take on all the generic properties of its class. The user needs to complete the definition by specifying values for dimensions, location, and so on (instance parameter). Some modelers allow users to define new generic classes; others restrict them to predefined generic classes.

2.2.4. Feature Relationships

The major function that features serve is to create associativity between entities in a product definition database. This association of entities makes it possible to encapsulate design or geometric constraints and do geometric reasoning. Some generic relationships typically needed are discussed next.

CUTOUT −5 (TRAPEZOID)	In sheet metal an opening whose shape is in the form of a quadrilateral where only two sides are parallel.
	Applications: 1. Cutouts in sheet metal for connectors
	Part Type: 1. Sheet Metal

ILLUSTRATION

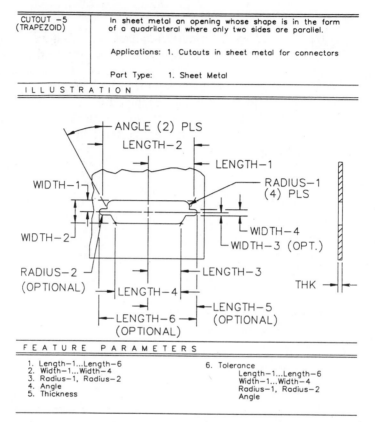

FEATURE PARAMETERS

1. Length−1...Length−6
2. Width−1...Width−4
3. Radius−1, Radius−2
4. Angle
5. Thickness

6. Tolerance
   Length−1...Length−6
   Width−1...Width−4
   Radius−1, Radius−2
   Angle

FIGURE 2.3. Dimension parameters of a feature. (Courtesy of Texas Instruments.)

Dependent and Independent Parameters. When one defines a new feature on a model some of the parameters of the new feature may already be fixed. For example, a through hole in a block must have a length equal to the depth of the block; the diameter of a recess feature or chamfer feature is fixed by the feature on which it is defined. Thus there needs to be a mechanism for features to derive certain parameters from the definition of other features. This is referred to as *parameter inheritance*, which is distinct from *property inheritance* between features of a family. Also, features that inherit parameters are referred to as *instance children features* and the source as *instance parent features*. Of course, an instance child feature can be the instance parent of other features. Features may need to derive parameters from several features, that is, a feature can have multiple parents. In simple cases, there might be a direct one-to-one link between a child feature parameter and a parent feature parameter. For example, the inside diameter of a torus to create a recess on a hole is equal to the diameter of the parent hole. The derivation of a dimension could be more

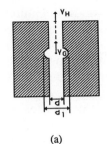

(a)

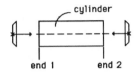

(b)   **FIGURE 2.4.** Parameter inheritance.

complicated, needing parameters from several parents, arithmetic computations, and even conditional statements. Some generic types of dependencies are the following:

1. *Direct Parameter Inheritance.* One or more dimensions of a feature must be calculated from parameters of others. An example is shown in Fig. 2.4a. The major diameter of the torus for the internal groove is equal to the diameter of the hole (d); the outside diameter (d_1) is

$$d_1 = d + 2t, \quad \text{where } t = \text{minor radius of torus (i.e., depth of recess)}.$$

2. *Orientation Dependencies.* The orientation of a feature is fixed by its instance parent. In the example of Fig. 2.4a the orientation of the groove (V_G) aligns with the orientation of the hole (V_H).

3. *Position Dependencies.* The location of a feature is fixed by the surface, edge, or corner of another feature. An example is shown in Fig. 2.4b, where one can only position the chamfer at either end of the cylinder.

4. *Feature Face Dependencies.* One or more geometric entities of a feature are constrained to lie on a specified face or faces of other features.

Adjacency Relationships. When a feature is deleted or modified, it may create problems for its neighbors. Some applications may need to have information regarding the surroundings of a feature (e.g., tool approach in machining). Also, feature validation, interference detection, compatibility of neighbors, and locating features relative to each other all require knowledge about what features are in the vicinity of a given feature.

Positioning/Placing of Features. When creating certain kinds of features, one might need to identify the entity on which the feature is to be placed and its location. For example, when defining a blind hole on a block, the fact through which it will go and the position of the center line need to be identified. To establish location/orientation on a given entity, some kind of a reference is needed on both the feature being positioned and the entity on which it is being positioned. Some systems use feature origins and feature coordinate systems that are predefined for every feature. One system allows users to pick reference entities for establishing a coordinate system. The datums are called "handles." This makes the system more flexible and also more convenient for defining position and orientation tolerances (29). However, if automatic feature recognition is used, handles are not available. This warrants the establishment of default coordinate systems for all features for use in locating and orienting of features. One may make provision for default coordinates to be overridden when defining features procedurally or interactively, but they are always used in automatic recognition.

Typical positioning methods may fall under the following categories:

1. Position a feature in world (global) coordinates.
2. Position a feature adjacent to (tangent to) face/edge/vertex.
3. Position feature faces/edges/axes on or along faces/edges of parent feature.

The use of relative positioning methods (belonging to classes 2 or 3 above) offers the advantage of encapsulating geometric constraints on location. Therefore, when one feature is modified or deleted, the change propagates to adjacent and child features. For example, in a stepped shaft each step may be positioned with respect to the adjacent step with all axes aligned; if the length of one of the steps were to be changed, all subsequent steps will be "pushed" outward, because of relative positioning. Thus relative positioning amounts to encoding geometric constraints that are maintained when the model is changed.

2.3. METHODOLOGIES FOR CREATING FEATURE MODELS

There are many alternatives for creating feature models in a geometric modeling context. In order to provide a framework for comparison, it is convenient to classify these methods into three broad groups:

1. *Interactive Feature Definition.* A geometric model is created first, then features are defined by human users by picking entities on an image of the part (Fig. 2.5a).

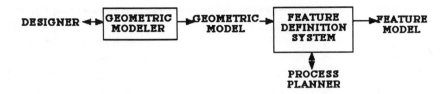

(a) Interactive feature definition

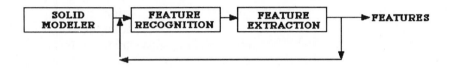

(b) Automatic feature recognition

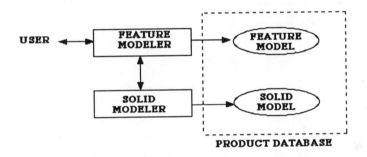

PRODUCT DATABASE

(c) Design by features

FIGURE 2.5. Schematics of feature definition approaches.

2. *Automatic Feature Recognition.* A geometric model is created first, then a computer program processes the database to automatically discover and extract features (Fig. 2.5b).

3. *Design by Features.* The part geometry is defined directly in terms of features; geometric models are created from the features (Fig. 2.5c).

The major characteristics of each of these categories and their subcategories are briefly discussed in the following sections.

2.3.1. Interactive Feature Definition

This methodology involves predefinition of the geometric model. Therefore the data structure of the geometric model is a major factor in the design of

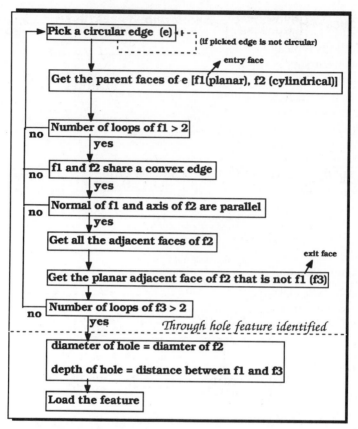

FIGURE 2.6. Interactive definition template.

the definition procedure. A 2D/3D wireframe or BRep solid model is created using a contemporary geometric modeling package. The database created is then read by a program that renders an image of the part on a CRT to allow the user to interactively pick topological entities (edges, faces) needed to define a feature. This information can be augmented with attributes such as tolerances, finish, or high-level nominal parameters (like hole diameter). This approach has been used largely for inputting data to programs for process planning and NC tool path generation. Figure 2.6 shows a typical algorithm for interactive definition supported by a template.

2.3.2. Automatic Feature Recognition

The input required for many application programs, such as process planning, NC part programming, and Group Technology coding, includes both the geometry and features. Various techniques have been developed in order to obtain this input directly from a geometric modeling database. This is popularly referred to as feature recognition, although the output of some techniques is not in the form of features but rather as machining volumes. These methods typically assume that all machining will be done by milling so it is not necessary to know the specifics of a feature other than its boundaries corresponding to final machined surfaces. For example, it does not matter if a machining volume is a rectangular pocket or an L-shaped slot because tool paths can be generated without this distinction. For this reason, *machining region recognition* and *feature recognition* are discussed separately in the following sections.

Machining Region Recognition. Much of the work in this area seems to have been focused on $2\frac{1}{2}$D milling. Grayer (16) and Parkinson (24) generated NC tool paths from recognized inner and outer boundaries, usually 2D profile curves, by offsetting the curves. Machining region recognition techniques may be classified into four categories:

- Sectioning techniques
- Convex hull algorithm
- Cell decomposition
- AI/Geometric reasoning

Sectioning Techniques. The part volume is sliced with a number of parallel planes that are perpendicular to the assumed tool approach direction. The intersection of the plane and the part model defines the boundary of the part at the plane. To generate NC code, Grayer decomposed 3D space into slices and passed each slice into an existing area clearance program. Parkinson used sectioning to deal with 3D faces (planes not parallel or perpendicular to spindle sections or a general curve surface). The 3D faces were intersected with planes parallel to the *XZ* or *YZ* planes. The intersection curves produced thus were split into straight line segments so as to keep within a given tolerance. These straight-line segments were offset by tool radius in the direction of the surface normal at their start and end points and at a convex edge (concave in the finished part) with a boundary face. CAM-I has used sectioning technique for volume decomposition of complex depressions (14). This is discussed later.

Convex Hull Algorithm. Woo (38) developed the decreasing convex hull algorithm, in which the volume to be removed was decomposed into machining volumes. Figure 2.7 shows how the convex hull algorithm works. The difference between the object and its convex hull is computed recursively, until the null set is obtained (i.e., until the object equals its convex hull). The object can then be represented as a sequence of convex volumes with alternating signs. The decomposition is not always useful because it could result in a removal volume that does not correspond to a single machining

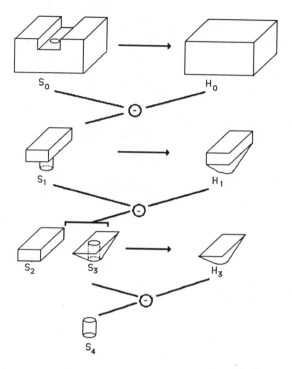

FIGURE 2.7. Convex hull algorithm [38].

operation (an odd-shaped feature). Also, the base stock is of awkward shape sometimes because it is the convex hull of the initial shape rather than standard bar stock.

Cell Decomposition. Cell decomposition techniques typically use a spatially enumerated model of the part or the part decomposed into a number of cells. The cells that are to be removed by machining are recognized. Armstrong (2, 3) used this strategy to produce a milled part for a given part and stock geometry. A spatially ordered cell decomposition of the part was produced by a lattice of planes parallel to the major axes as shown in Fig. 2.8. The cells and infinite planes were then positioned coincident with each planar face and tangential to each cylindrical halfspace. Each cell was classified either as a stock cell, a part cell, or a semipart cell. To derive roughing cuts, the algorithm selected cells accessible by the tool, and a path was simulated. If no collision with the desired part were detected, the path was concatenated with previous paths. If a collision were detected, then the tool was lifted to a safe plane. For finishing, the spatially ordered cells were scanned until one was found to have a face requiring machining. Then adjacent cells were examined for concatenation of that face or linked faces until a path became closed or a boundary was encountered.

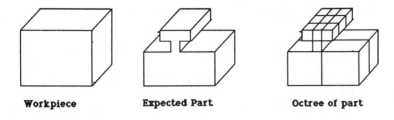

Workpiece   **Expected Part**   **Octree of part**

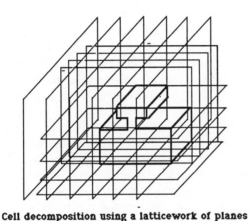

Cell decomposition using a latticework of planes

FIGURE 2.8. Cell decomposition.

AI/Geometric Reasoning. Preiss (28) has developed a rule-based algorithm that identifies individual operations by goal driven search. Using a boundary representation, the program starts searching at a vertex and follows an edge to the next connected edge and so on until it returns to the original edge. If the procedure succeeds, a profile cut is generated. If not, subloops are traversed for each disjoint section. Thus the program generates paths for clearing areas and pockets without actually recognizing features.

Feature Recognition. Feature recognition differs from machining region recognition in that portions of the geometric model are compared to predefined generic features in order to identify instances that match the predefined ones. Specific tasks in feature recognition may include the following:

- Searching the database to match topologic/geometric patterns
- Extracting recognized features from the database (remove a portion of the model associated with the recognized feature)
- Determining feature parameters (e.g., hole diameter, pocket depth)

- Completing the feature geometric model (edge/face growing, closure, etc.)
- Combining simple features to get higher level features

Both volume-based methods and surface-based methods have been devised. It is difficult to classify recognition methods into a clear taxonomy because there is considerable overlap between the various techniques. Because of space limitations, only two types of feature recognition are presented here: *boundary-based matching* and *volume-based decomposition*.

Matching. Generic features are first formalized in terms of their geometric and/or topologic characteristics. Then search algorithms are devised to determine which of these characteristics are present in the geometric model (or reconstituted or augmented model). Since solid model data structures are usually graph structures, *graph matching* has been a popular method for feature recognition. Pure graph matching done on unaugmented solid models amounts to topological matching; that is, the characteristics are based on the number of entities, topological type, connectivity, and adjacency. If matching were done this way, features of very different semantics would be classified as being the same. Therefore some subclassification using geometric relationships is necessary. An entity classification method, devised by Kyprianou (22), has been used widely. It is based on the magnitude of the angle of intersection. In this method, edges are classified either as convex, concave, smooth convex, or smooth concave as shown in Fig. 2.9. Smooth edges are reclassified as concave/convex on the basis of local curvature. Vertex classification is based on the vertex's incident edges; if two or more incident edges are concave, the vertex is deemed to be concave, otherwise it is classified as convex. Loops are classified as convex (all edges convex), concave (all concave), and hybrid (mixture). Faces of the object are marked "primary" if they contain a concave edge or an inner loop, and primary faces are ordered on the basis of the number of concave edge sets. A hierarchical faceset data structure is built by processing the geometric model, which now contains entities tagged by the above classification.

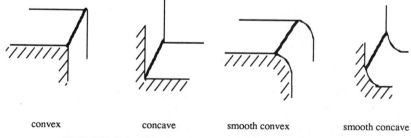

convex   concave   smooth convex   smooth concave

FIGURE 2.9. Kyprianou's edge classification criteria.

Kyprianou himself used a feature grammar to determine features from rules and faceset data structure. A metalanguage was also developed for specifying GT schemes (20); the faceset data structure was integrated to derive the DT code.

Another technique that has been applied to matching is *syntactic pattern recognition* adapted from vision systems. In these systems geometric patterns are described by a series of typically straight, circular, or other curved line segments. Simple patterns can be concatenated to give compound patterns. Languages have been developed for describing these sequences algebraically and manipulating them with operators that form a grammar. Features can be recognized by parsing the feature against the object's description in the grammar. Syntactic pattern recognition was applied to features by Kyprianou (22) and by Choi et al. (8). Jakubowski et al. (19) and Staley et al. (36) used strings of straight lines and curve segments to recognize 2D profiles of holes. Graph grammars and shape grammars have also been developed for matching feature shapes.

The most common matching method is based on rules. Features are formalized by templates that consisted of pattern rules. Templates are defined for both general features (like holes) and specific features (e.g., flat bottomed, constant-diameter hole). A general rule for a hole, for example, looks like this (17):

> The hole begins with an entrance face. All subsequent faces of the hole share a common axis. All faces of the hole are sequentially adjacent. The hole terminates with a valid hole bottom.

Rules such as these are expressed as a set of both geometric and topologic conditions, each of which had to be tested separately; all conditions have to be satisfied in order for the rule to be satisfied. Henderson (17) created, in effect, a graph grammar for feature recognition. Henderson's recognition and extraction algorithm involved the following steps: determine cavity volume (difference between stock and part), recognize general features in each cavity, classify general features into specific features, create and subtract the volume corresponding to each feature from the cavity, and repeat all the above steps until there are no residual cavities. The scope of the study was limited to sweep features.

Volume Decomposition. The purpose of volume decomposition is to identify material to be removed from a base stock and to break down this volume into units corresponding to distinct machining operations. Generally, the total material volume to be removed by machining is found by a boolean difference between the stock and the finished part. This volume must then be decomposed into units that correspond to practical machining operations that match machining features as shown in Fig. 2.10 for an example part. A well known work on volume decomposition is that done by General

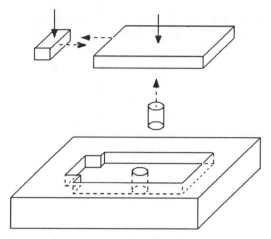

FIGURE 2.10. A part volume decomposed into its delta volumes.

Dynamics for CAM-I (14). The purpose of the project was to achieve a high degree of automation for generating NC programs for parts defined by "noncomplex" surfaces (planar, quadric, and cylindrical). An algorithm was developed for operating on BRep model of the total volume to be removed, augmented with tool accessibility codes for each face. A library of generic delta volumes existed in the system; new delta volumes could be added by users. This set of generic volumes was required to meet the criteria for completeness and richness as specified below:

- For every milled part the material to be removed can be decomposed into the union of disjoint delta volumes (completeness).
- For any volume of material to be removed there exists a delta volume contained in it (richness).

These two criteria guaranteed that any machining volume could be decomposed into a set of generic delta volumes. Decomposition was carried out in two major stages. First, the primitive (parameterizable) volumes were recognized and extracted by surface extension. Because surface extension could be done by interrogating the BRep model, considerable computations were saved. Complex depressions ($2\frac{1}{2}$ pockets) were recognized by sectioning. A set of cross sections was constructed by using a set of planes that were perpendicular to the cutter axis. Then relationships between adjacent cross sections were determined to decompose the volume into disjoint "super-delta volumes," which were decomposed further based on tool accessibility. All delta volumes had to have at least one accessible face. A face on a delta volume was inaccessible if it coincided with a face of the finished part. If a face were partly coincident with the finished part, then it was assigned connectivity to another delta volume with which the rest of the face was coincident. Finally, all deltas were compared with generic delta volumes and classified.

2.3.3. Design by Features

In this approach, features are incorporated in the part model from the beginning. Generic feature definitions are placed in a library from which features are instanced by specifying dimension and location parameters and various attributes. Two methodologies are common: *destruction by machining features* and *synthesis by design features*.

Destruction by Machining Features. This approach goes by various names in the literature, such as destructive solid geometry or deforming solid geometry. A part model is created by boolean subtracting features from a base stock model. The design and manufacturing plan are concurrently developed by transforming a base stock model into the final part model through the application of operations that correspond to stock removal. Prototype systems using this approach have been demonstrated at Standford (11) and at Purdue (37). The commercial system Pro-Engineer (from Parametric Technology) also supports this approach. All these systems use a set of predefined features that are subtracted from the base solid. Features are defined by attribute slots encompassing dimensions, tolerances, finish and starting (tool entry) face. In the Purdue and Stanford systems, process plans are generated and tested with each design change. In the Stanford system a "team" of expert systems works concurrently to generate, simulate, and verify plans. Expert systems included feature machining expert, fixturing experts, tooling experts, and collision checkers. Thus, when the design is complete, the process plans, tool designs, and NC programs are also complete.

Synthesis by Design Features. This includes systems that allow one to design by adding or subtracting features without a starting base stock, as shown in Fig. 2.11. Many research and commercial systems belong to this category. Generic features are predefined in terms of rules and procedures. Procedures may include methods for instancing, modifying, copying, deleting features, generating solid models, deriving certain parameters, and validating feature operations. For example, Chung (9) developed a prototype system for creating solid models in SDRC's GEOMOD via features defined in an expert system development package KEE from IntelliCorp. The ASU Features Testbed (29, 30) uses a parallel representation: a boundary model and a constructive model that represent the volume or cavity corresponding to the feature (referred to as the feature producing volume or FPV) with a Union or Difference operator. This solid representation is not specific to any particular solid modeler; through an "Evaluate" command all

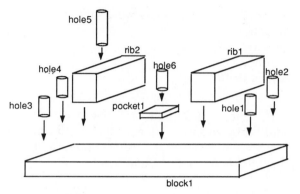

FIGURE 2.11. Synthesis by design features.

FPVs are combined and commands are translated through an interface to solid modeler specific commands.

There are many variations of the above general scheme. The features stored in libraries may be application oriented; Luby (23) defined casting features, Irani (18) defined injection molding features, and Chung (9) provides features needed in gating design for investment casting. Chung (9) demonstrated a general purpose template for adding new features to the library. The ASU Features Testbed provides an application-independent method for feature definition (30, 31). The system consists of two shells, one for part design and the other for concurrent mapping and applications. The shells contain mechanisms for defining generic features, adding features to the library, and using features in design but leave user organizations to customize the system, thus avoiding the difficulty of working with a hard-coded set of features (34).

2.3.4. Comparison of Definition Methods

The human assisted definition method is easy to implement and it can work off IGES (Initial Graphics Exchange Specification) or modeler specific database of contemporary systems. Only features needed for an application (e.g., process planning) need be identified. For models containing a large number of feature, this method can be time consuming. In many current implementations the burden of picking valid entities lies on the user. The job can be made easier with some changes; for example, if the user indicates he/she wants to identify a flat-bottom hole, then the system prompts that he/she must pick a cylindrical face and a planar face; when entities are picked the system will check if they are of the appropriate type. Procedures must exist for automatically deriving the diameter and depth of the hole from the geometric model. The number of topological entities is arbitrary and often depends on intersections performed in construction of the model.

For example, the cylindrical surface of the hole may be represented by several faces. Several systems allow one to merge topological entities lying on the same geometric entity, so this will always be a prerequisite.

Considerable progress in feature recognition has been made. Principal among the advantages of feature recognition is the use of current geometric modeler databases or even IGES. Another advantage is that recognition can be made application specific, allowing each application program to have its own recognition program. More work is needed in handling interacting features.

Sectioning techniques suffer from many inherent problems but are still the most commonly used. They are successful with simple $2\frac{1}{2}$D parts that do not have any undercut portions. The presence of undercuts, inclined surfaces, and nonplanar surfaces causes complications. When many features occur in the same plane, each feature is machined before moving to the next plane. This yields nonoptimal tool paths. Also, the slicing planes must be chosen appropriately, such that critical sections (i.e., portions where the cross section changes drastically) are not omitted.

Tree manipulation suffers a drawback due to the nonuniqueness of CSG. Convex hull decompositions algorithms often do not produce a usable decomposition because it can result in removal of volumes that do not correspond to a single machining operation (an odd-shaped feature). Also, the stock shape can be awkward because it is the convex hull of the initial shape. Only a handful of researchers have worked on tree manipulation or volume decomposition. Most of the work has been in the area of graph or pattern matching to locate features.

Design by features has the advantage that it allows designers to transfer to the database much of the information available at the design stage. This richer and higher level database is available for use by downstream applications. It is even possible to implement real-time manufacturability evaluation and concurrent design and process planning. However, the set of features used in design is not finite. One needs to determine how many features should be contained in the feature library and at what level of abstraction. Also, since features are application specific, the need for feature recognition by application does not go away when one designs by features. Finally, interactions between features can result in nongeneric shapes that do not exist in the database or they could make some generic dimension values obsolete.

Creating feature databases unaided by geometric modelers has the disadvantages that nonsense geometry can be created and the user cannot visualize the geometry. In DSG one destroys the stock model by removing features, which is a good way of providing input to CNC machines. However, the method requires the stock shape to be known *a priori* and it puts process planning responsibility on designers.

There have been some recent advances in unifying all three approaches (34). This requires "real-time" feature recognition, which is easier because

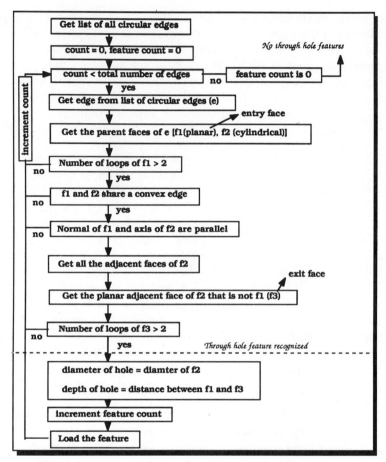

FIGURE 2.12. Automatic feature recognition algorithm.

the user can give the system some easy starting point. Figure 2.12 shows the algorithm for automatic recognition in such a system.

2.4. FEATURE REPRESENTATION

Features may be represented at various levels. For example, one could represent them by the process by which they may be created or by the resultant geometric model. The former representation has been termed *implicit* or *unevaluated* (27) and *procedural* (33); the latter has been termed *explicit or evaluated* (27) and *enumerative* (13, 33). Features may be defined more abstractly as a neutral description without any specification of how the feature is to become part of the geometric model. Explicit representations have commonly been used in interactive and automatic feature recognition and explicit representations in design by features. Some of the structures used for geometric representations of features are:

- Augmented graphs (21, 22, 26)
- Algebraic, syntactic (8, 36)
- Delta volumes (12, 14)
- Constraint-based BRep (6, 25)

Augmented graphs are usually based on face adjacency; the arcs of the graph are attributed with information on edge classification and geometric relationships. Syntactic languages have also been devised that encapsulate adjacency, connectivity, geometric orientation, and convexity/concavity of feature entities, though their use has been limited to 2D. Delta volumes are complete BRep models of closed spaces associated with tool accessibility codes for faces and connectivity information. Delta volumes are shown in Fig. 2.10; accessibility and connectivity codes for the delta volumes are indicated by solid and dashed arrows, respectively (14).

2.5. FEATURE VALIDATION

There are no universally applicable methods for checking the validity of features. It is up to the person defining a feature to specify what is valid or invalid for a given feature. This should not be confused with geometric or topological validity, which is based on rigorous mathematics. Features are invalid if any of the conditions declared in the generic definitions are violated. Such conditions could be based on size limits, shape, location, and so on. Therefore it is possible that some operations may result in valid (physically realizable) solids but may product invalid features. Typical checks that need to be done are compatibility of parent/dependent features, limits on dimension, and inadvertent interference with other features. There are situations in which the resulting features may be invalid. Intersection between feature volumes could:

- Make a feature nonfunctional
- Create nongeneric feature(s) from two or more generic ones
- Render feature parameters obsolete
- Give nonstandard topology
- Delete a feature by subtraction of larger feature
- Delete a feature by addition of larger feature
- Close an open feature

Several types of interaction are identified in Fig. 2.13. In Fig. 2.13a the hole is too close to the outside surface and so it breaks through. The passage

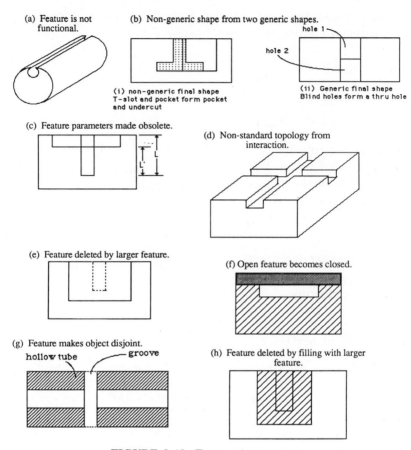

(a) Feature is not functional.

(b) Non-generic shape from two generic shapes.

(i) non-generic final shape T-slot and pocket form pocket and undercut

(ii) Generic final shape Blind holes form a thru hole

(c) Feature parameters made obsolete.

(d) Non-standard topology from interaction.

(e) Feature deleted by larger feature.

(f) Open feature becomes closed.

(g) Feature makes object disjoint.

(h) Feature deleted by filling with larger feature.

FIGURE 2.13. Feature interactions.

created does not qualify as a hole; its topology is not that of a hole and it may not be able to serve the function of a hole. In Fig. 2.13b, two legitimate generic features intersect and product a nongeneric shape in (i) and a generic shape in (ii); the topology of the final feature in (i) is nonstandard and the function of this shape will be different from that of either of the other two. In Fig. 2.13c, the creation of the pocket has eliminated the entry face of the hole; the depth of the hole has changed even though the user did not modify the hole parameters. In Fig. 2.13d two slots intersect, thus creating new topological entities. In Fig. 2.13e a new feature has completely deleted an old feature from the geometric model. In Fig. 2.13f a new feature has closed off a feature that was originally open. In Fig. 2.13g a groove placed on a tube is deep enough to make the part disjoint. In Fig. 2.13h the addition of a new volume deletes a cavity.

At the present time most features modelers do not perform automatic feature validation; it is the users responsibility. Research systems, however, have demonstrated the feasibility of automatic validation. Apart from checks on parent entities, parameter range, and position constraints, they can:

1. Detect intersections between feature entities using geometric modeler functionality.
2. Classify the type of interaction, when detected (like the classes given above).
3. Consult the rules/procedures specified by user or application to determine what action to take, which may include (a) disallow interaction, (b) send message to other features to alter affected parameters, (c) take no action, and (d) change feature type to match its current state.

2.6. CONCURRENT ENGINEERING WITH FEATURES

Concurrent engineering requires the sharing of product data between several tasks. How a part is modeled in terms of features varies from task to task. For example, if the part shown in Fig. 2.14 was synthesized by design features, it would not have the specifications of some of the machining features that do not exist in the mind of a designer—the step volume and the profiled step, in this case. Also, even in design, the method for synthesizing

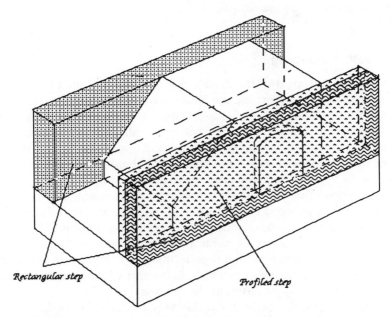

FIGURE 2.14. Examples of features specific to manufacturing.

parts from features is not unique; the same part could be viewed as having different features, depending on a person's viewpoint. Also, features vary from one type of part to another. For example, sheet metal parts have features like joggles or hems that are not relevant to machined parts; parts made from composites have lay-up features not found on homogeneous material parts. It is not possible to enumerate all possible features that everyone will ever need, although attempts to do so have been made (7). If a system supports a mechanisms for users to define their own features, it makes the system very flexible: The domain can be extended, features can be customized to the needs of a task or organization, and so on. However, there is also a desire to exchange data between systems from different vendors and between design and application programs. This need will be aided if the CAD community can agree on a standard set of features and their attributes. But a fixed set of features puts a constraint on designers and it is also unlikely that all product domains can be covered adequately with a fixed set; people will always need features that are not in the feature library.

Concurrent engineering with feature models could conceptually be done in one of two ways:

- By use of parallel, and equivalent, product models; this requires mapping of features from one application to another (feature mapping)
- By exchange of product data using a standard format (feature standardization)

These two possibilities are now discussed.

2.6.1. Feature Mapping

The theory of feature transformations can be found in Shah (33); some of the major points of interest are discussed here. Features depend on product type, application, and level of abstraction. A combination of these three factors defines a domain called a *feature space*. A feature space can be thought of as an *n*-dimensional vector space, where *n* is the maximum number of independent vectors in that space. Features can be considered analogous to vectors that are defined as linear combinations of the basis vectors. Suppose that the totality of information related to every kind of product, in all its aspects, over its entire life cycle, and for all applications conceivable defines a domain called a *feature hyperspace*. Then feature spaces for a given product or application are subsets of this hyperspace. These subsets are lower in dimension, that is, contain less information. Various kinds of relationships can exist between two subspaces that may be of the same dimension or different dimensions.

Feature spaces of the same dimension may be partially overlapping or be completely *disjount*. In the overlapping (*conjoint*) regions there are features with identical semantics; for example, a through hole has the same meaning on a machined part as on a sheet metal part so this can be regarded as a conjoint feature. In regions that do not overlap, features are meaningful in only one domain. For example, a blind hole can be machined but has no meaning for a sheet metal product; on the other hand, a joggle can only be formed on thin sheets but has no equivalent for machined parts.

Another type of relationship between feature spaces exists when the spaces are of unequal dimensions. A feature space is n-dimensional, where n is chosen such that the information level is just sufficient to carry out the tasks in the corresponding domain. Information in one domain may be selectivity abstracted to suit a different domain. This may be referred to as a *projection transformation* from n to $(n - m)$ space. The mapping from n to $(n - m)$ space is unique but the inverse is not. For example, in electronic design, a PCB can be a 2D compound feature, whereas for the mechanical designer it is a 3D compound feature.

Yet another relationship is that between "conjugate spaces," which are those subspaces that contain features that are composed of different variations of the same elements. For example, one may group elements in different ways to get different form features depending on one's point of view. The designer may place three rib features on a workpiece but together these result in two slot features that can be milled or shaped out from a workpiece; the ribs and slots are therefore complementary features.

Adjoint spaces are created by associating elements in one subspace to certain elements in another subspace. For example, "load" in structural analysis space can be associated with "boundaries" in geometric space.

When two subspaces are fully disjoint, there is no transformation available. For conjoint regions of overlapping spaces an identity transformation is used for the subspace formed by intersection. Conjugate transformations require geometric reasoning. Projection transformations are achieved by selective discarding of information (abstraction). Inverse projection transformations are generally not unique. Feature mapping is still a research topic; the types of transformation mechanisms have been identified, but methods for implementing general purpose transformations have not been found.

In a concurrent engineering environment a modeler needs to support multiple feature models that are stored in parallel. For example, the modeler must permit the user to define both ribs (design viewpoint of a feature) and slots (manufacturing viewpoint of the same feature) in the same model. There can be several problems if multiple viewpoint models coexist. First, if some changes are made to a feature in one viewpoint, it may destroy or invalidate features in other viewpoints. Second, if some application were processing the feature model, how will it know that all independent features have been processed? Since the purpose of multiple viewpoints is to meet the needs of specific applications, one solution that has been suggested is to designate only one viewpoint of features as the *primary viewpoint model* and to provide a method for transforming features from the primary to other viewpoints. Of course, any changes made to the primary model invalidate all the secondary models. Ideally, one would like to have a mechanism for making incremental modifications to the secondary model when the primary model is changed, but it is not clear if this is possible. Because of the role designated for the primary model, it is necessary for the primary model to be complete, that is, contain the complete and unambiguous definition of a part. On the other hand, it is not necessary to require secondary models to be complete.

2.6.2. Feature Standardization

The first version of a Form Feature Information Model (FFIM) has already been completed under the PDES/STEP umbrella (23). The FFIM appears to be adequate for static (snapshot) data exchange limited to shape information only. This is sufficient for applications that simply need to use the geometry features of a fully developed design, but it cannot be used for exchanging feature data between two feature modelers that each develop the model partially. The following comments will be made about the feasibility of feature data exchange, in principle, rather than specifically related to the FFIM work.

It should first be recognized that features technology is a fast developing field; the basic concepts and methodologies are still evolving. The development and acceptance of standards for features could adversely affect research. The feature concept presented in this chapter clearly identifies two components of a feature: the *geometric shape* and the *engineering semantics*. Neither component is limited to a finite number. Geometric/topological elements can be combined in various ways to yield virtually an unlimited number of shapes. These shapes are also not subject to any unique parametrization or procedural definition. The situation is even more complicated when one considers engineering semantics. The meaning, intent, or significance of various shapes depends on user viewpoint and convenience, the product type, the application, and so on. It does not appear that such information is standardizable. On the other hand, it is undesirable to develop systems that will work in isolation, unable to communicate with the outside world. Therefore some trade-offs will have to be made.

One other aspects that impacts the standardization issue is the extensibility of feature libraries. The ability to extend or modify the feature library is very desirable because it enhances flexibility, versatility, and convenience. However, it is detrimental to standardization. How can data-related user-defined features be exchanged?

The alternative approaches that could be taken are (a) standardize specific "common" shapes and semantics through consensus opinion, (b) standardize general classes of shapes only, not specific ones, (c) standardize specific shapes but incorporate attribute slots for conveying nonstandard information to application programs, and (d) standardize low-level shape and semantic elements from which all features can be synthesized. Option (a) will limit the number of features and make the system rigid, but both shape and semantics of features will easily be exchanged. Option (b) will allow users to extend the system to new features as long as they fall within standard taxonomical classes. Only limited amounts of semantic and specific information will be exchanged this way. The third option (c) provides a compromise by permitting company- or product-specific, nonstandard information to be passed on to programs that can be set-up to decode such data. The feasibility of option (d) is unknown at this time.

2.6.3. PDES Form Feature Information Model (FFIM)

It is perhaps appropriate to outline the current version of the FFIM, which will eventually be an international standard for feature data exchange. The FFIM defines a form feature as a portion of the skin of a shape that conforms to some stereotypical pattern and is considered a unit for some purpose. Nonshape information, such as functionality, processes, or surface finish, is not supported. FFIM assumes that, in the design process, the use of form features is optional, and every portion of the model need not be represented as a feature. Assemblies, joints, flexible members, and so on are not addressed by the FFIM.

The FFIM classifies features into several types; the two main classes are explicit and implicit features:

- *Explicit form features* are groupings of elements of the geometric model. In explicit features, the geometric shape elements that are necessary to define the feature are listed explicitly. For example, a hole may be represented as a list of two faces—a cylindrical face and a planar face forming the bottom of the hole. Explicit features therefore interpret the meaning of existing data rather than add new information.
- *Implicit form features* model shape information parametrically rather than geometrically. An implicit representation describes how the feature may be constructed. It is considered as a modification of some preexisting shape and, by the token, adds information to the model. For example, a hole can be defined as the axisymmetric sweep of a straight line about a specified axis. It is important to note that the actual faces/geometric elements corresponding to the implicit feature need not be physically present in the model.

A feature may have both an explicit representation and an implicit representation. It may also have a number of concurrent implicit representations. In the FFIM, information for supporting explicit features is treated very generally, while implicit features are specific types, divided into six classes, which were listed in Section 2.2.

A volume is represented either as a sweep or as a ruling. The further classification of feature volumes is shown in Fig. 2.15. Definition of a volume as a feature sweep requires the specification of a profile and a sweep path. The types of feature sweeps and sweep paths specified in FFIM is shown in Fig. 2.16. End conditions can be defined for the sweep. For example, a pocket can be defined as a sweep of a rectangular profile along a linear path. The fillet between the pocket walls and the bottom face of the pocket can be modeled by an end blend condition. A feature ruling defines a feature volume as a ruled surface definition. In addition to representing the feature volume, the faces of the shape that bound the implicit feature may be defined. For example, a hole feature in a block will have an entry face and an exit face that can be represented as bounds to the feature. The manner in which the feature blends into the preexisting shape can also be represented using edge blends. This enables the representation of chamfers, fillets, and so on.

In many cases, the implicit representation of a feature may depend on another implicit representation. For example, an internal groove may be

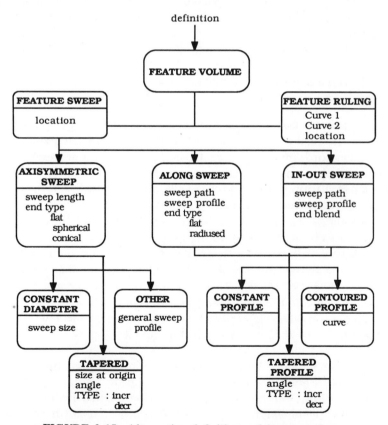

FIGURE 2.15. Alternative definitions of feature volume.

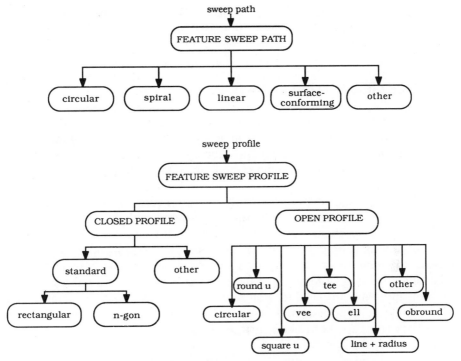

FIGURE 2.16. Specification of sweep path and profile.

defined on a hole, both of which are defined implicitly. In this case, the groove cannot exist without the hole, and hence, the FFIM provides a mechanism for ordering implicit features. Implicit features may be ordered in two ways. When one implicit representation refers to another representation, it automatically implies a precedence relationship. Another way of specifying precedences is by making a statement that one feature should precede another.

The class of the implicit feature describes its generic shape. The location of the feature can be defined in several ways. The most versatile way of representing volume-based features (depressions, protrusions, and passages) is by using a local coordinate system for the feature and positioning it in a known axis system that is generally the world coordinate system. In some cases, a geometric entity can be used to locate the feature. For example, a bendline can be used to represent the bend deformation feature. Other features, such as transition features and area features, can easily be located by reference to preexisting shape elements. For example, a fillet may be located by the surfaces it blends.

Since it is common for many features to be the same in all respects except location, the replication of features is supported by the FFIM. Replicate features are defined by declaring it to be identical to another feature and specifying its location. Feature patterns are also supported. A pattern feature is a replication of a feature in a geometric pattern such as a circular or a matrix pattern.

2.7. FEATURE-BASED APPLICATIONS

Feature-based applications must transform feature models from a design viewpoint into a manufacturing or analysis view. This transformation may include feature extraction, decomposition into lower level entities, reconstruction by geometric reasoning, and augmentation with new data or entities.

A basic requirement is access to the feature-geometry database. Sample queries for NC are (a) list all features associated with face n, (b) list the stock size (may correspond to the representation), and (c) list all pockets.

Applications should be able to augment the feature model with data private to them to create one or more secondary models. This includes attaching attributes to features or faces, for example, surface finish to a face or tool approach for a slot. For finite element analysis, the augmented model may include loads and restraints. Several applications may use the same data (perhaps both need surface finish).

To support the database queries, a static interface to the feature modeler and solid modeler is required. A dynamic interface is also needed for applications. For example, verification of NC programs includes tests gouging of a part and machining a fixture, both of which require a solid-modeler command interface for the boolean operations.

As the transformation occurs from the decision model to the feature-based application secondary model, a number of interdependencies arise. If a pocket has been transformed into its tool path, will a change in the design feature model affect the applications tool path? Since the augmented model is not accessed through the design model, the change will not be direct. However, design changes must propagate to the application, flagging situations that need human resolution. If an application changes the geometry of a part, these alterations will not propagate "up." Such situations should be noted as an attribute of the application feature model. Examples of the need to alter part geometry include building an idealized model for finite element analysis or changes needed for work holding.

REFERENCES

1. Ansaldi, S., DeFloriani, L., and Falcidieno, B. (1985). Geometric Modeling of Solid Objects by Using a Face Adjacency Graph Representation, *ACM SIGGRAPH*, Vol. 19, No. 3.

2. Armstrong, G. T. (1982). *A Study of Automatic Generation of Non-Invasive N.C. Machine Paths from Geometric Models*, Ph.D. Dissertation, Department of Mechanical Engineering, University of Leeds.

3. Armstrong, G. T., Carey, G. C., and de Pennington, A. (1984). Numerical Code Generation from a Geometric Modeling System. In: *Solid Modeling by Computers: From Theory to Application*, edited by M. S. Pickett and J. W. Boyse, Plenum Press, New York.

4. Bobrow, J. E. (1985). N.C. Machine Tool Path Generation from C.S.G. Part Representations, *Journal of CAD*, March, pp. 69–76.

5. Buchard, R. L. (1987). *Feature Based Geometric Constaints Applied to CSG*, M.S. Thesis, Purdue, May.

6. Bunce, P. G., Pratt, M. J., Pavey, S., and Pinte, J. (1986). Features Extraction and Process—Specific Study, *CAM-I Report*, R-86-GM/PP-01.

7. Butterfield, W., Green, M., Scott, D., and Stoker, W., (1986). Part Features for Process Planning, *CAM-I Report*, R-86-PPP-01.

8. Choi, B., Barash, M., and Anderson, D. (1984). Automatic Recognition of Machined Surfaces from a 3D Solid Model, *Journal of CAD*, March, pp. 81–86.

9. Chung, J. C., Cook, R. L., Patel, D., and Simmons, M. K. (1988). Feature-Based Geometry Construction for Geometric Reasoning, *ASME Computers in Engineering Conferences*, San Francisco, July/August.

10. Cunningham, J., and Dixon, J. (1988). Designing with Features: The Origin of Features, *ASME Computers in Engineering Conferences*, San Francisco, July/August.

11. Cutkosky, M., Tenenbaum, J., and Miller, D. (1988). Features in Process Based Design, *ASME Computers in Engineering Conferences*, San Francisco, July/August.

12. Dong, X., and Wozny, M. (1988). FRAFES, A Frame-Based Feature Extraction System, *Proceedings of International Conference on Computer Integrated Manufacturing*, Rensselaer Polytechnic Institutes, May 23–25, pp. 296–305.

13. Dunn, M. (1988). PDES Form Feature Information Model (FFIM), *PDES Form Features Group*, Mark Dunn (Coordinator).

14. General Dynamics Corporation (1985). Volume Decomposition Algorithm—Final Report, *CAM-I Report*, R-82-ANC-01.

15. Grayer, A. R. (1976). *A Computer Link Between Design and Manufacture*, Ph.D. Dissertation, University of Cambridge.

16. Grayer, A. R. (1977). The Automatic Production of Machined Components Starting from a Stored Geometric Description. In: *Advances in Computer Aided Manufacture*, edited by D. McPherson, North-Holland, Amsterdam, pp. 137–150.

17. Henderson, M. R. (1984). *Extraction of Feature Information from Three Dimensional CAD Data*, Ph.D. Dissertation, Purdue University.

18. Irani, R. K., Kim, B. H., and Dixon, J. R. (1989). Integrating CAE, Features, and Iterative Redesign to Automate the Design of Injection Molds, *ASME Computers in Engineering Conference*, Anaheim.

19. Jakubowski, R. (1982). Syntactic Characterization of Machine Part Shapes, *Cybernetics and Systems*, Vol. 3, No. 1.

20. Jared, G. (1984). Shape Features in Geometric Modeling. In: *GM Seminar on Solid Modeling*, edited by M. S. Pickett and J. W. Boyse, Plenum Press, New York.

21. Joshi, S., and Chang, T. C. (1988). Graph-Based Heuristics for Recognition of Machined Features from a 3-D Solid Model, *Journal of CAD*, Vol. 20, No. 2, pp. 58–66.

22. Kyprianou, L. (1989). *Shape Classification in Computer Aided Design*, Ph.D. Dissertation, University of Cambridge.

23. Luby, S. C., Dixon, J. R., and Simmons, M. K. (1986). Creating and Using a Features Database, *Computers in Mechanical Engineering*, Vol. 5, No. 3.

24. Parkinson, A. (1985). The Use of Solid Models in BUILD as a Database for N.C. Machining, *Proceedings of PROLAMAT 1985*, Paris, pp. 293–299.

25. Parks, R. D., and Chase, T. R. (1989). Representing Mechanical Parts Using Feature Specifications and Positional Constraints: A Contrast with PDES, *ASME Computers in Engineering Conference*, Anaheim.

26. Pinilla, J., Finger, S., and Prinz, F. (1989). Shape Feature Description and Recognition Using an Augmented Topology, NSF Design Eng. Conference, University of Massachussetts, Amherst, 1989.

27. Pratt, M. J., and Wilson, P. R. (1985). Requirements for Support of Form Features in a Solid Modelling System—Final Report, *CAM-I Report*, R-85-ASPP-01.

28. Preiss, K., and Kaplansky, E. (1983). Automatic Mill Routing from Solid Geometry Information, *Proceedings of the 1st International IFIP Conference on Computer Applications in Production and Engineering*, Amsterdam, The Netherlands.

29. Shah, J., Sreevalsan, P., Rogers, M., Billo, R., and Mathew, A. (1988). Current Status of Features Technology, Report for Task 0, R-88-GM-04.4, *CAM-I*, Arlington, Texas.

30. Shah, J., and Rogers, M. (1988). Functional Requirements and Conceptual Design of the Feature-Based Modeling System, *Computer Aided Engineering Journal*, Vol. 5, No. 1, February, pp. 9–15.

31. Shah, J. and Rogers, M. (1988). Expert Form Feature Modeling Shell, *Computer Aided Design*, Vol. 20, No. 9.

32. Shah, J. (1989). Feature Transformations Between Application Specific Feature Spaces, *Computer Aided Engineering Journal*, Vol. 5, No. 6, February, pp. 247–255.

33. Shah, J., Rogers, M., Sreevalsan, P., and Mathew, A. (1989). Functional Requirements for Feature Based Modeling, Report for Task I, R-89-GM-01, CAM-I, Arlington, Texas.

34. Shah, J., Rogers, M., Sreevalsan, P., Hsiao, D., Mathew, A., Bhatnagar, A., Liou, B., and Miller, D. (1990). The ASU Feature Testbed: An Overview, Technical Report, Department of Mechanical & Aerospace Engineering, Arizona State University.

35. Shah, J., and Mathew, A. (1991). An Experimental Investigation of the STEP Form Feature Information Model, *Computer Aided Design*, Vol. 23, No. 4, May, pp. 282–296.

36. Staley, S., Henderson, M., and Anderson, D. (1983). Using Syntactic Pattern Recognition to Extract Feature Information from a Solid Geometric Model Data Base, *Computers in Mechanical Engineering*, September, pp. 61–66.

37. Turner, G. P., and Anderson, D. C. (1988). An Object Oriented Approach to Interactive, Feature Based Design for Quick Turn Around Manufacturing, *ASME Computers in Engineering Conferences*, San Francisco, July/August.

38. Woo, T.C. (1982). Feature Extraction by Volume Decomposition, *Proceedings Conference on CAD/CAM Technology in Mechanical Engineering*, Cambridge, MA, March.

CHAPTER THREE

Moving Objects in Intelligent Design

TOSIYASU L. KUNII, YOSHIHISA SHINAGAWA, and HIRONOBU GOTODA

Department of Information Science, Faculty of Science, The University of Tokyo

3.1. INTRODUCTION

When designing any object, we can do better if the designing system can work intelligently to show us how the designed object operates or is used for us to be able to judge whether the design satisfies our needs. Such a design system can also be used effectively to design manufacturing processes and assembly processes of the designed object and to tell us intelligently whether the design is manufacturable and assemblable. The intelligent design system of this type can be realized only if the system is based on the knowledge of how the object designed and the related objects, such as machines used to manufacture and assemble, move and also on the knowledge of how the objects are viewed and displayed. The generalized knowledge of moving objects is presented here as a four-dimensional (4D) model, and the generalized knowledge of how to view and display the moving objects is presented as a part of the 4D model. The 4D model consists of the following elements:

1. The topology of shapes that can be either continuous, as in the case of analytical surfaces, or discrete, as in the case of volume rendered images and textures.
2. The geometry of the shapes that can also be either continuous or discrete.
3. The other attributes of the shapes that denote properties that are either visible or invisible and also those that are physical or chemical.
4. A set of topological shape operators to define, update, and search the topology.

Intelligent Design and Manufacturing, Edited by Andrew Kusiak.
ISBN 0-471-53473-0 © 1992 John Wiley & Sons, Inc.

5. A set of geometric shape operators to define, update, and search the geometry.

6. A set of view operators to define a viewer in terms of viewpoints and view angles to limit the scope of the topology and the geometry of the shapes that are to be viewed.

7. A set of display operators to map what is viewed to given (logical or physical) display devices.

8. A set of database operators to handle the other attributes.

Two case studies of the 4D model given above are shown in this chapter. In each case, topological or geometrical knowledge is adequately used to represent the shapes of objects. With such knowledge it is shown that the 4D model becomes compact and can be used in various scientific or engineering applications.

The first case study concerns the movement of a viewer inside an object. The automatic generation of the path followed by the viewer while going inside an object is described. One application is in medicine to simulate the effect of a gastroscope or a needle otoscope. Furthermore, in speleology, walking through caves reconstructed from topographic data is of great significance. The method can also be used for traversing a 3D maze. Such an animation is referred to as a "walk-through" animation. To generate the path, it is assumed that the Reeb graph of the object is known. The knowledge contained in the Reeb graph is used to reconstruct the surface of the object from a given set of contour lines.

The second case study concerns the representation of the shapes of soft objects such as garments. When a garment is deformed under dynamic constraints, wrinkles are formed or extinguished showing interesting patterns of shape changes. Studying such shape changes of wrinkles is crucial in developing fashion or interior CAD systems. While the knowledge used in the previous case is topological, here the geometrical knowledge deduced from singularity theory is applied. Singularity theory allows us to model qualitative aspects of the shapes of garment wrinkles. The display examples of this study show the usefulness of our 4D model.

3.2. AUTOMATIC GENERATION OF THE VIEW FUNCTION

The generation of the path of the viewer is necessary, for example, in simulating the effect of a gastroscope or a needle otoscope (18, 19, 20) and to create an animation. The problem is reduced to generating a "view function" that passes through the objects (24). In the rest of this chapter, the term "view function" refers to a function that gives the location of the viewpoint as a function of time.

This situation is totally different from ordinary path planning, where it is necessary to move among objects and the motion is restricted to a plane.

For a "walk-through," the path traverses the inside of complicated objects and is not restricted to a plane. Furthermore, it is difficult to constitute good cost functions (5) because the objects consist of a large number of surface patches.

The method for generating the view function depends on how the interior of the object is described. One of the commonly used descriptions is in the form of cross-sectional data, for example, as in the case of CT images (11) or photographs of serially sectioned celloidin specimens. In this chapter, the contour lines on each cross-sectional plane are used.

To generate the view function, it is assumed that the Reeb graph of the object is known. A Reeb graph is used to denote the topological structure of objects and is used to decide the locus of the view function. Then a method to compute the geometrical location of the viewpoint on each cross section is proposed. By combining this geometrical data with the topological information in the Reeb graph, the view function is generated.

3.2.1. Reeb Graph

In this section, a Reeb graph is introduced to determine the topological shape of the view function. A Reeb graph represents the topological "skeleton" of the 3D object. The locus of the function is decided according to the given Reeb graph. George Reeb first introduced this graph in his thesis (for details, see ref. 28).

Definition. Let $f : M \rightarrow R$ be a real-valued on a manifold M (2). The Reeb graph of f is the quotient space of the graph of f in $M \times R$ by the following equivalence relation: $(X_1, f(X_1)) \sim (X_2, f(X_2))$ holds if and only if $f(X_1) = f(X_2)$ and X_1, X_2 are in the same connected component of $f^{-1}(f(X_1))$.

First, the surfaces of the 3D objects to be walked through are considered to be two-dimensional manifolds in R^3. Then the Reeb graph of the height function $h(X)$ on these manifolds is considered. Here $h(X)$ gives the height of the point on the manifold $X = (x_1, x_2, x_3)$, where $x_1, x_2, x_3 \in R$ and the x_3 axis is set to be perpendicular to each cross-sectional plane of the objects, that is,

$$h(x_1, x_2, x_3) = x_3 .$$

For simplicity, we assume that the equation of the ith cross-sectional plane from the bottom is $x_3 = z_i$ and this plane is referred to as the ith cross-sectional plane. For example, the Reeb graph of the height function of the torus shown in Fig. 3.1a is as in Fig. 3.1c. This is easy to see when we consider the cross-sectional planes as in Fig. 3.1b; all the contour lines on each plane are represented by a representative point in the Reeb graph. The

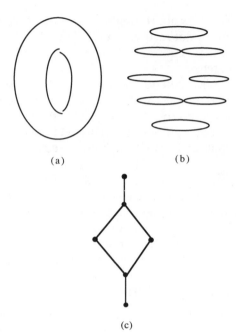

(a)

(b)

(c)

FIGURE 3.1. (a) A torus, (b) its cross sections, and (c) its Reeb graph.

graph shows the "skeleton" of the manifold and so the topological shape of the locus of the view function. The Reeb graph can also be used to reconstruct a surface between consecutive contours. That is, when there is a path from the equivalence class of one contour to that of the other, there is a surface patch between the two contours.

3.2.2. Choice of Representative Values

The Reeb graph loses its geometrical information on the $x_1 x_2$ plane. By the recovery of the geometrical information, the locus of the view function is completely decided. We recover this information by choosing a representative value from each equivalence class; we associate a point with each contour line on the plane and the locus of the view function passes through this point. For example, we can associate each contour line with its center.

The locus of the view function must lie in the interior of the object. Therefore the representative point of each equivalence class must lie in the interior of each contour line that it represents. We assume that the contours are approximated by polygons. The easiest choice is the center of gravity of each polygon. However, it is not always in the interior as shown in Fig. 3.2. The center of the kernel (22) of the polygon that approximates a contour line seems to be a reasonable choice because from the center we can view the whole contour. As illustrated in Fig. 3.3, there are, however, cases where the kernels do not exist.

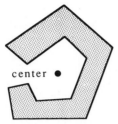
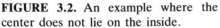
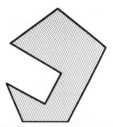

center ●

FIGURE 3.2. An example where the center does not lie on the inside.

FIGURE 3.3. The kernel is absent.

Therefore we choose the representative point as follows. In the following part, each cross-sectional plane $x_3 = z_i$, which is determined by the CT images or the slices of specimens, is referred to as the ith cross-sectional plane $(i = 0, 1, \ldots, n)$. First, the representative value of the bottommost cross-section plane (0th cross-sectional plane) is computed. Next, we compute the representative value of the contours on the upper cross-sectional plane. This process is iterated until the topmost cross-sectional plane and the representative values are determined on all the cross-sectional planes. Finally, the representative points on the cross sections are linearly interpolated to get the path of the viewpoint. In other words, the viewpoint moves along the line segments that connect each representative point on the cross sections. Instead of a linear interpolation, a cardinal spline (7) can be used, but this complicates the implementation.

Computation of each representative point of the contour on the cross-sectional plane is done as follows. First, we map the polygon that represents the contour on the ith cross-sectional plane onto a unit square centered at the origin by translating and scaling (see Fig. 3.4). This process is referred to as the normalization ν_i. The normalized image of the representative point of the previous cross-sectional plane [the $(i - 1)$th cross-sectional plane] by the map ν_{i-1} is also brought in to the unit square. This point is referred to as R. For the bottommost cross-sectional plane $(i = 0)$, R is set to the origin. Then R is tested to see if it lies inside the normalized polygon on the ith cross-sectional plane. If R is contained in the normalized polygon, it is

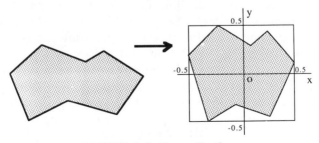

FIGURE 3.4. Normalization.

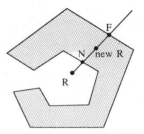

FIGURE 3.5. Selection of the representative point.

chosen as the representative of its equivalence class. Otherwise, the point inside the polygon near R is chosen. To be precise, we search for the edge of the polygon that is nearest to R first. Here, the distance between an edge E and a point R is defined as

$$\min_{p \in E} (p, R) \,,$$

that is, minimum distance between R and a point included in E. Then an open ray from R that passes through the middle of the edge referred to as the point N is computed. The intersection of the ray and each edge of the polygon is then calculated. The intersection second nearest to R is referred to as the point F. The representative of this contour is the midpoint of the line joining the point N and F (see Fig. 3.5). Finally, the representative R' is obtained by ν_i^{-1}.

3.2.3. View Direction and Other Implementation Details

Line of Sight, Twist Angle, and Velocity. The line of sight and the twist angle at the viewpoint are essentially independent of the view function. In this chapter, however, they are defined using the view function.

There are several ways to define the line of sight. One way is to use the tangent line of the view function, that is, df/dt, where $f : \mathbf{R} \to \mathbf{R}^3$ is a view function. Since the locus of the view function consists of linear line segments, the line of sight at the viewpoint is made to coincide with the line segment that the viewpoint is on. When the viewpoint is at an end point of a segment, the tangent line is not defined. In order to avoid a jump in the line of sight at each end point, the line of sight changes smoothly from the direction of the previous line segment to the present one (see Fig. 3.6). When a cardinal spline is used instead of a linear interpolation, the line of sight is just the tangent line of the curve at the viewpoint.

When the interval between the adjacent cross-sectional planes is small, the direction of the tangent line changes rapidly. In this case, the direction of the line of sight at the time t is set to be

$$f(t + \Delta t) - f(t) \,,$$

where Δt is a constant. This corresponds to "always looking ahead."

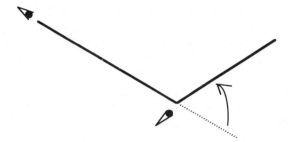

FIGURE 3.6. Movement of the line of sight.

FIGURE 3.7. Organs in the external, the middle, and the inner ear.

FIGURE 3.8. Walk-through in the organs.

The twist angle is defined to be the right-hand rotation about the line of sight. In our implementation, it is kept constant during the walk-through. The time needed to move from one cross-sectional plane to another is kept constant in the implementation. This defines the velocity of the viewpoint.

Display Examples. We now present examples of the finished product. Figure 3.7 shows the organs in the external, middle, and inner ear. Figure 3.8 shows the walk-through in the organs.

3.3. GENERATING SMOOTH SURFACES FROM CROSS-SECTIONAL DATA BASED ON HOMOTOPY MODEL

By using the topological information in the Reeb graph of the object, surface patches are generated between the adjacent contours. That is, when there is a path from one equivalence class of one contour to that of the other, there is a surface patch between the two contours. For the surface generation, the triangular tile technique has been popular: Triangular patches are generated between adjacent contours that are approximated by linear line segments (4, 6, 9). Spline approximation was also used: Surfaces are reconstructed with a spline approximation (30). The homotopy model (23, 25) presented here uses a homotopy and aggregates both the triangular tile technique and the spline approximation overcoming their drawbacks. The homotopy model consists of continuous toroidal graph representation and homotopic generation of surfaces from the representation. The existing methods considered each contour line as a set of line segments. In the homotopy model, contour lines are considered as shape functions and the surface generated is the locus of the homotopy that transforms the function of one contour to that of the other. This enables us to handle the problem continuously, not discretely. The continuous toroidal graph shows how each contour line is parametrized as a function.

As for the triangular tile technique, Fuchs et al. (9) used the toroidal graph representation and, based on graph theory, illustrated an example that minimizes surface area. This approach requires a great deal of computation to get the optimal solution. Christiansen et al. (6) provided a simpler triangulation scheme based on the shortest diagonal algorithm, which can be regarded as a kind of greedy algorithm. Their approach also included handling branches, which is adopted in our implementation. While this scheme works well when adjacent loops are similar in shape, it produces defective triangles if the consecutive contours vary widely. Ganapathy and Dennehy (10) improved the coherence between a contour pair by transforming the contours such that the perimeter of each is exactly equal to one. Their method was based on local constraints only. Kaneda et al. (13) proposed the addition of another condition to the shortest diagonal algorithm to remedy this problem. This condition, however, cannot always be satisfied and the algorithm is still greedy.

Essentially, triangulation algorithms generate defective triangles if the consecutive contours differ widely in shape, as in Fig. 3.9a (q_3–p_0–p_1). "Defective" means that the normal vectors of the triangles are perturbed and the surface generated appears to have artificial wrinkles or folds. In Fig. 3.9, for example, the normal vector N_A at the point A is much different from the normal vector N_B of the point B, which is in the neighborhood of A, while N_A is the same as the normal vector N_C at the point C, which is far from A. Therefore the surface appears to have a wrinkle around the point A. The wireframe representation of an example is as in Fig. 3.9b and its shaded image is shown in Fig. 3.9c, where the defective triangle is painted in red. When we have another contour that has the same shape as the contour p_0–p_1 above the contour q_0–q_1–$\cdots$–q_5, the surface generated is as in Fig. 3.9d, where the defect becomes more visible. The artificial wrinkle generated cannot be eliminated even when we use smooth shading, as shown in Fig. 3.9e. This problem cannot be solved by modifying the connection of

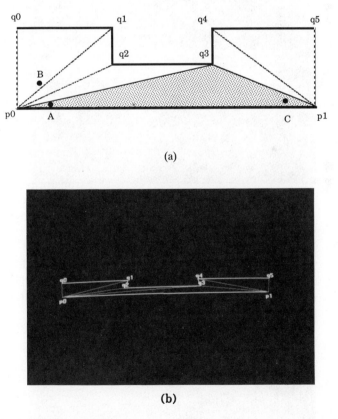

(a)

(b)

FIGURE 3.9. (a) Defective triangles that are inevitable when the triangulation method is used, (b) the wireframe representation of an example, (c) its shaded image, (d) the surface generated with the same contour shape as p_0–p_1 above the contour q_0–q_1–$\cdots$–q_5, and (e) its smoothly shaded image.

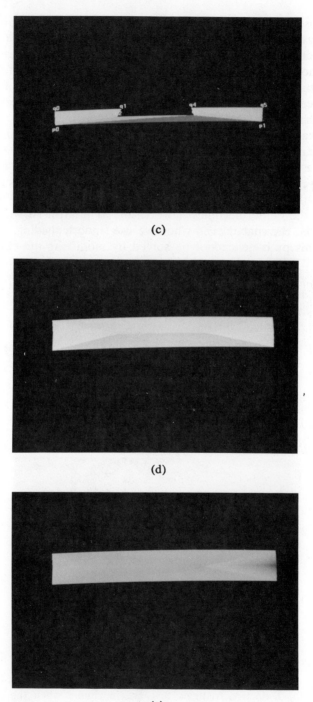

(c)

(d)

(e)

FIGURE 3.9. (*Continued*).

nodes because no node exists between p_0 and p_1. Therefore the surface generated by the triangulation method depends heavily on the choice of the nodes of each contour. We show that the problem is solved by using the homotopy model based on the continuous toroidal graph. There is another problem with the triangulation method. That is the smoothness of the surface generated. Although triangulated surfaces with smooth shading seem smooth, the surface normals are ambiguous and it is hard to understand the exact shape of the objects when the shape is complicated. It is necessary to obtain absolutely smooth surfaces when observation of the exact shape is required. This is also realized by using the homotopy model.

3.3.1. Homotopy

In this section, the concept of a "homotopy" (e.g., see ref. 2) is described. First, a homotopy is defined as follows.

Definition. Let f, $g : X \to Y$ be maps, where X and Y are topological spaces. Then f is homotopic to g if there exists a map $F : X \times I \to Y$ such that $F(x, 0) = f(x)$ and $F(x, 1) = g(x)$ for all points $x \in X$. Here $I = [0, 1] \subset \mathbf{R}$. This map F is called a homotopy from f to g and we shall write $f \cong g$. If for some subset A of X

$$F(a, t) = f(a) \quad \text{for all } a \in A, \text{ for all } t \in I$$

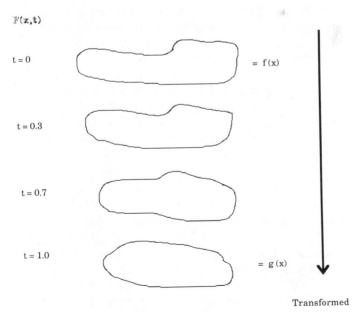

FIGURE 3.10. Homotopy.

holds, f is said to be homotopic to g relative to A and is written $f \cong g$ rel A. When F is defined by $F(x, t) = (1 - t)f(x) + tg(x)$, it is called a straight-line homotopy (e.g., see ref. 2).

We use the homotopy to generate surfaces connecting consecutive contours; that is, the surface connecting consecutive contours is represented by the locus of the homotopy that transforms one of the contours to the other (see Fig. 3.10). In Fig. 3.10, the upper contour is represented by the function f and the lower by g. The surface generated is the locus of the homotopy from $f \cdot u$ to $g \cdot v$, where (u, v) are functions on the continuous toroidal graph that indicates the correspondence between the contours. The next section discusses the toroidal graph in detail.

3.3.2. Toroidal Graph

Toroidal Graph Representation. The continuous toroidal graph shows how each contour line is parametrized as a function. Before discussing the continuous toroidal graph, it is necessary to present the original discrete version of the toroidal graph.

First, assume that a contour line is to be approximated by a string of linear line segments. Let one contour be defined by a sequence of m distinct contour points $P_0, \ldots, P_{m-1}$, and let the other contour be defined by $Q_0, \ldots, Q_{n-1}$. Let us assume that the orientations of both loops are the same. As Christiansen and Sederberg (6) pointed out, triangulation must satisfy two conditions: If two nodes of the same contour are to be defined as the nodes of the same triangle, they must neighbor each other on the contour line. Also, no more than two vertices of any triangle may be recruited from the same contour line. Fuchs et al. (9) reduced this rule to one in graph theory. They represented mutual topological relations of triangles in a toroidal graph, which is adopted as the basic representation in this chapter. In this graph, vertices correspond to the set of all possible spans between the points $P_0, \ldots, P_{m-1}$ and the points $Q_0, \ldots, Q_{n-1}$ and the arcs correspond to the set of all the possible triangles (see Fig. 3.11). The graph of an acceptable surface we consider here is connected and is in the following form: For every vertex of the graph, one arc is incident to it and one arc is incident from it.

Continuous Version of the Toroidal Graph. This section introduces a continuous version of the discrete toroidal graph. First, the lower and the upper contours must be represented by parameters: Points on the contours are designated by a function $f, g : I \to \mathbf{R}^3$, where $I \subset \mathbf{R}$ is an interval $[0, 1]$ and $f(0) = f(1)$ and $g(0) = g(1)$. In the continuous toroidal graph, horizontal and vertical distance between two vertices represents the difference of the parameter values between the two.

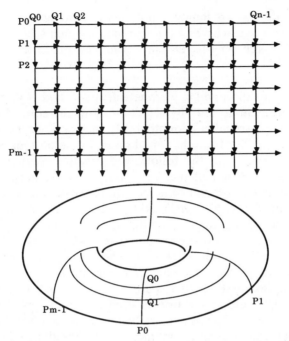

FIGURE 3.11. The toroidal graph representation (9).

Example 1. Arc length is used as the parameter. Let l_1 and l_2 be the length of the lower and the upper loops. For example, the horizontal distance between (P_{i1}, Q_{j1}) and (P_{i2}, Q_{j2}) represents the arc length between P_{i1} and P_{i2} over l_1 and the vertical length represents the arc length between Q_{j1} and Q_{j2} over l_2. The point (x, y) on the graph represents the pair of P and Q, where P is the point whose arc length from P_0 is xl_1, and Q is the point whose arc length from Q_0 is yl_2.

There are various ways to choose the parameter.

Example 2. When the shapes of contour lines are close to circles, the argument can be used as the parameter. In this case, it is essentially the same as with the cylindrical coordinate system.

The contour lines need not be approximated by linear segments. Parametric curves such as the Bezier curve, the B-spline curve, or the cardinal spline curve can be used instead.

Example 3. The parameter of the spline basis function is used as the parameter of the graph. To use the linear interpolation discussed later, arc-length representation is desirable. However, the conversion from the curve parameter to the arc length is not easy to compute. One remedy is to approximate the spline curves by linear line segments and use its length instead of actual arc length. Display examples presented later use this approximation.

When a path passes through (x, y), where $f(x)$ is P and $g(y)$ is Q, it means that P and Q are "connected" by a homotopy discussed in detail later.

   As long as the parameter monotonously increases as the point on the contour line goes farther from the initial node, it does not matter essentially what parameter is used in the following discussion. Without any loss of generality, it can be assumed that the parameters are normalized ($x, y \in [0, 1]$). The continuous version of an acceptable path is represented on this graph as a graph of $(x, y) = (u(s), v(s))$, where $u : I \to I$ and $v : I \to I$ are monotonously increasing functions with $u(0) = 0$, $u(1) = 1$, $v(0) = 0$, and $v(1) = 1$. Figure 3.12 shows an example. The discrete toroidal graph is the special case of its continuous version and the conversion is straightforward (see (Fig. 3.13): Its graph is the concatenation of that of $y = q_i$ and $x = p_i$. The surface represented by this path is as in Fig. 3.14. As for the connection, straight-line connection is used as an example. It represents triangles, all the points on the base being connected with the opposite vertex. The heuristic triangulation method proposed by Ganapathy and Dennehy (10) can be seen as the approximation of $y = x$ by a step function when we use the continuous toroidal graph.

   Next, it is necessary to decide the trail of the path in detail. In this chapter, "linear interpolation" of "closest pair vertices" is proposed to decide the trail. The closest pair vertices are defined as follows.

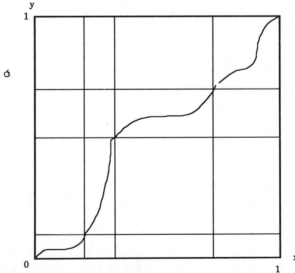

FIGURE 3.12. An acceptable path on the continuous toroidal graph.

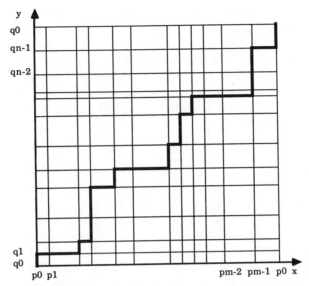

FIGURE 3.13. The natural expansion of a discrete toroidal graph to a continuous version.

Definition. A vertex (P_i, Q_j) is a *closest pair vertex* if

$$d(P_i, Q_j) = \min_{0 \leq k < n} d(P_i, Q_k)$$
$$d(P_i, Q_j) = \min_{0 \leq k < m} d(P_k, Q_j)$$

hold. Here $d(P, Q)$ is the distance between the points P and Q.

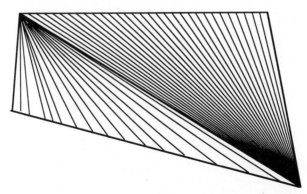

FIGURE 3.14. Correspondence between the two adjacent contours of a triangular mesh.

Acceptable paths that connects all the closest pair vertices do not always exist. Therefore connecting as many closest pair vertices as possible is examined next. Actually, we explored a simpler approach that groups closest pair vertices (23). We do not discuss it in detail here.

Finally, the details of the path are determined on this graph. Suppose the chosen closest pair vertices are (x_i, y_i) $(i = 0, 1, \ldots, k-1)$ and $0 = x_0 < , \ldots, < x_{k-1}$. The path is to be represented as functions $u(s) = s$ and $v(s); I \rightarrow I$, such that

$$v(s) = (y_{i+1} - y_i) \frac{(s - x_i)}{(x_{i+1} - x_i)} + y_i \qquad (x_i \leq s < x_{i+1}).$$

As stated previously, this is the linear interpolation of (x_i, y_i) and (x_{i+1}, y_{i+1}) (see Fig. 3.15). The surface represented by this path is as in Fig. 3.16. This example shows a loft surface. Comparing Fig. 3.9 and Fig. 3.16, it is obvious that the surface in Fig. 3.16 is smoother than that in Fig. 3.9 and the defective triangles are avoided. Figure 3.17 shows an example corresponding to Example 3. The homotopy model discussed later uses the toroidal graph representation expressed by $(u(s), v(s))$ to generate surface between contour lines.

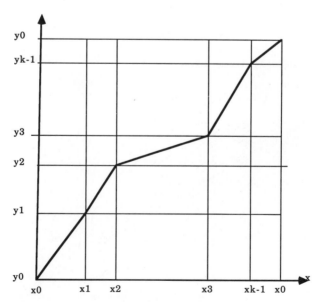

FIGURE 3.15. Linear interpolation of the closest pair vertices.

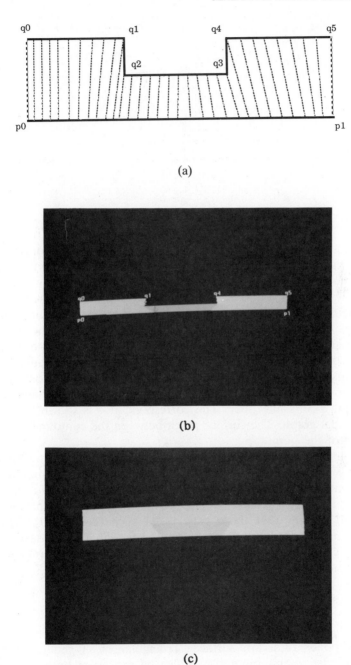

FIGURE 3.16. (a) A surface generated by a straight-line homotopy, (b) its shaded image, and (c) the surface generated with the same contour shape as p_0–p_1 above the contour q_0–q_1–$\cdots$–q_5.

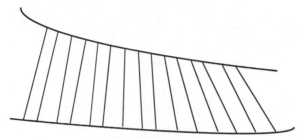

FIGURE 3.17. Correspondence between smoothly curved contours.

3.3.3. Surface Generation Based on the Homotopy Model

It has previously been stated that a vertex on the continuous toroidal graph represents the correspondence between the point of the upper contour line and that of the lower contour line and that corresponding points are connected by a homotopy. When the homotopy is a straight-line homotopy, it simply means that the two points are connected by a straight-line segment. This section discusses the surface generation in detail. In the homotopy model, the surface patch between the adjacent contours is generated by connecting corresponding points on each contour by a homotopy. In other words, in this model, all the points on a contour have their corresponding points on the adjacent contours and a homotopy is used for connecting the corresponding points like fiber. The correspondence is shown by the continuous toroidal graph. The surface patch between the contours is regarded as the locus of the homotopy.

To be more precise, let the lower and the upper contour lines be expressed by maps $f, g : X \to \mathbf{R}^3$, where $X = [x_0, x_1] \subset \mathbf{R}$. For simplicity, let the lower contour line be on the plane $z = 0$ and the upper $z = 1$ and $X = I$. In this chapter, "two-points $P(p_x, p_y)$ and $Q(q_x, q_y)$ connected by a homotopy F" means that there exists $s \in I$ such that $F(s, 0) = f(u(s)) = (p_x, p_y, 0)$ and $F(s, 1) = g(v(s)) = (q_x, q_y, 1)$ hold. That is, the surface to be generated is expressed by the homotopy between $f(u(s))$ and $g(v(s))$. When straight-line homotopy is used, it is expressed as

$$F(s, t) = (1 - t(f(u(s)) + tg(v(s)) ,$$

which corresponds to a loft surface. Triangular mesh generated by the Fuchs's or Christiansen's method can also be expressed by a straight-line homotopy. Cardinal splines (7) can also be used. Let the function representing a series of four contour lines be f_{-1}, f_0, f_1, f_2 and the respective acceptable paths be $(x, U_{-1}(x))$, $(x, U_0(x))$, $(x, U_1(x))$. Then the surface between f_0 and f_1 is

$$F(s, t) = w_{-1}(t)f_{-1}(U_{-1}^{-1}(x)) + w_0(t)f_0(x) + w_1(t)f_1(U_0(x))$$
$$+ w_2(t)f_2(U_1(U_0(x))) ,$$

where

$$[w_{-1}(t) \quad w_0(t) \quad w_1(t) \quad w_2(t)] = [t^3 \quad t^2 \quad t \quad 1] \begin{bmatrix} -a & 2-a & -2+a & a \\ 2a & -3+a & 3-2a & -a \\ -a & 0 & a & 0 \\ 0 & 1 & 0 & 0 \end{bmatrix}.$$

When each f_i is represented by a cardinal spline function, the surface is referred to as the cardinal spline surface and a display example is presented later. When each f_i is represented by a B-spline, this equation is similar to that of Wu et al. (30). However, the major difference is in the use of U_i, which makes it possible to handle the parameter values more precisely. This is useful for reconstructing complicated objects.

When two contours contain different numbers of loops, branch handling is necessary. Christiansen's (6) method is one case of this branching and is used in our implementation.

3.3.4. Display Examples

This section presents examples of the finished product. Figure 3.18 shows the three auditory ossicles (malleus, incus, stapes). The outline curves are approximated by the cardinal spline and a straight-line homotopy is used for surface reconstruction. As noted in Example 3, the arc length of the cardinal spline curve is approximated by the length of linear line segments. Figure 3.19 shows the same objects reconstructed by using a homotopy that corresponds to the cardinal spline surface. The outline curves are approxi-

FIGURE 3.18. Three auditory ossicles reconstructed by a straight-line homotopy.

FIGURE 3.19. Three auditory ossicles reconstructed by a homotopy corresponding to a cardinal spline surface.

FIGURE 3.20. Three auditory ossicles reconstructed using the triangulation method with Gouraud shading (20).

mated by the cardinal spline. Figure 3.20 shows the same objects reconstructed by Christiansen's triangulation method with Gouraud shading. As mentioned earlier, the shape looks ambiguous. In these figures, it is quite obvious that the homotopy model is superior to the others in generating a surface from a set of contour lines.

3.4. MODELING OF GARMENT WRINKLES

3.4.1. Fashion Design Systems

Fashion designing is a combination of art, industry, and commerce. The design process has a number of stages including initial sketch by a designer and extraction of patterns from the initial sketch by a pattern-making expert. The extracted patterns, when assembled, are expected to satisfy the original concepts of the designer. The traditional tools of the designer have long been limited to crayons and paper. But recently, several CAD systems were proposed to assist the designers with graphical editing and 3D previewing tools.

INFADS, proposed by Kunii and his group (15), was one of the earliest fashion design systems. Its attractive features are interactive design facility combined with a color picture database system and the introduction of a recursive concept of "picture" and "texture." Later, Terai (26) analyzed that interactive fashion design systems should meet the following three requirements: (a) design and manufacturing facilities are integrated; (b) draping behavior of garments can be simulated; and (c) display facilities of colors, patterns, and the feelings of the garment are supported. VIRGO, presented by Noma et al. (17) is a system designed to achieve the first requirement, and Terai (26) proposed an attribute grammar approach to the third requirement. There are also other works addressing the problem of (a) or (c), but little has been done to fulfill the second requirement.

Hinds and McCartney (12) presented a system that can display draped shape of garments. They used sinusoidal functions defined on quadrilateral patches to create folds or wrinkles. The resulting images, however, look rather stiff. Although sinusoidal functions may be useful to represent the distortion of "stiff" objects such as a steel plate, they fail to model the shape of garment wrinkles. Because wrinkles are created through the nonlinear interaction of the elements constituting "soft" objects, it is necessary to understand the nature of "softness' for modeling the shapes of wrinkles.

In the following subsections, we discuss the modeling of garment wrinkles. We first investigate the techniques for modeling soft objects, and then present the method to extract characteristic features of wrinkles. Finally, display examples are given, which show that our modeling primitives of wrinkles can suitably represent drastic shape changes observed in wrinkle formation processes.

3.4.2. Modeling of Soft Objects

Since the term soft objects was introduced by Wyvill et al. (31), modeling of soft objects has been studied intensively and now has grown to be one of the most attractive fields in computer graphics and animation. Because soft objects are structurally unstable and often consist of many components, it is extremely difficult to analyze their behavior under various constraints. When the force exerted on objects is not strong and is within a certain range, the shape change of a hard or rigid object occurs on a comparatively small scale and only a few factors, such as Young's modulus, Poisson's ratio, and Lame's constant, are sufficient to determine the behavior. In the case of soft objects, however, many elements interact with each other in a very complicated manner and sometimes even a subtle applied force results in a drastic shape change of the objects. Due to these difficulties in modeling soft objects, it was not until recently that the problem has become fully addressed.

Recently, two major modeling techniques—usually called physically based modeling and geometric modeling—have been devised to tackle these problems. Geometric models (3, 31) use geometric information to represent the shape of objects. A Jacobian (a coordinate transformation matrix) and a scalar field are examples of such geometric information. Although geometric models are simple and easy to program, dynamical constraints are relatively difficult to get incorporated into these models. Thus the resulting objects sometimes behave unrealistically.

Physically based models (21, 27, 29), while computationally more complex than geometric models, offer the realism. A set of differential equations derived from physical laws play a central role here. At first glance, these physically based models with the advent of high-power and low-cost workstations look superior to geometric models, because they seem to be able to simulate what is really occurring. However, physically based models have several difficulties. First, they are computationally too expensive. Since the number of elements making up a soft object is very large, real-time processing becomes impossible. Second, they tend to be unstable under widely changing dynamical constraints. Since differential equations are no more than local descriptions of an object, numerical simulation is necessary to determine the global structure. The resulting global structure depends on the parameter values of the simulation, but not always in a continuous manner. Sometimes it is drastically affected by a small change of the parameter values. To avoid such a drastic change, we have to determine the parameter values more accurately, which is very difficult because of the complex structures of soft objects.

From the above discussions, it follows that physically based models can certainly represent local structures of soft objects better than geometric models, but they are not always suitable to represent the global structure. In contrast, geometric models can represent the global structure but are not always suitable to represent local structures accurately. If we succeed in integrating both modeling techniques, we obtain a model that behaves well locally as well as globally. Our approach is such an attempt to achieve a better modeling of soft objects.

3.4.3. Characteristic Features of Wrinkles

As stated earlier, our major objective is to model the shapes of garment wrinkles. Since a garment is an object composed of many components, it is not practical to simulate its behavior by directly solving differential equations. We prefer to use some global information to reduce the amount of computation.

The existence of wrinkles can serve as such global information. When a garment is deformed, wrinkles are formed or extinguished. Shape changes are mainly observed around wrinkles while the other parts of the garment remain unchanged. In fact, the geometry of the wrinkles and of the other parts seems to be different: At the wrinkles both metric and curvature change, whereas at the other parts metric is preserved and only curvature changes. This difference may best be utilized by employing a mathematical method known as singularity theory (1). This theory provides a mathematical foundation to deal with qualitative geometric changes of a given system.

The basic idea of singularity theory is to consider the singular set of a surface-to-surface mapping. More specifically, a series of projections of a given surface are taken and analyzed. Figure 3.21 shows three typical types of projections. In the framework of singularity theory we are interested only in contours. The theory shows that the signs depicted in Fig. 3.21 are the only stable patterns in general, and the other types of signs are unstable; that is, if we take a projection from a slightly different direction, the pattern is decomposed into some combinations of the signs shown in Fig. 3.21. An important point here is that one can distinguish general types of signs that are stable from special types of signs that are unstable. If no a priori knowledge is assumed on the surface to be analyzed and if a projection is taken in an arbitrary direction, then the sign to be observed is almost always one of those in Fig. 3.21. The other types of signs are too rare to be observed. However, if the surface does have a special structure, one can expect that the rare types of signs are also observed.

In the case of garment wrinkles, the signs appearing in the projections are, in most cases, cusps, folds, or crossing lines. As shown in Fig. 3.22, however, there are special instances where the other types of signs emerge by changing the directions of viewing. This figure shows a situation where a cusp and a fold approach each other (a), then merge (b), and finally depart from each other (c). Note that (a) and (c) show different configurations: the lines that form a cusp and a fold are exchanged in the process of merging. The merged state (b) is classified as the p+ +c singularity (1, 14), which describes the structure of branching.

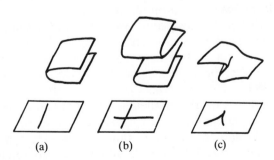

(a)   (b)   (c)

FIGURE 3.21. Typical signs: (a) fold, (b) crossing lines, and (c) cusp.

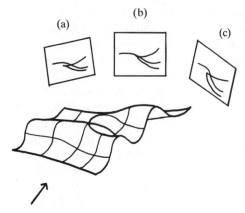

FIGURE 3.22. A case where a p+ +c singularity emerges: (a) cusp and fold, (b) p+ +c singularity, and (c) cusp and fold.

The same p+ +c singularity also describes the structure of vanishing. Branching and vanishing are complementary to each other. A point on a surface can be a branching point when it is viewed from one side and can be a vanishing point when viewed from the other side. Since this singularity is very rare, the behavior of the points corresponding to this singularity will produce far greater constraints than the behavior of the other points. Such points will be the characteristic points of garment wrinkles. By specifying the local structures around the characteristic points, more realistic garment wrinkle images can be synthesized.

3.4.4. Modeling Primitives of Wrinkles

As explained previously, the local structures around the characteristic points remarkably influence the global structures of garment wrinkles. Since the number of characteristic points is small, great data compression can be achieved by representing a wrinkle surface as a collection of characteristic points and the local structures around them.

The basic sets of modeling primitives are the positions and the types (branching or vanishing) or characteristic points. These sets of primitives alone, however, are not sufficient to reconstruct the original shape in a satisfactory way. Other primitives are necessary to specify the local structures around the characteristic points more precisely. There is a trade-off between the simplicity of the modeling primitives and the exactness of representation. Several choice can be considered.

Our choice is to add the contours that are associated with the special singular configurations (see Fig. 3.23) as another set of modeling primitives.

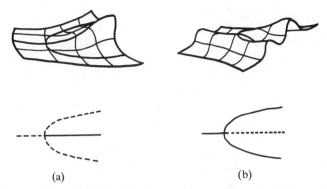

(a)   (b)

FIGURE 3.23. Modeling primitives: (a) branching and (b) vanishing.

This enables us to restrict the number of possible wrinkle shapes to be reconstructed. For specifying the contours, we choose several sample points and interpolate the coordinate values between those points by a spline function. Other choices are of course possible, but our method can produce a reasonably good appearance without requiring so much computation.

In summary, the modeling primitives of garment wrinkles are (a) positions and types of characteristic points and (b) contours associated with characteristic points. To represent the contours, several sample points need to be specified.

It is usually observed that several wrinkles appear on one garment cloth and become connected with each other or diminish at their ends. Our modeling primitives can describe such a complex situation: Connections can be represented by branching points and diminishings by vanishing points. Such a description will yield a graph, as shown in Fig. 3.24, where vertices correspond to the characteristic points and edges to the associated contours. This graph adequately illustrates the "backbones" of wrinkles.

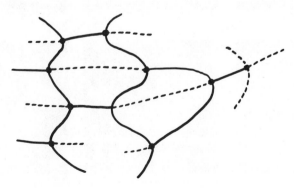

FIGURE 3.24. Global arrangement of wrinkles.

3.4.5. Display Examples

This section gives several examples of wrinkle formation processes. Here we consider the wrinkles formed around the arm of a jacket. The first basic parameter is the angle between the human forearm and the upper arm at the elbow. To facilitate the work of animation, we assume that the characteristic points are not newly created or destroyed during the process of wrinkle formation. Based on this assumption, we also include the initial configuration of characteristic points as the second parameter of this animation.

In the following, we examine two surface reconstruction methods from the modeling primitives (method 1 and method 2). For each case, two animation sequences are shown corresponding to the different initial configuration data: Configuration (a) has 13 branching points and 23 vanishing points; configuration (b) has 16 branching points and 24 vanishing points.

Method 1. The simplest way to reconstruct a surface from the modeling primitives is to reconstruct it cross sectionally by using spline functions. For each step of animation sequence, the positions of characteristic points are first computed, the associated contours are then generated, and finally cross-sectional reconstruction is performed to determine the positions of the points lying on the surface spanned by the associated contours. Figure 3.25 shows the results of this method.

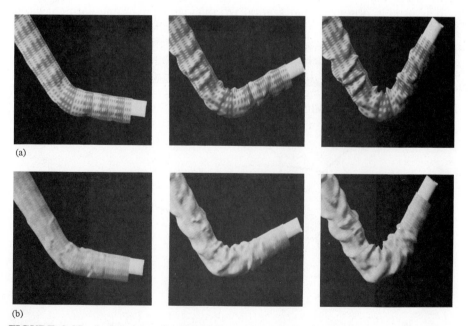

(a)

(b)

FIGURE 3.25. Animation of wrinkle formation process without preserving metric invariance.

Method 2. The spline method works reasonably well when the wrinkles are small. However, when the depth of the wrinkles increases above a certain threshold, the area of the reconstructed surface also increases and sometimes gives an artificial impression. This is mainly due to the fact that the curves generated by spline functions do not preserve their length.

Here we present an alternative method. There are two stages: the approximation stage and the relaxation (optimization) stage. In the approximation stage, we impose the constraint that the metric of the surface does not change. By this constraint we can eliminate the possibility of getting a bizarre pattern, as shown in Fig. 3.25. Of course garments are elastic in reality and at some parts (particularly around characteristic points) the metric may not be completely preserved. Thus in the relaxation stage, we remove this constraint and perform and dynamic simulation only with the constraint that the surface to be generated passes through the characteristic points. As shown in Fig. 3.26, much better garment wrinkle images can be obtained by this method.

Modeling of garment wrinkle formation processes has a certain similarity with modeling of the initial creation processes of the cosmos, where mass was formed from energy. Singularity theory has been applied in the two cases. Can material be modeled as cosmic wrinkles? Intelligent design of the cosmos is beyond our power, but is a challenging theme.

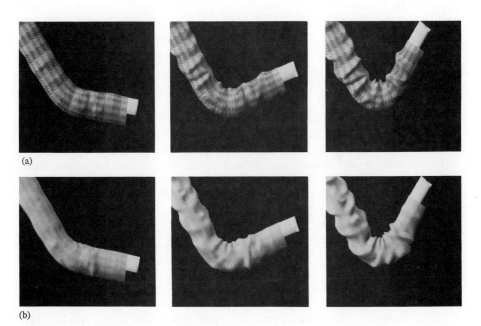

(a)

(b)

FIGURE 3.26. Animation of wrinkle formation processes partially preserving metric invariance.

REFERENCES

1. Arnold, V. I. (1986). *Catastrophe Theory*, 2nd ed., Springer-Verlag, New York.
2. Armstrong, M. A. (1983). *Basic Topology*, Springer-Verlag, New York, pp. 88, 103–105, 169.
3. Barr, A. H. (1984). The Global and Local Deformations of Solid Primitives, *Computer Graphics (Proceedings SIGGRAPH)*, Vol. 18, No. 3, pp. 21–29.
4. Boissonnat, J. D. (1988). Shape Reconstruction from Planar Cross Sections, *Computer Vision, Graphics, and Image Processing*, Vol. 41, No. 1, pp. 1–29.
5. Breen, D. E. (1989). Choreographing Goal-Oriented Motion Using Cost Functions. In: *State-of-the-Art in Computer Animation*, edited by N. Magnenat-Thalmann and D. Thalmann, Springer-Verlag, New York, pp. 367–380.
6. Christiansen, H. N. and Sederberg, T. W. (1978). Conversion of Complex Contour Line Definitions into Polygonal Element Mosaics, *Computer Graphics (Proceedings ACM SIGGRAPH)*, Vol. 12, No. 3, pp. 187–192.
7. Clark, J. H. (1981). Parametric Curves, Surfaces and Volumes in Computer Graphics and Computer-Aided Geometric Design, *Computer Systems Laboratory, Technical Report No. 221*, Stanford University.
8. Dobkin, D., Guibas, L., Hershberger, J., and Snoeyink, J. (1988). An Efficient Algorithm for Finding the CSG Representation of a Simple Polygon, *Computer Graphics (Proceedings ACM SIGGRAPH)*, Vol. 22, No. 4, pp. 31–40.
9. Fuchs, H., Kedem, Z. M., and Uselton, S. P. (1977). Optimal Surface Reconstruction from Planar Contours, *Communications ACM*, Vol. 20, No. 10, pp. 693–702.
10. Ganapathy, S., and Dennehy, T. G. (1982). A New General Triangulation Method for Planar Contours, *Computer Graphics (Proceedings ACM SIGGRAPH)*, Vol. 16, No. 3, pp. 69–75.
11. Herman, G. T., and Liu, H. K. (1979). Three-Dimensional Display of Human Organs from Computer Tomograms, *Computer Graphics and Image Processing*, Vol. 9, No. 1, pp. 1–21.
12. Hinds, B. K., and McCartney, J. (1990). Interactive Garment Design, *The Visual Computer*, Vol. 6, No. 2, pp. 53–61.
13. Kaneda, K., Harada, K., Nakamae, E., Yasuda, M., and Sato, A. G. (1987). Reconstruction and Semi-Transparent Display Method for Observing Inner Structure of an Object Consisting of Multiple Surfaces. In: *Computer Graphics 1987*, edited by T. L. Kunii, Springer-Verlag, New York, pp. 367–380.
14. Kergosien, Y. L. (1981). La fammille des projections orthogonales d'une surface et ses singularités, *Comptes Rendus*, Vol. 291, No. 1, pp. 929–932.
15. Kunii, T. L., Amano, T., Arisawa, H., and Okada, S. (1975). An Interactive Fashion Design System "INFADS," *Computer and Graphics*, Vol. 1, pp. 297–302.
16. Kunii, T. L., and Gotoda, H. (1990). Modeling and Animation of Garment Wrinkle Formation Processes. In: *Computer Animation '90*, edited by N. Magnenat-Thalmann and D. Thalmann, Springer-Verlag, New York, pp. 131–147.
17. Noma, T., Terai, K., and Kunii, T. L. (1986). VIRGO: A Computer-Aided Apparel Pattern-Making System. In: *Advanced Computer Graphics, Proceedings*

of Computer Graphics Tokyo '86, edited by T. L. Kunii, Springer-Verlag, Tokyo, pp. 309–401.

18. Nomura, Y. (1982). A Needle Otoscope, *Acta Oto-laryngo (Stockholm)*, Vol. 93, pp. 73–79.

19. Nomura, Y. (1982). Effective Photography in Oto-laryngology—Head and Neck Surgery: Endoscopic photography of the middle ear. *Oto-laryngology and Head and Neck Surgery, 1982*, Vol. 90, pp. 395–398.

20. Nomura, Y., Okuno, T., Hara, M., Shinagawa, Y., and Kunii, T. L. (1989). Walking Through a Human Ear. *Acta Oto-laryngo (Stockholm)*, Vol. 107, pp. 366–370.

21. Platt, J. C., and Barr, A. H. (1988). Constraint Methods for Flexible Models, *Computer Graphics (Proceedings ACM SIGGRAPH)*, Vol. 22, No. 4, pp. 279–288.

22. Preparata, F. P., and Shamos, M. I. (1985). *Computational Geometry: An Introduction*, Springer-Verlag, New York.

23. Shinagawa, Y., Kunii, T. L., Nomura, Y., Okuno, T., and Hara, M. (1989). Reconstructing Smooth Surfaces from a Series of Contour Lines Using a Homotopy. In: *New Advances in Computer Graphics*, edited by R. A. Eanshaw and B. Wyvill, Springer-Verlag, New York, pp. 147–161.

24. Shinagawa, Y., Kunii, T. L., Normura, Y., Okuno, T., and Young, Y.-H. (1990). Automating View Function Generation for Walk-through Animation Using a Reeb Graph. In: *Computer Animation '90*, edited by N. Magnenat-Thalmann and D. Thalmann, Springer-Verlag, New York, pp. 227–237.

25. Shinagawa, Y., and Kunii, T. L. (1991). The Homotopy Model: A Generalized Model for Smooth Surface Generation from Cross Sectional Data. *The Visual Computer*, Vol. 7, No. 2–3, pp. 72–86.

26. Terai, K. (1987). *Computer-Aided Design of Fabrics*. Master's thesis, Department of Information Science, Faculty of Science, University of Tokyo.

27. Terzopoulos, D., Platt, J. C., Barr, A. H., and Fleisher, K. (1987). Elastically Deformable Models, *Computer Graphics (Proceedings SIGGRAPH)*, Vol. 21, No. 4, pp. 205–214.

28. Reeb, G., (1946). Sur les points singuliers d'une forme de Pfaff completement integrable ou d'une fonction numerique, *Comptes Rendus Acad. Sciences Paris*, Vol. 222, pp. 847–849.

29. Weil, J. (1986). The Synthesis of Cloth Objects, *Computer Graphics (Proceedings SIGGRAPH)*, Vol. 20, No. 4, pp. 49–54.

30. Wu, S., Abel, J. F., and Greenberg, D. P. (1977). An Interactive Computer Graphics Approach to Surface Representation, *Communications ACM*, Vol. 20, No. 10, pp. 703–712.

31. Wyvill, G., McPheeters, C., and Wyvill, B. (1986). Data Structure for Soft Objects, *The Visual Computer*, Vol. 2, No. 4, pp. 227–242.

■■■■■■ CHAPTER FOUR

Qualitative Reasoning Methods in Design

D. NAVIN CHANDRA, S. NARASIMHAN and K. P. SYCARA
School of Computer Science, Carnegie–Mellon University, Pittsburgh, Pennsylvania

Design is the process of generating a description of a physical artifact that satisfies given specifications (including goals and constraints) that describe the desired function of the artifact. Design can thus be viewed as a transformation from the functional domain to the physical domain (26, 32, 36). In order to effect such transformations, a problem solver must be able to reason at different levels of abstraction from the functional to the physical. In the conceptual stages of a design, much of this reasoning is qualitative in nature: Traditional analysis methods cannot be applied to incomplete, conceptual designs. Of the two major phases of a design process, *conceptual design* and *detailed design*, tool-building efforts in CAD have concentrated mainly on the latter phase. Applications have been limited to tasks such as computer-aided drafting, solid modeling, numerical optimization, simulation, and analysis. Examples of such systems are STRUDL (Structural Design Language, ref. 33), SPICE (Electrical Network Analysis, ref. 34), and GDS (a drafting tool, ref. 24). Having made substantial contributions toward the development of tools for detailed design, researchers are now focusing their attention on the earlier phase: conceptual design.

Conceptual design is that part of the design process in which problems are identified, functions and specifications are laid out, and appropriate solutions are generated through the combination of some basic building blocks. Conceptual design, unlike analysis, has no fixed procedure and involves qualitative reasoning. Here are some observations on the role of qualitative reasoning in conceptual design:

Intelligent Design and Manufacturing, Edited by Andrew Kusiak.
ISBN 0-471-53473-0 © 1992 John Wiley & Sons, Inc.

- Conceptual designs are generated without committing to specific details. For example, a sketch of an aircraft's landing gear may not have any dimensions or material specifications. Primitive components are synthesized by reasoning about various combinations of the component behaviors and how they all fit together to provide the desired function.
- Designers often resort to "back-of-the-envelope" assessments to make design decisions. They are able to analyze an incomplete, nonnumeric specification (e.g., a sketch) of a design.
- Designers can interpret and explain the behavior of a design without carrying out elaborate numerical simulations.
- Designers can draw on their own experiences to relate a new design problem to similar designs they have encountered before. This requires an ability to recognize analogical similarities among designs even if they are in completely different domains.

Qualitative physics and qualitative reasoning are areas of artificial intelligence (AI) aimed at automating the reasoning about the behavior of physical systems without relying on numbers. By using a representation that captures the physical characteristics of a device, qualitative physics-based programs can simulate and predict the behavior of the device. This kind of reasoning is called *envisioning*. Most of the research in this branch of AI is aimed at envisioning the behavior of a given device. In design, however, we are interested in doing just the opposite—given a behavior description, one has to come up with a device that will display the required behavior. In conceptual design, qualitative reasoning (QR) tools play a role similar to that of traditional analysis tools in detailed design. As we shall see later, QR methods are being used to evaluate and guide the generation of conceptual designs.

The principles underlying an analysis tool can sometimes be used to guide design decisions. For example, if the design problem can be expressed as a collection of parameters, and if the desired behavior can be expressed mathematically, then numerical solution and optimization techniques can yield appropriate values for the parameters. This approach, however, cannot easily be used for conceptual design where the specification of the artifact to be designed is expressed in terms of a behavioral description. We can, however, use qualitative reasoning. In this chapter we examine approaches to using qualitative reasoning methods in design. The rest of the chapter is organized in two parts. The first part provides a brief overview of qualitative reasoning methods and the second part is a survey of some design systems that use QR methods. Readers who are familiar with QR methods may skip the next part of the chapter and go directly to Section 2.

4.1. BACKGROUND ON QUALITATIVE REASONING

In his "Naive Physics Manifesto," Hayes (14) attempted to formulate a theory on how human beings reason naively about the events occurring in the physical world. Humans don't solve differential equations to predict that a ball thrown up will eventually come down. How are they able to reason in this fashion? That is the question both naive physics and commonsense reasoning attempt to answer. Most of the research in qualitative physics, an offshoot of naive physics, has been centered around the domain of analysis of physical devices and processes.

4.1.1. Qualitative Physics

The basic idea in qualitative physics is that the complexity of a physical system can often be reduced by considering only certain abstractions of its behavior. If we model the system as a set of parameters that can assume real values and represent its behavior as the set of changes that these parameters undergo with respect to time, then one can say that qualitative physics deals with an abstraction of these parameter values. In other words, values of the parameters that model the behavior of the device are supposed to assume only a certain set of values, which are an abstraction of the quantitative real number line. These values are called qualitative values and represent certain intervals on the real number line. The transition points between these intervals are called landmarks.

Quantity Space. Qualitative variables can assume one of a set of qualitative values called the quantity space. A quantity space can be defined as a set of symbols representing qualitative and landmark values. A useful abstraction of a real number is whether it is negative, zero, or positive. These can be represented by the symbols $[-]$, $[0]$, and $[+]$, respectively.

Definition 1. The quantity space, denoted as Q, can be defined as the finite ordered set $\{[-], [0], [+]\}$, where $[-] < [0] < [+]$.

Definition 2. If x. is a quantitative variable taking any real value, then the qualitative value of x is defined as

$$
\begin{aligned}
[x] &= [+] \quad \text{iff } x > 0 \\
&= [0] \quad \text{iff } x = 0 \\
&= [-] \quad \text{iff } x < 0 .
\end{aligned}
$$

The condition that Q should be ordered and finite is what makes the quantity space an abstraction of the real number line.

[x] + [y]

[y] \ [x]	-	0	+
-	-	-	
0	-	0	+
+		+	+

[x] . [y]

[y] \ [x]	-	0	+
-	+	0	-
0	0	0	0
+	-	0	+

FIGURE 4.1. Definitions of qualitative addition and multiplication.

Qualitative Operations. Similar to the arithmetic operations on the real number line, one can define qualitative operations like addition, subtraction, multiplication, and division on the quantity space. The table in Fig. 4.1 lists the definitions of the addition and multiplication operations. Subtraction and division operations can be similarly defined.

Representation Based on Confluences. In qualitative physics, the model of a physical device is a set of qualitative differential equations called confluences (5). Confluences are the qualitative abstraction of the differential equations that model the system and can be derived from the differential equations for the model. Consider the differential equation representing the behavior of the pressure regulator given in the Fig. 4.2:

$$\frac{\partial P}{\partial t} + \frac{\partial A}{\partial y} - \frac{\partial Q}{\partial t} = 0 \, .$$

This equation can be written as the following confluence: $\partial P + \partial A - \partial Q = 0$, where for any quantitative variable x, $\partial x/\partial t$ is written as ∂x for brevity.

According to DeKleer and Brown (5), confluences alone cannot

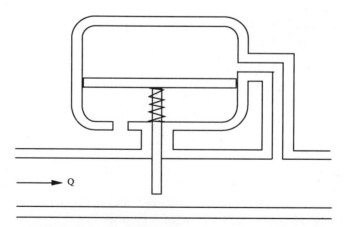

FIGURE 4.2. Schematic diagram of a pressure regulator.

adequately model the behavior of the device. In the above example, the confluence $\partial P + \partial A - \partial Q = 0$ is inadequate because when the valve is closed no flow (Q) is possible, the area (A) is unchanging, and the pressure (P) across the valve is unconstrained. But the confluence states that if $\partial A = 0$ and $\partial Q = 0$, then $\partial P = 0$. The above confluence violates the "no-function-in-structure" principle.

The no-function-in-structure principle states that the laws of parts of the device may not presume the functioning of the whole. For example, consider a light switch. The model of the switch states that "if the switch is on, the current flows and if the switch is off, no current flows." This violates the no-function-in-structure principle because the model is not a universal truth.

Coming back to the pressure regulator model, the contradiction is avoided by including in the model the notion of qualitative states. Qualitative states divide the behavior of a component into different regions, each of which is described by a different set of confluences. The full model for the valve can be expressed as the following:

Open: $[A = A_{MAX}], [P] = 0, \partial P = 0$;

Working: $[0 < A < A_{MAX}], [P] = [Q], \partial P + \partial A - \partial Q = 0$;

Closed: $[A = 0], [Q] = 0, \partial Q = 0$.

Qualitative Process Modeling. Engineering design does not only deal with physical devices but also with physical processes. Next, we discuss the qualitative process theory (QPT) as developed by Kenneth Forbus (12), which deals with such processes.

Any physical process involves a set of objects that are related to one another in a certain way. For the process to occur, certain preconditions in terms of the relations between objects and certain object properties have to be satisfied. A process model should include these in its description of the process.

In QPT, an object is represented in terms of individual views. A given object might behave differently in different contexts. For example, when the surrounding temperature is less than the freezing point, water behaves as a solid, while at temperatures above its boiling point it behaves as a gas and can take various shapes depending on the shape of the container. These differences are captured in the individual views. An example of an individual view is given below.

Individual View: Contained-liquid
Individuals:
   con: a container
   sub: a liquid

Preconditions: con can contain sub

Quantity Conditions: The amount of sub in con is greater than zero

Relations: There is a p such that: p is a piece-of-stuff, the substance of p is sub, p is in con, and the amount of $i(p)$ is equal to the amount of sub in con

Thus an individual view consists of the following (15):

1. *Individuals.* This refers to the set of objects participating in the physical process.
2. *Quantity Conditions.* This refers to the relations that must exist between various parameters of the objects for the process to take place. The relations can be expressed qualitatively.
3. *Preconditions.* This refers to the state of the objects that must be present for the physical process to take place. Preconditions differ from quantity conditions in that they need not be quantitative parameters. For example, for the container to contain water, the precondition is that the container should be able to hold liquids.
4. *Relations.* These refer to physical laws that hold good because of the state of an object. For instance, when the individual view is an ideal gas, the relation $PV = nRT$ holds good.

In QPT, processes are assumed to be the only cause of the change occurring in the system and are represented somewhat like the objects themselves. For example, the boiling process can be represented as follows:

Individuals:

   w: a contained liquid. The boiling temperature of w is temperature$_b$

   hf: an instance of the heat-flow process, such that the destination of hf is w

Quantity Conditions:

   hf is active

   $\sim($temperature$_w <$ temperature$_b)$

Relations:

   There is g such that:

   g is a piece-of-stuff

   g is a gas

   g is of the same substance as w

   temperature$_b$ = temperature$_g$

   Let generation-rate be a quantity such that:

   generation-rate > 0

   generation-rate is proportion to flow-rate$_{hf}$

Influences:

$I - (\text{heat}_w, \text{flow-rate}_{hf})$; counteracting the heat flow's influence

$I - (\text{amount}_w, \text{generation-rate})$

$I + (\text{amount}_g, \text{generation-rate})$

$I - (\text{heat}_w, \text{generation-rate})$

$I + (\text{heat}_g, \text{generation-rate})$

In this process definition, w and hf are variables that are bound to specific instances of individual views or other processes when an instance of the process is created. In QPT notation, for any two variables x and y, $I + (x, y)$ means that x increases with y and $I - (x, y)$ means that x decreases with y.

Qualitative Simulation Systems. Qualitative simulation provides a way of simulating devices. Two approaches that are found in the AI literature have been implemented as ENVISION and QSIM.

ENVISION (5) uses confluences to model a physical system. These confluences can be systematically derived from the ordinary differential equations. Solving these confluences using the various predefined operations, and some domain heuristics, yields the results of the simulation. The qualitative values that any qualitative variable can possess has a direct mapping to its quantitative value. This kind of abstraction is not without side effects. One main side effect is the large number of alternatives arising due to ambiguity. In the other hand, its main advantage is that it allows us to specify values less precisely.

QSIM (19) is another system based on envisionment but it uses an extended quantity space. QSIM simulates by generating possible behaviors and selecting the one that satisfies the qualitative constraints. The system generates new landmark values at which state transitions occur.

4.1.2. Relating Geometry and Behavior: The Configuration Space Approach

We now examine a qualitative reasoning method that interprets behavior based on the geometry of physically interacting objects. The ideas draw on an abstraction of the geometry and kinematics called the *configuration space*. The term configuration space was introduced by Lozano-Perez (20, 21) in the domain of motion planning for robots. Recently, this representation has been applied to the analysis and design of mechanisms by Faltings (8) and Joskowicz (16).

The c-space describes every possible placement of the links of a mechanism. A legal placement is one where links do not physically interfere. A point in the c-shape is a vector of the values of the position and orientation parameters of each link. Since all placements are enumerated in the c-space, any motion can be described as a curve in the legal region of the c-space.

A configuration of a single object is a vector of six parameters, three positions and three orientations. Consider a mechanism with two links. If regarded individually, the two links have a total of 2 times 6, that is, 12 degrees of freedom. However, as two objects cannot overlap in space, some configurations become illegal. Hence the c-space is partitioned into subsets corresponding to legal and illegal placements of the links. The illegal or forbidden region is shaded in all the figures. Regions corresponding to legal configurations, shown as unshaded areas in the figures, are where all motions of the mechanism must occur. (The c-spaces shown in this chapter are all hand-drawn approximations.)

Some simple mechanisms have two links and each link has exactly one degree of freedom. Figure 4.3 (top half) shows such a mechanism where the disk is constrained to rotate about its center and the rod is constrained to translate horizontally as indicated. The configuration vector for this mechanism thus has two elements, namely, Ω, the angular position of the disk, and X, the linear position of the rod. Taken separately, the ranges for X and Ω would have been $[0, +\infty]$ and $[0, 2\pi]$. However, since the rod cannot overlap the disk, its range of motion becomes confined to the intervals $[X_o, +\infty]$. The range of rotations for the disk remains unaffected. The c-space is thus divided into regions corresponding to the free and forbidden placements of the links.

The next example, shown in Fig. 4.3 (bottom half), is slightly more complicated. The addition of the projection on the disk introduces a change in the behavior of the device. The modified c-space now has a notch in the legal region. To understand why this change occurred, let us trace the sequence of events as the cam rotates in a counterclockwise direction. As the left upper tip of the projection touches the follower face, it pushes the

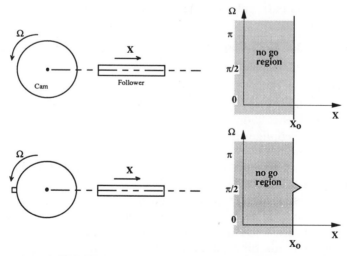

FIGURE 4.3. Simple configuration spaces.

follower. The X value of the follower then begins to increase. This is reflected by the beginning of the notch in the c-space. This continues until a position is reached where the top face of the projection and the bottom face of the follower are parallel and touch. The sequence of events is then reversed and the follower descends. If the height of the projection is gradually reduced and the c-space repeatedly redrawn, it will be seen that the depth of the notch gradually reduces until the original c-space is reached in the limiting case where the projection has a height of zero.

It is usually impractical to calculate the entire c-space of a mechanism in a single step because of the high dimensionality of the space ($6n$, where n is the number of links). A useful approach is to consider the mechanism to be composed of small functional units. The c-space for each individual unit is calculated and these are composed to give the c-space for the entire mechanism. Methods for doing this are suggested by Joskowicz (16), who rigorously proves the validity of this procedure.

With this we have come to the end of the first part of the chapter. We have briefly discussed various qualitative reasoning methods that have been used in engineering design systems. We now review some specific design systems.

4.2. APPROACHES TO QUALITATIVE REASONING IN DESIGN

The design systems discussed here have been placed under three major headings:

1. Systems that synthesize primitives using qualitative methods to guide the search for and evaluation of solutions.
2. Systems that use qualitative methods to reason about analytic equations.
3. Systems that reason about geometry and its relation to kinematic behavior.

4.2.1. Qualitative Methods in Synthesis

Synthesis is the process of configuring a new design from a given set of components such that the final artifact satisfies a given set of functional specifications. The components used in synthesis may either be predetermined primitives or pieces of other designs that are extracted and reused in a new context. Often, these components have to be modified to suit the new context. Examples of work in AI-based synthesis include building and architectural design (13, 22, 28, 35), circuit design (39), and the design of mechanical devices (2, 7, 25, 37). In this section we examine systems that use qualitative reasoning approaches in design synthesis.

Three research efforts are presented here: (a) motion synthesis by connecting the inputs and outputs of primitive mechanisms; (b) Ibis, a system that connects a given set of primitives to satisfy a given goal, Ibis goes beyond motion synthesis by including functional parameters such as pressures and flow rates, and (c) CADET, a system that synthesizes entire designs, or pieces of designs, that are drawn from a database of prior designs by recognizing useful analogies.

Qualitative Motion Synthesis. Mechanical designs can be viewed as being synthesized from conceptual building blocks that perform specific kinematic functions (18). Given the spatial relations among the inputs and outputs of a device, it is possible to find a configuration of the basic building blocks that will realize the desired function.

Some of the primitive mechanisms are (a) a gear pair that provides a rotation–rotation transformation, (b) rack and pinion that provides a rotation–translation transformation, and (c) the cam and follower system that provides a rotation–translation transformation. Kota (18) views these primitives as specific relationships between inputs and outputs. He expresses these relations as matrices relating the basic six degrees of freedom of the input and the output.

Consider, for example, the slider crank mechanism (Fig. 4.4). There are two ways in which one could use the mechanism: One could either rotate the crank and treat the slider movement as an output or one could move the slider and treat the crank's movement as output. In this case, the crank can be rotated only about the x axis and the slider can be moved along the y and z axes.

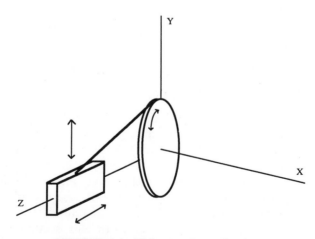

FIGURE 4.4. Slider crank mechanism.

The resulting outputs are indicated in an incidence matrix called the *motion transformation matrix* (MTM). The matrix for the slider and crank mechanism is

$$
\begin{array}{c}
\begin{array}{cccccc} Rx & Ry & Rz & Tx & Ty & Tz \end{array} \\
\begin{array}{c} Rx \\ Ry \\ Rz \\ Tx \\ Ty \\ Tz \end{array}
\left[\begin{array}{cccccc}
0 & 0 & 0 & 0 & 1 & 1 \\
0 & 0 & 0 & 1 & 0 & 1 \\
0 & 0 & 0 & 1 & 1 & 0 \\
0 & 1 & 1 & 0 & 0 & 0 \\
1 & 0 & 1 & 0 & 0 & 0 \\
1 & 1 & 0 & 0 & 0 & 0
\end{array}\right].
\end{array}
$$

The column elements are inputs and the rows are outputs. Consider the first column: A crank in the yz plane can be rotated about the x axis (Rx), and the slider can move anywhere in the yz plane (Ty and Tz). This relationship is indicated by nonzero elements. The matrix represents the basic notion of a slider and a crank for various orientations of the crank: rotation about x, y, and z axes.

In addition to the MTM, the system has several other matrices to capture motion constraints, cost considerations, and physical parameters of the design. The matrix of constraints indicates the directionality of relationships and reciprocating behavior (if any). For every incidence in the MTM, indicating an input–output relationship, the constraint matrix states whether the relationship is bidirectional or unidirectional, reciprocating or not.

A Design Example. Let us consider a synthesis example. The system is given the task of converting a rotation about the x axis (Rx) to a rotation and translation about some axis a, that lies in the xz plane. Let us call this vector Ha. This specification can be written as follows: $Rx \rightarrow Ha$. The specification can be converted into the following decomposition: $Rx \rightarrow Ty \rightarrow Ta \rightarrow Ha$, which corresponds to the following three matrices:

$$
\left[\begin{array}{cccccc}
0 & 0 & 0 & 0 & 0 & 0 \\
0 & 0 & 0 & 0 & 0 & 0 \\
0 & 0 & 0 & 0 & 0 & 0 \\
0 & 0 & 0 & 0 & 0 & 0 \\
1 & 0 & 0 & 0 & 0 & 0 \\
0 & 0 & 0 & 0 & 0 & 0
\end{array}\right]
\left[\begin{array}{cccccc}
0 & 0 & 0 & 0 & 0 & 0 \\
0 & 0 & 0 & 0 & 0 & 0 \\
0 & 0 & 0 & 0 & 0 & 0 \\
0 & 0 & 0 & 0 & 1 & 0 \\
0 & 0 & 0 & 0 & 0 & 0 \\
0 & 0 & 0 & 0 & 1 & 0
\end{array}\right]
\left[\begin{array}{cccccc}
0 & 0 & 0 & 1 & 0 & 1 \\
0 & 0 & 0 & 0 & 0 & 0 \\
0 & 0 & 0 & 1 & 0 & 1 \\
0 & 0 & 0 & 1 & 0 & 1 \\
0 & 0 & 0 & 0 & 0 & 0 \\
0 & 0 & 0 & 1 & 0 & 1
\end{array}\right].
$$

The decomposition is done using heuristics that know how to convert a vector into an ordered set of orthonormal vectors. The heuristics are based on the existing building blocks the system knows about. This prevents the

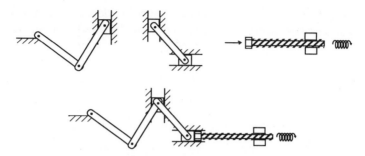

FIGURE 4.5. Building blocks.

system from producing spurious unrealizable alternatives. It does, however, make a closed-world assumption on the primitives it knows about; assuming all new designs are a combination of known primitives. The matrices shown about correspond to various building blocks that can be configured into a design as shown in Fig. 4.5. The top half of the figure shows the three primitives corresponding to the three matrices. The synthesized design is shown in the bottom half of the figure.

Discussion. The motion synthesis approach provides a method for recognizing a given behavior in terms of known primitive behaviors. This is one of the first formalized ways of viewing design as the synthesis of kinematic processes. This approach, however, is not suited to intermittent mechanisms, where the connectivity of kinematic pairs changes during the operation of the device.

Qualitative Synthesis with Interactions. The notion of synthesizing devices from known components is extended beyond basic kinematics in the Ibis system (40, 41). In Ibis, components are represented as sets of interactions among behavioral parameters of the component. This approach allows one to use any aspect of a behavior, not restricting behavior descriptions to just one domain (e.g., qualitative motion synthesis).

Let us start by examining the representation used in Ibis. A simple pipe, for example, is a device that takes some input flow (Qin) and produces some output flow ($Qout$). The flow rates depend on the resistance of the pipe (R) and the pressure difference across the pipe (Pd). The relationship among these parameters is called an *interaction*. The interaction model of the pipe is shown in Fig. 4.6.

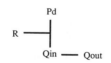

FIGURE 4.6. Simple pipe.

In addition to primitive devices, the interaction-based model is used to capture domain principles. Consider the examples in Fig. 4.7. A container for fluids can be modeled with two interactions: (a) the rate of change of fluid in the container (dV/dt) is determined by the rate of flow of fluid into the container ($Qtop$) and the rate of flow of fluid out of the bottom of the container ($Qbottom$); and (b) the pressure at the bottom of the container ($Pbottom$) is determined by the height of fluid in the container H and the atmospheric pressure ($Patm$). The container represented in the figure is actually a very large *vat* in which the height of the liquid does not change over time ($dH/dt) = 0$).

Figure 4.7 also shows two domain rules. The conservation of mass rule states that the inflow (Qin) into the outflow ($Qout$) from a node are equal. The compatibility rule states that the pressure difference at a node (Pd) is the difference of the pressures at the nodes $P1$ and $P2$; that is, $Pd = P1 - P2$.

A Design Example. Given a large vat and a punch blow, one has to design a device that will maintain the height of the fluid in the bowl even when fluids are removed from the bowl. The specification is given as

$$Hvat - Hbowl = d(Hbowl)/dt \, .$$

This means that the rate of change of the height of fluid in the bowl is a direct function of the height difference between the bowl and the vat. The bowl and vat are defined as containers. The vat, being very large, has a predetermined parameter: $dH/dt = 0$. In other words, its level does not change.

The nodes on the interaction graphs are treated as terminals of the components. The system designs by connecting compatible nodes to produce a working design. The program starts by building a graph of all possible connections among the given device models. This is done by placing *links* between all the terminals that can be connected. For example, all flow rates could be connected to all other flow rates. Figure 4.8 shows all the known interactions and links among them. The links are shown in fine lines.

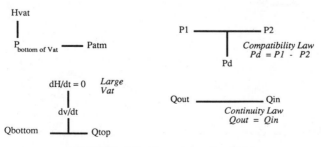

FIGURE 4.7. Containers for fluid.

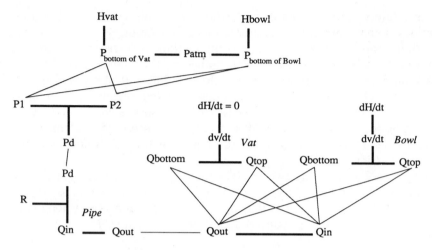

FIGURE 4.8. Interactions and links.

The design is generated by finding a path through the interactions and links that will satisfy the specified behavior. In this case, the system has to find a relationship between the rate of change of height of fluid in the bowl [$d(Hbowl)/dt$] and the difference in the heights of the vat and the bowl. Ibis looks for a path that terminates only on constants (e.g., *Patm and Qtop* of the bowl), unless the variable is a desired variable. In this design example, the desired variables are $d(Hbowl)/dt$, *Hbowl*, and *Hvat*. The grey line in Fig. 4.9 represents a path through the given interactions. The resulting design is the pipe attached between the bottom of the bowl and the bottom of the vat.

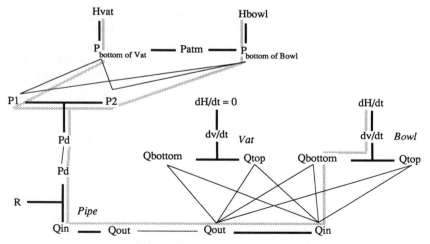

FIGURE 4.9. Finding a legal path.

If several legal paths are found, each minimal path is treated as a candidate topology. Ibis uses the topology of existing interactions when proposing candidates; in this way, it achieves function sharing (38).

Discussion. The interaction approach goes beyond motion synthesis by including higher order relations such as pressure and rates of change. The program also explicitly incorporates physical laws and principles. A major advantage of Ibis is that it can handle feedback mechanisms. The other two synthesis systems discussed in this chapter cannot, as yet, handle feedback.

Ibis, however, has one major drawback. Its ability to solve a problem depends on the syntactic form of the interactions it is given. For example, the bowl problem could have been solved by attaching the *Qbottom* of the bowl directly to the *Qbottom* of the vat. Ibis will not find the solution. This is because the goal has the term *Hvat* − *Hbowl* in it and the only way the program knows how to achieve a pressure difference is through the compatibility law. The only other interaction that "knows" about a pressure difference is the pipe.

The program is thus sensitive to the nature of the goal statement. The real goal of this design is to have a punch bowl automatically restore its level every time some punch is taken out of it. The corresponding goal could have looked like this: $d(Hbowl)/dt = 0$, while *Qtop* is nonzero (i.e., some punch is being drained off the top of the bowl). Given the same interactions as before, Ibis will not be able to solve the problem in its new form. This is because the program suffers from a *functional fixedness* (6, 23): It cannot view the given components directly because it does not recognize behavioral equivalence between the given goal specification and what can be achieved by the pipe, the vat, and the bowl.

Behavioral Synthesis. New designs can be synthesized by drawing ideas from a database of prior designs instead of relying only on a given set of primitives. We now examine CADET, a tool that generates designs by using parts of prior designs (28). Relying on prior designs to solve new problems is an attractive idea, however; because there is no one-to-one correspondence between the desired behavior of a device and the individual component behaviors, it is often not possible to find relevant cases by using the given overall behavioral specification as an index into the case database. This problem is approached in CADET through behavior-preserving transformation techniques to transform an abstract description of the desired behavior of the device into a description that can be used to find relevant designs in memory.

The CADET system uses a transformational approach to convert functional specifications of a device into behaviorally equivalent forms. The transformational approach decomposes given behavior specifications into "sub-behaviors" without altering the overall desired behavior. The decompositions do not impose any a priori structure or topology on the physical realization of the design. The decompositions are collections of sub-behaviors with information on how the sub-behaviors must interact to produce the overall device behavior. In effect, the program is not sensitive to the form of the given behavior specification. Even if a given behavior specification is not compatible with the representation of the cases in memory, the program is able to apply transformations to find behavior equivalence between the specification and relevant design cases in memory.

Behavior Representation and Transformation. In CADET, influence graphs are used to represent the behavior of devices. When this representation is incorporated in a device case base, it becomes possible to retrieve cases that match given behavior specifications. If retrieval using the design specification fails to retrieve relevant cases, the system should be able to recognize how a combination of component behaviors could produce the required effect.

Most existing case-based systems use a predefined set of indices to access cases in memory. This indexing strategy is limiting, since salient features of the current case, which could constitute good indices, may not directly match the predefined index set. Index transformation is a way to change the given salient features of the current cases in ways to match the indices under which previous cases have been stored, thus making accessible to the problem solver previously inaccessible cases. The transformation technique described here is applicable to any domain in which behavior can be modeled as a graph of influences.

The transformation technique in CADET is based on two qualitative rules involving influences. The two elaboration rules used in CADET are:

Design Rule 1. If the goal is to have x influence z, and if it is known *a priori* that u influences z, then the goal could be achieved by making x influence u.

Design Rule 2. If the goal is to have x influence z and if it is known *a priori* that some two quantities p and q influence z, then the goal could be achieved by making x influence p or q, or both.

The two design rules transform a given influence into a more detailed set of influences that are behaviorally equivalent to the original influence. As transformation operators, the first rule yields a serial transformation and the second one yields parallel transformations.

The design rules are used to transform goal influences into more elaborate influence graphs. This is done by first applying the given domain laws. If the domain laws are unable to generate elaborations that can be realized by the cases in memory, one can try to hypothesize variables. The idea is to hypothesize new influences and then find cases that may be used to achieve those influences. As new variables are introduced, corresponding new influences are hypothesized. In addition, as influences are all supposed to be based on physical laws or principles, the introduction of new variables implies that unknown laws or principles (unknown to the program) are being hypothesized. After hypothesizing influences, the case base is used to find prior designs that may embody some physical law or principle that matches the hypothesized influence. Using this approach, CADET often retrieves cases from outside the current design domain that are analogically related to the current design problem. It is for this reason that CADET's solutions are innovative (for definition of innovation see ref. 28). The approach is able to solve design problems by drawing analogies to prior designs and by exploiting physical laws and effects other than those included in the given domain description.

Design Example. Consider the design of a hot/cold water faucet. The function of the faucet is described as:

A device that *mixes* hot water at temperature T_h and cold water at temperature T_c with flow rates Q_h and Q_c, respectively, and allows the control of the mixed water temperature T_m by a mechanical signal S_t and its flow rate Q_m and by a mechanical signal S_f. In addition, the two controls should be independent: S_t should not influence Q_m and S_f should not influence T_m.

This specification is input to CADET as goals, constraints, and a qualitative description of the governing physical laws and principles.

Goals. The goal of the system is to select a set of components that take a certain input and produce a certain output. These goals can be represented using influences:

$$\left[\frac{\partial T_m}{\partial S_t}\right] = [+] \quad \text{and} \quad \left[\frac{\partial Q_m}{\partial S_f}\right] = [+].$$

Constraints. For the faucet, the behavior specification says that the signal S_t should not influence Q_m and the signal S_f would not affect T_m. For CADET the above constraints are input as

$$\left[\frac{\partial T_m}{\partial S_f}\right] \neq [+], \quad \left[\frac{\partial T_m}{\partial S_f}\right] \neq [-], \quad \left[\frac{\partial Q_m}{\partial S_t}\right] \neq [+], \quad \left[\frac{\partial Q_m}{\partial S_t}\right] \neq [-].$$

Process Description. The process controlled by the faucet is the mixture of fluid flow. The behavior of the mixture can be described in terms of the following equations:

$$Q_m T_m = Q_h T_h + Q_c T_c \quad \text{(conservation of energy)}, \tag{1}$$

$$Q_m = Q_c + Q_h \quad \text{(conservation of mass)}, \tag{2}$$

where $T_h > T_c$.

The Design Process. CADET first tries to find cases that directly match the given goal. If it fails, it starts elaboration by alternatively applying the serial and parallel operators. Before it starts hypothesizing variables, it tries to elaborate using the known domain laws.

Here are some of the alternatives generated[1]:

1. $(T_cS_t+)(Q_cS_f+)(Q_cS_t-)(Q_cS_f+)$.
2. $(Q_mS_t-)(T_mS_t+)(Q_mS_f+)(T_mS_f+)(Q_mS_t+)(T_mS_t+)(Q_mS_f+)(T_mS_f-)$.
3. $(T_cS_t+)(Q_cS_f+)(T_cS_t+)(Q_hS_f+)$.

The first solution is rejected because it violates the given constraints. In the second solution we find that S_t influences T_m positively but influences Q_m both negatively *and* positively. We can assume that S_t will not influence Q_m if the quantitative increase and decrease are equal[2]. Similarly, we find that S_f will influence Q_m positively and can be made not to influence T_m. The potential design, based on the cases retrieved for this transformed index, is given in Fig. 4.10. The third solution is shown in Fig. 4.11.

Discussion. Through the process of influence hypothesis and matching, the system is able to use physical laws and principles embedded in prior design cases to achieve its current goals. In this way, CADET is opportunistic about the principles it exploits in a design. This is unlike other approaches, which assume that all the relevant principles have been identified *a priori*, as in the Ibis system (41). Because CADET hypothesizes influences, it does not limit itself to the given knowledge. Consequently, it can generate elaborations that represent designs that have never been conceived of before. Furthermore, CADET's ability to recognize behavioral equivalences reduces its sensitivity to the form of the problem description.

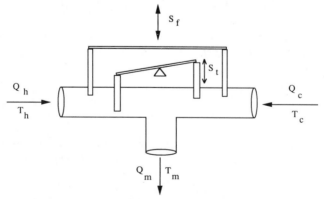

FIGURE 4.10. Potential final design for alternative 2.

[1]In CADET the influence $[\partial Q_m/\partial Q_h]=[+]$ is represented as the list (Q_mQ_h+).
[2]The interpretation is that the design *could* potentially satisfy the constraints.

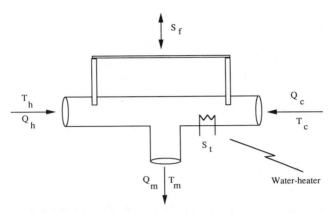

FIGURE 4.11. Potential final design for alternative 3.

The CADET system has two limitations. (a) It cannot handle feedback mechanisms like Ibis does. CADET's influence graph representation could be used to represent feedback, but the transformation heuristics will fail to preserve behavioral equivalence. (b) CADET cannot handle intermittent mechanisms. This limits the program from handling a very wide variety of engineering systems (31). The motion synthesis method, however, is able to handle such mechanisms (e.g., dwell mechanisms) quite naturally.

4.2.2. Qualitative Analytic Methods in Design

Design is often classified into two categories: routine design and nonroutine design. According to Cagan and Agogino (4), nonroutine designs are characterized by the creation of new variables and thus an expansion of the design space. Routine designs, on the other hand, are restricted to a fixed set of variables and thus a predefined design space. In this section we examine two systems that deal with nonroutine design: 1^{st}PRINCE and PROMPT.

Nonroutine Design by Dimensional Variable Expansion. 1^{st}PRINCE is a nonroutine design system (3). It innovates structural designs by reasoning from first principle knowledge to discover new design prototypes. Design knowledge is input to 1^{st}PRINCE in the form of equations pertaining to the domain of interest as a symbolic optimization problem, which represents a good initial design prototype. The system then uses qualitative techniques of monotonicity analysis as derived from Karush–Kuhn–Tucker (KKT) conditions of optimality to select a critical integral. The design space is then expanded by what is called the "Dimensional Variable Expansion" (DVE), which essentially divides the integral into a set of smaller ranged integrals. Based on the following definitions, 1^{st}PRINCE has a set of rules that allow it to manipulate the design space.

Definitions

1. The monotonicity of a continuously differentiable function $f(x)$ with respect to variable x_k is the algebraic sign of $\partial f / \partial x_k$.
2. A constraint $g_i(x) \leq (\geq) 0$ is active at x_o if $g_i(x_o) = 0$. Otherwise it is inactive.
3. A positive variable x_k is said to be bounded above by a constraint if the variable is maximum when the constraint is active. It is said to be bounded below if the variable is minimum when the constraint is active.

Based on these definitions, Papalambros and Wilde (30) give three rules of monotonicity analysis that define well-constrained optimization problems without overconstrained cases.

Rule 1. If the objective function is monotonic w.r.t. a variable, then there exists at least one active constraint that bounds the variable in the direction opposite to the objective.

Rule 2. If a variable is not contained in the objective function, then it must be either bounded from both above and below by active constraints or not actively bounded at all.

Rule 3. The number of nonredundant active constraints cannot exceed the total number of variables.

In 1^{st}PRINCE, the design space is expanded by splitting a critical integral into a summation of smaller ranged integrals:

$$\int_0^x X = \int_0^{x_1} X_1 + \int_{x_1}^x X_2.$$

This division creates the new dimensional variable x_1 and the new vectors of variables X_1 and X_2. The division takes place because monotonicity analysis determines that x_1 might have a critical effect on the objective function. This division does not take place indefinitely. 1^{st}PRINCE uses inductive inference to identify trends in the constraint activity and uses that information to extrapolate the current design space to the final design space.

A Structural Design Example. Consider a circular solid beam under a transverse load, as shown in Fig. 4.12. This design problem is input to 1^{st}PRINCE as the following optimization problem.

(a) $W = \displaystyle\int_0^L \pi \rho r_1^2 \, dx = \pi \rho L r_1^2;$

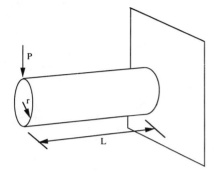

FIGURE 4.12. A circular beam under transverse load.

(b) $\sigma = \dfrac{Mc}{I} = \dfrac{4Px}{\pi r_1^3}$;

(c) $\tau = \dfrac{PQ}{It} = \dfrac{4P}{3\pi r_1^2}$;

(d) $\tau \le \tau_y$;

where ρ is the mass density, x is the distance along the beam of length L, P is the transverse load, M is the bending moment (equal to Px), c is the maximum distance from the neutral axis (equal to r_1), I is the moment of inertia, Q is the first moment, and t is the width.

Consider the optimization of W based on the transverse load only. The integral (a) can be divided into two integrals:

$$W = \int_0^L \pi \rho r_1^2 \, dx = \int_0^{x_1} \pi \rho r_1^2 \, dx + \int_{x_1}^L \pi \rho r_1^2 \, dx \, .$$

Solving the optimization problem for x_1, 1stPRINCE gets $x_1 = L/2$ (Fig. 4.13b). Further division of the integral leads to Fig. 4.13c. At this point, 1stPRINCE inductively infers the trend and comes up with the final design as in Fig. 4.13d.

Prompt. PROMPT (27) is another innovative design tool that discovers new design members to satisfy the given constraints whenever none of the existing prototypes satisfies the constraints. The structure of the existing prototype is modified by reasoning qualitatively about the domain principles and applying some heuristics called modification operators. For instance, let us say that PROMPT is given the following design constraints for the design of a cylindrical beam. Weight $W \le w_1$, torsional stiffness $T_s \ge T_1$, length $= L_1$.

Given the structure of a cylindrical beam under torsional stress, for example, PROMPT can retrieve from its knowledge base the set of equations that describe the behavior of the structure as follows:

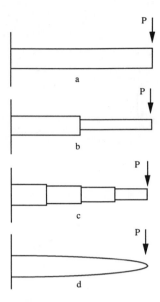

FIGURE 4.13. Inductive inference of the optimal beam.

$$\tau_{\theta z} = G\gamma_{\theta z} \, ,$$

$$\gamma - (\theta z) = (d\phi/dz)r = (\phi/L)r \, ,$$

$$df = \tau_{\theta z} \, dA \, ,$$

$$dM = r \, df \, ,$$

$$M_t/\phi = (1/L) \int^{A} rGr \, dA \, ,$$

$$W = \int D \, dA \, dz \, ,$$

where, $\tau_{\theta z}$ is the shear stress, $\gamma_{\theta z}$ is the shear strain, $d\phi$ is the angle between two cross sections a distance dz apart, r is the distance from the center of an area dA, df is the force exerted by the area dA, and dM is the torque exerted by the area dA.

The system focuses attention on the fifth and sixth equations because they describe the stiffness and the mass of the beam in terms of its underlying structure. Examining these equations shows that the contribution of a given area dA to stiffness varies proportionally to r^2 and the contribution of the same area to mass is invariant with respect to r. This leads to the conclusion that moving mass from areas of low r to areas of high r increases the stiffness of the beam but does not affect the mass of the beam.

Reasoning heuristics of this type are implemented in PROMPT as operators of modification. An example of such an operator is the mass redistribution operator (MRO). The MRO captures the heuristic "move mass from areas of low load to areas bearing high load." Examples of other operators included in PROMPT are (a) redistribution of mass that includes removal and addition of mass, (b) changing the material distribution, (c) changing the shape, and (d) additional and removal of elements.

In the current example PROMPT chooses the first operator based on the above-mentioned qualitative reasoning. The final design that satisfies the constraints is a hollow cylinder, shown in Fig. 4.14.

PROMPT's ability to analyze problems in such detail is based on a large knowledge base called the "Graph of Models" (1). This database uses a graph structure to store physics knowledge about principles, such as bending loads, buckling, torsion, and vibration. Using this knowledge, the program can reason about an artifact from first principles and can derive its behavior from its structure.

Discussion. The PROMPT and DVE approaches cannot easily be applied to the design of a devices that involve behaviors arising from the interaction among physical members, in other words, mechanisms.

PROMPT's use of modification operators gives the program the ability to generate designs that are innovative variables of the original design. This approach has not yet been applied to problems other than member design. In problems where one has to both modify and synthesize objects to achieve required kinematic behaviors, one may adopt an approach that combines the abilities of PROMPT and DVE with the synthesis methods described in the previous section. The introduction of domain knowledge and heuristics into these first principles based methods could yield interesting results.

The next section presents a first, but definitive, step toward the integration of design for kinematics, and the design of the shape of the objects involved in the kinematic mechanism. The approach, however, is very different from those described previously.

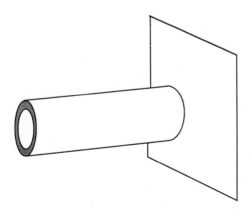

FIGURE 4.14. Innovative design of a hollow circular beam.

4.2.3. Qualitative Shape Design

Qualitative reasoning using configuration spaces provides us a way of qualitatively relating behavior and geometry. This allows us, for the first time, to design the very shape of physical objects in order to satisfy given behavior specifications. Qualitative shape design goes beyond a purely synthetic approach that just configures primitive building blocks. Described here are two approaches to shape design: parametric design using place vocabularies and feature-based shape design.

Qualitative Parametric Design. The geometry of a device's configuration space can be represented as a place vocabulary: a graph whose nodes are kinematic states and whose edges are transitions between the states. A kinematic state is a unique and qualitatively distinct contact between the various boundaries of two physically interacting objects (9).

Faltings (10) presents a design methodology that parametrically modifies the boundaries of objects to achieve a desired behavior. The boundary of an object is represented as smooth segments (lines and arcs) and discontinuities—points at which the smooth segments are connected. The configuration space can be abstracted to a kinematic topology (11) that is useful in conceptual design. In his latest work, Faltings has shown how place vocabularies can be generated directly from geometry without having to calculate the configuration space. This results in substantial savings in computation. In addition, the place vocabularies can be related causally to dimensions of the geometric entities involved. For example, consider the sketch of a pair of gearwheels. At an abstract level, there are only five different qualitatively different topologies (Fig. 4.15). The arrows indicate legal transitions among the five distinct states.

It turns out that very little addition information is needed to determine whether the two gears will mesh. All this information is obtained by considering only convex and concave curvatures of an object, without requiring precise geometry. This result is particularly relevant to conceptual

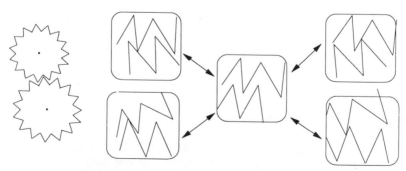

FIGURE 4.15. Place vocabulary for two cogs.

design. In conceptual design it is very important to be able to relate behavior and geometry, when the geometry is specified only in terms of a rough sketch.

Feature-Based Shape Design. Shape design can be viewed as a process of deleting, adding, or modifying physical features of a given artifact in order to achieve a given behavior specification. In one of the first research efforts in shape design, Joskowicz and Addanki (17) have approached the problem using knowledge that relates physical features to boundaries in the configuration space.

A Design Example. Figure 4.16 shows the initial geometry and its configuration space. As the cam rotates, the follower can go as far left as $X = 0$. The desired behavior is specified in terms of a c-space (Fig. 4.17). At two locations along the rotation of the cam, the follower can go farther left than originally.

The problem is solved by first evaluating the difference between the original and desired c-spaces. The missing or differing features in the configuration space are then related to features in the physical space. To facilitate this kind of reasoning, the authors have classified different types of boundaries that arise from the possible pairwise contacts between feature types. This table specifies, for each type of contact and design space, the type of configuration space boundary produced, together with the set of equations that define it. This allows the program to add or delete edges and arcs to the contours of the existing geometry.

In this example, an edge is removed and three new edges are added to accommodate each difference between the original and the desired configuration space. The resulting modified design is shown in Fig. 4.18.

The design procedure is applied iteratively. After physical changes are made to the artifact, the new configuration space is computed and compared to the desired c-space. If there are still some differences, the design procedure is repeated.

Discussion. The c-spaces and place vocabulary approaches are the only way in which we know how to qualitatively relate kinematic behavior and the

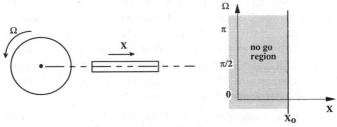

FIGURE 4.16. Initial configuration.

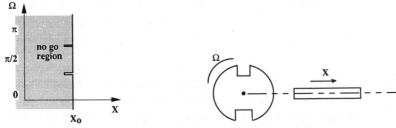

FIGURE 4.17. Desired behavior in the form of a configuration space.

FIGURE 4.18. Final design.

geometry of objects. Its use in design allows us to go beyond mere configurations to the actual manipulation of shape. This is a significant step for computer-aided design. The second approach can produce new geometries that satisfy a given specification. It goes beyond making parametric changes to a given geometry. Joskowicz and Addanki, however, have not addressed the problem of feature sharing and interaction in more complex systems.

These approaches have one major limitation: They cannot easily handle high degree-of-freedom mechanisms. Consider a simple table lamp mechanism with one moving axis of rotation. The corresponding c-space is three dimensional and is very difficult to generate and to reason about. Consequently, these approaches have been applied successfully only to kinematic pairs (i.e., two interacting objects with only one degree of freedom each). This does not, however, mean that these approaches cannot be applied to complex mechanisms. A complex device might be composed of many kinematic pairs, where each subsystem can be designed using the c-spaces approach.

4.2.4. Summary

Design systems that use qualitative methods can be viewed under the following three headings:

1. Systems that synthesize design primitives using qualitative methods to guide the search for and evaluation of solutions. We examined three such systems: (a) Kota's system that reasons about motion and synthesizes designs by transforming one type of motion into another; (b) *Ibis*, a system that assembles a collection of lumped-parameter models; and (c) CADET, a program that recognizes behavioral similarities between a given problem and prior designs stored in a design memory.

2. Systems that use qualitative methods to reason about analytic equations. We examined two approaches: (a) 1st PRINCE, a system that proposes

new variables to split a design into discrete elements whose shapes can be optimized locally to yield a globally optimized shape; and (b) PROMPT, a system that reasons analytically to guide the application of certain modification operators that mutate a given design into something new and innovative.

3. Systems that reason about geometry and its relation to kinematic behavior. Two approaches have been discussed here: (a) a place vocabulary-based approach, which allows the system to access the behavior of interacting objects even when the design is merely a conceptual sketch; and (b) a configuration spaces approach, which allows the system to draw a correspondence between physical features of the design and its behavior.

ACKNOWLEDGMENT

This research is being supported by DARPA and AFOSR under contract number F49620-90-C-0003.

REFERENCES

1. Addanki, S., and Davis, Ernst S. (1985). A Representation for Complex Domains. In: *Proceedings of the Ninth International Joint Conference on Artificial Intelligence*.
2. Brown D. C., and Chandrasekaran, B. (1986). Expert Systems for a Class of Mechanical Design Activity. In: *Proceedings of the First International Conference on AI Applications in Engineering*, edited by D. Sriram and B. Adey, Computational Mechanics, Southampton, UK.
3. Cagan, J. (1990). *Innovative Design of Mechanical Structures from First Principles*, Ph.D. dissertation, Department of Mechanical Engineering, University of California, Berkeley.
4. Cagan, J., and Agogino, A. M. (1987). Innovative Design of Mechanical Structures from First Principles, *Artificial Intelligence in Engineering Design and Manufacturing*, Vol. 1, No. 3, pp. 169–189.
5. Dekleer, J., and Brown, J. (1984). A Qualitative Physics Based on Confluences, *Artificial Intelligence*, Vol. 24, pp. 7–83.
6. Dunker, K. (1945). On Problem-Solving, *Psychological Monographs*, Vol. 58, No. 270.
7. Dyer, M. G., Flowers, M., and Hodges, J. (1986). EDISON: An Engineering Design Invention System Operating Naively, *Proceedings of the First International Conference on Applications of AI to Engineering*, April.
8. Faltings, B. (1986). *A Theory of Qualitative Kinematics in Mechanisms*, Technical report UIUCDCS-R86-1274, Department of Computer Science, University of Illinois at Urbana–Champaign.

9. Faltings, B. (1990). Qualitative Kinematics in Mechanisms, *Artificial Intelligence*, Vol. 44, pp. 89–119.

10. Faltings, B. (1990). Qualitative Representations in Conceptual Design. In: *Proceedings of IFIP Conference on Modelling the Innovation*, North-Holland, Amsterdam.

11. Faltings, B., Baechler, E., and Primus, J. (1989). Reasoning About Kinematic Topology, *Proceedings of the IJCAI 1989*.

12. Forbus, K. (1984). Qualitative Process Theory, *Artificial Intelligence*, Vol. 24, pp. 85–168.

13. Gero, J. S., Maher, M. L., and Zhang, W. (1988). Chunking Structural Design Knowledge and Prototypes. In: *Artificial Intlligence in Engineering: Design*, edited by J. S. Gero, CMP/Elsevier, New York.

14. Hayes, P. (1985). Naive Physics 1: Ontology for Liquids. In: *Formal Theories of the Commensense World*, edited by J. Hobbs, B. Moore, Ablex Publishing Corporation, Norwood, NJ.

15. Iwasaki, Y. (1989). Qualitative Physics. In: *Handbook of Artificial Intelligence No. IV*, edited by A. Barr, P. R. Cohen, and E. A. Feigenbaum, Addison-Wesley, Reading, MA.

16. Joskowicz, L. (1986). *Reasoning About Shape and Kinematic Function in Mechanical Devices*, Technical report TR No. 402, Courant Institute of Mathematical Sciences, New York University, New York.

17. Joskowicz, L., and Addanki, S. (1988). From Kinematics to Shape: An Approach to Innovative Design, *Proceedings of the Seventh National Conference on Artificial Intelligence (AAAI88)*, pp. 347–352.

18. Kota, S. (1990). A Qualitative Matrix Representation Scheme for the Conceptual Design of Mechanisms, *Proceedings of the ASME Design Automation Conference (21st Biannual ASME Mechanisms Conference)*.

19. Kuipers, B. J. (1986). Qualitative Simulation, *Artificial Intelligence*, Vol. 29, pp. 289–338.

20. Lozano-Perez, T. (1983). Spatial Planning: A Configuration Space Approach, *IEEE Transactions on Computers*, Vol. C-32, No. 2, pp. 560–570.

21. Lozano-Perez, T., and Wesley, M. (1979). An Algorithm for Planning Collision-free Paths Among Polyhedral Obstacles, *Communications of the ACM*, Vol. 22, pp. 560–570.

22. Maher, M. L. (1984). *HI-RISE: An Expert System for the Preliminary Structural Design of High Rise Buildings*, Ph.D. dissertation, Department of Civil Engineering, Carnegie–Mellon University, Pittsburgh.

23. Maier, N. R. F. (1931). Reasoning in Humans: II. The Solution of a Problem and Its Appearance in Consciousness, *Journal of Comparative Psychology*, Vol. 12, pp. 181–194.

24. McDonnell Douglas Corporation (1984). *Graphic Design System*.

25. Mittal, S., Dym, C., and Morjaria, M. (1985). PRIDE: An Expert System for the Design of Paper Handling Systems. In: *Applications of Knowledge-Based Systems of Engineering Analysis and Design*, edited by C. Dym, American Society of Mechanical Engineers, New York, pp. 99–116.

26. Mostow, J. (1985). Toward Better Models of the Design Process, *The AI Magazine*, pp. 44–57.

27. Murthy, S. S., and Addanki, S. (1987). PROMPT: An Innovative Design Tool, *Proceedings of the Sixth National Conference on Artificial Intelligence*, pp. 637–642.

28. Navinchandra, D. (1991). *Exploration and Innovation in Design: Towards a Computational Model*, Springer-Verlag, New York.

29. Navinchandra, D., Sycara, K. P., and Narasimhan, S. (1991). Behavioral Synthesis in CADET, A Case-Based Design Tool. In: *Proceedings of the Seventh Conference on Artificial Intelligence Applications*, IEEE Press, New York.

30. Papalambros, P., and Wilde, D. (1988). *Principles of Optimal Design*, Cambridge University Press, New York.

31. Pu, P. (1989). *Qualitative Simulation of Ordinary and Intermittent Mechanisms*, Ph.D. dissertation, Department of Computer Science, University of Pennsylvania, Philadelphia.

32. Rinderle, J. R. (1982). *Measures of Functional Coupling in Design*, Ph.D. dissertation, Massachusetts Institute of Technology, Cambridge.

33. Roos, D. (1966). *ICES System Design*, MIT Press, Cambridge, MA.

34. Spice Group (1975). *SPICE2, A Computer Program to Simulate Semiconductor Circuits*, Technical report ERL-M520, University of California, Berkeley.

35. Sriram, D. (1986). *Knowledge-Based Approaches for Structural Design*, Ph.D. dissertation, Carnegie–Mellon University, Pittsburgh.

36. Steinberg, L. I. (1987). Design as Refinement Plus Constraint Propagation: The VEXED Experience, *Proceedings of the Sixth National Conference on Artificial Intelligence*, pp. 830–835.

37. Steinberg, L., Langrana, N., Mitchell, T., Mostow, J., and Tong, C. (1986). *A Domain Independent Model of Knowledge-Based Design*, Technical report AI/VLSI Project Working Paper No. 33, Rutgers University, New Brunswick, NJ, March.

38. Suh, N. P., Bell, A. C., and Gossard, D. C. (1978). On an Axiomatic Approach to Manufacturing and Manufacturing Systems, *Journal of Engineering for Industry*, Vol. 100, No. 2, pp. 127–130.

39. Tong, C. (1986). *Knowledge-Based Circuit Design*, Ph.D. dissertation, Stanford University, Stanford, CA.

40. Williams, B. (1989). *Invention from First Principles via Topologies of Interaction*, Ph.D. dissertation, Massachussetts Institute of Technology, Cambridge, MA.

41. Williams, B. (1990). Interaction-Based Invention: Designing Novel Devices from First Principles, *Proceedings of AAAI-90*, pp. 349–356.

■■■■■ CHAPTER FIVE

Integration of Expert Systems, Databases, and Computer-Aided Design

KENNETH F. REINSCHMIDT and GAVIN A. FINN
Stone & Webster Engineering Corporation, Boston, Massachusetts

5.1. INTRODUCTION

Computer-aided engineering (CAE) began some 30 years ago with the use of computers to perform engineering calculations. The 1960s saw the introduction of matrix analysis, the beginnings of the finite-element method, and the development of the first large-scale software packages.

In the 1970s, engineering applications achieved widespread use in practice, in part through service bureaus. Finite-element analysis became established for the static and dynamic stress analysis of complex mechanical parts and structural systems. Computer-aided design and drafting (CADD), which at that time was largely two-dimensional drafting, was adopted by leading design firms. Computer-aided design and manufacturing (CAD/CAM) and numerically controlled (NC) machines were introduced in manufacturing.

In the 1980s, there was an explosion in the amount of software available and in the use of computers in design and on the plant floor. Mature methods for finite-element analysis, nonlinear analysis, dynamic analysis, and simulation have improved the engineering and design process and have made possible the design of more economical manufactured products with greatly improved confidence in their performance.

The application of computers to engineering design has become common practice in most of today's engineered projects. The use of computer-aided design (CAD) software spans many engineering functions, ranging from

Intelligent Design and Manufacturing, Edited by Andrew Kusiak.
ISBN 0-471-53473-0 © 1992 John Wiley & Sons, Inc.

simple drafting aids to full-fledged three-dimensional modeling and design tools. Many of the latest design packages allow for the inclusion of attribute information in the database, assigning physical meanings to the graphical representation of components, subsystems, and systems. In addition to advances in software architectures, hardware developments have made CAD more accessible to the majority of engineers. A great deal of engineering design work is now being accomplished by means of CAD systems.

Also in the 1980s, the length of the engineering and design development process became a competitive factor. Reduction of product lead times has become a significant competitive advantage in many industries. For example, it has been estimated that each additional day of lead time in the introduction of a new model automobile costs the manufacturer about $1 million in lost profits (not merely lost revenues) (3). Therefore an automobile manufacturer with a lead time to market of 4 to 6 months longer than the competition could lose hundreds of millions of dollars in profits.

In the 1990s, the response to these competitive conditions is concurrent engineering (9). In concurrent engineering, the requirements of assembly and manufacture are considered simultaneously with other design requirements in order to reduce product development lead times. That is, products are explicitly designed for the manufacturing process. Manufacturing issues drive the design process to a greater extent than before.

Concurrent engineering requires the integration of the design process with the manufacturing process. In the competitive world of the 1990s, CAD will become too important to the organization's competitive strategy and survival to be left to the traditional CAD support group. Those organizations that cannot adapt will be in competitive difficulties, whereas those that do adapt will find that their CAD strategy and their market strategy must be integrated. The traditional fragmented, specialized design process leads inevitably to suboptimization—the optimal design of subsystems rather than the optimization of the complete system or product. Such suboptimized products will not be competitive with products optimized for manufacturability, and the present global marketplace for manufactured products will tend to eliminate inefficient producers.

These competitive factors have led to the demand for CAD systems specifically oriented toward concurrent engineering. The ability to make the design process more intelligent by considering manufacturing and other considerations during design has become a major step toward the realization of concurrent engineering.

It has long been the desire of engineers to include design expertise within the CAD software, but conventional software technologies have not been well suited to this task. Using the technology of expert systems, it is now possible to incorporate judgmental design knowledge within the CAD environment. This ability allows for the distribution of relatively rare expert design and manufacturing knowledge to a wide range of designers and engineers, thus improving overall design performance and reducing the design time.

The integration of computer-aided design systems and expert system shells allows expert systems applications to interact with CAD systems to provide advice regarding the feasibility and desirability of proposed designs. In the design review mode, this architecture allows designers to propose configurations, have the expert system review the design and provide modification advice, and then graphically modify the design. Expert systems integrated with computer-aided design have provided practical, convenient, and cost-effective solutions to a number of common problems associated with engineering and manufacturing.

The implementation of expert systems integrated with computer-aided design can provide significant advantages when a firm is faced with any of the following situations:

- Experienced engineers and designers have been lost or are about to be lost due to retirements, promotions, transfers, relocations, reductions in force, or other reasons.
- Available engineers and designers are unfamiliar with the requirements of the manufacturing process.
- Product design requires highly skilled specialists with specific and scarce know-how.
- Designs are complex, with tight manufacturing tolerances, special finishes, complex surface geometry, or complex subassemblies.
- Product design requires a large, multidisciplinary team of engineers and designers.
- Corporate standards for product design are very specific and detailed.
- It is desirable to increase the productivity of engineers and designers by automating part of the design process, such as the production of details or the development of specific customer modifications to basic generic designs.
- It is necessary to respond quickly to customer orders to keep business.
- Manufacturing, machining, assembly, or other production requirements have a significant impact on the design.
- There is considerable iteration in the design process due to manufacturing engineering review and comment on designs.
- Manufacturing, machining, or assembly is to be preformed by automated machinery, such as numerically controlled tools or robots.
- Production scheduling, procurement of materials or components, group technology, or reduction of work-in-process inventories are important considerations.

- Competitive factors make it necessary to reduce the engineering and production lead time in the development and introduction of new or modified products.

This chapter describes expert systems technology and the structure of the engineering design expert system and discusses how expert systems can provide higher levels of expertise and knowledge to designers and engineers. The accessibility of specialized knowledge allows them to make sound design decisions, thus improving design quality and reducing design time.

5.2. SUBJECT DESCRIPTION

5.2.1. Three-Dimensional Design

Because products are manufactured in a three-dimensional world, the use of three-dimensional computer-aided design is essential to concurrent engineering and to the application of expert systems in design. The use of three-dimensional solids modeling for the design of entire products, previously impractical, has been made practical by the rapid advances in computer graphics hardware. Reduced Instruction Set Computer (RISC) workstations now permit the display of full three-dimensional CAD models with hidden lines removed or in image (shaded) mode, and the improvements in the price/performance of computer memory permit very large three-dimensional solid models to be stored at acceptable cost.

Only with three-dimensional design are dimensions (lengths, areas, volumes) maintained in true one-to-one scale, without the distortions characteristic of orthographic or isometric projections on the drawing plane. With solid modeling, the computer can automatically compute true surface area, volume, weight, center of gravity, moments of inertia, and other properties directly from the three-dimensional representation. That is, these three-dimensional models are not merely pictorial visualizations of a project but are databases of the engineering and manufacturing properties of the design.

The three-dimensional database is organized around physical objects or parts. These objects constitute the natural language of expression for engineering, design, and manufacturing. To ensure the integrity of the design database, each object should be represented uniquely whenever possible. If multiple representations of an object are necessary, consistency between the representations in the database must be assured.

The three-dimensional models provide the benefits of physical scale models while eliminating many of their deficiencies. The three-dimensional computer models are built by the designers of the facility in the process of design, rather than by specialist model makers. Not only does this eliminate the expensive and time-consuming step of building a physical model from the completed engineering drawings, it means that the model is available to other users earlier in the design process, for coordination with mating parts, for manufacturability review, and so on.

5.2.2. Database Integration and the Addition of Intelligence to the Design Models

Design data must be interactively accessible to all authorized users during the design process, while changes to the database are controlled. Design change control and design data management are central issues in the use of computerized design systems. With developments in database management systems, particularly relational databases, object-oriented databases, and data modeling, integrated product databases have become practical. Design data in these integrated databases are available to all authorized users on an open basis, not merely by transfer of files, but by immediate on-line access to data through the CAD system.

The integrated CAD database represents a computer model of all stages of product development, from conceptual design through engineering analysis, detailed design, material requirements planning, purchasing, and manufacturing. Whether the database resides on a single server or is distributed over a number of computers, all participants have access to the necessary data through compatible database management systems. The database management system also provides facilities for maintaining the security and integrity of all data in the system.

Now engineers are learning to model design processes using techniques such as entity-relationship diagrams and design structuring (4, 11). The purpose of data modeling is to determine how to represent design and manufacturing data elements and how these data elements are interrelated, in order to develop applications that consciously improve work methods and engineering productivity.

The implementation of expert design systems requires a reduction in paper drawings as the master design documents. Paper drawings are useful as media for human communications but not for computer communications. Even drawings scanned into computer imaging systems are useless for this purpose, because these raster images are not intelligent and cannot be understood by expert systems or other concurrent engineering applications. The CAD database must become the reference design and the basis for materials procurement, manufacturing resource planning, and manufacturing process design.

To create drawings when needed, the three-dimensional computer models can be sliced and compressed into two-dimensional projections for final drafting. This extraction and projection onto two-dimensional drawings can in many cases be accomplished or assisted by expert systems, which contain the knowledge needed to create standard two-dimensional views from the three-dimensional database. Design changes must of course be made only on the three-dimensional model, rather than on the two-dimensional drawings, to preserve the integrity of the database; new drawings can be regenerated as necessary.

In addition, graphical design information can be distributed on inexpensive personal computers such as PS/2s. Simplified, menu-driven programs open up the process by permitting users who are not familiar with the use of CAD systems to view and mark up drawings on PS/2s linked to the CAD design models and to the database of design and manufacturing information. These developments increase the value of the design data and extend the usability of the CAD models, by making design data accessible on personal computers at all engineering locations and on the plant floor.

Three-dimensional design models are also closely linked to engineering application programs for stress analysis, vibration analysis, and so on. By integration of the data, the design basis and the engineering analyses are kept consistent. Designs developed in this way are of higher quality, less information is lost in the interfaces between specialized engineering functions, and engineers have greater job satisfaction because tedious transcription of data is eliminated.

5.3. THEORETICAL BASIS

5.3.1. Expert Systems Defined

Expert systems are a part of the field of artificial intelligence (AI). *The Handbook of Artificial Intelligence* (1) defines artificial intelligence to be a part of computer science that is concerned with designing computer systems that exhibit the characteristics normally associated with intelligence in human behavior, such as understanding language, learning, reasoning, and solving problems.

Expert systems are not concerned with understanding language, or other aspects of intelligence, but are only concerned with solving problems. As distinct from other methods of artificial intelligence, expert systems are not based on any general theory of intelligence or universal science of cognition. Expert systems are based on the engineering premise that intelligence lies in knowledge; that is, people who know more exhibit more intelligence (can solve more problems) than those who know less. Consequently, it follows from this premise that, if enough knowledge can be represented in a computer, the computer will display intelligent-like behavior.

One of the first expert systems was the program MYCIN, developed in the early 1970s at Stanford University to assist medical doctors in the diagnosis of bacterial infections in human patients (2). From this beginning, many diagnostic expert systems have been developed in engineering and manufacturing, as it turns out that the reasoning involved in diagnosis of the causes of problems is suitable to representation by expert systems.

Whether an expert system in fact replicates the reasoning process used by a human expert is primarily of interest to computer scientists; as a practical matter it is only necessary that the expert system replicate the results that the human expert would have obtained under the same circumstances.

A report by the National Academy of Sciences defines expert systems to consist of two parts: a body of knowledge and a mechanism for interpreting this knowledge (8). The body of knowledge in turn consists of two parts: heuristics or rules that represent the knowledge to solve problems in a particular domain, and facts about the specific problem to be solved. The objective of expert systems development is to capture the knowledge of an expert in a particular area, to computerize this knowledge, and to transfer it to others.

An expert system then contains all of the following elements:

- An inference engine
- A knowledge base separate from the inference engine
- A database
- An explanatory facility

The inference engine draws logical conclusions, or inferences, from the expert knowledge contained in the knowledge base and the specific problem conditions contained in the database. One of the discoveries that made expert systems useful was the realization that, by separating the inference engine form the knowledge base, the inference engine may be used in any number of expert systems.

The knowledge base is a computer representation of the knowledge of the domain expert, expressed in one of several forms, for example, as rules. A knowledge base is specific to each expert system application.

The database is a set of input data, pertinent to a particular problem, that is used by the inference engine to reach a conclusion. These data may be obtained from the user of the expert system, from a computer database, from a CAD system, or from the output of another computer program.

Expert systems also include explanatory facilities, which explain to the user why or how the expert system has arrived at a particular conclusion. That is, if the user asks the expert system to explain its results, the explanatory facility determines the specific knowledge (from the knowledge base), combined with the user's input data, that led to this conclusion and displays these to the user. The explanatory facility makes the expert system easier to use, easier to understand, and easier to accept. The existence of an explanatory facility is one of the features that distinguishes an expert system from any other computer program.

A rule base is the most common kind of knowledge base. In a rule base, the domain expert's knowledge is expressed in the form of production rules, each of which consists of a set of conditions and set of consequents. These production rules are commonly expressed in a form as close as possible to natural language (e.g., English). A typical rule is of the form:

```
IF        ⟨Condition A is true⟩
    AND ⟨Condition B is true⟩
    .

    .

    .

    AND ⟨Condition N is true⟩
THEN       ⟨Conclusion X can be drawn}
ELSE       ⟨Conclusion Y can be drawn⟩
```

If the conditional statements are all evaluated to be true, then the rule is said to fire, and the consequent statements are asserted to be true as well. The consequents of one rule may be used as the conditions of other rules. This produces a chain reaction effect, as the firing of one rule leads to the firing of others. This chain reaction is carried out by the inference engine.

A practical expert system does not consist of a few rules, each with a very large number of conditions and consequents, but rather a large number of rules, each with a few conditions and consequents.

The chain reaction behavior of rule-based systems is of three types: forward chaining, backward chaining, and mixed. In forward chaining, or data-driven, systems, the inference engine uses the given data (from the user or from the database) and determines whether any rules can fire. If so, the chaining process continues, until one or more conclusions, or terminal consequents, are reached or no more rules can fire. If rule-firing stops before a conclusion is reached, the inference engine asks the user for more information, which may cause new rules to fire.

In backward chaining, or goal-driven, systems, the inference engine posits a possible conclusion and then chains backward through the rules from this conclusion to the data. If the conditions required to sustain the hypothesis are incompatible with the actual data, then the conclusion is rejected and a new conclusion is hypothesized. This continues until one or more conclusions have been validated by the data or all possible hypotheses have been rejected. Mixed chaining is a composite of forward chaining and backward chaining.

Not all expert systems are based on rules. Some are based on other knowledge representational schemes, such as frames or objects. However, rules are the easiest for the design or manufacturing engineer to understand and are the basis for most available expert system shells.

Figure 5.1 shows how an expert system is developed. The domain expert, who possesses the knowledge or expertise, interacts with a knowledge engineer, who is experienced in asking questions and converting the expert's answers into rules. The knowledge engineer is also experienced in selecting which inferencing method to apply and in structuring the rule base efficiently. The rules resulting from this collaboration of domain expert and knowledge engineer then form the knowledge base.

Figure 5.2 shows a user interacting with a completed expert system. The

FIGURE 5.1. Expert system development.

knowledge base in the computer has been created earlier by the collaboration of the domain expert and the knowledge engineer. The user supplies the input data specific to his problem, or problem data may be extracted from a computer database or CAD system. The inference engine compares the user's input with the rules in the knowledge base to determine if it can arrive at any conclusion. If not, the inference engine will ask the user for more information. The user may ask the inference engine why it needs this information, or why it has arrived at a certain conclusion, through the explanatory facility.

After the knowledge base is first developed, it should be tested by the users, in order to refine the knowledge base and to make the system reliable and easy to use. An iterative process is preferred. It is generally more efficient to start with a few rules, test them in practice or in a prototype

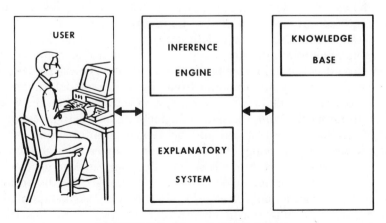

FIGURE 5.2. Use of expert system.

system, and then to add rules to increase the depth of knowledge and to handle situations not originally contemplated.

Since the rules in the rule base are expressed in English, or something like English, rules may be added over a period of time by the original domain expert or by others. This evolutionary improvement is characteristic of expert systems, as compared to other types of computer programs. If many domain experts contribute their knowledge to a rule base, the expert system over time may become more knowledgeable and capable of problem-solving than any one of the experts who contributed to its development. This is particularly true of expert systems that make manufacturing knowledge available to designers through CAD systems: The rule base may become a synthesis of manufacturing know-how beyond the capabilities of any single person.

The demonstrated benefits of expert systems include making design and manufacturing expertise more widely available in the organization, improved design decision-making, and enhanced performance of design personnel and manufacturing equipment. Design and manufacturing engineering are fertile fields for the application of expert systems because much expertise is heuristic—that is, based on experience and judgment rather than on theory—and is therefore generally suitable for representation by rules. The English-like nature of the rules lends itself to knowledge base development by practicing industrial engineers and designers, and experience has shown that practicing engineers can readily express their expertise in rule form.

Many early expert systems were stand-alone applications, but the full potential of expert systems in engineering and design can only be achieved by integration of the knowledge bases with the manufacturing database and the computer-aided design system. Stand-alone expert systems require excessive manual data input to be truely useful in practical design situations. By integrating an expert system shell with the database and the CAD system, the expert systems can automatically interrogate the design database for any required information to satisfy their rules. The results of the expert system sessions can also be automatically inserted into the design model.

With this integration, designers can execute expert systems interactively while in the process of design using the CAD system. The expert system can retrieve any necessary information from the CAD model or from the database. Advice provided by the expert system is displayed on the workstation screen during the computer-aided engineering and design session. The output of the expert system can also be geometric information, which is inserted directly into the computer-aided engineering and design models. In this way, expert systems can advise the designer, make suggestions to improve the design for manufacturability (6), change the dimensions of objects in the design model, or even produce entire designs automatically. Like the familiar engineering analysis programs, expert systems free engineers from tedious details in order to give them more time to deal with the significant (and interesting) issues of engineering and design.

5.3.2. Expert System Shells

Before 1985, it was generally believed that the development and use of expert systems required specialized computer languages, such as LISP or PROLOG, and specialized computer hardware, such as LISP machines. With the development of commercial expert system shells, this position changed. Today, many commercial expert system shells are available for mainstream computers from mainframes to personal computers.

An expert system shell is a computer program that combines an inference engine, an explanatory facility, and knowledge base development tools. A shell is essentially an expert system without the knowledge base, which must be created by the expert system developers. A shell can be applied to a number of different knowledge bases, to create different expert systems. That is, different knowledge bases can be developed without having to rewrite the inference mechanism each time.

An analogy with spreadsheets may be useful to those familiar with these programs. The expert system shell is analogous to the spreadsheet program as it comes from the box; it has no knowledge in it. The knowledge base is analogous to the rules or macros that the spreadsheet developer inputs to the spreadsheet program to define how to solve a particular type of problem. The database is analogous to the data that the spreadsheet user inputs to the spreadsheet program to solve a specific problem. The explanatory facility has no analog in the spreadsheet program.

There are now many commercially available expert system shells. Different shells use different forms of knowledge representation and inferencing. The selection of the proper form of knowledge representation requires experience and training on the part of the knowledge engineer. Issues such as problem-solving strategies, knowledge representation methods, development environments, and ability to interact with existing CAD systems and databases must be considered.

5.3.3. Hardware and Software Environments

Initial expectations of the memory requirements and the amount of central processing unit (CPU) time that would be required for expert systems led to the belief that only specialized AI workstations (LISP machines) would be feasible for delivery and development of expert systems. It is now possible, however, for expert systems to be developed and delivered on standard mainstream computer equipment used for CAD applications.

5.4. PROCESS AND PRODUCTION

5.4.1. The CAD System

Only three-dimensional computer-aided design systems are suitable for concurrent engineering, because only with three-dimensional models are true dimensions and engineering properties preserved. The essential factors required for interactive three-dimensional design include (a) true solids representation (not just wireframes), (b) true boolean operations (union, addition, subtraction), (c) fast display and hidden-line removal at the workstation, (d) shading with multiple light sources, and (e) very large data space for very large models.

Many computer-aided design systems have provided some of these features, but typically the response has been too slow for practical use in interactive design, or the model size is too small for realistic designs, or both. Many computer-aided design systems derived from two-dimensional drafting systems provide three-dimensional views for display purposes but do not support truly interactive design in the three-dimensional space. Large memory capacity and high-speed processing, along with high-resolution engineering graphics workstations, provide the bases for practical three-dimensional design and engineering.

The CAD system should be a commercially supported system, capable of mechanical design and a wide variety of applications.

In an open, integrated system, it must be possible for the user to temporarily exit from the CAD system to execute an application program, a query to the database, or an expert system, and then to return to the CAD system at the point from which the user left. The user wishes to see an intelligent CAD system, not the interface between a CAD system and an expert system shell; therefore the execution of the expert system from within the CAD session must be as transparent and seamless as possible.

Based on these requirements, the CAD systems CATIA and CADAM were used for the examples presented here. These two systems are probably the most widely used high-end CAD systems for mechanical, automotive, and aircraft design.

5.4.2. The Database Management System

For the integrated database concept to work, data transfer from the CAD system to the relational database is not sufficient: The link must be intimate and interactive. That is, the designer at a CAD workstation must be able to interactively insert and retrieve data from the database without delays, and a user of the database must have immediate access to the geometric information. Conversely, the inclusion of attribute information in a proprietary CAD database does not support the integrated database concept, because such information is available only through the CAD system. In the integrated database, any user, not only CAD users, must have the ability to access all data to which the user is entitled.

The most suitable available database model is the relational data model. The relational data model provides the maximum flexibility, because this data structure is easily modified and expanded, for all participants in the project to develop new uses for the data. Some of the inherent advantages of a relational data model for this application are:

- Simple and readily understandable data structures (i.e., tables)
- Data structures independent of programs, so that both the database and the application programs can be independently modified and maintained
- Use of a standard, universal, easy-to-use database language (e.g., SQL—Structured Query Language)
- Rapid prototyping and development of new applications
- Access to data both on-line and through user-written application programs
- Availability of multiple user views
- Easy modification, expansion, and additions to views, tables, fields, and indices, so that it is relatively easy to extend the data structure to meet new needs as they are identified
- Support of very large data structures

References to the database must be processed very rapidly, in order to provide acceptable response time to the users. Large memory capacity and high processing speed are essential to the operation of the database.

In order to find the data necessary to satisfy the conditions of the rule base, the expert system must have access to more than graphical design data. Specifically, some CAD systems, databases store graphical elements in the form of lines and points, relating the geometry of the design rather than the engineering information. In order to accommodate engineering data relating to the geometrical elements, it is most convenient to use a relational database structure. By assigning attributes to the graphical objects, the relational database can contain almost any type of information. In this scheme, database management functions (such as searches) can be applied to the database, allowing the system user (or expert system) to obtain necessary information or to make changes easily.

For the reasons itemized previously, the DB2 database management system was selected for integration with the CAD systems and the expert system shell.

5.4.3. The Expert System Shell

To complete the intelligent design system, a commercial expert system shell is needed. This shell should be capable of both backward and forward chaining, should be easy to use, and should support rules in a language very close to English. It should be able to read and write data from the relational database management system. It must be possible to execute the run-time version of the knowledge base from within the CAD system; that is, the expert system shell cannot create its own environment in the computer. It should be possible to develop and test knowledge bases independently of the CAD system and then to execute these knowledge bases from within the CAD session without modification.

For these reasons, the expert system shell ADS (Aion Development System) was selected for integration with CATIA, CADAM, and DB2.

5.5. APPLICATIONS

5.5.1. Interactive Design Advisors

In the interactive mode, the expert system shell, or a run-time module of the inference engine and knowledge base, is executed from within the CAD system. Figure 5.3 shows the general structure of the on-line designer's assistant. The designer, using the CAD system, may decide to invoke an expert system, or the expert system may be invoked automatically as a consequence of some action or decision of the designer. The run-time inference engine executes the rule base; forward chaining is most often used in these applications. To fulfill the conditions of the rules, the inference engine requires data. These data are obtained in the following hierarchical order:

If the required data are contained in the CAD model currently in use by the designer, these data are used to satisfy the rules.

If all the required data are not contained in the current CAD model or in the relational database, the inference engine asks the user for the required data.

For this integrated system to work, it is clear that the entities or objects must be defined consistently in the CAD model, in the relational database, and in the rule base. Attributes must be consistently defined in the relational database and in the rule base. (Attributes are not stored in the CAD model to prevent the possibility of inconsistencies between attributes in the CAD

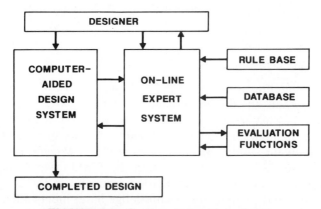

FIGURE 5.3. On-line designer's assistant.

model and attributes in the relational database.) Consequently, a data model and data dictionary are needed to control the definitions or entities and attributes.

If all the data needed to satisfy the rule base and to reach a conclusion are available to the inference engine from either the CAD model or the relational database, then no questions will be asked of the designer, who simply sees the resulting advice or suggestions displayed on the screen. If all the data necessary to reach a conclusion are not available from the CAD model or the relational database, then the user sees questions displayed on the CAD screen. This process continues until a conclusion is reached.

Example. As a very simple example of how this process can be applied, consider the case of the design of a circular rod that fits into a hole drilled into a part. One of the first steps is to select the material of the rod, attribute ROD.MATERIAL, which may be equated to the part material or selected by a rule such as the following:

```
IF      ROD.ENVIRONMENT IS CORROSIVE
THEN    ROD.MATERIAL IS STAINLESS-STEEL
ELSE    ROD.MATERIAL IS CARBON-STEEL
```

This rule specifies the rod material uniquely, depending on the value of the given condition. It is also possible to select the rod material from a list of available materials contained in the database.

The rod diameter may be determined from strength requirements, to satisfy a given allowable stress. Rules can be used to select the applicable code or other service requirements, such as:

```
IF       〈Condition A is true〉
     AND 〈Condition B is true〉
     .
     .
     .
     AND 〈Condition N is true〉
THEN     CODE IS ASME

IF       PART.SERVICE IS CRITICAL
     AND CODE IS ASME
     AND ROD.MATERIAL IS STAINLESS-STEEL
     .
     .
     AND 〈Condition N is true〉
THEN     ROD.SAFETY-FACTOR IS 3
```

The minimum rod diameter to meet the given loads, safety factors, and so on may be determined from familiar strength of materials formulas, which may be written in the form of rules. If the stainless steel rods available in the inventory are stored in the database by the manufacturing inventory management system, then additional rules and database queries can immediately select the rod size that is available and meets the strength requirements.

In addition to setting attributes uniquely, rules can also be used to propagate constraints on attributes. The rod selected by the above rules, with diameter ROD.DIAMETER, must fit into a hole with diameter HOLE.DIAMETER. For ease in manufacturing, there must be some finite gap, or play, between the rod and the hole. For manufacturability, this gap cannot be less than ROD.PLAY.MINIMUM. This minimum value may be assigned by a rule such as:

```
IF      ROD.MATERIAL IS STAINLESS-STEEL
    AND ROD.ASSEMBLY IS MANUAL
    .
    .
    .
    AND ⟨Condition N is true⟩
THEN    ROD.PLAY.MINIMUM IS 0.0001*ROD.DIAMETER
```

For satisfactory functioning of the part, the play between the rod and the hole cannot be too loose. This constraint can be controlled by rules such as:

```
IF      ⟨Condition A is true⟩
    AND ⟨Condition B is true⟩
    .
    .
    .
    AND ⟨Condition N is true⟩
THEN    ROD.PLAY.MAXIMUM IS 0.005*ROD.DIAMETER
```

Additional rules may be used to modify the constraints, for example, making them tighter:

```
IF   ROD.PLAY.MAXIMUM >0.01
THEN ROD.PLAY.MAXIMUM IS 0.01
```

After these rules have been applied, the expert system must select a hole diameter between the minimum value ROD.DIAMETER plus two times ROD.PLAY.MINIMUM and the maximum value ROD.DIAMETER plus two times ROD.PLAY.MAXIMUM. The hole diameter should be a standard drill diameter. Suppose that standard drill sizes are stored as attribute SIZE in the table DRILL.TABLE. Then a database query is used to find the hole diameter equal to a standard drill diameter lying between these constraints:

```
EXECUTE SQL SELECT FROM DRILL.TABLE 'SIZE' WHERE 'SIZE' >
'ROD.DIAMETER'+2*'ROD.PLAY.MINIMUM'               AND'SIZE' <
'ROD.DIAMETER'+2*'ROD.PLAY.MINIMUM' INTO 'HOLE.DIAMETER'
```

Finally, a rule can be used to create the geometric feature for the hole or to write an error message to the user if there is no standard drill diameter within the tolerances:

```
IF      HOLE.DIAMETER IS NOT NULL
THEN    CALL   CREATE.GEOMETRY   (HOLE,   HOLE.DIAMETER,
        HOLE.DEPTH)
ELSE    WRITE MESSAGE (''There is no standard drill diameter
        within the prescribed tolerances for the rod and
        hole. Do you wish to use a nonstandard drill or to
        override              the              standard
        tolerances?'')
```

In this case, if the database query returns the value NULL, there is no standard drill bit to drill a hole within the constraints set by the rod diameter and the minimum and maximum play. In an interactive CAD session, the expert system displays a message on the CAD screen, asking the designer to resolve the problem, which may involve overriding one of the manufacturing standards or redesigning the rod in such a way that the standards can be met. In a batch process, the design could be rejected for resolution by a designer, or additional rules could be generated to resolve the difficulty automatically. Alternatively, a more complex set of rules could be written to size the rod and the hole simultaneously, to assure a feasible solution.

If the result of the database query is not NULL, then a standard drill bit has been found that meets the derived constraints, and a geometric construction routine is called to create the hole part in the model geometry.

Part Size Advisor (Landing Gear). To illustrate the interactive design process with a simple graphical example of the use of expert systems, Fig. 5.4 shows a plot taken from a CATIA screen showing a three-dimensional model of an aircraft landing gear. At this point, the designer invokes a knowledge base to assist the designer in selecting the proper size for the landing gear strut. In Fig. 5.4, the CAD system asks the user to indicate the desired piston casing on the display. The selection is made by positioning the cursor (shown as an arrow).

This screen shows how the user, in the interactive mode, inputs data to the knowledge base (the identification of the part geometry) by simply pointing to the geometry on the screen. With this identification, the expert system now has available, from the CATIA geometric model, all the geometric properties (length, diameter, etc.) and other attributes (material, etc.) of the piston casing, for use in further inferencing.

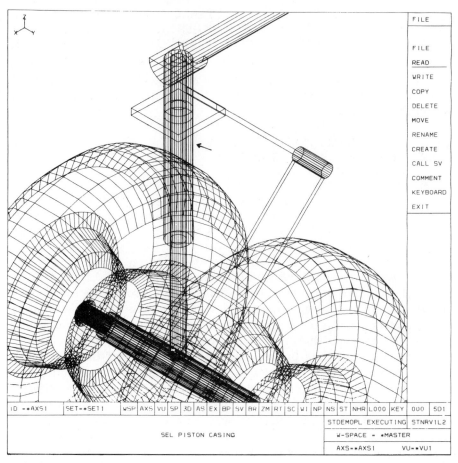

FIGURE 5.4. CATIA screen for strut analyzer expert system: selection of piston casing.

In Fig. 5.5, the ADS inference engine finds that forward chaining stops before a conclusion is found, due to lack of data. One missing item of necessary information to size the strut is the weight of the aircraft. The expert system then asks the user to input the weight of the aircraft. This question is asked because the required information is not available from the CAD model (which shows only the landing gear) or from the database (the weight of the aircraft is not an attribute to the landing gear), although the database could have been designed to provide this information if it were known. The user enters the value 19000.

This screen demonstrates how the user can enter real numerical values during the CAD session, in response to the question posed by the expert system. If the value of the aircraft weight had been available as a CATIA attribute or from the DB2 database on the aircraft under design, the rule

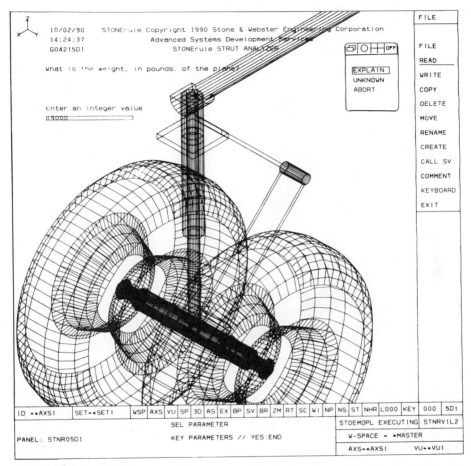

FIGURE 5.5. CATIA screen for strut analyzer expert system: input of aircraft weight.

condition would have been instantiated automatically and the question would not have been asked of the user.

In Fig. 5.6, the ADS inference engine finds that forward chaining again stops before a conclusion is found, again due to lack of data. One missing item of necessary information to size the strut is the location of this landing gear on the aircraft. The inference engine then asks the user to input this location by selecting one of a set of multiple-choice options. This question is asked because the required information is not available from the CAD model or from the database.

Although the location might be a preassigned attribute of the landing gear, the user in this case is using a generic model and is currently designing the landing gear for a specific location; that is, the user is defining this

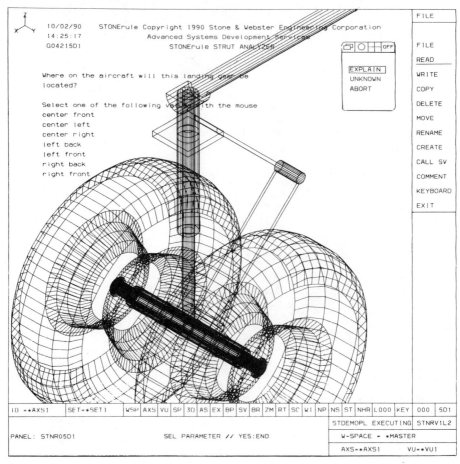

FIGURE 5.6. CATIA screen for strut analyzer expert system: input of landing gear location.

attribute. The user selects one of the possible landing gear locations from the list displayed. This screen demonstrates how the user can make multiple-choice selections from a list displayed by the expert system.

The inference engine processes a number of rules to design this strut, but in this case only two rules require data that cannot be retrieved or inferred from the CAD model or from the relational database, or from the consequents of other rules. The inference engine derives a safety factor for the piston casing from the rules and calculates the required size from the safety factor, the weight of the aircraft, and other considerations. In Fig. 5.7, the expert system creates a new geometric object and causes CATIA to display the suggested geometry of the piston casing on the CAD workstation screen. The recommended piston casing, which is larger than the original piston

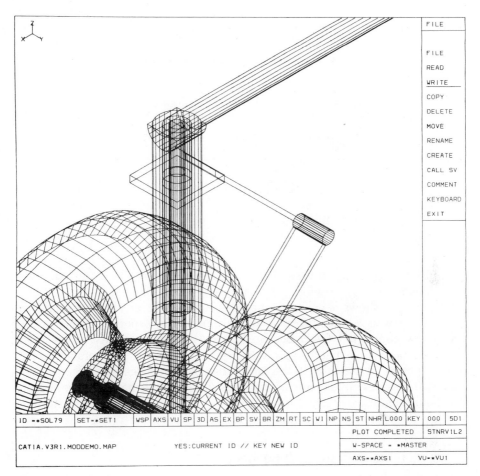

FIGURE 5.7. CATIA screen for strut analyzer expert system: display of recommended piston casing geometry.

casing, is displayed on top of the original and in a different, blinking, color. That is, the ADS knowledge base has created new geometry in the CATIA database, which is displayed along with the original geometry. In this case, the result of the knowledge base is only a suggestion or recommendation; the user can accept it, or reject it and accept the original geometry for the piston casing, or generate the user's own new geometry.

5.5.2. Design Review

Design review can be performed by expert systems as an interactive process or as a batch process. The example in the previous sections may be considered to be an interactive design review, in which the user's design for

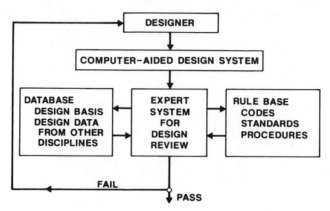

FIGURE 5.8. Expert system for design review.

the piston casing was reviewed by the expert system and a new size was recommended.

Figure 5.8 shows a general schematic for the use of expert systems for design review as a batch process. In this application, expert systems are used to review completed CAD design files for manufacturability or for other considerations as a batch operation, without direct user interaction. If a design passes all the rules in the knowledge base, the design is accepted and released; if a design fails any of the rules, the design is held and the reasons for failure and suggestions for resolving the problem are written as overlays on the CAD model or drawing.

5.5.3. Automatic Drawing Production

Expert systems can be used for the generation of two-dimensional engineering drawings from three-dimensional design models. In this case, the knowledge base represents the rules for generation of standard drawings, such as cross sections or isometrics, from three-dimensional models. Given these rules, the two-dimensional projections of cross sections can be created by the CATIA system, and the appropriate notes and other annotation can be added.

5.5.4. Automatic or Parametric Design

By extension of the interactive use of expert systems discussed earlier, entire engineering design processes can be expressed in the form of rules, so that certain design steps can be made completely automatic. In this way, expert systems can be used for parametric design or the automatic generation of complete designs from a set of parameters input by the designer.

Structural Steel Connection Advisor. As an example of parametric or automatic design of details, a steel connection detailing expert system has been developed. This expert system uses a database of geometric data from a three-dimensional CATIA design model, created during the process of design of a steel framework. This geometric database contains information on the layout and sizes of the steel members. After the steel framework has been designed, it is analyzed by a three-dimensional stress analysis program. The DB2 database contains forces and moments from the results of the stress analysis. Code requirements are contained in the knowledge base. The expert system then generates connection detail designs automatically, using these data, either interactively for single connections selected by the user or as a batch process for all connections in the steel framework. The results of the expert system are output through CATIA as detail drawings ready for fabrication.

This type of expert system, for automatic design, increases engineering productivity by eliminating design errors and by eliminating completely the man hours required for this design step.

5.6. FUTURE DEVELOPMENTS

The power of expert systems lies in their ability to separate knowledge about specific areas of application (the knowledge base) from general problem-solving knowledge (the inference engine). This separation is accomplished by the expert system shell.

Much of what is considered to be expertise is actually experience distilled in the form of heuristic rules. Experts are those people who have accumulated more rules, in certain specific areas of interest, then have nonexperts. These heuristic rules, if expressed verbally, can be coded into expert systems that embody the superior problem-solving capabilities of the human experts.

The value of expert systems lies not in the replacement of experts by computer programs. There is no known instance of this ever occurring, because no expert system is as good as a human expert, if one is available. The value of expert systems lies in their ability to replicate the performance of experts on judgmental tasks in situations in which there are not enough human experts, because they are retired, sick, on vacation, in another location, or otherwise not available.

In the proper situations, expert systems can improve productivity, reduce errors, and improve quality. Many expert system applications in manufacturing have proved to be valuable, but these have barely scratched the surface of the possibilities in this field.

REFERENCES

1. Barr, Avron, and Feigenbaum, Edward A. (1981). *The Handbook of Artificial Intelligence*, William Kaufmann, Los Altos, CA, p. xi.

2. Buchanan, Bruce G., and Shortliffe, Edward H. (1984). *Rule-Based Expert Systems: The MYCIN Experiments of the Stanford Heuristic Programming Project*, Addison-Wesley, Reading, MA.

3. Cusumano, Michael A., and Nobeoka, Kentaro (1990). *Strategy, Structure, and Performance in Project Development: Observations from the Auto Industry*, Working Paper 3150-90-BPS, Sloan School of Management, Massachusetts Institute of Technology, Cambridge, MA, April.

4. Eppinger, Steven D., Whitney, Daniel E., Smith, Robert P., and Gebala, David A. (1990). *Organizing the Tasks in Complex Design Projects*, Working Paper 3183-89-MS, Sloan School of Management, Massachusetts Institute of Technology, Cambridge, MA, June.

5. Feigenbaum, Edward A., McCorduck, P., and Nii, H. P. (1988). *The Rise of the Expert Corporation*, Random House Times Books, New York.

6. Jakiela, Mark J. (1989). *Intelligent Suggestive CAD Systems: Research Overview*, Paper LMP-89-021, Laboratory for Manufacturing and Productivity, Massachusetts Institute of Technology, Cambridge, MA, February.

7. Madnick, Stuart E., and Wang, Y. Richard (1987). *A Framework of Composite Information Systems for Strategic Advantage*, Working paper 1937–87, Sloan School of Management, Massachusetts Institute of Technology, Cambridge, MA, September.

8. National Academy of Sciences—National Academy of Engineering (1983). *Report of the Research Briefing Panel on Computers in Design and Manufacturing*, p. 60.

9. Nevins, James L., and Whitney, Daniel E. (1989). *Concurrent Design of Products and Processes: A Strategy for the Next Generation in Manufacturing*, McGraw-Hill, New York.

10. Robertson, David C., and Allen, Thomas J. (1990). *Evaluating the Use of CAD Systems in Mechanical Design Engineering*, Working Paper 3-90, International Center for Research on the Management of Technology, Sloan School of Management, Massachusetts Institute of Technology, Cambridge, MA, January.

11. Smith, Robert P., and Eppinger, Steven D. (1990). *Modeling Design Iteration*, Working Paper 3160-90-MS, Sloan School of Management, Massachusetts Institute of Technology, Cambridge, MA, June.

CHAPTER SIX

Management of Product Design: A Strategic Approach

DEBASISH N. MALLICK and PANAGIOTIS KOUVELIS
Department of Management, College of Business Administration,
The University of Texas at Austin

6.1. INTRODUCTION

To be successful, a business organization has to identify the needs of its customers, develop appropriate products to satisfy those needs, and [barring "hollow corporations"—those companies that do little or no manufacturing (16)] should have a manufacturing system to produce those products in a profitable manner. Three functional areas that support these activities of a business organization are marketing, design, and manufacturing. To be competitive, each one of these three areas needs to be managed strategically (Fig. 6.1). During the past two decades, marketing has been playing an important role in the business strategy of a corporation. In contrast, until recently manufacturing has been playing only a reactive role in the strategy formulation process. Attempts are being made now to use manufacturing as a strategic resource to gain competitive advantage (13, 15). Manufacturing and marketing, however, are only two of the three links necessary to complete the design–manufacturing–marketing chain, the chain that connects the product, the producer, and the customer. In the long run, a business organization can compete only by offering products superior to its competitors' products. The business strategy of a corporation is not complete without a design strategy to develop the product in the first place. Yet, the concept of design as a strategic weapon has been almost nonexistent in the corporate strategic debate.

This chapter begins by examining the strategic nature of the product design decisions in Section 6.2. In Section 6.3, it identifies various issues that

Intelligent Design and Manufacturing, Edited by Andrew Kusiak.
ISBN 0-471-53473-0 © 1992 John Wiley & Sons, Inc.

FIGURE 6.1. Design strategy in corporate strategy framework.

have made management of product design a challenging task. In Section 6.4, it presents recent developments in the area of strategic management of product design. In Section 6.5, it reviews various organizational structures and design support systems as alternatives available to the product design managers. Concluding remarks from the authors are found in Section 6.6.

6.2. PRODUCT DESIGN AS A STRATEGIC ACTIVITY

In an ever demanding global marketplace, customers are insisting on higher quality, greater reliability, faster delivery, and more product variety. To compete successfully, an organization needs to be profitable. However, that no longer necessarily implies the lowest cost operation. To be successful, an organization must compete simultaneously along many of the following dimensions: *price*, *quality*, *speed*, *flexibility*, and *service*. These dimensions are known as "content variables" in the manufacturing strategy literature (1). Here they are referred to simply as the *dimensions of competition*. The following discussion is intended to demonstrate how product design affects an organization's ability to compete on any of the above dimensions of competition.

6.2.1. Price

Price is the amount of money that customers have to pay to exercise ownership rights over a product. Market demand sets an upper limit to the price that a company can charge for its products. To remain profitable, corporations need to charge a price that covers their entire cost of development, manufacturing, and marketing of the product, including a fair return on their investment. Therefore price positioning of a product is limited by its costs. In most cases, the total cost of a product is dominated by its development and manufacturing costs. Researchers have shown that a large portion of the manufacturing cost is again determined at the design stage. Studies reveal that, for some products, up to 80% of the production costs can be determined at the design stage (5). Consequently, product design can and should be used strategically to achieve a desired price positioning. For example, in the early 1970s, the Japanese automobile-makers entered the low-end subcompact segment of the U.S. automobile market primarily based on their ability to design automobiles that cost less to manufacture.

6.2.2. Speed

"Speed" is generally used, in place of "deliver speed," to refer to the time it takes from the development of a product concept until the product reaches the market. Competition on speed, also known as *time-based competition* and *time-based strategy*, is the ability to deliver new products to the market fast. Positioning along the speed dimension is becoming increasingly popular in the United States and abroad. A recent survey of 50 major U.S. companies found that practically all companies put "speed" at the top of their priority lists. An increase in speed in the product development process often pays off even if it means going over budget. From Ref. 8: "An economic model developed by the McKinsey & Co. management consulting firm shows that high-tech products that come to market six months late, but on budget, will earn 33 percent less profit over five years. In contrast, coming out on time and 50 percent over budget cuts profit only by 4 percent." When competing on speed, it is desirable to have products with simple design to facilitate ease of manufacture. However, achieving a desired level of performance with a simple and easy to manufacture product might require a high level of innovation. Decisions made at the product design stage, particularly those affecting manufacturability of the product, significantly influence a company's ability to compete on the speed dimension (8).

6.2.3. Flexibility

Flexibility is the ability to respond to unexpected changes in the marketplace. There are three major types of flexibility a company may like to have.

These are *volume flexibility* (i.e., the ability to respond to demand changes), *product flexibility* (i.e., the ability to produce a variety of products), and *lead-time flexibility* (i.e., the ability to introduce new products to the market fast). Higher levels of all three types of flexibility are easier to achieve with simple products. Product complexity increases the burden on the manufacturing system, because more components, strict tolerances, or tighter assembly requirements demand more sophisticated controls (e.g., inventory, quality) and entail higher chances of costly mistakes. For high volume this usually results in the need for more dedicated (i.e., mass production) processes. For low volume products with tight specifications, computer-controlled systems such as flexible manufacturing systems (FMS) are required to achieve flexibility. Both options are very expensive. Mass production and computer integrated manufacturing (CIM) systems require a significant capital investment and a long planning horizon to justify their acquisition, which makes them unattractive options to managers in turbulent markets. Also, it is much easier to set up a manufacturing facility for products with simple design. Sometimes simple product designs can be accommodated within existing manufacturing facilities with little or no modifications.

6.2.4. Quality

The concept of competing on quality is slightly more complex. This is because the word "quality" has different meanings to different people. According to Garvin (11), there are eight separate dimensions of quality. These are performance, features, reliability, durability, aesthetics, serviceability, and perceived quality.

Performance refers to the primary operating characteristics of a product. For an automobile, these would be traits like acceleration, handling, fuel consumption rate, and comfort. *Features* are the secondary characteristics that supplement the product's basic functions. For an automobile, these would be sunroof, power windows, customized wheel-cover, and so on. The performance and features are primarily determined by its design. *Reliability* is an engineering and technology related product characteristic that reflects the probability of a product failing within a specified period of time. Reliability of a product is influenced, in general, by the decisions on number and types of components to be used in a product since each component represents a potential source of failure. *Durability* is the amount of abuse a product can absorb during its economic life and still meet performance expectations. Increased level of durability is usually associated with increased level of innovation, because it involves innovative use of materials and processes specified by the product designers. *Aesthetics* is the combination of product attributes that best matches the preferences of a specified consumer. Aesthetic appeal is very much dependent on individual preferences. Yet the mass appeal of the aerodynamic revolution of the 1930s and the more recent success of the Sony Walkman make us believe that there is a general preference for certain types of design. It is the designers job to find such a design. *Serviceability* is the speed, courtesy, and competence of repair that a company provides for its products in the field. The serviceability of a product is inherent in its design. Many products are so complex that they require user training, periodic maintenance, or both. It is much easier and faster to carry out repair work on a simple product. The serviceability can also be improved by a modular design. For a product with modular design, a component failure can be confined within the module where the component belongs. It is very clear that performance, features, reliability, durability, serviceability, and aesthetics are established at the design stage. This is why they are often grouped together as *design quality*.

Conformance is the degree to which a product's operating characteristics match preestablished design standards. Loose specifications are easier to meet. However, relaxation of these specifications without sacrificing performance of other dimensions is a challenge to the designers. Product design in general, not just tolerance setting, affects the level of conformance that can be achieved and passed on to the customer.

In many instances, consumers do not always possess complete information about a product's attributes. Hence they rely on indirect measures, defined as *perceived quality*, when comparing different brands. A very carefully managed design, manufacturing, and marketing strategy is required to affect consumers' perceptions positively. However, it should be noted that reputation, a primary contributor to perceived quality, can only be built by selling well-designed products over a very long period of time.

6.2.5. Service

From the discussion on serviceability, it is clear that decisions made at the product design stage affect a business organization's ability to provide after-sales service to the customer in a cost effective way.

Depending on the product and the market segment of interest, any one or many combination of these dimensions can be used strategically to achieve a product positioning and market differentiation as required to gain competitive advantage. "Product design is a strategic activity, by intention or by default" (26), because major commitments to these dimensions are usually made during the design stage of the product. Corporations not making their product design decisions strategically are competing from a position of disadvantage.

6.3. WHY IS DESIGN MANAGEMENT A DIFFICULT TASK?

From the existing literature on the subject and practices followed by business organizations, it has been found that not many business organizations manage their products strategically and most managers are not very comfortable with the concept of "management of design." The following discussion examines issues that are primarily responsible for making strategic management of product design a difficult task.

6.3.1. Lack of a Precise Definition of Design

In general, people in business and outside recognize the importance of design. However, there is no agreement on how to define it. The word design is often used in a very broad context. "Design" typically refers to products, packaging, facilities, and so on. However, sometimes it is being used to refer to the elements of style, such as "designer hairstyle" or "designer jeans." From Ref. 19: "At the opposite extreme, philosophical explorations of design often begin with the prehistoric invention of stone tools, thereby listing under the rubric of design, at least by implication, all scientific and technological advancements." Unfortunately, neither approach yields a valid concept that can be used by business or by the general public to define "design" in a useful way because these types of generic concepts tend to dilute focus (19).

6.3.2. Problems Caused by Division of Labor

Product design in its present form emerged during the Industrial Revolution. Earlier, a typical craftsperson would design and manufacture a product based on a customer's request. This used to be an expensive process that not many could afford. Therefore customers started to value standardized products, which could be manufactured at low cost. To achieve cost reduction, work became increasingly compartmentalized through division of labor. The simple job shop process used by craftspeople to manufacture customized products was replaced by complex flow processes devoted to the mass production of standardized products. The tasks of designing a product became separated from the task of manufacturing and marketing it. "A new type of individual, the designer, devised two-dimensional patterns and three-dimensional prototypes to be copied with automatic precision by workers and machines" (19). Designers became separated from the customers and had to rely on marketing's interpretation of customers' needs based on market research of dubious validity.

As divisions and departments separated from one another, functionally and physically, communication problems and interdepartmental conflicts began to emerge. The goal of good design, as measured by appropriateness to the job it has to do (3), was not always compatible with the goals of other functions such as manufacturing, marketing, and finance. Due to the relative ease of measurement, manufacturing effectiveness became synonymous with low cost production in most business organizations. Marketing experts tried to increase market share using strategies emphasizing more on the price, promotion, and distribution rather than on the well-designed products. Designers were sidelined with limited power in the corporate hierarchy. The new corporate order forced them in the back seat during the corporate debate for strategy formulation, letting marketing and finance executives steal the show.

6.3.3. Overemphasis on the Marketing Concept

In the past, when there was little or no competition, companies sold whatever they made. When the competition began to intensify, business organizations tried to solve their problems with the "marketing concepts." The marketing concept, in most cases, prescribes a two-step process. First, companies need to identify their customers' needs. Second, they should find a cost effective way to satisfy those needs. This is a very powerful concept because it focuses on customer satisfaction. However, it is extremely difficult, in some cases, to identify what the customers' needs are. This is because customers are not always able to articulate their needs and desires properly. In general, they tend to define their needs in terms of existing products. When dealing with innovative new products such as the Sony Walkman or Selective Laser Sintering Machines, in most cases customers will not be able to imagine or predict the product even at a conceptual level, let alone tell the market researchers that they wanted it. Therefore over-reliance on existing market research methodology forces designers to become less innovative, which forces the business to become vulnerable to aggressive and innovative competitors.

6.3.4. Lack of Design Education

It is rather common to find business executives or managers to make strategic design decisions without a broad and in-depth understanding of product design. This is because they are from a background (e.g., law, accounting, finance) that is far removed from the product design activity. Design is not an integral part of most major business school curricula, with few exceptions. "Of the 250-odd engineering schools in the U.S., design is still generally taught only in the context of a few 'machine design' courses, which traditionally stress analytical design techniques more than actual design" (7). The problem of design education is exacerbated by the fact that the field of design has been deficient of explicit and logical knowledge that can be generalized. Recent activities supported by some of the business organizations in this country and abroad and the efforts by the Corporate Design Foundation of Boston, Massachusetts, to educate business leaders of today and tomorrow have been a major step forward. However, much still needs to be done.

6.3.5. Design Perceived as an Unmanageable Activity

There is a general perception that design is a complex and creative activity that is difficult to manage. The notion of industrial design triggers in the mind of most business people vague artistic forms and talents, which cannot be translated into solid business arguments. Engineering design reminds one of blueprints and tolerances, which most business people find an unnecessary detail to understand, and even more so to consider in their decision-making process.

In general, managers believe that any attempt to manage design effectively would require too many changes that are organizationally destructive. Consequently, design in most organizations is treated as a jobbing task. Companies use it when they need it for one-off projects or it is treated as a function separated from the mainstream business activity, such as manufacturing, marketing, and finance. However, design is a business function and businesses are able to compete because of it. "Good design is good business." It must be recognized that managing design is conceptually no different from managing other corporate functions, such as finance or information systems (20).

6.4. DESIGN STRATEGY

Fitzsimmons et al. (9) define product design as the entire process and combined effort of engineering, industrial, and human factor design specialists that finally results in coming up with the product(s), often in a prototype form. *Engineering design* is the process that validates the technical feasibility of a product concept and determines how to provide the required "functions" to a product. *Industrial design* refers to the process of providing "forms" to these functions, and *human factor design* includes such processes as ergonomic design and psychological factor analysis, which make a product easier to use (Fig. 6.2).

They also define "design strategy," using the language of business, and present a conceptual framework to facilitate development, implementation, and control of design strategy as an integral part of the overall business strategy. Design strategy is defined "as a pattern of decisions to gain competitive advantage through design. This can happen either through the design of new products that create new markets, or through supporting existing market needs better than the design function of the competitors" (9). Their proposed framework defines two *basic design variables* to facilitate a strategic insight into the multidimensional design management problem (Fig. 6.3). These variables are product complexity and innovation level used in a design. Complexity of a system is measured by the number of elements in that system and the level of interactions among these elements (4). For example, a product with a larger number of components is more

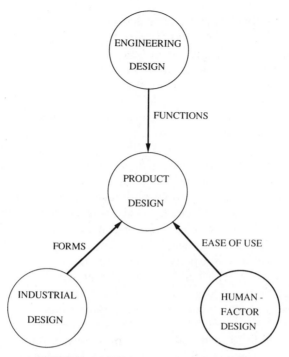

FIGURE 6.2. Elements of product design.

complex than a similar product with fewer components. Also, the complexity is increased if the components have multifunction capabilities rather than single functions. *Innovation* is defined as the incorporation of new ideas and technology (10). The level of innovation is measured by the degree of

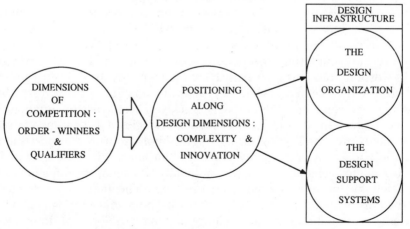

FIGURE 6.3. Design strategy framework.

change an innovation brings in comparison to the existing level of "performance" of a product. Complexity in itself is not necessarily undesirable. It is often a necessary component in building, or maintaining, competitive advantage (12). Similarly, innovation in itself is not always desirable. In certain product–market segments it may be necessary to pursue a low innovation strategy. For example, a high innovation level that leads to short product life cycles may not always be desirable from a business perspective.

The implementation of design strategy requires selection of an appropriate design infrastructure (Fig. 6.3). The design infrastructure decisions consist of selecting the *design organization*—that is, the organization structure necessary to support a specific design strategy best—and the *design support systems*—that is, the tools and techniques that can be used to manage the design activity efficiently. This is an entirely different approach than the traditional reactive mode of design management, because it forces the manager to make a strategic choice by considering advantages and disadvantages of various alternatives. This research work is a significant step forward because it provides a framework that can be used to facilitate participation of design in the corporate debate as an equal partner with other functions, such as manufacturing and marketing.

Depending on the strengths and weaknesses of a business organization, and the opportunities and threats posed by its environment, the available strategic alternatives can be classified into three general categories. The first category can be classified as *market driven* strategies: Here marketing acts as a sensor of the market needs and transmits them to manufacturing and design so that the organization as a whole can respond to these needs in the best possible way. Take the example of Chevy-Geo. Here the company identified the market needs for a fuel efficient subcompact car. Then it went to design and manufacture. The second category can be classified as *design driven* strategies: This refers to the situation where the design function takes up the leadership role in the strategic planning process by acting as an innovator of new products. The organization as a whole needs to find the best possible way to manufacture and market these new products. A new product can be the result of a technological breakthrough such as the Selective Laser Sintering System designed, manufactured, and marketed by DTM Corporation, Austin, Texas. Here the company itself was triggered by the invention of a new technology that will enable us to convert CAD-generated concepts into solid models quickly. These types of situations can also be characterized as *technology driven* strategies. Or it can be the result of an innovative application of existing technology, with the most suitable example being that of the Sony Walkman, an innovative product based on existing audio technology that was originally designed, manufactured, and marketed by the Sony Corporation of Japan. The third category can be classified as *manufacturing driven* strategies: The manufacturing systems are typically capital intensive and require a long development time, which makes them less responsive to quick changes. If not planned strategically, they have the potential to constrain the business strategy of an organization. The manufacturing driven strategy refers to the situation where an organization has to design and market products that can be supported best by a given manufacturing system. This is a very important short-term strategy, because in the short run only the products that can be manufactured by the existing system can be marketed and these should be designed accordingly: for example, when steel industry designers at Nippon Steel of Japan came up with VDS (vibration damping steel), which could be manufactured with minor modification of their existing facility. Here marketing went out to find new applications such as quiet washing machines and better soundproofing of automobile doors.

The underlying assumption for the above three general categories of design strategies is that a misfit between design, manufacturing, and marketing strategies will lead to suboptimal utilization of the organizational resources. And conversely, a good fit will produce positive synergy, which is not possible to achieve by focusing on any of the three functions.

6.5. DESIGN INFRASTRUCTURE

Design infrastructure refers to the organization structure and design support systems that can be used to deliver a particular design strategy. The design infrastructure choices made by a company reflect the market requirements and positioning desired along the competitive dimensions. Setting up the design infrastructure is the crucial step in implementing any design strategy.

6.5.1. Structuring the Design Organization

Product Development Process. The product development process generally consists of four iterative stages. These are concept design, technical design, detail design, and manufacturing process design (Fig. 6.4). The new

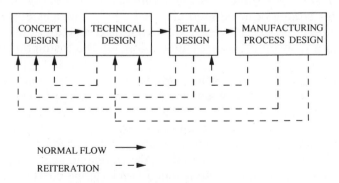

FIGURE 6.4. Stages of product design.

product concept and rough product specifications are developed at the *concept design* stage by matching a company's technological strengths with marketing's assessment of the customer's need. Technical feasibility of the initial concept is analyzed and engineering knowledge is used to meet the increasingly complete and accurate specification at the *technical design* stage. *Detail design* is the process of documenting the complete description of an almost designed product with exact drawings and bill of materials. This stage is usually followed by prototype building and testing to fine-tune a design before releasing a complete "production-intent" design. The *manufacturing process design* refers to the process of designing the molds and tools, jigs and fixtures, work-cell configurations, and process planning necessary in manufacturing the product. A number of passes through the first three stages are usually made before a design can reach the manufacturing process design stage (7).

Product Development Strategies. Depending on the product and the market segment of interest, a business may opt to follow a strategy through these four stages in a sequence, concurrently, or as a combination of both. Traditionally, companies have been following the *sequential product development strategy*, which is often referred to as the "throwing it over the wall" strategy. Here each group of specialists does their work and then passes it on to the next group. Due to the sequential nature of the process, decisions made at an upstream stage constrain the downstream stages to overall suboptimal options. Also, the sequential execution of different stages requires longer lead time. Poor interface between different stages causes communication problems and misunderstandings, which increase the number of iterations needed to complete a design. This makes it very difficult to forecast project completion accurately, causing frequent delays and late starts. However, it is important to recognize that sequential product development strategy may be the only option for a large, complex, precedence-constrained project. A sequential product development strategy is easier to implement because it places much less demand on the organizational resources and requires less organizational efforts for coordinating the project. This strategy can be used effectively by companies enjoying technological superiority in a stable market with little competition; that is, where speed is not a relevant competitive dimension and the customer is willing to wait.

Concurrent product development strategy refers to the simultaneous design of the product and its manufacturing and support processes. For the case where attention is restricted to the engineering design, but not on the industrial and human factor design, this approach is also referred to as concurrent engineering. Under the concurrent product development strategy, all four stages of the product development need to proceed in parallel. It is intended to force designers, in their design process, to consider all strategic elements of the product life cycle from conception through disposal, including product quality, facility design, and production control and scheduling. Simultaneous processing of all four design stages is expected to reduce the time it takes for a product, from the concept stage, to reach the market and to reduce the total cost of doing business by improving the interface between different stages and eliminating costly iterations through different stages. In addition to shortened lead time and reduced cost, concurrent product development strategy is expected to lead to better quality of design by elimination of the suboptimization that is inevitable under a sequential product design approach. Concurrent product development strategy is an appropriate strategy when companies need to introduce new products fast in a competitive environment. However, it is important to recognize that the concurrent product development strategy usually places an increased demand on the organizational resources and requires increased efforts in communication and coordination. The costs of such efforts must be weighted against the anticipated benefits of using such an approach.

The *team approach* is a relatively new product development strategy, which tries to achieve some of the advantages of the concurrent approach while keeping investment in organizational resources to a minimum. Under this strategy, all stages of the product development need not be carried out simultaneously, but individuals working on different stages are brought together to provide a system perspective through constant interaction from the very beginning of the project. This strategy avoids unnecessary iterations through increased communication and results in shortened lead time. Team approach attempts to encourage innovation through informal coordination, which requires less effort, and by giving a free hand to the members of the team. It is appropriate when a higher level of organizational flexibility is needed to meet the demand for new and innovative products. However, for companies introducing a large number of products over a short period of time, the problem of resource allocation may cause conflicts among different teams championing different products. Also, this approach produces slow organizational learning because experience gained by a team is not shared by the entire organization.

Product Development Organization. There are three general types of organization structures that are commonly used by companies engaged in product development activities, as shown in Fig. 6.5. Under *functional organization*, individuals are grouped together by their work, knowledge, and specialties. The interest of each function is represented to the top management by a senior functional manager. The work of the different functional areas is coordinated through detailed specifications of the guidelines to be followed by each function and by occasional meetings to resolve differences among different functions. Functional organization minimizes duplication of organizational resources. However, this form of organization tends to overspecialize and to develop a narrow corporate viewpoint, thus making conflict resolution difficult.

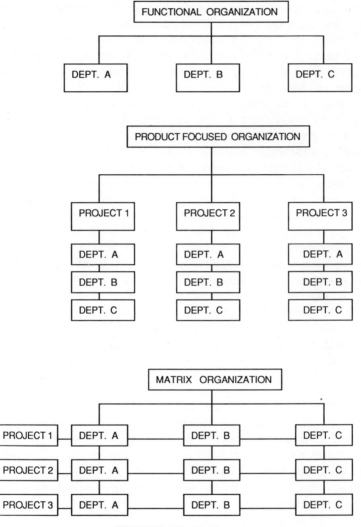

FIGURE 6.5. Design organizations.

Product-focused organization is a relatively self-contained group formed around a particular product development project by dedicating individuals from different functional areas on a temporary basis. The advantage of this approach is that it allows more attention to each product development project. Informal coordination, requiring less effort, can be used to achieve interaction of functional activities. Product-focused organizations can be used to encourage innovation by assigning greater responsibility to their team members. However, this form of organization tends to duplicate expensive organizational resources and is not suitable for centralized control.

In uncertain business environments, lateral coordinating approaches perform better than traditional vertical chains of command. *Matrix organization* attempts to capture the benefits of different specialists through lateral coordination, while maintaining specialized organizational units and thus avoiding duplication of efforts. Under matrix organization structure, each functional specialty is represented by individuals from different functions in the product development project under a project manager who is responsible for coordinating the project. Depending on the power and the position of the project manager in the organizational hierarchy, Hayes et al. (14) classify the matrix organization into two types. "Lightweight project manager" is a middle-level or lower-level manager having considerable expertise but little status or influence in the organization. Under this organizational structure, functional managers retain the "line authority" over the people working on the project. The project manager who is responsible for coordinating the project activities can only advise as a "staff." Heavyweight project manager" is a senior person in the organization who has expertise and experience and often outranks functional managers. Except for long-term career development, individuals from different functions but working on a product development project are under direct control of the project manager. Matrix organization is found suitable for coordinating efforts on large complex projects involving a number of product lines. It is also useful for projects that have a limited time span and that need different people for each phase of the work. However, this type of organization often forces people to work for more than one boss. Collaboration and constructive conflict resolution are essential for successful implementation of matrix organizations.

From the discussion on product development strategies and product design organization, it appears that there exists a one-to-one relationship between a product design strategy and an organization structure. That is, a functional organization is suitable for sequential product development strategies, matrix organization is suitable for the concurrent product development approach, and product-focused organization is appropriate for the team approach. Fitzsimmons et al. (9) point out that the design of organizational structure to manage a product development process is, and should be treated as, a strategic issue. Yet this has not been a well researched area. The above authors have emphasized that like most strategic issues there is no one general organizational structure that is suitable for managing all levels of product complexity and challenges in product innovation. For example, when an organizational structure that has been designed to deal with a high level of complexity (or innovation) is employed to deal with a low complexity (or innovation) situation, it can result in an inefficient utilization of expensive resources. A "good" strategy is the one that attempts to ensure a proper fit between a choice of product development strategy and the organization structure to be used to implement it, taking into account the market characteristics and any existing limitations on available organizational resources.

6.5.2. Design Support Systems

The *design support system* refers to the selection of tools and techniques to be used to support the design strategy. An extensive set of tools is available that can be used to manage the product design process. A partial list of these tools and techniques is presented in Table 6.1.

Axiomatic Approach. The axiomatic approach to design attempts to manage design activities through all four stages of the product development process by specifying a set of axioms. These axioms are the basic laws and fundamental principles that have been found to govern all good designs in practice. Axiom 1, the *independence axiom*, requires independence of functional requirements. It states that in a good design each functional requirement of a design problem is satisfied independently by some aspect of the solution. Integration of design features in a single physical part is recommended only if functional requirements can be satisfied independently without any compromise. Axiom 2, the *information axiom*, requires minimization of information content. It states that a good design must be as simple as possible. That is, a good design should attempt to satisfy all functional requirements while keeping the complexity of the product to a minimum level. According to this approach, a design problem is to be defined in terms of functional requirements and constraints. The solution to the design problem is to be arrived at by translating these functional requirements into design parameters in accordance with the axioms of good design (22).

Design for Manufacture Guidelines. The *design for manufacture guidelines* try to lead to product designs that are easy to manufacture. They are based on the belief that manufacturing success cannot be assured at the production stage after a product has been designed; therefore product design and process planning must be integrated. These guidelines encourage reduction in the number of parts and part variations across product families through design of parts for multiple use to achieve cost reduction. Some of these guidelines encourage the design of components that are easier to fabricate. Others require designs that will eliminate separate fasteners and ensure foolproof (so that a product cannot be assembled in a wrong way) and simple (so that it can be used with automated machines) assembly processes. They also attempt to improve manu-

TABLE 6.1. The Design Support Systems

Axiomatic approach (22)
Design for manufacture guidelines (21)
Design for analysis principle (23)
Taguchi method (17, 24)
Group technology (18)
Quality function deployment (12a)
Computer-aided design (1a, 25)

facturing efficiency through modular design and elimination of time-consuming adjustments during assembly. These "guidelines are systematic and codified statements of good design practice that have been empirically derived from years of design and manufacturing experience" (21). These guidelines can be used to stimulate creativity and encourage good design practice that will result in products that are inherently easier to manufacture (21).

Design for Analysis Principle. The *design for analysis principle* attempts to achieve reduction in concept-to-market time by reducing the time spent for analysis at the technical design stage. It proposes that "designers should be constrained to work with only those designs [of products or systems] that can be analyzed easily and quickly by simple tools" (23). The principle is based on the belief that a rapid exploration of many simple to analyze designs is better than spending time analyzing a single complex design. Under this principle, it is possible to restrict the complexity of a product design to be a desired level by specifying suitable analysis tools. For example, by restricting the designers to the use of simple primitive solids (e.g., blocks, cones, and cylinders) of different dimensions, it is possible to achieve a simple design quickly. Research based on case studies of various product development projects has shown that the design for analysis principle leads to the design of products that are simple and easy to operate, and a constrained process provides strategic focus to the designers during the product development process by allowing them more time to focus on the overall improvements, instead of being involved in many complex details (23).

Taguchi Method. The *Taguchi method* extends quality improvement activities to product and process design. It is based on the premise that "quality is a virtue of design. The 'robustness' of products is more a function of good design than of on-line control, however stringent, of manufacturing processes. Indeed—though not nearly so obvious—an inherent lack of robustness in product design is the primary driver of superfluous manufacturing expense" (24). *Robustness* of a design refers to a design that ensures a product will never fail to perform its intended functions during its useful life. Robust designs are associated with reliable products. The Taguchi method provides a way to develop specifications for robust design by using the statistical design of experiments theory. This method attempts to determine the settings of product design parameters that make the product's performance insensitive to environmental variables, product deterioration, and manufacturing imperfections. The Taguchi method is a very cost effective technique for improving quality, because it attempts to achieve reduction in performance variation by reducing the influence of the sources of variations at the product design stage. It also increases manufacturability and reduces the product development and life-cycle costs of a product (17).

Group Technology. The *group technology* approach to design attempts to decompose a system into subsystems by classifying parts into parts families based on their design features. The classifications can be carried out either by a *visual method* or by a *coding method*. In the visual method, parts are grouped based on their similarities of geometric shape in a systematic way. However, this method is dependent on personal preference. Hence it can only be used effectively when dealing with a design that has a limited number of parts. In the coding method, each part is given a numerical or alphabetical code based on its geometrical shape and complexity, dimensions, type and shape of raw material, required accuracy, and so on. Group technology can achieve significant reduction in production lead time, work-in-process, labor and tooling requirements, rework and scrap materials, and reduced paper work by eliminating duplication of efforts if the current requirements can be satisfied with an existing part with minor design changes (18).

Quality Function Deployment. In most product development projects marketing inputs are used only during the concept design stage. After a concept is developed and approved for technical design, little interaction takes place between the design engineers and marketing experts. This forces designers to make decisions with little or no customer input at the later stages of the design process. When the marketing experts or the customers finally get to see the prototype design, it is usually too late to make any changes to the design without getting involved in costly redesign work, because the ability to make changes to a design declines very rapidly as it nears completion. This may cause cost overruns and long delays in product introduction. Sometimes this may even lead to products that customers never wanted in the first place. The quality function deployment (QFD) attempts to overcome these problems by forcing interaction between marketing experts and design groups throughout the product development cycle, assuring that design decisions are made with the full knowledge of all technical and market trade-off considerations (8a).

The *house of quality* is one of the basic design tools of this approach. It was first introduced by the Mitsubishi Corporation of Japan in 1972. Through a set of planning and communication routines, the house of quality forces designers to design products that customers want. It is based on the belief that marketing, design, and manufacturing personnel must work together throughout the entire process of product design to design products that reflect customers' desires and needs. It begins by asking: "What do customers want?" These requirements are identified as CAs (i.e., customer attributes) and ranked by their order of importance along a vertical axis. Comparison with the competition is made with respect to these CAs at this stage. In the next stage, those engineering characteristics (ECs) are identified that can influence the CAs along a horizontal axis. It is possible for some ECs to affect more than one CA. The body of the matrix known as the "relationship matrix" is filled up by a multifunctional team, indicating the degree of influence of each EC over each CA. This matrix can then be used to evaluate the effect of engineering changes over CAs (12a).

Computer-Aided Design. Traditionally, computer-aided design refers to a hardware–software combination that allows two and three-dimensional graphic design on CRT monitors. Besides allowing quick development of three-dimensional models, which provide a better insight than the two-dimensional blueprints, these systems can be used to visualize complex mathematical functions and computer simulation of designs. However, new software-based systems are available to facilitate implementation of the design support systems discussed earlier in this section.

The *Mechanical Computer-Aided Engineering System* (MCAE) can be used to test a design analytically, thus eliminating the need for prototype building at the early stage of design. It allows engineers to play "what if" games by varying assumptions about materials, size, and other operating conditions. The system allows the designers to see the effects of hypothetical working environments and estimate the production cost while making design decisions (25).

Burling et al. (1a) point out that for successful introduction of a product into the factory, computer-aided systems must be integrated with the design aids and the factory specific information so that the product coming out of the design stage can be manufactured by the available system. They describe the *AT&T computer aids* that can be used to design products with manufacturing guidelines.

The *Design for Assembly System* by Boothroyd and Dewhurst and the *Assemblability Evaluation Method* by Hitachi can "calculate producibility scores for nearly any product, based on the number of parts, the number of its standardized parts, the simplicity of couplers, the motion involved in its assembly, and so on" (6). Some of these systems can even provide estimates of production time and manufacturing costs (6).

Group technology databases are usually computerized and presented to the designers so that they can have the full knowledge of the parts design inventory and retrieve them as necessary. *Concurrent Design Environment (CDE) software*, marketed by Mentor Graphics of Oregon, allows loading of design criteria to the computers used by the designers. The system will signal a warning if the designer violates any of the specified guidelines (2).

Most of the tools and techniques that were discussed in this section are based on much broader technological and scientific principles, which would have made a more in-depth discussion of these tools and techniques beyond the scope of this chapter. The interested reader is referred to the cited literature for further information. The purpose of this exposition is to emphasize the ways in which these design support systems become a part of the design management strategy.

6.6. CONCLUSION

Product design is a strategic activity that affects the efficient operation of the entire business. There exists very little awareness of design as a strategic business activity among the academic community and practitioners of business. To break down these barriers between design and the rest of the business, a common language is needed. Design is not just an engineering activity taking place behind high walls far away from the corporate headquarters and manufacturing facilities. It should be managed strategically like any other business resource. In spite of its importance, design strategy should not dominate the other functional strategies. Design must work together with manufacturing and marketing to serve the business leaders of today and tomorrow in order to remain competitive in an uncertain environment.

REFERENCES

1. Adam, E. E., and Swamidass, P. M. (1989). Assessing Operations Management from a Strategic Perspective, *Journal of Management*, Vol. 15, No. 2, pp. 181–203.
1a. Burling, W. A., Bartels, L., Barbara, J., O'Neill, L. A., and Pennine, T. P. (1987). Product Design and Introduction Support System, *AT&T Technical Journal*, Vol. 66, No. 5, pp. 21–37.
2. Caminiti, S. (1990). Products to Watch, *Fortune*, Vol. 122, No. 9, p. 154.
3. Caplan, R. (1985). Good Design Is, *Across the Board*, Vol. 22, No. 5, pp. 23–26.
4. Cooper, W. W., Sinha, K. K., and Sullivan, R. S. Measuring Complexity in High Technology Manufacturing: Indexes for Evaluation, *Interfaces* (Special Issue on Strategies in High-Tech Industries), Forthcoming.
5. Corbett, I. J. (1986). Design for Economic Manufacture, *Annals of International Institute for Production Engineering Research*, Vol. 35, No. 1, pp. 93–97.
6. Dean, J. W. Jr., and Susman, G. I. (1989). Organizing for Manufacturable Design, *Harvard Business Review*, Vol. 68, No. 1, pp. 28–36.
7. Dixon, J. R., and Duffy, M. R. (1990). The Neglect of Engineering Design, *California Management Review*, Vol. 32, No. 2, pp. 9–23.
8. Dumaine, B. (1989). How Managers Can Succeed Through Speed, *Fortune*, Vol. 119, No. pp. 54–59.
8a. Fine, C. T. (1989). Developments in Manufacturing Technology and Economic Evaluation Models. Unpublished manuscript, Cambridge, MA: Sloan School of Management, MIT.
9. Fitzsimmons, J. A., Kouvelis, P., and Mallick, D. N. (1990). The Design Strategy and Its Interface with Manufacturing and Marketing Strategy: A Conceptual Framework, *Proceedings of the 21st Annual Meeting of Decision Sciences Institute*, Vol. 2, pp. 1507–1509.
10. Foster, R. N. (1986). *Innovation, the Attacker's Advantage*, Summit Books, New York.
11. Garvin, D. A. (1984). What Does "Product Quality" really mean, *Sloan Management Review*, Vol. 26, No. 1, pp. 25–43.
12. Hagel, J. (1988). Managing Complexity, *McKinsey Quarterly*, Spring, pp. 1–23.
12a. Hauser, J. R., and Clausing, D. (1988). The House of Quality, *Harvard Business Review*, Vol. 66, No. 3, pp. 63–73.
13. Hayes, R. H., and Wheelright, S. C. (1984). *Restoring Our Competitive Edge: Competing Through Manufacturing*, Wiley, New York.
14. Hayes, R. H., Wheelright, S. C., and Clark, K. B. (1989). *Dynamic Manufacturing: Creating the Learning Organization*, Free Press, New York.

15. Hill, T. (1989). *Manufacturing Strategy: Texts and Cases*, Irwin, Boston.
16. Jones, N. (1986). The Hollow Corporation, *Business Week*, No. 2935, pp. 56–59.
17. Kackar, R. N. (1985). Off-line Quality Control, Parameter Design and Taguchi Method, *Journal of Quality Technology*, Vol. 17, No. 4, pp. 176–209.
18. Kusiak, A. (1990). *Intelligent Manufacturing Systems*, Prentice-Hall, Englewood Cliffs, NJ.
19. Meikle, J. (1989). Design in the Contemporary World, a paper prepared from the Proceedings of the Stanford Design Forum 1988, *Pentagram Design*, pp. 15–16.
20. Olins, W. (1985). Management by Design, *Management Today*, February, pp. 62–69.
21. Stoll, H. W. (1988). Technical Report: Design for Manufacture, *Manufacturing Engineering*, Vol. 100, No. 1, pp. 67–73.
22. Suh, N. P., Bell, A. C., and Gossard, D. C. (1978). On an Axiomatic Approach to Manufacturing and Manufacturing Systems, *ASME Journal of Engineering for Industry*, Vol. 100, No. 2, May.
23. Suri, R., and Shimizu, M. (1989). Design for Analysis: A New Strategy to Improve the Design Process, Technical Report No. 89-3, University of Wisconsin, Madison.
24. Taguchi, G., and Clausing, D. (1990). Robust Quality, *Harvard Business Review*, Vol. 68, No. 1, pp. 65–75.
25. Villers, P. (1988). Designing for Predictability: New MCAE Tools, *Harvard Business Review*, Vol. 66, No. 4, p. 86.
26. Whitney, D. T. (1988). Manufacturing by Design, *Harvard Business Review*, Vol. 66, No. 4, pp. 83–91.

CHAPTER SEVEN

Neural Networks in Design of Products: A Case Study

HORMOZ ZAREFAR and JOHN. R. GOULDING

Department of Mechanical Engineering, School of Engineering and Applied Science, Portland State University, Portland, Oregon

7.1. INTRODUCTION

Consider the design conflict of two interconnecting shafts rotating at different speeds in a typical computer-aided design and computer-aided manufacturing (CAD/CAM) environment. Also, consider novice engineers who are often assigned projects, which to them incorporate new or vaguely familiar design skills and practices. What can be done to increase the intelligence and productivity of CAD/CAM?

Some experienced managers and supervisors argue that product background and design expertise can be learned only through years of practice; design handbooks are consulted for basic design equations and rules of thumb, and supervisors are in the loop for how-to expertise (24). If this is so, the novice engineer must spend less time physically producing the *best* design and more time creatively researching the possible design alternatives. The conventional approach is to automate the operations of drafting, simulation, and planning using CAD/CAM systems. While this approach reduces the time it takes to reiterate a design, more can be done to ensure that the *best* design is also the first design.

CAD/CAM should be an environment through which a novice engineer accesses handbooks, supervisors, and design equipment. Most important, it should aid the novice engineer in selecting the best mechanism, from a knowledge base of proven designs, in a known design conflict. This chapter discusses an implementation of a neural network-based hybrid CAD/CAM design environment for mechanical power transmissions, illustrated in Fig. 7.1.

Intelligent Design and Manufacturing, Edited by Andrew Kusiak.
ISBN 0-471-53473-0 © 1992 John Wiley & Sons, Inc.

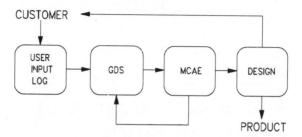

FIGURE 7.1. GDS environment overview.

Mechanical power transmissions, gears, belts, chains, hydraulics, pneumatics, and electrical design solutions are linked with the design conflict of the above interconnected shafts rotating at different speeds. Functional feature relationships and decision-based design, manufacturing, and customer requirements represent the design conflict. The knowledge base links the conflict to proven mechanisms for the *best* solution. In this research, a 171-question Gearbox Design Supervisor (GDS) program is used to extend a knowledge-based CAD design system. The GDS selects the *best* CAD scaling engine most applicable to the design conflict, illustrated in Fig. 7.2. The primary contribution of the GDS is to tackle "back to the drawing board" paradigms using decision-based criteria.

7.2. EXPERT SYSTEMS TO DESIGN SPUR GEARBOXES

7.2.1. Overview of Mechanical Design

Mechanical design has long been considered an ill-defined domain that does not lend itself to a rigorous computerization formalism. The design process has been divided into various categories by different researchers and authorities. In general, the collective knowledge that leads an engineer from

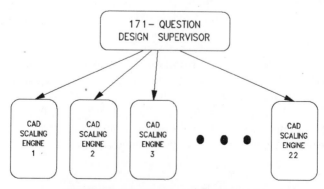

FIGURE 7.2. Supervisory approach.

a set of requirements and guidelines to the realization of a final product can be termed *design*. Moreover, designing can be considered an endeavor in problem solving, both at the conceptualization and the detail levels. The problem-solving strategies can be classified as formal and informal. Formal category of the design deals with the utilization of scientific tools in mathematics, materials, physics, continuum mechanics, and so on. These tools are readily available to the designers via text and handbooks and are employed to perform *synthesis* and *analysis*. Informal knowledge, on the other hand, relies primarily on human designer's insight and expertise and the intangible *heuristics* or rules of thumb.

Pahl and Beitz (18) propose a broad classification of design as *original*, *adaptive*, and *variant*. Original design is associated with elaboration of original solution principles to the design of a system and contemplation of new assemblies or machines and involves heavy participation of cognitive and intuitive knowledge. In adaptive design, known and recognized systems and solution principles are adapted to a given problem. It can be considered a blend of redesign (size variation) of some components of the selected system and original design of other subassemblies. Variant design, on the other hand, is merely concerned with resizing (scaling) of previously originated components to conform to the requirements of the new system design. As one might expect, there is no universal agreement as to where the boundary of one design type ends and another one begins. The majority of designing activities, however, fall under the latter two categories.

7.2.2. Background

Attempts in the automation of the design process via computer modeling have been ongoing for the past 7 years. Proliferation of artificial intelligence (AI) programming fueled the interest; in particular, expert systems were thought to be a mechanism for capture and dissemination of design knowledge (1, 4, 9). However, it is now apparent that the entire design cycle cannot be readily computerized, due in part to the highly personal and ill-structured nature of the creative abilities of the designers. The intriguing results reported by Ullman et al. (28) suggest the narrow focus of the design engineer. These findings are supported in a study by Waldron et al. (29) in which a designer's conceptual knowledge increases by exposure to more products and varieties of designs. It has therefore become apparent that a more productive designer should consider an array of possible design strategies and then attempt to prune the matrix of possibilities using various requirements and constraints.

To tackle the automation of the design requires means of developing appropriate methodologies for computer formalization. One of the important features of a systematic design methodology is its ability to facilitate application of the known iterative solutions to a known problem domain.

Techniques to provide such facilities must include guidelines to formalize, preserve, and invoke the known solutions to the problem at hand.

Although emulation of original design is beyond the reach of any computer formalism, the following sections demonstrate a step forward in automation of adaptive design by incorporating a hybrid approach. This approach addresses decision-making processes that have traditionally been treated as a separate stage in the product development cycle. These are less demanding tasks on human creativity; tasks that can be dealt with by employing prudent computer paradigms. The method combines the elements of variant design with decision-making abilities of a lead engineer or a design manager to bring about (from a set of alternative designs) an acceptable design. We call this a *design team approach* and propose a collaborative (hybrid) computational strategy to address the design-related issues (31).

7.2.3. Parallel Axis Gear-Drive Expert System (PAGES)

The proposed hybrid approach for a variant-based mechanical design was tested by developing a prototype parallel axis spur gear-drive system. The prototype is based on the knowledge captured from texts and handbooks (6, 7, 24). The knowledge-based subprograms consisted of gear arrangement, lubricant selection, and housing design. These subsystems were augmented with algorithmic data processing routines for gear tooth and shaft sizing to form the complete hybrid system. Flow of information in the interactive system is shown in Fig. 7.3.

Two coordinator modules were designed to enhance knowledge and data communication among the modules. Based on the information provided from the gear arrangement and gear tooth modules, the gear-shaft coordinator is responsible for determining the types and number of shafts to be designed. The system coordinator is responsible for the integrity of the

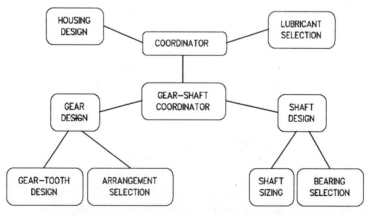

FIGURE 7.3. Spur Gearbox MCAE: CAD scaling.

design hierarchy, communication of information among subsystems, and providing explanation facilities at the rule-based level upon request. The system modules and their module types are listed in Table 7.1. The input set required to start the program is shown in Table 7.2.

The PAGES system is primarily a mechanical redesign (variant design) system with the ability to *resize* an existing array of components forming a generic spur gear drive. It was designed as a working prototype for spur gear drives and as such the system functioned as expected, emulating a computerized model of the variant design with limited heuristic input. It successfully produced variant designs of a generic system time and again. The heuristic knowledge employed in the system was primarily extracted from design texts and handbooks (6, 7, 24). Examples include fillet size selection in the shaft design and choice of lubricant for extreme ambient temperatures. Although the system was designed to emulate a "design team" approach, it became apparent that for a more robust design and close approximation of adaptive design, there is a need for integrating more decision-making paradigms into the system. In other words, PAGES was not able to handle adaptive designs in which a designer may wish to consider different variant design options.

TABLE 7.1. Parallel Axis Geardrive Modules and Types

Module	Type
Gear arrangement	Rule-based
Gear tooth design	Algorithmic
Gear-shaft coordinator	Rule-based
Shaft design	Algorithmic
Lubricant selection	Rule-based
Housing design	Rule-based and algorithmic
System coordinator	Rule-based

TABLE 7.2. Input Set for Gear-Drive Expert System

Input Requirements
Application environment
Horsepower transmitted
Axis orientation
Relative direction of rotation
Speed reduction
Pinion speed
Gear finishing process
Required life cycles
Ambient temperature
Shaft overhung length
Overhung load

7.3. DRAWBACKS IN CURRENT MCAE TECHNOLOGY

Two major drawbacks exist in current CAD/CAM software. First, embedded design alternatives needed by engineers during the mechanism conception and rework stages are lacking. Second, the software operator needs a thorough understanding of the intended design and the how-to expertise needed to create and optimize the design alternatives.

First generation CAD/CAM software reduced labor costs involved in reiterating product designs. Second generation CAD/CAM software, recently introduced, incorporates knowledge-based expert rules, equations, and proprietary languages to *scale* previously designed products to satisfy new design requirements (19, 20, 22). Third generation CAD/CAM software, or Mechanical Computer-Aided Engineering (MCAE) software, addresses both drawbacks (31).

To facilitate the conceptual design process, a *design supervisor* must be integrated into the MCAE system. To link the operator with the MCAE software, an interactive *intellectual* go-between must exist to translate design conflicts (or requirements) into the proven mechanisms of design. The *design supervisor* also manages dependencies and the independent attributes, imbedded in the knowledge base, to reduce the time needed to bring the product from conception to market. Factors considered include engineering analyses, sales and marketing influences, facilities resource uses, and customer and corporate relationships in procedure. The foregoing describes the GDS. Real-time concurrent engineering, response to changing design requirements, and adapting to the engineering experience of the user are all goals of *design supervisor* research.

A working GDS has been developed for an existing mechanical power transmission MCAE software system. Figure 7.4 illustrates the block overview of the GDS system environment. Basic mechanical requirements are

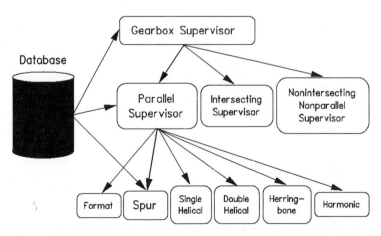

FIGURE 7.4. Block GDS: Supervisory approach.

necessary; however, *intellectual* design requirements are also needed to identify a particular gearbox as the *best* design. Furthermore, interactively combining the MCAE with the GDS solves the "back to the drawing board" paradigm.

7.4. AN INTRODUCTION TO THE GDS PARADIGM

To infer the parallelness of two lines from perpendicularity, the basic strategy involves three rules. Start with two lines, *A* and *B*, and draw a third line, *C*, that is normal. From this relationship, we ask whether *A* is perpendicular to *C*, and whether *B* is perpendicular to *C*. When both of these conditions are true, we can infer that *A* is parallel to *B*. Stated succinctly:

Rule 1. IF *A* is perpendicular to *C*;
      THEN Goal I is satisfied.
Rule 2. IF *B* is perpendicular to *C*;
      THEN Goal II is satisfied.
Rule 3. IF Goal I and Goal II are satisfied;
      THEN *A* is parallel to *B*.

The three rules form an expert system. However, the expert system asks too much of the user. Should the user determine perpendicularity among the lines by eye, by using a go/no-go gauge, or by using some type of scale? The only *correct* method is to measure the angles with a scale and let the program determine perpendicularity.

When we measure the angle between two lines we must allow for some tolerance to define that we mean by "perpendicular." Let us assume that our finest gradient of measure is 0.01°; then the word "perpendicular" means $90 \pm 0.1°$. Figure 7.5 illustrates a zone of tolerance applied as a

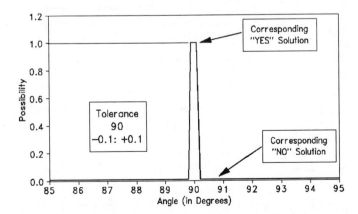

FIGURE 7.5. Binary logic: perpendicularity in a tolerance expert system.

(traditional) binary expert system. Therefore the expert system is:

Rule 1'. IF $|\angle AC - 90| \leq 0.1$;
    THEN Goal I is satisfied.
Rule 2'. IF $|\angle BC - 90| \leq 0.1$;
    THEN Goal II is satisfied.
Rule 3'. IF Goal I and Goal II are satisfied;
    THEN A is parallel to B.

A fault exists with this binary expert system, however. None of the rules compensates for the error if $\angle AC$ is exactly 90° and $\angle BC$ has a tolerance error of ±0.11°. For a large expert system (with hundreds of rules), the process of such error checking is a major problem because it is unlikely that we would know the results for all possible combinations of input. To minimize such problems, adopt the methodology of fuzzy logic.

Fuzzy logic maps hidden relationships between rules. A relationship exists between Rule 1' and Rule 2' because they both deal with the tolerance between a common line. It is possible to embed a third rule into the knowledge base if we express the rules as continuous functions. A Yes/No output is desired; so the expert system will require a rule-based quantifier (defuzzifier). Figure 7.6 illustrates a fuzzy expert system. Gauss' function of standard distribution will represent the paradigm of physical measurement, and one standard deviation will map all correct answers as the set of solutions ≥0.9 using the tolerance of 0.1°. This mapping represents a partial function of the desired result. Figure 7.7 illustrates Fig. 7.5 as a continuous fuzzy map. The fuzzy logic hybrid expert system consists of the following:

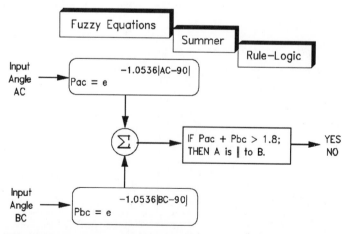

FIGURE 7.6. Fuzzy expert system: continuous equation expert.

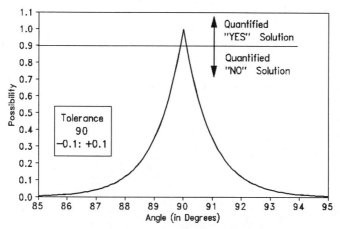

FIGURE 7.7. Continuous logic equations: perpendicularity in a fuzzy expert system.

Rule 1″. $P_{AC} = e^{-1.0536|\angle AC - 90|}$
Rule 2″. $P_{BC} = e^{-1.0536|\angle BC - 90|}$
Rule 3″. IF $\frac{1}{2}(P_{AC} + P_{BC}) \geq 0.9$;
            THEN A is parallel to B.

Suppose the user measures $\angle AC = 90.0°$. Then Rule 1″ gives possibility $P_{AC} = 1.00$. Thus $\angle BC$ can range between $90.0 \pm 0.2°$; and Rule 3″ still yields "A is parallel to B." Figure 7.8 illustrates a 3D solution surface for 85–90° measurement (space). By choosing the underlying function and the method of quantification, rules may be mapped to suit any linear or nonlinear solution. The significance of this paradigm cannot be over-emphasized.

7.5. THE CASE FOR FUZZY KNOWLEDGE REPRESENTATION

Prehistoric cave drawings depict hunting tactics. Later, Euclid's *Elements* (circa 300 B.C.) contained 465 geometric propositions embodying axiomatic methods and formal deductive reasoning. Then Aristotle, Descartes, Leibniz, and Spinoza organized knowledge through logic sequences. More recently, George Boole developed the symbolic operators AND, OR, and NOT used in *boolean* logic (circa 1806). Today, computer programmers model knowledge using symbolic operators and structured heuristic languages. The common thread between prehistoric hunters and computer programmers is the desire to apply an acquired knowledge.

Knowledge, in the formal sense, is mathematical equations, procedural algorithms, and heuristic arguments that automate the mechanistic processes that humans perform. For example, stress can be calculated in a mechanical

3D Fuzzy Equation Surface for Parallelleness

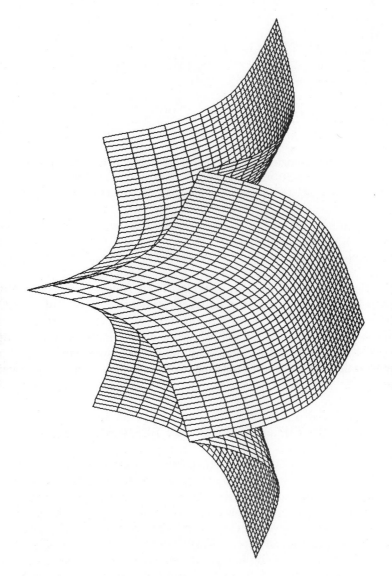

Angle AC vs Angle BC for 85–90°

FIGURE 7.8. Parallel logic surface.

Possibility—"Yes" for > 0.9

part to optimize a design, formulas can be developed to schedule a process, and rules can be written to simulate the action of procedure. Since 1883, Kelvinists have sought to advance the state of science through knowledge expressed in numerical reasoning (26). Still, from more than 100 years of research, the paradigm of the glass being either half-full or half-empty has yet to be realized.

Knowledge, in the formal sense, is also truth, and truth does not contradict itself. Lord Kelvin was not concerned with human judgment, perception, or emotion when he spoke of expressing knowledge through numbers. Yet human intellect deals with abstract phrases such as "cost is low," "quality is high," "it is beautiful," and so on. The words *low*, *high*, and *beautiful* have definite meanings, but such words cannot be described through crisp numbers or mechanistic representations. It is the human equation that has yet to be fully addressed in formal logic.

In 1945, Vannevar Bush suggested that diverse knowledge should be grouped using intellectual associations (3). For example, consider the phrase "somewhat short life." What does this phrase mean? In the context of cycle fatigue of a machine part, it could mean years. In the context of a process schedule, it could mean days. In the context of a business decision, it could mean hours. Often, real-world problems are *fuzzy* problems (33).

Intellectual knowledge was considered irreducible before the mid-1960s. That is, the traditional method of applying black and white rules to real-world problems becomes unreasonable as the number of associations increases. In 1965 Zadeh introduced the concept of fuzzy sets. A fuzzy set associates intellectual knowledge through continuous functions (33). Fuzzy truths, fuzzy connectives, and fuzzy rules map the judgment, perceptions, and emotions behind the mechanistic processes of knowledge. In fuzzy terms, the glass is both half-full and half-empty, and the context of the question determines the result. So the nature of the glass takes on different meanings in different contexts.

The application of fuzzy logic, the combination of fuzzy set theory and traditional logic, to an intellectual paradigm results in a fuzzy answer. A glass recently filled halfway is *usually* considered half-full. The word *usually* describes the possibility of "half-full" being the correct answer in the context of recent filling. Thus fuzzy logic extends traditional Yes/No logic to bridge intellectual paradigms in design and manufacturing.

7.6. THE CASE FOR NEURAL NETWORK KNOWLEDGE REPRESENTATION

The GDS system is designed to (a) provide expertise when human expertise is lacking, (b) exhaustively evaluate the design conflict (without combinatorial search explosions), and (c) arrive at solutions when given incomplete or uncertain information (unlike MCAE programs which perform numerical design) given a fuzzy *intellectual* knowledge base. Several programming methodologies exist: fuzzy logic expert systems, nondeterministic polynomial time complete (NPC) solutions (13), and neural networks (N-Nets). For this work, a unique back-propagation N-Net was implemented.

The capabilities of fuzzy expert systems are inherently well-suited to contribute to solutions of the kinds of problems described (e.g., expert systems have been used for medical diagnoses since the mid-1970s). The major drawback, however, is that the programmer is required to define the functions underlying the multivalued, or ranked, possibility optimization. While much is gained from fuzzy rules, system development time is comparable to conventional expert systems. Furthermore, expert-type rules use a comprehensive language system that may have built-in biases, embedded goals, and hidden information structures, which may result in errors (21). NPC programs, unlike expert systems, utilize mathematical relationships, look-up tables, and empirical knowledge to model functional, physical, and geometric attributes. Because NPC programs optimize the part-at-hand and not the design concept, this technique is incompatible with fuzzy intellectual knowledge bases.

N-Nets, like NPC programs, use mathematical relationships to optimize systems. Unlike NPC programs and rule-based expert systems, N-Nets evaluate all the design conflict constraints simultaneously. Like fuzzy logic programs, N-Nets are capable of statistical decision-making given incomplete and uncertain information. Unlike rule-based fuzzy logic, N-Nets are developed using black-box techniques that model (or learn) the knowledge base. That is, N-Nets do not use rules, in the formal sense; so development time is greatly reduced from that of rule-based modeling. Additionally, N-Nets can be designed to adapt to the user's requirements in the field (12, 25). The decision criteria are summarized in Table 7.3.

It is more practical to think of N-Nets as a technique for building continuous equations than as a tool for modeling the brain (as implied in the name). It must be conceded, however, that rule-based expert systems are much easier (for humans) to error-check than an ensemble of continuous equations. It appears that the fuzzy associative memory technique should provide the fast and robust data-based learning of N-Nets. For this research,

TABLE 7.3. Systems Compared

Design Criteria	Rules	NPC	N-Nets
Adaptation in the field			√
Data-based learning			√
Fuzzy knowledge bases	√		√
Incompleteness and uncertainty	√	√	√
Possibility theory	√		√
Simultaneous evaluations		√	√

and for all of the above-mentioned reasons, it was determined that the GDS should be modeled using neural networks.

7.7. MODELING DESIGN AND MANUFACTURING EXPERIENCE

Generally, senior design engineers can design complex mechanical parts better than novice designers because, through their broader knowledge and experience, they can recognize similarities in known components of the overall design. This is a process of pattern recognition. The N-Net implementation of the design supervisor is, foremost, a pattern recognizer and will classify problems that cannot be expressed as algorithms, calculations, searches, or configuration matrices.

For the design supervisor to model intellectual design and manufacturing *experience*, input and output pairs of known causal relationships, homomorphic heurisms, and cogent observations of design intent must be assembled as a *training-set* knowledge base. The N-Net, as a black box, is trained on this knowledge base until the error between the desired output and the actual output is reduced below an acceptable threshold (for the quantitizer). Figure 7.9 illustrates the *pupil–teacher* relationship involved with black-box (supervised) training. When an N-Net *learns*, the structure of the N-Net ensemble equations are changed to map a continuous relationship (in N-dimensional hyperspace) of the knowledge base. For a robust knowledge base, N-Nets have the ability to generalize a solution (having desired results, called *care space*) to an unknown problem. That is, the training set provides discrete points on a global surface, which is described by a single equation (the N-Net).

It is this generalization, as in the parallel-line example, that is utilized in the GDS. For this research, the structure of the N-Net/GDS was designed to provide 5.128×10^{27} possible mappings (called *performance space*), and 619 relationships were assembled into a knowledge base. Design intent is synthesized in the form of multiple-choice questions. The N-Net/GDS was trained on 10,000 variations within the knowledge base to form the training set. From the structure of the N-Net and training set, it was calculated that the GDS would have a care set of 12,990 possible mappings. Figure 7.10 illustrates the architectural relationship of an untrained network, the per-

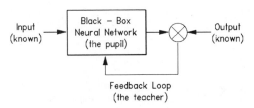

FIGURE 7.9. N-Net/Teacher feedback: error correction training.

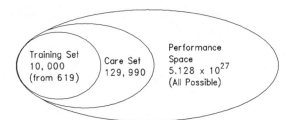

FIGURE 7.10. GDS solution space.

formance space, to a trained network, the care space. The marriage of N-Nets with the design supervisor paradigm has thus resulted in a robust pattern recognizer capable of generalizing and resolving ambiguity in comprehensive power transmission models with an intuitive feel.

7.8. BACK-PROPAGATION NEURAL NETWORKS

A back-propagation neural network program simulates a parallel distributed processing system containing a matrix of artificial neurons, or nodes. Figure 7.11 illustrates a basic node capable of summing weighted input signals and producing output via a threshold transfer function. The GDS uses a sigmoid threshold transfer function; however, the binary threshold transfer function is illustrated for clarity.

Nodes are connected together in a feed-forward fashion to form layers. The input layer (all input nodes) and the output layer (all output nodes) are accessible to the user; the middle layers remain *hidden*. The N-Net stores information in the weighted input lines that connect the nodes. Weights may either reinforce or suppress, with positive or negative bias, the amount of input signal. The need for hidden layers is best illustrated in the exclusive OR (XOR) example in Fig. 7.12. Figure 7.13 demonstrates how the XOR N-Net can be used as an expert system to determine if a gear can be

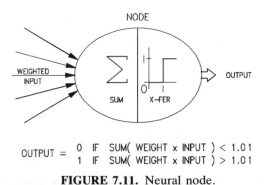

$$\text{OUTPUT} = \begin{array}{l} 0 \quad \text{IF} \quad \text{SUM(WEIGHT} \times \text{INPUT)} < 1.01 \\ 1 \quad \text{IF} \quad \text{SUM(WEIGHT} \times \text{INPUT)} > 1.01 \end{array}$$

FIGURE 7.11. Neural node.

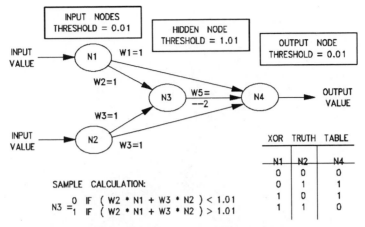

FIGURE 7.12. XOR neural network schematic.

manufactured with straight teeth based on axial geometry. On the surface, this network appears to be doing little more than a rule-based expert system. By changing the transfer function from binary to sigmoid, for example

$$OUTPUT = \frac{1}{1 + e^{-\alpha[\Sigma(WEIGHT \times INPUT) + \beta]}},$$

fuzzy relationships can now be encoded, as in the parallel-line example. Alpha (α) determines the degree of fuzziness, and β determines the threshold level. Generally, α is a fixed value during training, and β is varied by connecting it through a weight to an input having a fixed value, called a bias.

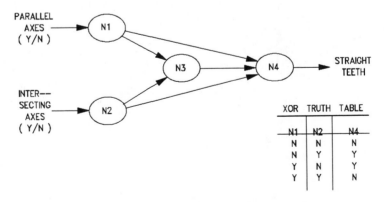

FIGURE 7.13. XOR neural network expert system.

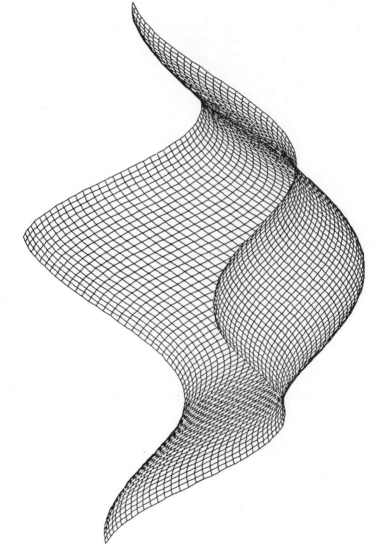

Error scape for a two-weight node

Weight #1 vs weight #2—showing (local) minimum

Error from desired output

FIGURE 7.14. Three-dimensional (3D) error surface.

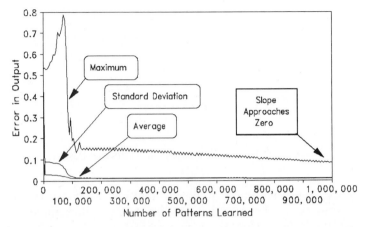

FIGURE 7.15. The 3×4 SN-GDS training history: tested on 106 random patterns.

The training set data pairs, expressed as numerical values, are applied to the input and output layers during a training session. The weights are adjusted, and a solution is mapped, using the back-propagation of errors method (17, 25, 30). Back-propagation of errors, during training, is similar to the feedback loop of a control system; however, instead of discrete compensation, the equation describing the output is modified. Back-propagation employs a gradient descent method in which the weight adjustment is determined from the (back-propagated) error at each node. Before training, the weight values are random. After training, weight values approach an averaged minimum error for the entire training set. Figure 7.14 illustrates a hypothetical error-scape for a node having two weights. During training, patterns are applied across the node, and the error is reduced stepwise. Convergence is guaranteed if the architecture of the N-Net structure is robust for the training set. However, maximum error may never approach zero because error-scapes are generally multidimensional with many (acceptable) local minimums. Figure 7.15 shows the training history for the GDS implementation of a parallel-axis expert system.

To design an expert system, different groups of closely related patterns must be presented during a training session. Generally, upper, midrange, and lower pattern values (describing the care set) will internally force the network to synthesize the embedded relationships between different problems. A *well*-defined and well-trained N-Net will produce correct solutions for untrained problems (30).

7.9. IMPLEMENTING NEURAL NETWORKS IN HYBRID SYSTEMS

In this research, 22 different gearbox designs were modeled through 171 questions from a literature search of gear handbooks (2, 5–8, 14, 15, 16a,

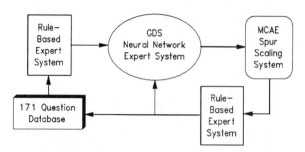

FIGURE 7.16. Parallel supervisor network design.

26, 27). Each question has a nonlinear multiplication associated with it inside the network. However, no single question can determine the overall outcome of the GDS. That is, each question carries roughly the same weight. So key questions, such as the determination of axis orientation, were handled as rules. The basic supervisory approach is to use the 171 questions as input to a hybrid system of rules and GDSs to select the *best* MCAE to *scale* the design, as illustrated in Fig. 7.16.

Three feedforward back-propagation N-Nets of unique structure implement parallel, intersecting, and nonintersecting nonparallel axis GDSs, as shown in Fig. 7.17. The knowledge base was defined by categories: mecha-

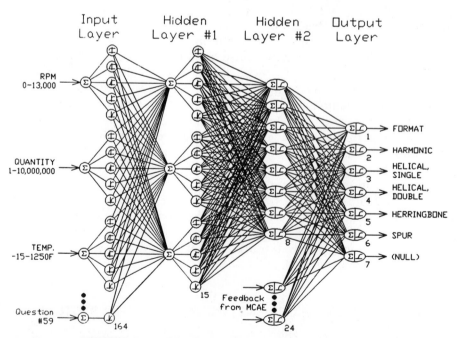

FIGURE 7.17. GDS/Superneurode architecture: 3 × 5.

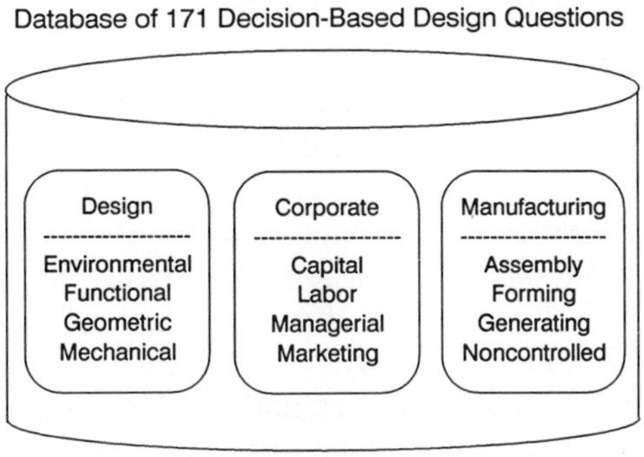

Database of 171 Decision-Based Design Questions

Design	Corporate	Manufacturing
Environmental Functional Geometric Mechanical	Capital Labor Managerial Marketing	Assembly Forming Generating Noncontrolled

FIGURE 7.18. Intellectual requirements.

nism design, corporate practice, and manufacturing methods. Figure 7.18 illustrates the three categories.

Because N-Nets are example-based rather than rule-based, they do not require conventional programming; thus several interactive, but not user modifiable, N-Nets are trained for a similar class of problems at the software house, and a second set of untrained adaptive N-Nets is also distributed. The trained N-Nets supervise the (transparent) learning of the untrained nets based on user interaction and feedback loops. Thus the design supervisor software will adapt itself to the needs of the user, as illustrated in Fig. 7.19. In this fashion, once the MCAE program has completed *scaling* a proven gearbox design, the adaptive N-Net/GDS is used to check the results from the first N-NET/GDS, as detailed in Fig. 7.17. The final result is a recursive program environment that identifies the *best* possible solution and then verifies the design intent.

Because the gearbox is treated as a simple component of some larger design, the GDS is only required to select and coordinate the design of the *best* gearbox. The corporate practice and manufacturing methods input are stored as permanent databases. The mechanism design input is controlled by the user, who responds to 21 questions, which have 84 possible results. The 21 queries of the GDS encompass the design information required to build any of the 22 gearboxes. A gearbox can be designed within minutes.

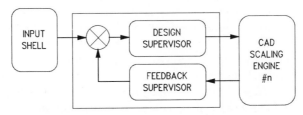

FIGURE 7.19. Interactive design supervisor.

7.10. CONCLUSION AND FUTURE WORK

The Gearbox Design Supervisor (GDS) program was run interactively with an MCAE program capable of scaling four spur gearbox designs. In each test case, the GDS successfully coordinated the design process. When the overall system was used by novice engineers (senior design students in the mechanical engineering curriculum), several aspects became apparent. Designers who had little knowledge about mechanical power transmissions were able to identify the gearbox systems. Those who supplied ambiguous data were able to play what–if games and foster creative abilities. Finally, designers who used the system reduced development time considerably.

It proved necessary to use as many design questions as possible to define the gearbox design problem in the earliest stages of development. The hypothesis was to provide multiple initial designs and then select from these the *best* design for engineering analysis. Then the outcome of the engineering analysis would be checked against the initial design parameters to prove that the *best* design was selected. In addition, it was determined that the system must be capable of addressing incomplete and conflicting demands and constraints found in actual design problems.

References for this research included gear design handbooks, technical articles, and expert rules of thumb to build a database of more than 170 design queries. These queries spanned the disciplines of engineering, manufacturing, marketing, production, and management. A match of proven solutions to value ranges within the queries then formed the knowledge base.

From the known solutions, it was evident that all design queries would need to be formulated as ranges of continuous values. That is, there were no discrete "yes" or "no" answers to the design queries. Furthermore, the data contained many conflicting and overlapping values. A secondary requirement of the system was to have it manage missing responses to the queries by a novice user.

It was determined that fuzzy logic was necessary to build the inference engine and would be modeled through neural network techniques. A training set was developed from the ranges of values of the known problem design conflict and solution pairs; real values and unknowns (missing responses) were randomly selected to form more than 10,000 patterns in the training set. The neural network was then trained from these data to map distributed solutions throughout the care set.

By testing various neural network architectures, a hierarchy of intermediate design solutions, or supervisors, proved beneficial to parsing the design problem. In addition, piecewise decomposition of individual solutions was available at each layer in the hierarchy. The most successful neural network learning algorithm was back-propagation. By using a new and unique five-(weight) layer, fully interconnected modified three-layer architecture, the (highly nonlinear) intellectual knowledge base was successfully modeled. Furthermore, the use of supervisory layers in a hybrid expert

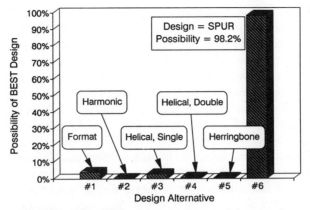

FIGURE 7.20. GDS super gearbox training results.

system/neural network knowledge base inference engine is essential to program efficiency and piecewise decomposition of possible solutions.

The GDS performs very well in producing the Yes/No type of decisions required to identify the *best* MCAE *scaling* program. Figure 7.20 is the spur gearbox output from the GDS training verification session. Note the *Yes* confidence level attained for the spur gearbox design and the excellent *No* rejection of the other 21 gearbox designs. Similar results were obtained for the other 21 gearbox designs.

In our research, novices were supplied with outlines to design a gearbox; however, their outlines were incomplete and contained ambiguous data. Figure 7.21 is an example of the worst-case output from the GDS during interaction with novices whose (unknown) objectives were to design a double helical type gearbox. Note the excellent *No* rejection of the other 19 gearbox designs.

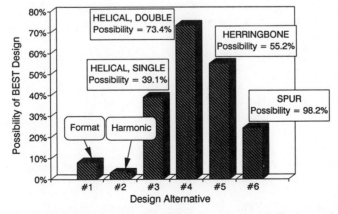

FIGURE 7.21. GDS novice user results. Data: one-quarter helical, one-half herringbone, and one-quarter unanswered.

Although the GDS-based system interacted with only one MCAE program, the results of interacting with the 21 other MCAE programs were simulated for each test case. The primary contribution of the GDS was to expand knowledge-based CAD software systems to tackle "back to the drawing board" paradigms. In addition, better response to changing design requirements, better use of capital equipment and direct labor costs, and faster new product development have been achieved. The result of implementing the design conflict methodology is to make decision-based reasoning information accessible to the user throughout the engineering process.

The GDS has demonstrated its viability in mechanical engineering design. Future work will link the GDS interactively with several independent MCAE programs. An interactive process, similar to the role of the traditional *design supervisor*, should emerge. Yesterday, engineering software was used to create drawings. Today, MCAE is used to *scale* products. Tomorrow, diverse product designs will be the end result of the design supervisor methodology.

REFERENCES

1. Brown, D. C., and Chandrasekaran, B. (1983). An Approach to Expert Systems for Mechanical Design, *Proceedings IEEE Design Automation Conference*, pp. 173–180.
2. Buckingham, E. (1960). *Design of Worm and Spiral Gears*, Industrial Press, New York.
3. Bush, Vannevar (1945). As We May Think, *Atlantic Monthly*, July, pp. 101–108.
4. Dixon, J. R., and Simmons, M. K. (1984). Computers that Design: Expert Systems for Mechanical Engineers, *Computers in Mechanical Engineering*, Vol. 2, No. 3, pp. 101–18.
5. Drago, R. J. (1988). *Fundamentals of Gear Design*, Butterworth, Boston.
6. Dudley, D. W. (1954). *Practical Gear Design*, McGraw-Hill, New York.
7. Dudley, D. W. (1962). *Gear Handbook: The Design, Manufacturing, and Application of Gears*, McGraw-Hill, New York.
8. Dudley, D. W. (1984). *Handbook of Practical Gear Design*, McGraw-Hill, New York.
9. Elias, A. L. (1983). Computer-Aided Engineering: The AI connection, *Aeronautics and Astronautics*, July/August, pp. 10–18.
10. Goulding, J. (1991). Adaptive Transfer Functions in Back-Propagation Networks, *Proceedings 1991 International Joint Conference on Neural Networks*, Seattle, WA.
11. Goulding, J., and Zarefar, H. (1990). Power Transmission Recognition System, *Proceedings 1990 International Society of Mini and Microcomputers Conference*, New Orleans, LA.
12. Hecht-Nielson, R. (1990). *Neurocomputing*, Addison-Wesley, Reading, MA.
13. Hoeltez, D. A., and Chieng, W. (1989). Designing Mechanisms with Expert Systems, *Machine Design*, September 12, pp. 163–168.

14. Houghton, P. S. (1952). *Gears: Spur, Helical, Bevel, and Worm*, Technical Press, London

15. Lynwander, P. (1983). *Gear Drive Systems, Design and Application*, Marcel Dekker, New York.

16. Mead, C. (1989). *Analog VLSI and Neural Systems*, Addison-Wesley, Reading, MA.

16a. Merritt, H. F. (1971). *Gear Engineering*, Halsted Press, New York.

17. NeuralWear, Inc. (1989). *NeuralWorks: An Introduction to Neural Computing, NeuralWorks User's Guide, NeuralWorks I, NeuralWorks II, Rev. 2.00.* Neural-Ware, Sewickley.

18. Pahl, G., and Beitz, W. (1984). *Engineering Design*, Springer-Verlag, New York.

19. Rouse, N. E. (1989). Getting Concepts Down in the Computer, *Machine Design*, January 12, pp. 50–66.

20. Rouse, N. E. (1989). Designers Gain Insight into the Factory, *Machine Design*, June 8, pp. 56–64.

21. Rumelhart, D. E., McClelland, J. L., and The PDP Research Group (1986). *Parallel Distributed Processing*, 2 Vols., MIT Press, Cambridge.

22. Santalla, R. W. (1988). Smart CAD Builds Large Plants, *Computer-Aided Engineering*, June, pp. 56–60.

23. Shigley, J. E., and Mischke, C. R. (1989). *Mechanical Engineering Design*, 5th ed., McGraw-Hill, New York.

24. Shigley, J. E., and Mitchell, L. D. (1986). *Mechanical Engineering Design*, 4th ed., McGraw-Hill, New York.

25. Simpson, P. K. (1990). *Artificial Neural Systems: Foundations, Paradigms, Applications, and Implementations*, Pergamon, New York.

26. Thompson, Sir William (1891). *Popular Lectures and Addresses*, Macmillan, London.

27. Trautchold, R. (1955). *Gear Design and Production Rules and Working Formulas*, Columbia Graphs, Columbia, CT.

28. Ullman, D. G., et al. (1987). Toward Expert CAD, *Computers in Mechanical Engineering*, November/December, pp. 56–70.

29. Waldron, M. B., et al. (1987). A Study of Visual Recall Differences Between Expert and Naive Mechanical Designers, *Proceeding 1987 International Conference on Engineering Design*, Vol. 1, pp. 86–93.

30. Wasserman, P. D. (1989). *Neural Computing Theory and Practice*, Van Nostrand Reinhold, New York.

31. Zarefar, H. (1986). *An Approach to Mechanical Design Using a Network of Interactive Hybrid Expert Systems*, Ph.D. Thesis, University of Texas, Arlington.

32. Zadeh, Lotfi A. (1987). Fuzzy Sets. In: *Fuzzy Sets and Applications: Selected Papers by L. A. Zadeh*, edited by R. R. Yager, S. Ovchinnikov, R. M. Tong, and H. T. Nguyen, Wiley, New York, pp. 105–146. [Reprinted from *Information and Control*, 8: 338–353 (1965).]

33. Zadeh, L. A. (1987). Outline of a New Approach to the Analysis of Complex Systems and Decision Processes. In: *Fuzzy Sets and Applications: Selected Papers by L. A. Zadeh*, edited by R. R. Yager, S. Ovchinnikov, R. M. Tong, and H. T. Nguyen, Wiley, New York, pp. 105–146. [Reprinted from *IEEE Transactions on Systems, Man, and Cybernetics*, SMC-3: 28–44 (1973).]

PART II

DESIGN OF MANUFACTURING SYSTEMS

CHAPTER EIGHT

Flexible Fixturing for Intelligent Manufacturing

BOPAYA BIDANDA and C. K. MURALIKRISHNAN
Department of Industrial Engineering, University of Pittsburgh, Pittsburgh, Pennsylvania

8.1. INTRODUCTION

The past decade has witnessed a concerted thrust toward flexibility and integration in manufacturing operations. CIM, CAD, FMS, and so on are no longer intimidating buzzwords but practical realities on the road to intelligent manufacturing systems. Smaller batch sizes and a group technology orientation have led to the development of flexible work cells and systems. There is increasing awareness of the need to evaluate manufacturing operations from a systems perspective.

In the traditional model, manufacturing is a set of sequential activities from product design to after-sales service. The design of a product is based on market needs and aesthetics. This is translated into a set of functional requirements by the product designer, who develops a set of detailed dimensioned and toleranced engineering drawings. Next, the process planning department determines the individual machining, assembly, and inspection operations required to produce the part. Based on the engineering part drawings and the process plan, the tool design department designs the jigs, fixtures, and special cutting tools needed to implement the process plan.

After raw material to produce the part material is available, schedules are drawn and the first production run, sometimes referred to as prototyping, is used to iron out wrinkles in the production process. Any change in product design at this point means that the sequential chain of activities must be repeated. Changes in jig and fixture design are especially time consuming due to their complexity. The role of fixture design in an intelligent manufacturing environment is shown in Fig. 8.1.

Intelligent Design and Manufacturing, Edited by Andrew Kusiak.
ISBN 0-471-53473-0 © 1992 John Wiley & Sons, Inc.

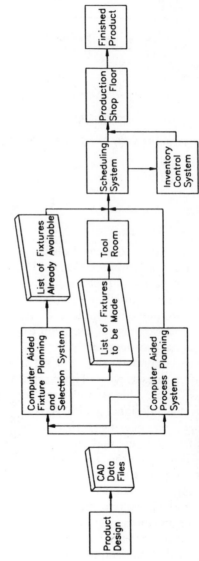

FIGURE 8.1. The role of fixture design in the manufacturing sequence.

It can be seen from the above model that product and process design must run in parallel and be closely coordinated to achieve matching of needs and requirements. This ensures that process constraints are considered as part of the product design and consequently eases the fabrication phase (20).

Jig and fixture costs can account for up to 20% of total product cost (28). The design and manufacture of jigs and fixtures are tedious and time consuming processes due to the following reasons:

1. Jigs and fixtures are manufactured to close tolerances. As a rule of thumb, jig and fixture tolerance is about 30–50% of part tolerance and thus their manufacture needs considerable setup time.
2. Since jigs/fixtures are mostly made of hardened steel, the kinds of machining operations that can be used for their manufacture are constrained.

The need for integration between CAD and CAM (i.e., design to manufacturing) has been well documented. Research efforts toward this integration have primarily focused on selected areas such as computer aided process planning (CAPP), automated NC tool path generation, off-line programming, and manufacturing resources planning. On the other hand, the automation of fixture design and selection has not been sufficiently addressed. Automated design can reduce cost by standardizing design and reducing lead time in the design process. While the concept of dedicated fixturing (i.e., a dedicated fixture for a single part and operation) is acceptable in traditional mass manufacturing environments, lead time and cost considerations have accentuated the need for more versatile fixtures in flexible and cellular manufacturing environments. Examples in the literature (12, 19, 24) suggest that adaptable fixturing could reduce the costs of traditional fixturing by as much as 80%. Modular fixtures provide the opportunity to automate the design and assembly of fixtures.

The objective of this chapter is to introduce the reader to current trends in industry and also to research developments in the area of fixturing, with a focus on machining fixtures.

8.2. PRINCIPLES OF FIXTURE DESIGN

Fixtures are workholders that hold, locate, and support the workpiece during the machining cycle. The objective is to ensure that the workpiece is located precisely with respect to the cutting tool or machine tool during the operation. Since cutting forces tend to vary with the type of machining operation, fixtures are generally classified according to the type of machine on which they are used (e.g., milling fixtures, turning fixtures).

Efficient fixture design involves the consideration of a number of factors related to the machine, product, and process. The fixture should provide adequate support to prevent distortion and deflection of the workpiece under the influence of machining forces. The design should also aid the cutting force in clamping down the part. Typical machine considerations that affect fixture design are table travel, spindle swing, spindle size, cutting tool dimensions, and method of mounting the fixture. Typical cutting considerations include type of material, material removal rate, speed, and feed. This translates into a complex set of variables in machining operations where multiple operations are performed in a single setting. Product information such as surface finish, tolerances, locating surfaces, material allowances, and batch sizes are no less important. The sophistication of the fixture is dependent to a large extent on economic considerations.

8.2.1. Principles of Location

It is necessary to distinguish between the functions of *location* and *clamping*. By location, we refer to the positioning of the workpiece in three-dimensional (3D) space at a desired orientation and position with respect to the machine. Clamping refers to the act of maintaining this fixed reference.

Method of Locating. An unrestrained solid body in 3D space possesses 12 degrees of freedom (dof) or 12 modes of movement (see Fig. 8.2). These consist of movements along the positive and negative directions of the X, Y, and Z axes and clockwise and counterclockwise rotations about the three axes. These dof must be restricted by locators and clamps. By the 3-2-1 method of location, nine dof are eliminated by appropriate placement of six locators on mutually perpendicular planes, as shown in the Fig. 8.3a. If the locating planes are not mutually perpendicular, the result is a wedging action that tends to lift the workpiece off the fixture and machine tool. The remaining dof (movement along positive X, Y, and Z) can be restricted by means of forces applied as shown (see Fig. 8.3a). These forces are supplied by clamping devices in a fixture. Clamps cannot be substituted by more locators because it would prevent loading and unloading of parts (16).

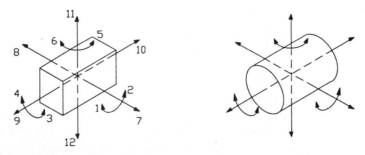

FIGURE 8.2. Degrees of freedom of rotational and prismatic workpieces.

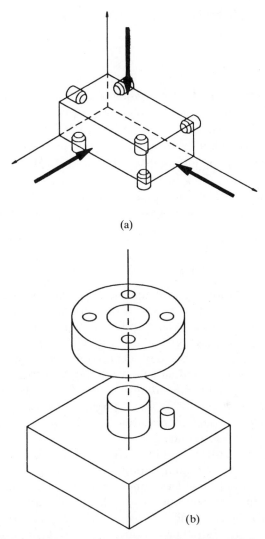

(a)

(b)

FIGURE 8.3. Restriction of degrees of freedom with locators.

Concentric and Radial Locating Methods. A slightly different approach
has to be adopted in dealing with parts that provide cylindrical surfaces or
holes for location. The purpose again is to restrict the dof of the workpiece
with the minimum number of locators. Figure 8.3b shows an arrangement
that is often used in radial location. The base plate and central locator arrest
nine dof. By using an additional radial locator (the small pin on the base
plate in Fig. 8.3b), rotation about the Z (vertical) axis is prevented. The
movement along the positive Z axis can be restricted by the clamping
device.

Curtis (8) details the following factors to be considered in selecting a locating scheme:

Locating Plane. Primary locating planes are established by three and only three locating points. This is directly derived from the fact that any plane can be defined by a minimum of three noncollinear points.

Part Features. Permissible variations in part size should not affect the location of the part. Also, the locators should be designed and placed to ensure repeatability of location from one workpiece to the next.

Stability. The locating points should be chosen as far apart as possible on any one surface. This not only ensures stability of the part but also reduces the error due to deviations in locator dimensions.

Cutting Forces. Locators should be placed so that the cutting forces aid in pushing the part against the locators.

8.2.2. Clamping Principles

Clamps provide the holding function once the part has been located. They supply the forces needed to restrict those dof that have not already been restricted by the locators. Many different types of standard clamps are available. Figure 8.4 shows a standard strap clamp that has been adapted to make it more flexible; it can now clamp jobs of varying heights. Clamp design should be simple and sufficient for the purpose. The clamps should not hinder loading and unloading of the workpiece. They should be positioned over rigid sections as far as possible. Excess clamping force may damage the part. Production requirements determine if power clamps or quick acting clamps are required.

A discussion of fixtures would be incomplete without a comparison of fixturing for rotational and nonrotational workpieces. Based on their geometry, workpieces are classified as either rotationally symmetric or prismatic. Rotationally symmetric workpieces are characterized by a similar clamping geometry, which varies only in diameter. The clamping elements can accomplish the dual functions of locating and clamping, resulting in a simpler fixture. Prismatic workpieces, on the other hand, usually require that the position determination be separate from and precede the clamping operation. Thus they create greater demands on the fixture design (26).

8.3. TRADITIONAL ROTATIONAL AND NONROTATIONAL FIXTURES

Here we begin by describing flexible fixturing issues of rotational components. Then flexible fixturing methodologies for nonrotational components are briefly addressed.

Flexible fixtures for rotational components have been in use for the last

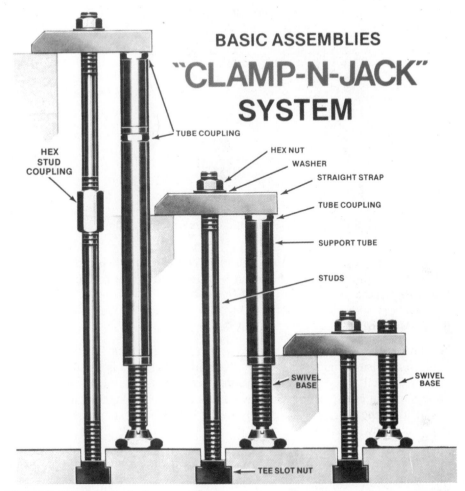

FIGURE 8.4. Standard clamp modified to increase flexibility. (Courtesy of Tobin Corporation.)

few decades. The following types of workholders can be called flexible since they are used to hold more than one type of component: centers, mandrels, collets, and chucks. Each is briefly described.

8.3.1. Centers

A plain center has a standard taper ground on one end while the other end is ground at a 60° taper. The workpiece to be machined also has a 60° hole drilled at each end. The workpiece is then mounted between the two centers and using a plate and a dog (see Fig. 8.5a) as a drive mechanism, the workpiece rotates along with the machine spindle, so that it can be

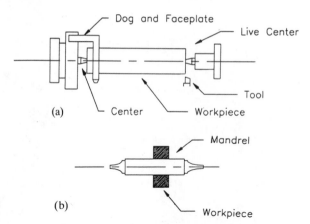

FIGURE 8.5. Schematic of center and mandrel.

machined. The difference between a plain center and a live center is that in a live center the 60° end is fitted on bearings and hence rotates along with the workpiece.

8.3.2. Mandrels

A mandrel is used to hold workpieces with through holes. The workpiece is held on an inside diameter and the machining operation is usually restricted to longitudinal type turning operations (see Fig. 8.5b). Standard mandrels are circular in cross section and have a taper of 0.003–0.005 inch per inch to ensure that the workpiece is rigidly held (27). Mandrels can be subdivided into the following classes:

1. *Plain, solid mandrels* are usually 4–12 inches long; here workpieces may be machined on both ends and the outside diameter.
2. *Gang mandrels* hold multiple workpieces together on a single mandrel, usually in production work.
3. *Expanding mandrels* consist of a spring-hardened, tapered sleeve that fits on a tapered pin that is longer than the sleeve. When the sleeve is pushed toward the pin, by means of a nut, it expands and holds the workpiece firmly.
4. *Cone mandrels* are conical in shape and hold workpieces with large internal holes.
5. *Nut mandrels* are threaded and are used to hold nuts that are being faced or chamfered on the end.

8.3.3. Collets

Collets are used for gripping premachined work and are designed to hold bar stock of circular, square, and hexagonal cross sections. Unlike chucks,

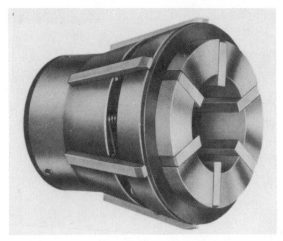

FIGURE 8.6. Multisize collet. (Courtesy of Pratt Burnerd America.)

collets exert a uniform pressure on the part and thus thin-walled parts can be machined without distortion from the holding forces. Figure 8.6 shows a multisize collet; the collet shown has a gripping range of 0.125 inch with an extended range of 0.010 inch above and below each nominal size. These collets usually come in sets with multiple sizes.

8.3.4. Chucks

Chucks are the most versatile of all workholding devices used on lathes. It is impossible to cover the entire gamut of chucks used in machine shops since there are a very large number of special chucks.

Chucks are normally prefixed by the terms "universal," "independent," or "combination." A universal or self-centering chuck is one in which each jaw is equidistant from the chucking axis and the jaws do not move independently. In an independent type chuck, the jaws can be moved independently. In a combination type chuck, the jaws can move either in unison or independently. Chucks are also classified by the number of jaws available on the chuck. Three jaw chucks are the most common and can hold circular or hexagonal cross sections, while four jaw chucks can hold circular, square, octagonal, and other cross sections.

1. Counter centrifugal chucks are used in high-speed turning operations, and shown in Fig. 8.7. Wedges or counterweights offset the weight of each jaw and counteract the centrifugal force developed when the spindle turns.
2. Diaphragm chucks use the principles of hydraulics to hold premachined components. In this type of arrangement, the chuck jaws are attached to a diaphragm; when pressure is applied to the diaphragm, the jaws open out (see Fig. 8.8). After the workpiece is in place,

FIGURE 8.7. Counter-centrifugal power chuck. (Courtesy of Autoblok Corporation.)

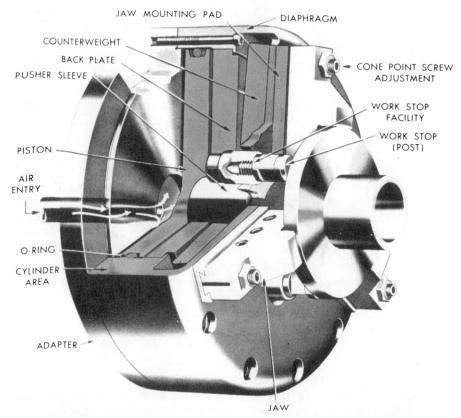

JAW MOUNTING PAD

DIAPHRAGM

COUNTERWEIGHT

BACK PLATE

PUSHER SLEEVE

CONE POINT SCREW ADJUSTMENT

WORK STOP FACILITY

WORK STOP (POST)

PISTON

AIR ENTRY

O-RING

CYLINDER AREA

ADAPTER

JAW

FIGURE 8.8. Diaphragm chuck. (Courtesy of ITW Woodworth.)

FIGURE 8.9. Two jaw automatic indexing chuck. (Courtesy of Autoblok Corporation.)

pressure is removed and the jaws clamp the workpiece tightly. These chucks are used extensively for grinding operations on lathes (15).

3. Spring jaw chucks are similar to a diaphragm chuck in that pressure is used to push a draw bar and force open the jaws of the chuck. However, each jaw is independent of the other, enabling the chuck to accommodate asymmetrical diametral variations in the workpiece. Moreover, the squaring action is better than in diaphragm chucks. These chucks are often used to hold gears (15).

4. Indexing chucks are used when a workpiece has multiple intersecting axes and machining must be carried out on different faces of the workpiece. Here the workpiece is mounted on an indexing chuck and more than one machining operation may be carried out in one setting (see Fig. 8.9).

Fixtures for prismatic (or nonrotational) parts are traditionally designed for a specific operation. These fixtures usually hold only one part type during a given operation. Fixtures consist of a steel baseplate with individual locating and clamping devices that are welded. If the part production volume is sufficient to justify the expense, these fixtures are premachined, carburized, hardened, and then ground to the final dimensions.

8.4. ADVANCES IN FIXTURING

8.4.1. Rotational Parts

Since flexible fixtures for rotational machining operations already exist, the focus of recent developments has been to automate the loading and unloading of workpieces and to reduce the changeover time during setup.

The first step toward automating the operation of chucks is to actuate them pneumatically or hydraulically. Power chucks have now replaced manually operated chucks in almost all production applications. This results not only in faster loading and unloading but also in the adjustment of gripping force. The clamping force can be varied depending on the application. Power chucks also provide follow-up force if chuck jaws bite into the workpiece during machining. In a similar situation, manual chucks will need to be tightened by the operator before continuing production. Chucks may be actuated by means of a cylinder mounted on the rear of the lathe spindle or one that is built into the chuck (self-contained hydraulic and pneumatic chucks).

Chucks are essentially multipurpose workholders accommodating different sets of jaws to hold different workpieces. While power actuation automates the material handling operation in loading and unloading, there is still a need to reduce unproductive time consumed in setup changes. This has been accomplished by quick change jaws. An average jaw changeover on a conventional chuck by an experienced operator takes about 20 minutes (5). With a quick change design this will take no more than 2 minutes. There are two basic quick change designs that exist. In the first, the top jaw disengages from the master or base jaw, and in the second, both disengage from the chuck body as a single unit. Figure 8.10 is a schematic explaining the operation of one type of quick change jaws.

Another approach is to change the chuck itself between setups. This is achieved by using a type of coupling between the spindle face and the chuck. The two halves of the coupling quickly and accurately interlock with each other. One such system developed by Illinois Tool Works Co. (see Fig. 8.11) delivers a positional repeatability of 0.0003 inch TIR (22). These advances facilitate the use of robot manipulators in chuck changing and loading/unloading operations.

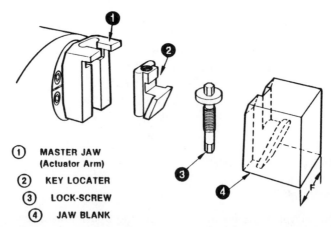

① MASTER JAW
   (Actuator Arm)

② KEY LOCATER

③ LOCK-SCREW

④ JAW BLANK

FIGURE 8.10. Quick change jaws. (Courtesy of ITW Woodworth.)

FIGURE 8.11. Quick change chuck. (Courtesy of ITW Woodworth.)

8.4.2. Nonrotational Parts

Fixtures for nonrotational components are traditionally designed for a single part and operation. However, the drive toward flexibility has led to the development of the following fixture types.

Modular Fixtures. Modular fixturing systems have been enjoying a resurgence of interest in the move toward flexibility on the shop floor although the concept of using a workholding system with modular, interchangeable parts has been prevalent since the 1940s (17). However, until recently, the need for cost effective workholding was not fully realized. Modular fixtures allow the user to develop many different fixtures from the basic set of fixturing elements comprising a modular fixturing kit. Flexibility is derived from the various configurations that are possible within the system. Components are easily built-up for one setup and then stripped and combined for a different operation. These systems are also refered to as "erector set tooling." The advantages are in reduced fixture stock and thus capital engagement, the reduction of labor cost in fixture design and fabrication, and the reduction in lead time (12). On the average, the design and assembly of a modular fixture can be completed in approximately 5% to 20% of the time required for a dedicated workholder (17).

System Types. Several systems are available commercially; these can be classified under two basic styles—the T-slot systems and the dowel pin systems. The difference is in the connecting principle used to connect individual modules. The T-slot system appears to be the more popular of the two: it uses a series of base plates with machined and ground T-slots that are used to mount fixturing elements and other accessories (see Fig. 8.12a). The T-slots are machined exactly perpendicular and parallel to each other to ensure precise alignment of elements. These systems are made to tolerances of approximately ± 0.0002–0.0004 inch in parallelism, flatness, and dimension. The advantage of the T-slot system is its adaptability in positioning:

(a)

(b)

FIGURE 8.12. Sample modular fixtures. (Courtesy of Flexible Fixturing Systems, Inc. and Fritz Werner Machine Tool Corporation.)

Components can be moved anywhere along the slots. The drawback, of course, is that repeatability from one setup to the next could suffer. Each setup is either photographed or stored in a CAD system to aid in subsequent rebuilding.

The dowel pin system has a similar philosophy except that the base plate has a pattern of holes that are used to locate and mount the accessories (see Fig. 8.12b). The advantage of repeatability is obtained at the cost of a slight loss in flexibility of position. In both systems, "hardened precision liners are used in the base plate and other components to ensure high accuracy. In addition all locating and chucking elements are hardened and ground in order to ensure precise shape, accurate dimensioning and durability" (14).

System Elements. The elements of a modular fixturing system typically comprise:

1. *Base elements* are the main carriers of a fixture, where all supporting, locating, and guiding elements are attached. The most common shapes of mounting bases are rectangular, square, and round.
2. *Supporting elements*, such as V-blocks and angle plates, that serve to support the component.
3. *Locating elements* include locating pins, centers, discs, and plates. They are responsible for properly locating the component.
4. *Guiding elements* such as drill bushes.
5. *Clamping elements* include all kinds of clamps, T-bolts, and clamping screws.
6. *Subassemblies* or *combined units* include indexing units, sine bars, angle tables, magnetic chucks, and vises (19).

By the very nature of their configuration, modular fixtures would seem to lend themselves to automated design and assembly. They can also be manually assembled. Estimates of build time of a new modular fixture design range from as little as 2 hours to 2 days (18, 19). Once the design has been stored in a CAD database or the setup has been photographed and documented, a repeat setup can be accomplished in a significantly lesser amount of time. Also, as setup personnel become more familiar with the system, the time required for the first setup is found to decrease considerably as experienced at GCA Corporation's Chelmsford manufacturing facility (18).

Shortcomings of Modular Fixtures. Modular components must have dimensions that are multiples of a relatively large factor. To fixture workpieces with intermediate dimensions, parts must be slid and sometimes even screwed and locked in the desired position. This leads to problems in repeatability in subsequent setups. Modular fixtures are also susceptible to greater tolerance stackup owing to their piecewise design.

Robotic assembly of modular fixtures is currently limited to very simple designs. Woodwark and Graham (29) question the suitability of existing designs for robotic assembly. They stress the need for an approach where accuracy, resolution, and rigidity will be provided by the intrinsic nature of the elements and the manner in which they locate with one other. Gandhi and Thompson (14) have proposed a methodology for automated design and assembly of modular fixtures based on the integration of state of the art methodologies from several disciplines.

Conformable Fixtures. Conformable fixtures are workholding devices that conform to the shape of the component to be machined. One approach to conformable fixtures is to use phase change materials ("authentic" or "pseudo") to encapsulate the workpiece (13). The material could be a low melting point alloy (authentic phase change) or a fluidized bed (pseudo phase change). In either case, the material in the fluid state is allowed to surround the block to be fixtured. A phase change that solidifies the material around the block is then induced. This has the effect of clamping the workpiece. This type of fixture is shown in Fig. 8.13. These methods are especially suited to irregular shapes. However, they suffer from the drawback that removing the low melting alloy from the workpiece could pose a problem. Also, a separate means of location of the workpiece in the workholder needs to be considered. The holding forces generated in the fluidized bed fixture are not sufficient for most metal cutting operations.

Flexible Mechanical Fixtures. Flexible mechanical fixtures that use modifications of traditional supporting and clamping techniques are in the developmental stage. Petal collets (25) consist of a set of tetrahedral "petals" connected to a common base, such that the entire fixture can open and close like the petals of a flower (see Fig. 8.14). This design can only hold and support a limited set of shapes. Another fixture with a similar approach is the multileaf vise (25) that consists of two jaws. The movable jaw is like a normal jaw; the stationary jaw, however, is made up of a series of coincident subjaws that can rotate about a common axis (Fig. 8.15).

   Both of the above approaches (conformable fixtures and flexible mechanical fixtures) take into account only the supporting and clamping requirements of a part. The location requirement is ignored and so they need to be

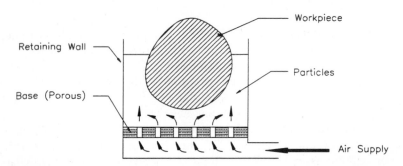

FIGURE 8.13. Fluidized bed conformable fixture. (Adapted from ref. 13.)

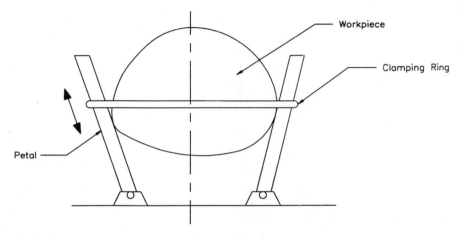

FIGURE 8.14. Flexible mechanical fixture: petal collet type. (Adapted from ref. 25.)

used in conjunction with an auxiliary metrological system that can locate the part with reference to the fixture.

Programmable Conformable Clamps. These clamps were designed to clamp and machine turbine blades. The system is flexible enough to accommodate a variety of blades. The clamps consist of hinged octagonal frames that can be opened to accept a blade and then closed. The lower half of each clamp employs plungers that, when released, are free to conform to the profile of the part above. A high strength belt is wrapped over the blade to hold it down against the plungers. The plungers are then forced against the blade using air pressure and mechanically locked in position (9).

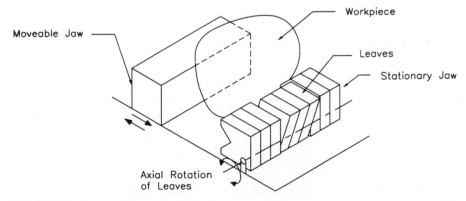

FIGURE 8.15. Flexible mechanical fixture: multileaf vise. (Adapted from ref. 25.)

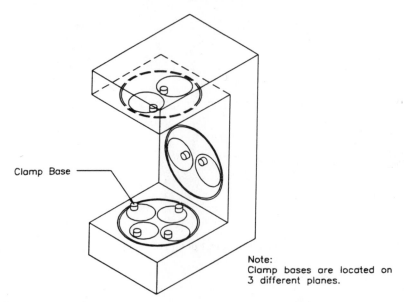

Clamp Base

Note:
Clamp bases are located on
3 different planes.

FIGURE 8.16. Double revolver type numerical control clamping machine. (Adapted from ref. 26.)

NC Fixturing Machines. Two versions of a numerically controlled clamping machine have been developed: the double revolver type (Fig. 8.16) and the translational type. The double revolver type consists of three sets of vertical cylindrical pins that rotate about two separate axes—the axis of a disk on which they are mounted and about their own axis. The translational type is a modified version and consists of locators that move in the X, Y, and Z planes and thus have six degrees of freedom. The locating points on each of these fixtures are able to reach any point in a given 3D workspace. Both types of fixture system are programmable and clamp a wide variety of part types (26).

8.5. AUTOMATED DESIGN OF FIXTURES

The objective of this section is to discuss and evaluate current research issues involved in the automated design of fixtures. The process of fixture design is a combination of rule-based logic and appropriate expert input. The major work packages in fixture design include the identification of clamping and locating points, and then the selection of specific standard elements from a set of available base plates, locators, and clamps. Traditionally, expert input is provided by the tool designer. An AI-based approach is relevant not only in the context of computerization to improve efficiency but also as an integrating link between CAD and CAM. One of the reasons for the relatively slow progress in automated fixturing is that CAD databases containing product information are ill-suited to fixturing needs. The kind of information needed for fixturing decisions (features, tolerances, surfaces needed to be machined, etc.) are seldom in a form that aids manufacturing analysis/planning. Also, while most drafting systems represent 2D or $2\frac{1}{2}$D models, fixture design involves 3D interpretation of objects. For rotational parts, a 3D interpretation from a 2D model is not very complicated; the process is more complex when considering prismatic components.

Early efforts to computerize fixture design focused on providing a graphical interface and a library of standard components, from which a designer could choose elements to build a fixture. These decisions are made by the fixture designer and designs often vary with the style of the individual designer.

The focus of research in this area has shifted to the application of AI-based techniques to automate a larger portion of the design process. Typical inputs to an automated fixture design system are workpiece description including geometry, features, tolerances, machining allowances, machining operations, and machine information. Outputs to be expected are fixture configurations for each machining operation. The processing involved includes interpretation of part features from the drawing, generation of workpiece orientation, calculation of fixturing and clamping coordinates, generation of alternative fixture configurations, and analysis of fixtures for sufficient location and stability. These problems are nontrivial; thus individual research efforts typically tend to address a subset of this area. The issues involved in automated design of fixtures can be classified as (a) schemes for representing the shape and structure information of a workpiece, in a form that will aid manufacturing analysis, and (b) establishment of rules for workholding, development of computer-based inference systems for fixturing decisions, and implementation in the form of automated assembly.

8.5.1. Schemes for Workpiece Representation

CAD systems available today include wireframe systems, boundary representation (B-Rep) systems, constructive solid geometry (CSG) systems, and the more recent analytic solid modelers (ASM).

Dong and Soom (10) have developed a Machining Process-Oriented Data Format (MPODF) to represent rotational parts in a wireframe system. Surface features, feature positions tolerances, surface finish, and so on are all represented as part of this format. The procedure used by them for automatic recognition of the 2D CAD database follows:

Uniformity of CAD Drawing Data. Since the construction of the 2D CAD model is system and user dependent, the drawing of the same part may have different coordinate data on two different systems or on the same system through two different users. Appropriate modifications have to be made to ensure that the drawing will always be represented by the same data.

Identification of Cylindrical Elements. Cylindrical geometric elements of various sizes were found to be the most frequently appearing, so cylindrical element recognition was used as the foundation for feature recognition. Since these appear as rectangular elements in 2D, this translates to identifying rectangles in the drawing.

Relational Analysis of Cylindrical Elements. Once the rectangles (or cylindrical elements) have been identified, the relationship between each of these is sought. To detect the presence of a shaft or hole, the algorithm determines which rectangles are contained inside others based on coordinate data.

Geometric Feature Recognition. Once rectangular elements have been classified as inner and outer, recognition of workpiece features is based on a special manufacturing based data format. A rotational surface is represented by horizontal edges of a rectangle with diameter and length provided. A hollow shaft or hole end plane is identified by the vertical edges of the two rectangles with outside and inside diameter.

Bidanda and Cohen (3) extend the concept of a CAD data storage format suited to manufacturing planning and analysis. Their application is again limited to rotational components. External surfaces are first numbered left to right in a clockwise direction; next, internal surfaces are numbered in the same way. A character "y" or "n" is used to indicate if a surface needs machining or not. In Fig. 8.17, surface 1 does not need machining and

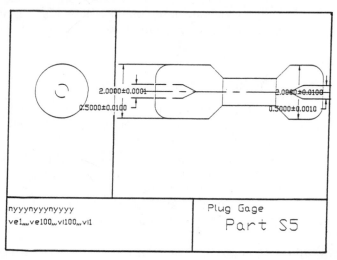

FIGURE 8.17. Sample part for entity and feature extraction. (From ref. 2, with permission.)

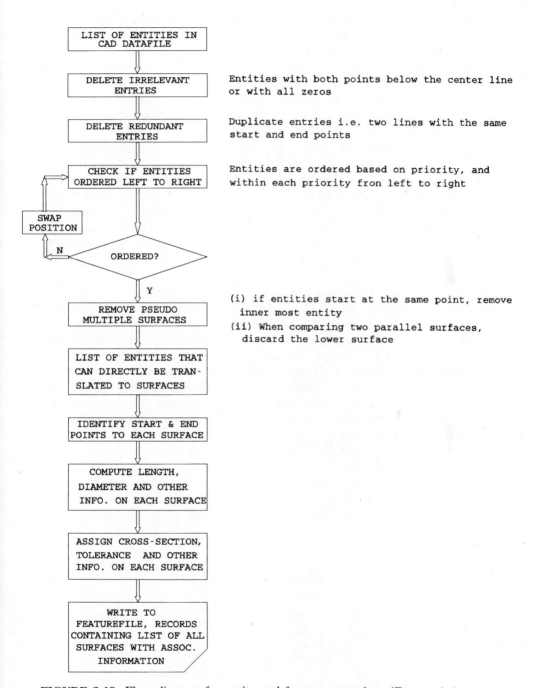

FIGURE 8.18. Flow diagram for entity and feature extraction. (From ref. 2, with permission.)

surfaces 2–4 do. This string of information and the tolerance information are stored in the title block of the drawing file. Tolerances are entered in the same order as surfaces; adjacent surface information is separated by a comma delimiter; multiple tolerances are separated by an asterisk. Once the format for the database has been established, feature extraction is facilitated. The procedure for the geometry-based feature extraction is shown in Fig. 8.18; it results in a feature file containing the set of features for the part as shown in Fig. 8.19.

Chou and Barash (6) outline an approach for representing prismatic parts. The topology of the body of the part is represented by a n-fold torus. This is done in a way that the workpiece is enclosed by a polyhedral envelope "such that the polyhedral surfaces provide both a reference frame for feature machining and candidate directions for supporting, locating and clamping the workpiece." Part features are depicted in $2\frac{1}{2}$D. Shape information is derived from the areas and directions (normals) of the main body surfaces. Principal orientations of the part are then identified by ranking the outward surfaces by orientation. To fix a part securely, it has to be supported, located, and clamped from certain combinations of the so-called principal orientations. The domain of any such fixture element is restricted to the "half-part" of the workpiece visible from that direction. During the design process, once the base surface has been determined, locating and clamping towers are constructed along the circumference of the workpiece.

Gandhi and Thompson (14) propose the use of mathematical modeling to represent a real object by its abstractions, which include geometric, physical, and kinematic descriptions and physical constraints. They suggest the use of solid models, where objects are represented as a combination of primitives, to represent the geometric description.

Ferreira and Liu (11) use a combination of boundary representation and feature-oriented schemes. The B-Rep scheme is used to describe the geometry and topology of the part along with descriptive information on the current state of a face (whether a face has been machined, whether it is available for locational reference, etc.). In addition, a feature description of

Fea.No.	Sur.No.	StartX	StartY	End X	End Y	Length	Dia.	Ht.	M?	Tol.	C.S.
1 fce.	lft.	5.500	6.000	5.500	6.500	0.00	0.00	0.50	n	vel	cir
2 cur.	ext	5.500	6.500	6.000	7.000	0.50	0.00	0.50	y		frm
3 hor.	ext	6.000	7.000	7.500	7.000	1.50	2.00	0.00	y		cir
4 tap.	ext	7.500	7.000	8.000	6.500	0.50	0.00	0.50	y		cir
5 hor.	ext	8.000	6.500	10.000	6.500	2.00	1.00	0.00	n	vel00	cir
6 tap.	ext	10.000	6.500	10.500	7.000	0.50	0.00	0.50	y		cir
7 hor.	ext	10.500	7.000	11.250	7.000	0.75	2.00	0.00	y		cir
8 cur.	ext	11.250	7.000	11.750	6.500	0.50	0.00	0.50	y	vil00	frm
9 fce.	rgt	11.750	6.500	11.750	6.000	0.00	0.00	0.50	n		cir
10 hor.	int	5.500	6.250	6.500	6.250	1.00	0.50	0.00	y		cir
11 tap.	int	6.500	6.250	7.000	6.000	0.50	0.00	0.25	y	vil	cir
12 tap.	int	10.500	6.000	11.000	6.250	0.50	0.00	0.25	y		cir
13 hor.	int	11.000	6.250	11.750	6.250	0.75	0.50	0.00	y		cir

FIGURE 8.19. Feature file for sample part. (From ref. 2, with permission.)

entities such as holes and slots is associated with faces of the boundary representation.

8.5.2. Establishment of Rules for Workholding

Bidanda and Cohen (3) classify the knowledge base required for fixturing decisions (for rotational components) into three distinct types of rule category:

1. Qualitative rules.
2. Quantitative rules.
3. Heuristics or rules of thumb.

Examples of qualitative rules used include:

1. If the holding surface is hexagonal, then a four jaw chuck cannot be used.
2. A part can be held between centers only if it has no through hole.

Quantitative rules relate machining and workpiece parameters. For example, if the diametral error due to tool and workpiece deflection (a function of cutting force, length of overhang, modulus of elasticity, etc.) is greater than the tolerance, then the workholding scheme is not feasible.
Examples of heuristic rules used include:

1. If process plan specifies a speed greater 400 rpm, then evaluate feasibility of using a counter centrifugal chuck (to prevent slippage).
2. When turning between centers, if the length/diameter ratio of the workpiece is greater than 6 and less than 10, add one steady rest to minimize sag and deflection.

Once a set of rules have been developed and stored in a knowledge base, they are applied depending on workpiece characteristics. Rule-based systems can be developed for such problems. Software programs based on recursive languages such as LISP and PROLOG are especially well suited for these tasks. Other benefits offered by such languages include mix of integer and real variables, extensive virtual memory, incremental compilation, and efficient search procedures (1).

In the case of rotational parts, the problem is predominantly one of fixture selection rather than of design. An algorithm for optimal selection of workholding devices for rotational parts was developed by Bidanda and Rajgopal (4). This algorithm is based on dynamic programming and chooses a fixture for each operation of the manufacture of a part (based on the process plan). Optimal fixtures are chosen from a set of feasible fixtures for each operation, such that setup time is minimized.

The problem of fixturing prismatic parts poses a greater challenge because the parts are more varied in geometry and the expert rules need to have more of a design and analytical basis. Chou et al. (7) developed a mathematical theory for the analysis and synthesis of fixtures for prismatic parts based on screw theory. Forces applied by locators and clamps and those generated by machining operators are modeled as wrenches (a load vector and a couple along the same axis). Fixturing solutions are then tested for four functional requirements: locational stability, deterministic work-piece location, clamping stability, and complete restraint.

A similar approach is also suggested by Gandhi and Thompson (14). Here machining, clamping, frictional forces, and torques are expressed by a resultant wrench. The system of forces is solved for static equilibrium to determine reactions at the locators. The choice of the support points is dictated in part by these solutions. Fixture design is then evaluated to calculate the deflections of the part and the result of this analysis may necessitate redesign of the fixture if necessary. Modular fixtures provide an opportunity to configure and automatically assemble fixtures for prismatic parts from a combination of standard elements. Much of the AI work on prismatic fixture design is centered around the use of modular fixtures, although the practical realization of automatic assembly to a reasonable degree of accuracy is not yet economically feasible.

Nnaji and Lyu (23) developed rules for expert fixturing of regular polygonal prisms for face milling operations. The primary input to the program is the workpiece description (which is constrained due to the assumption of a polygonal prism); system output includes a datafile containing the coordinates of fixturing points and a list of the modular fixture elements. The 3-2-1 method of locating and clamping is adopted and equations to calculate the six fixturing locations and three clamping points are derived. The first locating plane (3 locators) is constructed such that the area of the triangle formed by the three points is then large enough to ensure stability. For prisms with an even number of faces, the plane that is opposite and parallel to the surface to be milled is chosen. However, the prisms with an odd number of faces have an edge that is opposite to the surface to be machined. In this case, the first plane is constructed based on the two adjacent faces of the opposite edge. This ensures that the surface to be milled remains perpendicular or parallel to the milling cutter axis. The second plane having 2 locators is chosen in a manner that maximizes the distance between the locators. Also, these locators have the same X-axis coordinate as two of the locators in the first set (see Fig. 8.19). Their Y and Z coordinates are selected so that they do not interfere with the tool path. The third plane is related to the second in that the third locator should be at the same height as those in the second plane (as explained in Fig. 8.20). Finally, clamping is effected against the locators in the second and third planes.

Markus et al. (21) developed an expert system to assist in planning the

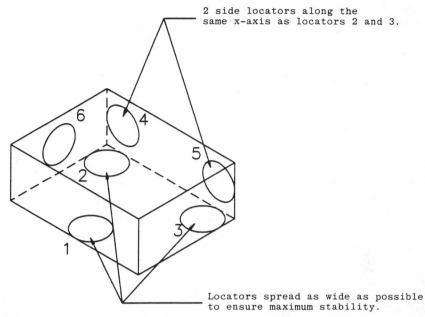

FIGURE 8.20. Procedure for identifying locator positions. (Adapted from ref. 23.)

initial configuration of fixture modules for simple prismatic parts. Workpiece description and clamping/support points are input to the system and result in the generation of fixture configurations. The fixture is composed of seats (rest buttons) and towers (a functional unit of the fixture holding one or more seats) and clamps. The base of the fixture is a pallet with slots machined in the form of an *X-Y* grid. Towers and clamps are positioned along these slots in a manner similar to commercial modular fixtures. Tower types are differentiated based on whether they locate in the horizontal direction or the vertical direction or are used for clamping. A tower is characterized by its type, orientation, and direction of the slot (*X* or *Y*) where it is fixed. In the first step of the design process, a collection of towers is identified based on the clamping and support coordinates entered. Subsequent design is completed using these towers only. A designer directing the search process can not set priorities and discard alternative solutions. Once designed, the fixture is assembled manually.

Fixture configuration involves realization of the relative positions of the modules and the workpiece. Gandhi and Thompson (14) demonstrate that spatial relationships between two objects, such as "fit," "against," and "coplanar," can be modeled mathematically. Geometric features such as face, edge, vertex, shaft, and hole are associated with the contact locations between the workpiece and the fixture and the spatial relationship between them is defined analytically (e.g., an edge is against a face if it lies completely in the plane of the face). Once the locator positions are determined using the force analysis described earlier, this knowledge of spatial relationships is used to select appropriate fixture modules. For example, locating a cylinder on a V-block can be considered to be satisfying the relationship "face (of the V) against shaft." Fixture modules are selected such that they satisfy the spatial relationships determined.

Feasibility of fixture layout depends on the absence of element interference. While this is true even of manually assembled designs, it is especially significant in an automated assembly approach, where, in addition to the interference between fixture elements, there may be interference between the setup hardware (typically the robot) and the fixture elements. Youcef-Toumi et al. (30) suggest a "holistic" approach to determine layout feasibility for modular fixtures. Workpiece layout requirements are transformed to element constraints. Their analysis centers around the physical characteristics of the modular elements of the fixturing system with particular attention on the vertical support modules used for location. For a given fixturing system, parameters such as "minimum distance between support points that can be supported by noninterfering modules" and "furthest possible distance between two support points where their candidate modules could interfere" are defined. Candidate modules for any support point are analyzed for interference with candidates for other support points. Modules are classified as interacting if there is a possibility of their interfering with other modules based on comparisons with the parameters defined. For example, if two support points are closer than the furthest possible distance described above, then they would be termed interacting. Actual interference analysis of interacting modules is accomplished by looking for intersections at boundaries of the respective CAD models. All such interactions are considered and the layout is deemed infeasible if there is at least one support point for which there is no single noninterfering module. Although this approach has been used in the analysis of sheet metal fixturing, it is general enough to be applied to other modular fixturing applications.

8.6. CONCLUSIONS

In this chapter, recent advances in fixturing as applicable to intelligent manufacturing systems were reviewed. These included developments in industry as well as research issues. Research and development in flexible fixtures are growing because of the promise of substantial productivity increase in manufacturing systems with improved workholders. The future will probably see a standardized manufacturing database that can be used for manufacturing planning and analysis. As can be seen from the chapter, there is also a need for a standardized graphical interface for manufacturing. The work by individual researchers is a first step toward alleviating this need. Rule-based systems have already demonstrated, in concept, that large-scale expert systems can be built to automate the process of fixture design, selection, and assembly. The next few years will hopefully see these systems being widely used in production planning departments at manufacturing plants.

ACKNOWLEDGMENTS

The authors thank Yi-Shin Deng for the many hours he spent in developing the illustrations.

REFERENCES

1. Badiru, A. B. (1990). Artificial Intelligence Applications in Manufacturing. In: *The Automated Factory Handbook—Technology and Management*, edited by D. I. Cleland and B. Bidanda, TAB Books, Blue Ridge Summit, PA.
2. Bidanda, B. (1987). *An Integrated CAD-CAM Approach for the Selection of Workholding Devices for Concentric Rotational Components*, Unpublished Ph.D. dissertation, Pennvania State University.
3. Bidanda, B., and Cohen, P. H. (1990). Development of a Computer Aided Fixture Selection System for Concentric, Rotational Parts, ASME Winter Annual Meeting, Dallas, TX.
4. Bidanda, B., and Rajgopal, J. (1990). Optimal Selection of Workholding Devices for Rotational Parts, *IIE Transactions*, Vol. 22, No. 1.
5. Burgam, P. M. (1984). Quick-Change Chuck Jaws—How Quick is Quick?, *Manufacturing Engineering*, pp. 43–45.
6. Chou, Y. C., and Barash, M. M. (1986). Computerized Fixture Design from Solid Models of Workpieces, *Proceedings of the Winter Annual Meeting of the ASME*, pp. 181–192.
7. Chou, Y. C., Chandru, V., and Barash, M. M. (1989). A Mathematical Approach to Automated Configuration of Machining Fixtures: Analysis and Synthesis, *Journal of Engineering for Industry*, Vol. 111, pp. 299–306.
8. Curtis, M. A. (1986). *Tool Design for Manufacturing*, Wiley, New York.
9. Cutkosky, M. R., Kurokawa, E., and Wright, P. K. (1982). Programmable Conformable Clamps, *AUTOFACT 4 Conference and Exhibition*, Society of Manufacturing Engineers, Dearborn, MI, pp. 11/51–11/58.
10. Dong, Z., and Soom, A. (1986). Computer-Automated Interpretation of 2D CAD Databases for Rotational Parts, *Proceedings of the Winter Annual Meeting of the ASME*, pp. 133–142.
11. Ferreira, P. M., and Liu, C. R. (1988). Generation of Workpiece Orientations for Machining Using a Rule Based System, *Robotics and Computer Integrated Manufacturing*, Vol. 4, No. 3/4, pp. 545–555.
12. Friedmann, A. (1984). The Modular Fixturing System, a profitable Investment, *Proceedings of the International Conference on Advances in Manufacturing*.

13. Gandhi, M. V., and Thompson, B. S. (1985). Phase Change Fixturing for Flexible Manufacturing Systems, *Journal of Manufacturing Systems*, Vol. 4, No. 1, pp. 29–39.

14. Gandhi, M. V., and Thompson, B. S. (1986). Automated Design of Modular Fixtures for Flexible Manufacturing Systems, *Journal of Manufacturing Systems*, Vol. 5, No. 4, pp. 243–252.

15. Gettleman, K., and Beaver, D. (1981). Getting the Grip on Workholding, Parts 1 & 2, *Modern Machine Shop*.

16. Hoffmann, E. G. (1984). *Fundamentals of Tool Design*, Prentice-Hall, Englewood Cliffs, NJ.

17. Hoffmann, E. G. (1985). Hoffmann on Tooling: Flexible Fixturing, Parts I and III, *Modern Machine Shop*, p. 102.

18. Krauskopf, B. (1984). Fixtures for Small Batch Production, *Manufacturing Engineering*, pp. 41–43.

19. Lewis, G. (1983). Modular Fixturing Systems. In: *Proceedings of the 2nd International Conference on Flexible Manufacturing Systems*, IFS Publications North-Holland, Amsterdam.

20. *Manufacturing Engineering* (1988). August, p. 43.

21. Markus, A., Markusz, Z., Farkas, J., and Filemon, J. (1984). Fixture Design Using Prolog: An Expert System, *Robotics and Computer Integrated Manufacturing*, Vol. 1, pp. 167–172.

22. Nastali, W. F. (1986). Workholding in Transition, *Manufacturing Engineering*, February, pp. 40–44.

23. Nnaji, B. O., and Lyu, P. (1990). Rules for an Expert Fixturing System on a CAD Screen using Flexible Fixtures, *Journal of Intelligent Manufacturing*, Vol. 1, No. 1, pp. 31–48.

24. Quinlan, J. C. (1984). New Ideas in Cost-cutting, Fast-change Fixturing, *Tooling and Production*, pp. 44–48.

25. Thompson, B. S. (1984). Flexible Fixturing—A Current Frontier in the Evolution of Flexible Manufacturing Cells, *ASME Paper #84-WA/Prod-16*, ASME Publications, New York.

26. Tuffentsammer, K. (1981). Automated Loading on Machining Systems and Automatic Clamping of Workpieces, *Annals of the CIRP*, Vol. 30, No. 2, pp. 553–558.

27. Wagener, A. M., and Arthur, H. R. (1941). *Machine Shop Theory and Practice*, Van Nostrand, New York.

28. Waldish, M. (1982). On the Edge of a Snake Pit—FMSs Are a Boon Used with Careful Planning, but Few Achieve This, *Engineer*, Vol. 255, pp. 32–33.

29. Woodwark, J. R., and Graham, D. (1983). Automated Assembly and Inspection of Versatile Fixtures. In: *Proceedings of the 2nd International Conference on Flexible Manufacturing Systems*, IFS Publications/North-Holland, Amsterdam.

30. Youcef-Toumi, K., Bausch, J. J., and Blacker, S. J. (1989). Automated Setup and Reconfiguration for Modular Fixturing, *Robotics and Computer Integrated Manufacturing*, Vol. 5, No. 4, pp. 357–370.

■■■■■■ CHAPTER NINE

Selection of Manufacturing Equipment for Flexible Production Systems

ULRICH A. W. TETZLAFF*

Department of Decision Sciences and MIS, School of Business Administration, George Mason University, Fairfax, Virginia

9.1. INTRODUCTION

The selection of manufacturing equipment for flexible production systems is of considerable importance for a manufacturing company, since it usually involves large capital investment. Therefore the selection has to be done carefully, applying a sophisticated concept and appropriate tools to ensure a smooth and economical working system.

The objective of this chapter is to analyze the process of equipment selection and to give an overview of applicable tools. Furthermore, it offers a new conceptual framework that can be used as a guideline for system planners and as a basis for further research. It is limited to the process of equipment selection for flexible production systems, which are computer controlled production systems capable of processing a variety of different part types. Their major components are computerized numerical controlled machines (CNC machines) and load/unload stations, handling systems for parts and tools, and a computerized planning and control system (for a more in-depth discussion on the taxonomy of flexible production systems see refs. 30 and 55).

Equipment selection is an integrated part of the whole design process of production systems. Consequently, first the design process itself is analyzed (Section 2) and then a conceptual framework for the design process is given. Next, different procedures for equipment selection are discussed (Section 3). In Section 4, applicable tools for equipment selection are presented.

*This work was completed when the author was at the University of California, at Berkeley.

Intelligent Design and Manufacturing, Edited by Andrew Kusiak.
ISBN 0-471-53473-0 © 1992 John Wiley & Sons, Inc.

Besides discussing specialized techniques, such as group technology, consideration is also given to how techniques with a usually broader scope like mathematical programming and knowledge-based systems are, or can be, applied to equipment selection. The chapter concludes with a summary and an outlook for future research.

9.2. THE DESIGN PROCESS OF PRODUCTION SYSTEMS

In order to examine the characteristics of the design process, its decisions, objectives, and restrictions are stated next. Then a conceptual framework for the design process is given.

9.2.1. Decisions

Decisions and their related variables must incorporate all those decisions that lead to an operational system. They comprise (see also refs. 26, 30, and 53):

- Selection of production system(s) used to produce the required products.
- Selection of parts and a quantity of them to be produced on a system over its whole lifetime.
- Selection of the equipment for the production system. This includes selection of the CNC machines (i.e., their type and the number of each); defining the number of load/unload stations in the system; selection of the transportation, material handling, and tool handling systems; defining the number of buffers and the inventory system; and defining the number of pallets and fixtures needed.
- Defining the type and structure of the planning and control system.
- Defining the layout of the system.
- Defining the number and skills of personnel needed.
- Defining the demand for and scope of flexibility needed.

It should be noted that these decisions are interdependent. This can easily be demonstrated by a simple model of causal relationships between the beginning of the design process (i.e., the given or expected gap of missing production capacity for some parts) and the result of the design process (i.e., the new installed production system(s) to fill this gap). The model consists of a causal chain that depicts the path of relationships between object groups to obtain one single alternative for the assignment of a part to a production system. These object groups are parts, process plans, operations, tools, machine tools, other machines, material handling systems, and other equipment needed (see Fig. 9.1).

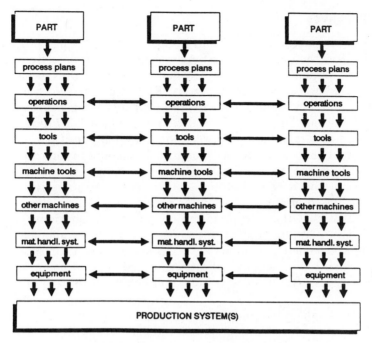

FIGURE 9.1. Model of causal relationships.

First, the part's (alternative) process plan(s) is examined; then each process plan is further subdivided into single operations. For each operation a specific tool or set of tools is required. To perform the operation, however, a machine tool is also necessary. Just to obtain a *feasible* solution for our part under consideration, the machine tools needed to produce the part must be supported by further machines (load/unload stations, washing machines, etc.), a material handling system, and all the other supplementary resources (see list above) to obtain an operational system. To obtain an efficient solution for a part, rather than just a feasible one, the inter-dependencies with other parts and with their causal chains must be taken into consideration. This is due to the fact that as soon as tools, machine tools, or other pieces of equipment are selected for the system, there are consequences for other parts. This is shown in Fig. 9.1 by horizontal arrows between members of one object group (i.e., between tools, machine tools, other machines, the material handling system, and other resources). Thus the vertical arrows show technical feasibility, whereas the horizontal arrows symbolize capacity restrictions, performance considerations, and cost or profit influences.

From this model of causal links between the part and the resulting production system, it can be concluded by observing the length of the chain that we are faced with an overwhelming array of possibilities. This would be the case even if only a few alternatives existed at each link of the chain and interdependencies to other parts are neglected.

9.2.2. Objectives

In the previous subsection, decisions necessary for the design process were discussed. Now the question arises: According to which objectives should a favorable or optimal solution be selected? In other words, what criteria should be used in making these decisions? Kalkunte et al. (26), for example, name three different objectives: cost, productivity, and product quality.

In the following the analysis is restricted to productivity. Productivity can be described as a relationship between input and output of production and thus it allows one to measure the economic feasibility of production (1, 11).

In order to define input and output of a production system, those decision variables that have an effect on the input or output of the system must be considered. Based on the analysis in the previous subsection, one can define output as the parts that are produced on the system and input as all the resources necessary to produce these parts (e.g., the equipment for the system, the personnel, raw material). To allow a comparison of differently measured inputs and outputs, the transformation to a common unit of measure is necessary. This can be accomplished by measuring in monetary units.

Based on the definition of productivity, three objectives can be derived, where each of them can be chosen depending on the circumstances surrounding the design process:

- Minimize only input (costs, present value of cash outflows), while output is constant.
- Maximize output (part production), while input is constant.
- Optimize the relationship between input and output.

To decide which of the three above-mentioned ways is to be used to optimize the system during the design process, the results of the preceding planning processes have to be considered. The planners might wish to configure the production system(s) in such a way that a given capacity gap is closed. Consequently, they choose the first objective. However, they might discover that the solution found consists of a system that causes initial investment costs to exceed the available budget. Alternatively, the planners might then either try to make adjustments on the preceding planning processes (i.e., the budget) or try to find a solution by using the second objective and produce as much as possible within their available budget. If, however, there is neither a limited budget nor a restriction on output, the last objective is applicable.

9.2.3. Restrictions

Some restrictions on the design process can easily be derived from the environment of the design process. From a long-range production plan a capacity gap is given, and from the finance plan the available budget to fill this gap is obtained.

Further restrictions are given by:

- Technological constraints due to the physical properties of the parts to be produced.
- Technological constraints due to the properties of the potential equipment.
- Limited space for the layout of the system.
- Necessary throughput times because of customer service requirements or because of necessary synchronization with other downstream or upstream production processes.

9.2.4. A Concept for the Design Process

The principal aim is to construct a decision model that enables one to solve the design problem. This could be done within a single model incorporating all the causal relationships between the decision variables in an adequate way, while considering all the given restrictions. However, for three reasons it is questionable whether such a simultaneous procedure is applicable.

1. *Complexity Problem.* As already shown by a simple model of causal relationships, the design problems are very complex and the set of alternatives grows rapidly with the number of decision variables.
2. *Method Problem.* It is doubtful whether the available tools are powerful enough to construct and solve a satisfactory homomorphic model of reality.
3. *Data Problem.* To explore all alternatives data requirements are extensive.

A way of circumventing these problems is to further structure the design process as a sequential procedure. This allows the reduction of the problem to a set of subproblems, which are simpler to solve and require less data. However, two disadvantages are inherent in this concept:

1. It is possible that favorable solutions or even the optimal solution are discarded at early steps in the sequential procedure, because they do not yet appear promising.
2. With the solution obtained from the first steps it might not be possible to generate a feasible solution in the following steps.

Nevertheless, these disadvantages can be avoided if the procedure is structured and modeled in such a way that the neglected causal relationships between decision variables of different subproblems are of minor impor-

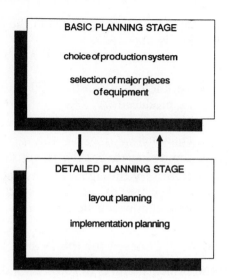

FIGURE 9.2. A planning concept for the design.

tance. This can be achieved, for example, by performing first those decisions that have a major impact on the planning objective. Moreover, the sequential planning procedure can be performed repeatedly with a feedback to the initial planning steps to ensure the generation of a feasible solution.

For the design of production systems a two-step planning procedure is suggested. The separation of decision variables is done by first selecting those variables that have a major influence on the objective and allow system costs or cash outflows for the system (input) and the system behavior (output) (e.g., throughput) to be estimated as accurately as possible during the first planning step. This allows coordination with other planning areas, that is, the investment plan and the long-range finance plan. Consequently, the first step consists of basic decisions including the choice of production system(s) and the selection of major pieces of equipment for each system. The second step comprises a more detailed planning stage. Here the final layout and the timing for the implementation of the system are considered. Figure 9.2 summarizes the concept once more. In the next two subsections both planning stages are described.

The Basic Planning Stage. Through sequencing the design process only a subset of decision variables is considered, while the objective remains the same and the number of restrictions might increase. As decisions for the basic planning stage, the following are considered:

- The part assignment to production systems.
- The type of production system(s).
- The type and number of machines, load/unload stations, and so on.

- The material handling system, especially the transportation system and its number of vehicles.
- The number of pallets and fixtures.
- Flexibility requirements.

However, the number of pallets and fixtures might also be subject to the following layout planning phase, depending on how much the planning objective is influenced by it. For example, if costs for pallets and fixtures play a negligible role compared to the other equipment costs, consideration during the detailed planning phase might be more appropriate.

The Detailed Planning Stage. If the basic planning stage is finished, a few problems still have to be solved before the design of the system is completed. If not already done in the basic planning phase, the number of pallets and fixtures has to be determined. Furthermore, the number of buffers and their location must be specified. The design of the material and tool handling system has to be completed. The structure of the required planning and control system and the amount and skills of personnel needed are additional aspects to be solved. Finally, the physical layout, considering restrictions on space and other resources, must be determined.

After the final physical layout of the system has been established, the planning of the different tasks for the implementation of the system and their timing is made. In addition to considering the physical implementation of the system, any necessary organizational adaptations and changes must be planned.

9.3. THE SELECTION PROCESS OF EQUIPMENT

Through analyzing the planning concept of the previous section it can be observed that the selection process of major equipment takes place during the basic planning stage. Here the major components for the production system(s) are selected. Thus a concept of how to perform the selection process of equipment is primarily given by the procedures of the basic planning stage. These procedures, which include (a) data collection, (b) linking, and (c) selecting, are described in the following subsections.

9.3.1. Data Collection

The purpose of data collection is to obtain all data necessary for the selection process. Considering the decisions to be made during the basic planning stage, this involves data for the following object groups: part types, process plans, operations, tools, machine tools, other machines, and the material handling system. The data collected can be technical data as well as cost data.

9.3.2. Linking

The objective of linking is to generate *feasible* links between the object groups. Considering the model of causal relationship in Fig. 9.1, this is identical to generating vertical links between the members of different object groups. For linking it is important to find all possible or at least all important links between the objects under consideration. This guarantees that a favorable feasible solution can be found at the end of the basic planning stage.

A possible way of linking is to generate vertical links from the part down to the machine tool or even further down to material handling systems. Thus a number of different assignments of machine tools, other machines, and material handling systems to a certain part type is obtained. In Fig. 9.3, one example of such a chain is shown.

Example 1. Part type 1 with three alternative process plans shall be manufactured on one/several production system(s). One vertical chain might have the following character: part type 1→ with alternative process plan 1b→ and its operation 1b.1→ can be performed with cutting tool of type aa→ using as a machine tool of type A (Fig. 9.3).

9.3.3. Selecting

The purpose of selecting is problem reduction and problem solving. Problem reduction is necessary, since, despite the fact that a two-stage planning procedure is applied, the relationships between the decision variables of the basic planning stage are still too complex to allow the application of a simultaneous solution procedure.

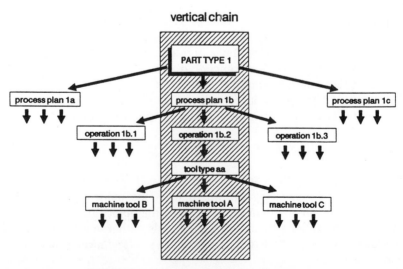

FIGURE 9.3. Vertical chains through linking.

During problem reduction, data are filtered by selecting members within one object group. Thus, considering the model of causal relationships, it consists of selecting objects with horizontal interdependencies within their object group. It is important that during this selecting process no dependencies get discarded, which could endanger the finding of a favorable, feasible solution. Thus selecting has a much greater impact on subsequent planning steps than does linking. The latter searches only for feasible vertical links. Selecting has to consider interdependencies between the members of an object group and what consequences the selecting result has on the interdependencies between members of object groups further down the vertical chain.

One way of selecting is to group part types together into part families, so that each part family is produced on one production system. However, the assignment of a certain part type to a part family already has a strong impact on the appearance of the *not yet configured* production system. It will influence the kind of operations that have to be performed by the system and what kinds of tools and machine tools are selected. Thus it will have a considerable impact on the interdependencies within object groups like tools and machine tools further down the vertical chain.

Example 2. Consider three part types—part type 1, part type 2, and part type 3. For simplicity assume that each part type has only one process plan consisting of one operation, that is, for part type 1 operation 1a.1, for part type 2 operation 2a.1, and for part type 3 operation 3a.1. Through linking it is observed that operation 1a.1 can be performed on one of the machine tools out of the set {mill A, mill B}, for operation 2a.1 on one out of {mill A, lathe C}, and for operation 3a.1 on {drill D}. As can be observed in Fig. 9.4, a different grouping of parts through selecting results in a different set of possible configurations, with each configuration consisting of different types of machines.

Another possibility for selecting is given in the next example. Here machines are selected in such a way that possible routes for a part through a not yet configured system are generated.

Example 3. Consider one part type 1 having only one process plan consisting of operations 1a.1 and 1a.2. The first operation 1a.1 can be performed with a tool of type aa on either milling center A or milling center B. The second operation 1a.2 can be performed with a tool of type bb at drilling machine C or drilling machine D. The possible routes for part type 1 are {mill A, drill C}, {mill A, drill D}, {mill B, drill C}, and {mill B, drill D}, assuming that the order in which the two operations take place is not relevant. Thus the selection procedure can create one, several, or all of these routes. Note that if all possible routes are generated, no data reduction occurs. However, in general, the number of possible routes is quite large and a reduction is unavoidable (e.g., see Section 9.4.2 for

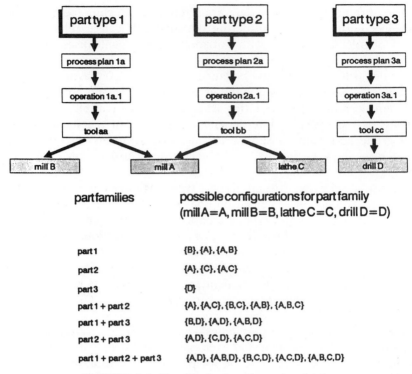

FIGURE 9.4. Data reduction through selecting.

applicable tools). On the other hand, since the selecting procedure takes place at the lower end of the vertical chain, the impact of this data reduction is not as strong as in the previous example.

9.3.4. Finding a Solution

Since the relations between decision variables of the basic planning stage are still too complex to allow for a simultaneous solution procedure, the suggested strategy for generating a favorable solution in accordance with a given objective is to apply a procedure that combines one or several linking and selecting procedures. However, when and how to apply linking or selecting is problem specific and depends in particular on the size of each object group.

One possibility is to start with selecting and group the part types first into part families, such that for each part family a production system must be configured. (The complications involved with this step were already outlined in the previous section.) Next, links must be generated for each part type of a family down to applicable machine tools or further down to the material handling system. According to a specific objective, a selecting procedure then has to select from this set of applicable machines those machines used for configuring the production system (see Fig. 9.5). A final linking step then provides the configuration with the complementary equipment necessary.

Example 4. Continuing Example 2, a first selecting step is performed by grouping part type 1 and part type 2 together to part family 1 and let part family 2 consist only of part type 3. Next, linking is accomplished by generating vertical links for all three part types. On the set of possible configurations for part family 1, that is, ({mill A}, {mill A, lathe C}, {mill B, lathe C}, {mill A, mill B}, {mill A, mill B, lathe C}), selecting is applied and configuration 1 with {mill B, lathe C} is obtained. For part family 2 only one possible configuration is available, that is {drill D}, which yields configuration 2. Finally, linking is again applied to complete the configuration by assigning all the necessary complementary equipment to the two

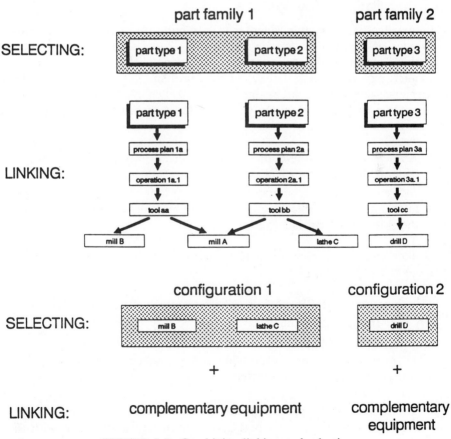

FIGURE 9.5. Combining linking and selecting.

systems. Note that selecting must be applied again, if during this last linking step several alternative choices for complementary equipment are available for a configuration and when each choice has a different impact on the planning objective.

9.4. TOOLS

This section presents tools applicable during the basic planning stage. The discussion is thereby restricted to generative tools, that is, tools that are intended to reach a favorable decision and thus can be applied for linking and selecting. Evaluative tools, which are intended to evaluate a given set of decisions, such as computer simulation, queueing networks, or petri nets, are not considered here (see ref. 54 for a discussion of evaluative versus generative tools).

9.4.1. Knowledge-Based Systems

A knowledge-based system denotes a computer program that can store knowledge of a particular domain and use that knowledge to solve problems from this domain in an intelligent way (22). Knowledge-based systems are also called expert systems if they refer to problem solving at that level of performance and in those areas that are usually achieved by human experts.

According to Kriz (29), a knowledge-based system contains at least the following two components:

1. *Knowledge Base.* Here the domain-specific knowledge is stored, in general, in the form of facts or rules.
2. *Inference Engine.* It operates on the knowledge base, performs logical inferences, and deduces new knowledge by applying rules to facts until the posed problem is solved.

Besides these basic components, a knowledge-based system should provide the following features:

- Explanatory capabilities of its behavior on request by the user.
- A user-friendly dialogue.
- An application-oriented knowledge representation language.
- The possibility of easily modifying and extending the knowledge base during the lifetime of the system.

A knowledge-based system with the above listed features, but an empty knowledge base, is called an expert system shell. By supplying it with the particular knowledge, a specific knowledge-based system can be built. (For an overview on shells, see refs. 4 and 12; on already existing systems in the development, implementation, and operation of automated manufacturing systems, see ref. 31.)

Until now, knowledge-based systems have been applied to perform selecting during the basic planning stage by grouping machine tools and equipment. The general concept is to evaluate the system behavior through either closed queueing networks or simulation and then to apply a set of rules. According to the model of causal relationships, these rules try to capture horizontal interdependencies and aim at improving a given configuration. Mellichamp et al. (35, 36) developed an expert system that suggests where modifications on the design of a flexible manufacturing system are appropriate and uses simulation to evaluate the design. Floss and Talavage (16) apply a two-stage procedure. First, to improve the system design, they apply an expert system on the estimates of the system behavior obtained through a closed queueing network model. Then, during the second step they use simulation for system evaluation and another expert system to fine-tune the design. However, these improvements are restricted to capacity changes (i.e., the number of machines, vehicles, etc.) of each type and do not consider the selection of alternative types of equipment.

A major difficulty of these approaches is that, for a given objective, the interdependencies between the decision variables and their sensitivity toward the objective are far more complex than can be modeled by a simple set of rules. Thus emphasis on more quantitative approaches for machine and equipment selecting is more appropriate.

To date, for linking purposes knowledge-based systems have only been used for the generation of process plans (a survey is given in ref. 20). However, the generation of feasible vertical links down to machine tools or even further seems to be a promising application. This is due to the fact that rules can capture quite well the technical constraints that have to be considered to ensure the feasibility of a link.

9.4.2. Group Technology Methods

Group technology comprises a number of methods for analyzing and arranging the part spectrum and the relevant manufacturing processes according to similarities (e.g., geometry, size, volume, type and sequence of operation). The goal is to establish a number of production cells, each of which is capable of producing a group or family of components (10). If group technology is applied to the design of production systems, each group technology cell can be interpreted as a system.

Ballakur (6) classifies the techniques in group technology into three classes:

1. *Form part families and then group machines into cells* (*part family grouping*). Here classification and coding schemes, such as the one of

Opitz (41), can be subsumed. An overview on coding systems can be found in Hyer and Wemmerlöv (24). Further cluster analysis can be applied to find part families (9).

2. *Form machine cells based on similarity in part routings and then allocate parts to cells (machine grouping).* Again, cluster analysis, such as the single linkage cluster analysis (33), or the approach of Gupta and Seifoddini (21), can be applied. Graph-theoretical methods like the one of Rajagopalan and Betra (42) or Askin and Chiu (2) are approaches for machine grouping. Olivia-Lopez and Purcheck (40) perform machine grouping based on combinatorial analysis.

3. *Form part families and machine cells simultaneously (machine–part grouping).* Manual techniques, such as the production flow analysis of Burbidge (8) and the component flow analysis of El-Essawy (15), perform such a simultaneous grouping procedure of parts and machines. Furthermore, a number of algorithmic techniques exist that are based on matrix reordering of a machine–part matrix, for example, the rank order clustering algorithm of King (28) or the bond energy algorithm of McCormick et al. (34). Some newer approaches also include cost considerations. Here, for example, those of Seifoddini (47), of Askin and Subramanian (3), and of Mutti and Semeraro (38) can be listed.

For a more in-depth review of group technology methods refer to theses by Ballakur (6) and Seifoddini (47) and papers by Huang and Houck (23) and Mosier and Taube (37). A more recent comparison of some techniques is given in Shafer and Meredith (48) and a survey of current practices in U.S. manufacturing industry in Hyer and Wemmerlöv (25).

The linking capabilities of group technology depend on the procedure used, whereas its selecting capabilities are, in general, strongly developed in all procedures. However, these selecting capabilities are restricted by the fact that most of the algorithms were originally developed for the redesign of existing job shops with the aim of combining the job shop with the advantages of a flow shop. Therefore cost objectives that play an important role during the basic planning stage (see Section 2.2) were neglected in the earliest approaches, such as production flow analysis, component flow analysis, or matrix reordering techniques like the rank ordering algorithm. Only in some newer techniques, for example, those developed by Seifoddini (47), Askin and Subramanian (3), and Mutti and Semeraro (38), are cost considerations incorporated. The latter, in particular, explicitly considers a new production system and not the reorganization of an old one.

Another drawback of these techniques is that they do not consider, with the exception of some newer approaches (39, 43), alternative process plans during grouping. Due to the flexibility of modern manufacturing equipment, this is, however, a very common option that allows for better utilization of resources and hence less equipment.

9.4.3. Mathematical Programming

Mathematical programming is concerned with the determination of the optimum of a function of several variables, which also have to satisfy a number of constraints. These constraints can be inequalities and/or equalities (59).

Depending on the kind of function and constraints (e.g., linear or nonlinear) and the values the variables can have (e.g., continuous or integer), different methods for finding an optimum solution exist. An overview of standard algorithms and methods is given by Eiselt and von Frajer (14).

In the past, several mathematical programming models were developed to optimize the design of flexible production systems. In the following two subsections, a classification of these models is given according to two criteria: (a) system behavior modeling and (b) decision variables (a more comprehensive classification that includes flexibility aspects is given in ref. 55). An in-depth discussion of each criterion gives the planner useful hints about the appropriate model for planning problems. Afterward an overview of existing models is given by the introduction of a model matrix.

System Behavior Modeling. The different modeling of system behavior has important consequences on the evaluation of required system parameters, for example, the throughput. Two classes of models can be distinguished: models that consider static system behavior and those that consider dynamic system behavior.

Static modeling, which neglects queueing processes and important aspects of the part flow, is achieved by comparing the available capacity of a machine or transportation system to its workload. The throughput of a station (set of identical machines or the transportation system) is then determined by the workload W_m at station m divided by the number of servers s_m (CNC machines, load/unload stations, vehicles, etc.), or equivalently the number of visits v_m divided by the service rate μ_m and the number of servers s_m. Moreover, the system's throughput can be determined by the bottleneck station b, that is, the cell or transportation system with the highest workload per server.

$$\frac{W_b}{s_b} = \frac{v_b}{\mu_b \cdot s_b} = \max_m \frac{v_m}{\mu_m \cdot s_m} \,.$$

The fact that the system's maximal throughput is given by the capacity of the bottleneck station allows simple linear programming models to be constructed.

Models that consider dynamic system behavior and thus queueing processes and part flow interactions comprise nonlinear queueing network formulas. However, the question arises as to what extent the throughput evaluation by static modeling differs from dynamic modeling of system behavior. If dynamic modeling is achieved by a standard closed queueing network of the Jackson type, it can be observed that the system's throughput T reaches a saturation point if the number of customers (pallets) increases toward infinity. It can be shown that this saturation point of the closed queueing network is identical to the throughput of the bottleneck station (46):

$$\lim_{N \to \infty} T = \frac{e_b}{v_b} \cdot \frac{G(M, N-1)}{G(M, N)} = \frac{\mu_b \cdot s_b}{v_b} = \frac{s_b}{W_b} \; ;$$

and it can be concluded that, when a very large number of pallets is in the system, dynamic throughput behavior is almost identical to static throughput behavior modeling. Note that $G(M, N)$ is the normalization constant derived from closed queueing network theory having M stations and N customers in the system. The relative flow at the bottleneck cell is given by e_b. See Gordon and Newell (17) or Bruell and Balbo (7) for a more comprehensive discussion of closed queueing networks of the Jackson type.

Decision Variables. Another distinction between models is due to the capacity for performing different design decisions based on the use of different decision variables incorporated in the models. Three basic categories of decision variables can be identified:

1. Variables for part routing/operation assignment. Here the part routing or the assignment of operations is determined. Based on given production requirements for each part type, it is necessary to decide how much of each part type is produced on each of its alternative routes, or which operation is performed on which resource.
2. Variables for server decisions. These decision variables describe how many CNC machines, load/unload stations, vehicles for the transportation system, and so on are to be implemented in the system.
3. Variables for pallet/part decisions. In some models with a limited number of pallets or parts allowed in the system, decisions about the number of pallets/parts are possible.

Note that some of the basic categories of decision variables can be aggregated, for example, server and pallet/part decisions, to a system decision variable. Further aggregations in each category are possible, for example, operations of a part type to operations of a part family.

Based on the capability of the models and their decision variables the following classification of optimization models is derived.

1. *Routing Optimization.* Here the routing of parts in a given production system is optimized. Therefore in this category only decision variables concerning the part routing/operation assignment are found. Since the

system is already configured, models of this type are not directly applicable for design optimization. However, routing optimization is integrated in quite a few more complex models.

2. *Capacity Optimization.* The necessary types of equipment must already have been selected in order to optimize capacity. Furthermore, the part routing through the system is given. Now the optimal quantity of each type of equipment must be found satisfying a certain objective and simultaneously obeying certain restrictions. As a result, variables for CNC machines/load–unload stations/transportation system decisions and/or variables for the number of pallets/parts decision are used.

3. *Equipment Optimization.* The models in this class are not only able to determine the optimal amount of each type of equipment but also allow the selection of equipment, that is, the inclusion or exclusion of different types of equipment. This makes it necessary to have decision variables for the part routing/operation assignment as well as for the CNC machines/load–unload stations/transportation system selection and possibly also for the number of pallets/parts decision implemented.

4. *Part and Equipment Optimization.* Besides equipment optimization, the part types to be produced on the system can also be selected in an optimal way. Here the same decision variables as for equipment optimization are required. However, part routing/operation assignments are now extended in such a way that part selection is possible.

5. *System and Equipment Optimization.* Here the type and number of production systems necessary to fulfill production requirements and their equipment are selected. Here all four of the above-listed categories of decision variables are applicable.

6. *Part, System, and Equipment Optimization.* These models comprise all of the above-mentioned decisions. For that reason it can be considered as being identical to the issue of the basic planning stage.

The classes of optimization models described above can be positioned as decision models in a hierarchical tree comprising an upper class and several lower classes (see Fig. 9.6) according to the concept introduced by Schneeweiß (45).

The Model Matrix. According to the two classification criteria, the optimization models of flexible production systems can be positioned in a two-dimensional matrix. This matrix consists of two columns, one for models with an unlimited number of pallets [UP], that is, static system behavior modeling, the other one for those with a limited number of pallets [LP], that is, dynamic system behavior modeling (see Fig. 9.7). Furthermore, the rows are assigned to different classes, derived from the decision variables used [RO, CA, EQ, PAEQ, SYEQ, PASYEQ]. Thus each position in the matrix represents a certain combination of the two classification criteria. According

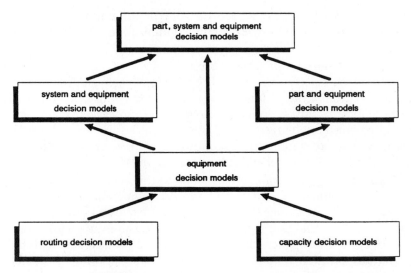

FIGURE 9.6. Hierarchy of decision models.

system behavior → ↓ decision problems	unlimited number of pallets (static system behavior modelling) [UP]	limited number of pallets (dynamic system behavior modelling) [LP]
routing optimization [RO]	ROUP-Secco-Suardo (1978) -Kimemia/Gershwin (1978) -Avonts et al. (1988)	ROLP-Kimemia/Gershwin (1978) -Yao/Shanthikumar (1987) -Tetzlaff-1 (1990) -Tetzlaff-2 (1990)
capacity optimization [CA]		CALP-Vinod/Solberg (1985) -Solot (1987) -Shanthikumar/Yao-1 (1987) -Shanthikumar/Yao-2 (1988) -Dallery/Frein (1988) -Shanthikumar/Yao-3 (1989) -Lee et al. (1989) -Tetzlaff (1990)
equipment optimization [EQ]	EQUP-Graves/Whitney (1979) -Graves/Lamar (1980) -Tetzlaff-1 (1990) -Tetzlaff-2 (1990)	EQLP-Tetzlaff-1 (1990) -Tetzlaff-2 (1990)
part and equipment optimization [PAEQ]	PAEQUP-Whitney/Suri (1985)	
system and equipment optimization [SYEQ]	SYEQUP-Sarin/Chen (1986)	
part, system and equipment optimization [PASYEQ]		

FIGURE 9.7. Model matrix.

to its position in this matrix, each model has a name showing its position and the author's name. The year of publication is added in parentheses to facilitate the citing of references. (For a complete discussion of these models see ref. 55).

With the models listed in Fig. 9.7, selecting can be performed primarily, for example, by performing selecting for parts, machines, and material handling systems to obtain part families and production system(s); a part, system, and equipment optimization is achieved. However, through the introduction of constraints, it is also possible to consider feasible vertical links, which are not given as input data. On the other hand, due to the large number of technical constraints, mathematical programming has only a restricted capability for linking.

9.5. SUMMARY AND OUTLOOK

To analyze the selection process of equipment for flexible production systems, the design process, during which the selection takes place, was analyzed first. Based on a planning concept derived from this analysis, it was possible to pinpoint three characteristic procedures that take place during selection: data collection, linking, and selecting. Finally, an overview of available generative tools for linking and selecting processes was provided.

Based on this overview, some suggestions for future research can be stated:

- Since the potential to apply knowledge-based systems for linking purposes has so far been neglected, further research is desirable in this area.
- As can be seen by quite a few white locations in the model matrix for optimization models based on mathematical programming, further developments are desirable, especially for models that consider dynamic system behavior and thus incorporate queueing aspects and part flow interactions.
- It is desirable to do further research on how to combine selecting and linking as well as their associated tools to obtain a favorable, feasible solution for the basic planning stage.

REFERENCES

1. Adler, P. S. (1987). A Plant Productivity Measure for "High-Tech" Manufacturing, *Interfaces*, Vol. 17, No. 6, pp. 75–82.
2. Askin, R. G., and Chiu, K. S. (1990). A Graph Partitioning Procedure for Machine Assignment and Cell Formation in Group Technology, *International Journal of Production Research*, Vol. 28, No. 8, pp. 1555–1572.

3. Askin, R. G., and Subramanian, S. P. (1987). A Cost-Based Heuristic for Group Technology Configuration, *International Journal of Production Research*, Vol. 25, No. 1, pp. 101–113.

4. Assad, A. A., and Golden, B. L. (1986). Expert Systems, Microcomputers, and Operations Research, *Computers & Operations Research*, Vol. 13, No. 2/3, pp. 301–321.

5. Avonts, L. H., Gelders, L. F., and Van Wassenhove, L. N. (1988). Allocation Work Between an FMS and a Conventional Jobshop: A Case Study, *European Journal of Operations Research*, Vol. 33, pp. 245–256.

6. Ballakur, A. (1985). *An Investigation of Part Family/Machine Group Formation for Designing Cellular Manufacturing Systems*, Ph.D. Thesis, University Wisconsin, Madison.

7. Bruell, S. C., and Balbo, G. (1980). *Computational Algorithms for Closed Queueing Networks*, North-Holland, New York.

8. Burbidge, J. L. (1963). Production Flow Analysis, *The Production Engineer*, Vol. 42, No. 12, pp. 742–752.

9. Carrie, A. S. (1973). Numerical Taxonomy Applied to Group Technology and Plant Layout, *International Journal of Production Research*, Vol. 11, No. 4, pp. 399–416.

10. Chakravarty, A. K., and Shtub, A. (1984). An Integrated Layout for Group Technology with In-process Inventory Costs, *International Journal of Production Research*, Vol. 22, No. 3, pp. 431–442.

11. Chew, B. W. (1988). No-Nonsense Guide to Measuring Productivity, *Harvard Business Review*, No. 1, pp. 110–118.

12. Citrenbaum, R., Geissman, J. R., and Schultz, R. (1987). Selecting a Shell, *AI Expert*, Vol. 2, No. 9, pp. 30–39.

13. Dallery, Y., and Frein, Y. (1988). An Efficient Method to Determine the Optimal Configuration of a Flexible Manufacturing System, *Annals of Operations Research*, Vol. 15, pp. 207–225.

14. Eiselt, H. A., and Frajer von, H. (1977). *Operations Research Handbook, Standard Algorithms and Methods*, W. de Gruyter, Berlin.

15. El-Essawy, I. F. K. (1971). *The Development of Component Flow Analysis as a Production Systems' Design for Multi-Product Engineering Companies*, Ph.D. Thesis, UMIST, UK.

16. Floss, P., and Talavage, J. (1990). A Knowledge-Based Design Assistant for Intelligent Manufacturing Systems, *Journal of Manufacturing Systems*, Vol. 9, No. 2, pp. 87–102.

17. Gordon, K. D., and Newell, G. F. (1967). Closed Queueing Systems with Exponential Servers, *Operations Research*, Vol. 15, No. 2, pp. 254–265.

18. Graves, S. C., and Lamar, B. W. (1980). A Mathematical Programming Procedure for Manufacturing System Design and Evaluation, *Proceedings of the IEEE International Conference on Circuits and Computers*, pp. 1146–1149.

19. Graves, S. C., and Whitney, D. E. (1979). A Mathematical Programming Procedure for Equipment Selection and System Evaluation in Programming Assembly, *Proceedings of the 18th IEEE Conference on Decision and Control*, Fort Lauderdale, pp. 531–534.

20. Gupta, T., and Ghosh, B. K. (1988). A Survey of Expert Systems in Manufacturing and Process Planning, *Computers in Industry*, Vol. 11, pp. 195–204.

21. Gupta, T., and Seifoddini, H. (1990). Production Data Based Similarity Coefficient for Machine-Component Grouping Decisions in the Design of a Cellular Manufacturing System, *International Journal of Production Research*, Vol. 28, No. 7, pp. 1247–1269.

22. Hayes-Roth, F., Waterman, D. A., and Lenat, D. B. (1983). *Building Expert Systems*, Addison-Wesley, Reading, MA.

23. Huang, P. Y., and Houck, B. L. W. (1985). Cellular Manufacturing: An Overview and Bibliography, *Production and Inventory Management*, No. 4, pp. 83–93.

24. Hyer, N. L., and Wemmerlöv, U. (1985). Group Technology Oriented Coding Systems: Structures, Applications, and Implementation, *Production and Inventory Management*, No. 2, pp. 55–78.

25. Hyer, N. L., and Wemmerlöv, U. (1989). Group Technology in the U.S. Manufacturing Industry: A Survey of Current Practices, *International Journal of Production Research*, Vol. 27, No. 8, pp. 1287–1304.

26. Kalkunte, M. V., Sarin, S. C., and Wilhelm, W. E. (1986). Flexible Manufacturing Systems: A Review of Modeling Approaches for Design, Justification and Operation. In: *Flexible Manufacturing Systems: Methods and Studies*, edited by A. Kusiak, North-Holland, Amsterdam, pp. 3–25.

27. Kimemia, J. G., and Gershwin, S. B. (1978). Multicommodity Network Flow Optimization in Flexible Manufacturing Systems. In: *Complex Materials Handling and Assembly Systems Final Report*, Vol. II, No. ESL-FR-834-2, Electrical Systems Laboratory, MIT, Cambridge, MA.

28. King, J. R. (1980). Machine-Component Grouping in Production Flow Analysis: An Approach Using Rank Order Clustering Algorithm, *International Journal of Production Research*, Vol. 18, No. 2, pp. 213–232.

29. Kriz, J. (1987). Knowledge-Based Systems in Industry: Introduction. In: *Knowledge-Based Systems in Industry*, edited by J. Kirz, Halsted Press, New York.

30. Kusiak, A. (1985). Flexible Manufacturing Systems: A Structural Approach, *International Journal of Production Research*, Vol. 23, No. 6, pp. 1057–1073.

31. Kusiak, A., and Heragu, S. S. (1987). Expert Systems and Optimization in Automated Manufacturing Systems, Working Paper No. 07/87, Department of Mechanical and Industrial Engineering, University of Manitoba, Winnipeg, Manitoba, Canada.

32. Lee, H. F., Srinivasan, M. M., and Yano, C. A. (1989). Algorithms for the Minimum Cost Configuration Problem in Flexible Manufacturing Systems, Technical Report 89-1, Department of Industrial Engineering & Operations Engineering, University of Michigan, Ann Arbor.

33. McAuley, J. (1972). Machine Grouping for Efficient Production, *The Production Engineer*, Vol. 51, No. 2, pp. 53–57.

34. McCormick, W. T., Schweitzer, P. J., and White, T. E. (1972). Problem Decomposition and Data Recognition by a Clustering Technique, *Operations Research*, Vol. 20, pp. 993–1009.

35. Mellichamp, J. M., and Wahab, A. F. A. (1987). An Expert System for FMS Design, *Simulation*, Vol. 48, No. 5, pp. 201–208.

36. Mellichamp, J. M., Kwon, O., and Wahab, A. F. A. (1990). FMS Designer: An Expert System for Flexible Manufacturing System Design, *International Journal of Production Research*, Vol. 28, No. 11, pp. 2013–2024.

37. Mosier, C., and Taube, L. (1985). The Facets of Group Technology and Their Impacts on Implementation—A State-of-the-Art Survey, *OMEGA The International Journal of Management Science*, Vol. 13, No. 5, pp. 381–391.

38. Mutti, R., and Semeraro, Q. (1987). A Heuristic Method to Group Pieces for Economical Production, Working Paper WP-1987-001-QS, Department of Mechanics, Polytechnic Institute of Milan, Milan, Italy.

39. Nagi, R., Harhalakis, G., and Proth, J.-M. (1990). Multiple Routings and Capacity Considerations in Group Technology Applications, *International Journal of Production Research*, Vol. 28, No. 12, pp. 2243–2257.

40. Olivia-Lopez, E., and Purcheck, G. F. (1979). Load Balancing for Group Technology Planning and Control, *International Journal of Machine Tool Design and Research*, Vol. 19, No. 4, pp. 259–274.

41. Opitz, H. (1966). *Verschlüsselungsrichtlinien und Definitionen zum werkstückbeschreibenden Klassifizierungssystem*, Essen, Federal Republic of Germany.

42. Rajagopalan, R., and Betra, J. L. (1975). Design of Cellular Production Systems: A Graph-Theoretic Approach, *International Journal of Production Research*, Vol. 13, No. 6, pp. 567–579.

43. Rajamani, D., Singh, N., and Aneja, Y. P. (1990). Integrated Design of Cellular Manufacturing Systems in the Presence of Alternative Process Plans, *International Journal of Production Research*, Vol. 28, No. 8, pp. 1541–1554.

44. Sarin, S. C., and Chen, C. S. (1986). A Mathematical Model for Manufacturing System Selection. In: *Flexible Manufacturing Systems: Methods and Studies*, edited by A. Kusiak, North-Holland, Amsterdam, pp. 99–112.

45. Schneeweiß, C. (1983). Elemente einer Theorie betriebswirtschaftlicher Modellbildung, *Zeitschrift für Betriebswirtschaft*, Vol. 54, No. 5, pp. 480–504.

46. Secco-Suardo, G. (1978). Optimization of Closed Queueing Networks. In: *Complex Materials Handling and Assembly Systems Final Report*, Vol. III, No. ESL-FR-834-3, Electrical Systems Laboratory, MIT, Cambridge, MA.

47. Seifoddini, H. (1984). Cost Based Machine-Component Grouping Model. In: *Group Technology*, Ph.D. Thesis, Oklahoma State University, Stillwater.

48. Shafer, S. M., and Meredith, J. R. (1990). A Comparison of Selected Manufacturing Cell Formation Techniques, *International Journal of Production Research*, Vol. 28 No. 4, pp. 661–673.

49. Shanthikumar, J. G., and Yao, D. D. (1987). Optimal Server Allocation in a System of Multi-server Stations, *Management Science*, Vol. 33, No. 9, pp. 1173–1180.

50. Shanthikumar, J. G., and Yao, D. D. (1988). On Server Allocation in Multiple Center Manufacturing Systems, *Operations Research*, Vol. 36, No. 2, pp. 333–342.

51. Shanthikumar, J. G., and Yao, D. D. (1989). Optimal Buffer Allocation in a Multicell System, *The International Journal of Flexible Manufacturing Systems*, Vol. 1, pp. 347–356.

52. Solot, P. (1987). Optimizing a Flexible Manufacturing System with Several Pallet Types, Working Paper O.R.W.P. 87/17, Department of Mathematics, Federal Polytechnic School of Lausanne, Lausanne, Switzerland.

53. Stecke, K. E. (1985). Design, Planning, Scheduling, and Control Problems of Flexible Manufacturing Systems, *Annals of Operations Research*, Vol. 3, No. 3, pp. 3–12.

54. Suri, R. (1985). An Overview of Evaluative Models for Flexible Manufacturing Systems, *Annals of Operations Research*, Vol. 3, pp. 13–21.

55. Tetzlaff, U. (1990). *Optimal Design of Flexible Manufacturing Systems*, Physica-Verlag, Heidelberg.

56. Vinod, B., and Solberg, J. J. (1985). The Optimal Design of Flexible Manufacturing Systems, *International Journal of Production Research*, Vol. 23, No. 6, pp. 1141–1151.

57. Whitney, C. K., and Suri, R. (1985). Algorithms for Part and Machine Selection in Flexible Manufacturing Systems, *Annals of Operations Research*, Vol. 3, pp. 239–261.

58. Yao, D. D., and Shanthikumar, J. G. (1987). The Optimal Input Rates to a System of Manufacturing Cells, *INFOR*, Vol. 25, No. 1, pp. 57–65.

59. Zoutendijk, G. (1976). *Mathematical Programming Methods*, Elsevier Publishing, New York.

CHAPTER TEN

Technology Selection and Capacity Planning for Manufacturing Systems

SHANLING LI and DEVANATH TIRUPATI

Management Department, College of Business Administration,
The University of Texas at Austin

10.1. INTRODUCTION

Capacity expansion decisions in most industries usually involve substantial capital investments and have received considerable attention from both academicians and practitioners. These decisions typically require an understanding of the trade-offs between several related factors and cannot be made in isolation. For example, process and manufacturing technologies in chemical, electric power, fertilizer, engineering, and communication industries are highly capital intensive and exhibit substantial scale economies. The availability of alternative technologies suggests that these choices should be made together with the expansion decisions. For firms facing demands distributed over geographical regions, plant location and expansion decisions are linked and the trade-off is between investment and transportation costs. The reader may note that any combination of these factors together with dynamics of product demands make technology choice and expansion decisions extremely complex.

In most process industries, automation and integration have been continuing trends for decades and the methodologies and models for supporting capacity and technology decisions accommodate their more important characteristics. However, in discrete goods manufacturing, recent developments in modern technologies such as flexible manufacturing systems (FMSs), computer integrated manufacturing (CIM), computer aided design (CAD), and flexible automation (FA) permit production of a wide variety of products with small changeover costs. Increased competition in the

Intelligent Design and Manufacturing, Edited by Andrew Kusiak.
ISBN 0-471-53473-0 © 1992 John Wiley & Sons, Inc.

marketplace, particularly from the overseas manufacturers, has resulted in short product cycles and has put a premium on flexibility in changing product mix in a dynamic fashion. Together, these factors encourage investments in facilities capable of producing several product families. The presence of economies of scale introduces additional complexity, making expansion decisions even more intricate.

Our objective is to provide the reader with an overview of the methodologies that are available for supporting technology and capacity decisions. The chapter is directed toward the practitioner with modeling interest and the researcher with application focus. The goal is to provide the reader with a flavor of the research in the area and describe its role in making technology choices and expansion decisions. Thus we concentrate on modeling issues and applications rather than on theoretical results. [For a comprehensive review of the literature focusing on the economic aspects of evaluation of flexible technologies, we refer the reader to a recent survey by Fine (15).] Since the resulting decision problems are complex and difficult to solve optimally, we elaborate on heuristic procedures and approximation methods that have been developed in this context.

The chapter is organized as follows: In section 10.2 we describe the major factors that play a key role in technology and capacity choices. Section 10.3 is devoted to models and methodologies that have been developed to support these strategic and tactical decisions. Several applications of these methods are discussed in Section 10.4. Finally, we conclude in Section 10.5 with a summary of outstanding issues.

10.2. CHARACTERISTICS OF THE TECHNOLOGY SELECTION AND CAPACITY EXPANSION PROBLEM

As indicated in the introduction, a number of factors influence strategic choices related to the technology selection and capacity additions. These include, among others, product mix characteristics, technology alternatives, cost parameters, and length of the planning horizon. In this section, we briefly describe the key factors of the capacity planning problem.

10.2.1. Product Mix

The range of products manufactured by a plant is a function of the production technology used and, to a large degree, determines the complexity of the planning problem. The simplest case is represented by a firm producing a single homogeneous product. Electricity generation in the power industry is a classic example that fits this case. The single product model can also be used for firms producing several variants of a basic product line. A stable product mix in such cases will permit aggregation of all variants into a single product. Several examples in the chemical and process industry fit such a model fairly well. In contrast to the fairly homogeneous products in the foregoing examples, the product mix in discrete part manufacture is quite diverse and dynamic. Between these extremes, a wide range of product mix patterns can be found. For example, in the oil refining industry the product mix is determined primarily by the choice of inputs and by the processing technology. Similarly, in the fertilizer industry, the product mix may consist of substitutable products (e.g., coal- and oil-based fertilizers) that are produced from entirely different sets of inputs and production processes. In such cases, the planning problem may be modeled as a single product case with alternate technologies.

Product mix plays an important role in technology selection because many factors related to product mix may complicate these decisions. In the case of a single product, the decision in technology choice focuses merely on the cost structure of different technologies since all of them have unique function to produce one product (6). When a firm produces multiple products, many factors, such as operational flexibility (dedicated, semiflexible, and flexible) and cost of alternative technologies, are involved in the decisions. Evaluation of the trade-offs between functions and costs of technologies becomes extremely difficult with a large number of products and a dynamic product mix.

10.2.2. Demand

Invariably, technology and capacity decisions are driven by product demands that may be characterized along several dimensions. This multidimensional attribute is perhaps the single most important factor that determines the investment levels and the amount of capacity additions. In most planning problems, product demands are assumed to be given (perhaps based on market analysis) and the objective is to develop medium- or long-term plans to meet this demand. However, in some integrative approaches, the demand levels are treated as decision variables to be determined in conjunction with choices related to production facilities. Since the former case is more common, in this chapter, we focus mostly on models and methods in which demand is considered as an exogenous parameter. Figure 10.1 provides a summary of the various factors that determine product demand characteristics. In what follows, we briefly elaborate on some of these aspects.

One of the key features that determine the complexity of the planning problem relates to the modeling of demand as either a stochastic or deterministic process. Clearly, actual demand is almost always uncertain and is stochastic. However, deterministic models are often easier to analyze and may be adequate for planning purposes. For example, for mature products with not very substantial random fluctuations in demand, it may be sufficient to make planning decisions based on average demands. The changes in demand can be met by intermediate-term production planning with inven-

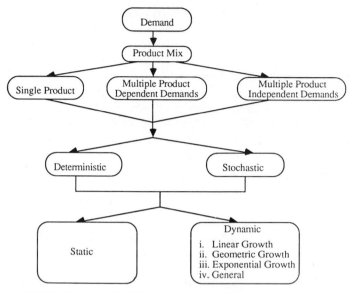

FIGURE 10.1. Multidimensioned attributes of demand.

tories and short-term capacity changes. This approach has an additional advantage that data requirements are consistent with the information normally available/collected by firms and with the decision hierarchy. We observe that while deterministic models are suitable for most applications, they do not capture some of the key flexibility benefits provided by modern integrative technologies such as CIM and FMSs. In such situations, it is desirable to model explicitly the effect of uncertainties in demand and the role of flexibility. (This aspect is discussed in detail in Section 10.3.)

A second aspect that influences capacity decisions is its behavior over time. In the simplest case, demand is time invariant and one-time decisions are sufficient (with provision for depreciation and loss of capacity). However, demands are often dynamic and the pattern of demand becomes a key factor. Figure 10.2 describes several patterns that are commonly encountered in practice. Clearly, well established growth patterns such as linear or geometric growth are easier to analyze. However, responsiveness to markets may give rise to general demand patterns as outlined in Fig. 10.2.

In environments with multiple products, the situation is further complicated by the interproduct influences on demand. In the simplest case, product demands may be considered independent with production technology as the common factor. However, it is more common for facilities to manufacture products with either positively or negatively correlated demands. The former case can be observed in firms producing automobile components for the same type of cars. An example of negatively correlated demands can be found in the consumer electronics industry producing a variety of substitutable products.

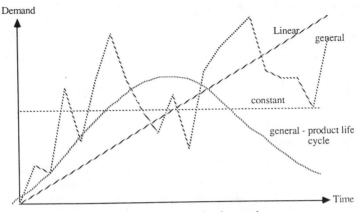

FIGURE 10.2. Some sample demand patterns.

10.2.3. Technology Choices

Traditionally, large volume mass production and low unit costs have been the prime motivators in the development of manufacturing technologies. These objectives have resulted in dedicated technologies with specialized equipment designed to produce a limited range of products in an efficient manner. Automobile assembly lines are classic examples of such production systems. However, in recent years there has been a growing trend in the development of integrated technologies that provide a great deal of operational flexibility. These are typically characterized by automation and computer controlled equipment to facilitate switch-over between products and thus permit production of a variety of products. Other benefits of these modern technologies include quality improvement and cost reduction.

As noted in the introduction, the development of integrative technologies has been a continuing phenomenon in process industries for several years and the methodologies for their evaluation incorporate all the relevant features. In contrast, in the fabrication and discrete part manufacturing industry, the characteristics of the modern technologies are somewhat different. Furthermore, these developments are fairly recent and their features have not been fully captured in the evaluation models. This, perhaps, is a major weakness of the methodologies that are currently available for supporting capacity and technology decisions.

We note that between the two extremes of dedicated and flexible technologies, a wide range of production processes are available. For example, it is quite common to find technologies with limited flexibility that are capable of producing a subset of products. Another variant of flexible technology consists of essentially dedicated plants that are amenable for conversion from one type to another. Such conversions typically involve one-time investment costs and, in some cases, may be irreversible. Examples of convertible technologies can be found in the communication industry (cable production) and in the energy sector (furnaces capable of conversion from coal to oil and vice versa).

In summary, it is important to understand and model the key features of technology alternatives as they relate to capacity and equipment choices. This modeling aspect requires considerable knowledge and judgment and is important since it determines the relevance of the results provided by the model.

10.2.4. Cost Function

Clearly, cost is one of the important factors in technology selection. Linear functions are most popular and continue to be used over a wide range of applications. The primary motivation for their use is ease of analysis. Also, they are appropriate in cases with no significant scale economies. However, it is well known that in several industries investments in plant and equipment exhibit substantial scale economies (42, 46). Concave functions, which permit modeling of several types of scale economies, have been extensively used in the development of analytical models for capacity planning. In practice, two specific forms described below have been very useful in capturing cost characteristics exhibited by several production technologies.

Power function

$$f(x) = Kx^{\alpha} ,$$

where $f(\cdot)$ is the cost function, x the amount of capacity addition $(x \geq 0)$, K a constant, and α the economies of scale parameter $(0 < \alpha \leq 1)$.

The power function has been found to be appropriate for process industries with significant scale economies such as chemicals, oil refining, fertilizers, and cement. A low value of α indicates large economies of scale and vice versa. For example, Manne (46) suggests a value of α between 0.6 and 0.8 for such cases.

Fixed Charge Cost Function

$$f(x) = \begin{cases} 0 & \text{if } x = 0 \\ F + vx & \text{if } x > 0 , \end{cases}$$

where F and v, respectively, are the fixed and variable costs in investment.

This form is particularly suitable for engineering and fabrication industries, where capacity may be added in relatively small increments. It is also useful in incorporating administrative, financial, and other expenses that are independent of the amount of capacity addition.

It should be mentioned that it is not necessary that the cost functions should be stationary over the planning horizon. This is an interesting and desirable feature in some industries since technological advances could lead to cost reductions in the future.

10.2.5. Planning Horizon

In the short run, technology and equipment are fixed and capacity planning relates to workforce planning and decisions such as overtime operation, subcontracting part of the demand, and production smoothing with inventory buildup. In contrast, technology and equipment choices are long-term decisions with the length of the planning horizon depending on many factors that include maturity of technology, level of investment, uncertainties in demands, costs, and dynamics of product mix.

Thus, for environments with stable, mature product mix and well defined capital intensive technology choices, long planning horizons are appropriate. For a rough cut analysis in such cases, continuous time, infinite horizon models provide quick solutions with little computational effort (details are described in Section 10.3). With more dynamic demands and several choices in capacity and technology, discrete time and finite horizon models that explicitly consider the alternatives are required to describe the relevant trade-offs. Typically, these problems are harder to solve and require specialized algorithms.

10.3. JUSTIFICATION APPROACHES FOR TECHNOLOGY SELECTION

In the discrete part manufacturing industry, recent developments in integrative technologies such as CIM and FMSs have made it possible to design plants with planning horizons longer than individual product life. This is particularly attractive since current trends suggest shortening product life in the market. As a result, manufacturing capabilities have become more important in defining the strategic position of the firm. However, a number of empirical studies indicate that while most managers have good conceptual understanding of the benefits of modern technologies, methods for their evaluation and adoption are rather inadequate. In this section we describe approaches for evaluation of alternative technologies and for adoption of investment and capacity plans. We focus on the role of analytic models in supporting these high-level managerial decisions. Accordingly, we start in Section 10.3.1 with a brief description of methods commonly employed in practice and point out major weaknesses in capturing the key trade-offs. The remainder of the section is devoted to the description of optimization models and solution approaches. Section 10.3.2 focuses on models for the single product case while Sections 10.3.3 and 10.3.4, respectively, describe models for two and multiple product families.

10.3.1. Practical Approaches

Discounted cash flow methods such as net present value (NPV) and internal rate of return (IRR) are still some of the most popular methods employed in practice for project selection and capacity planning. While these methods have been used successfully in ranking projects and in developing capital budgets, it is important to recognize that these approaches provide only marginal analysis. The marginal analysis focuses only on the values of incremental cash flows and does not reflect many benefits incorporated in advanced technologies. For example, flexible manufacturing systems (FMSs) have the benefits of reducing setup times, increasing production efficiency and improving manufacturing flexibility. Some of these benefits cannot be measured by cash flows and hence there is a need for more comprehensive methods that take into account all the relevant factors. In fact, much has been written recently (1, 7, 8, 33) on the difficulties of measuring manufacturing performance with traditional cost accounting methods. Researchers observed that many firms made their investments with either inadequate or no justification. Kaplan (34) calls this prevailing practice of justification of investment in modern manufacturing technology "justification by faith."

The need for appropriate approaches to measure the benefits of modern technologies has become critical. For example, based on an annual survey of large manufacturers in North America, Europe, and Japan, Tombak and Meyer (55) and Meyer et al. (50) conclude that American firms are lagging behind in their efforts to overcome the trade-offs between manufacturing flexibility and cost efficiency. Meredith and Suresh (48) observe that failure of justification of modern technologies leads to rejection of many worthwhile projects. These observations are consistent with those of Jaikumar (30), who found that, in comparison with the Japanese, U.S. manufacturers installed fewer flexible systems. This is partly due to the high costs of acquisition of flexible systems and partly due to lack of appropriate evaluation methods.

10.3.2. Modeling Approaches with Single Product

The issue of technology management has become particularly relevant to operations planners as a result of the introduction of new computer aided process technologies in recent years. In order to select from a set of available technologies, management must develop systematic procedures for analyzing the alternatives and determine the relevant trade-offs. In the remainder of this section we describe a number of models and paradigms that have been developed in the fields of operations management, management science, and economics to address related issues.

The best known single-product model is due to Manne (45, 46), who examined the planning problem with linearly increasing demand and stationary costs (except for discounting) over an infinite horizon. The basic trade-off is between economies of scale in investment costs (captured by concave cost functions) and the benefits due to delay in capacity additions. Manne's main result is to show that the optimal policy requires additions of capacity at regular intervals. This implies that capacity is added by building plants of fixed size and the planning problem reduces to determination of the optimal plant size (or, alternately, the frequency of capacity additions). Figure 10.3 shows the process of demand growth and capacity expansion over time. It is interesting to note the similarity between the amount of excess capacity and inventory in the EOQ (economic order quantity) model. Using Manne's characterization, the total discounted cost over the infinite horizon can be expressed as a function of the frequency of capacity addition in the following manner:

$$C(x) = \frac{f(xD)}{1 - e^{-rx}} , \qquad (1)$$

where D is the annual increase in demand (units/year)/year, x the time interval between successive plants (years), r the annual discount rate, compounded continuously (1/year), and $f(\cdot)$ the investment cost function.

The optimal plant size (xD) and the corresponding frequency of capacity additions (x) is determined by minimizing (1). For the special case in which investment cost function is the power function (described in Section 10.2), optimal x (x^*) is obtained as a solution to the following transcendental equation:

$$\alpha(e^{rx^*} - 1) = rx^* . \qquad (2)$$

It is interesting to note that the frequency of capacity additions depends only on the discount rate (r) and economy of scale parameter (α) and is

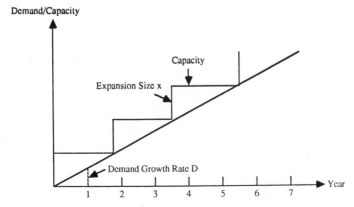

FIGURE 10.3. Process of demand growth and capacity expansion.

independent of the magnitude of demand growth (D). Solution of Eq. (2) can be simplified by tabulating the values of x^* as a function of r and α.

Srinivasan (54) extended Manne's results to the case in which demand growth is geometric rather than linear. Using an approach similar to that of Manne, Srinivasan showed that it is still optimal to construct plants at constant intervals of time. However, because of the geometric growth in demand, the plant sizes will not remain constant but will grow exponentially.

The continuous time models described so far suffer from two major drawbacks that make them appropriate only for a rough cut analysis. First, demand growth may not always follow a regular pattern such as linear or geometric. Second, standardization of plants and equipment may restrict choices to a few discrete alternatives. To address these and other related issues, a number of finite horizon, discrete period models have been developed in the literature. The choice of discrete time periods poses no problem because the interval lengths can be as small as needed. In fact, discrete time models with an interval length of a quarter or year are consistent with budget practices of most firms.

One of the earliest finite horizon models is due to Manne and Veinott (47), who consider the capacity planning problem with general demand growth. However, they assume that the size of capacity additions is unrestricted. This model is identical to that developed by Wagner and Whitin (58) in the context of lot sizing in production planning. Work related to capacity planning with discrete alternatives has been reported, among others, by Erlenkotter (11, 12) and Neebe and Rao (52). In what follows, we discuss in detail the model due to Neebe and Rao (52). Our objective is to describe the scope of such models and provide the reader with a flavor of the research in this area.

Notation

T	Planning horizon
I	Finite set of n expansion projects, $I = \{1, 2, \ldots, n\}$
z_i	Capacity of project i; $z_i > 0$ and may be brought on stream at the start of any period $t(t = 1, 2, \ldots, T)$
c_{it}	Investment cost for project i in period t
d_t	Demand in period t, $t = 1, 2, \ldots, T$
R_t	$\max_{\tau = 1, t} d_\tau$
r_t	Incremental demand in period t, $r_t = \max\{0, d_t - R_{t-1}\}$
x_{it}	$= \begin{cases} 1 & \text{if project } i \text{ is brought on stream at the start of period } t \\ 0 & \text{otherwise} \end{cases}$
y_t	Idle capacity in period t

Model [NR]

$$\text{Minimize} \quad \sum_{i=1}^{n} \sum_{t=1}^{T} c_{it} x_{it}$$

subject to

$$\sum_{t=1}^{T} x_{it} \leq 1 \qquad\qquad i \in I , \tag{3}$$

$$\sum_{i=1}^{n} z_i x_{it} + y_t - y_{t+1} = r_t \qquad t = 1, 2, \ldots, T , \tag{4}$$

$$\sum_{i=1}^{n} x_{it} \leq 1 \qquad\qquad t = 1, 2, \ldots, T , \tag{5}$$

$$x_{it} = 0 \text{ or } 1 \qquad\qquad i \in I, t = 1, 2, \ldots, T , \tag{6}$$

$$y_t \geq 0 \qquad\qquad t = 1, 2, \ldots, T . \tag{7}$$

In the formulation above, constraint (3) ensures that each project can be selected at most once over the planning horizon and constraint (4) guarantees that demand is satisfied in each period. Constraint (5) restricts the number of projects chosen in each period. Note that this model is useful not only in project selection but also in choosing technologies. In the latter case, a project can model technology choice with a specified capacity. For technologies with alternative capacity choices, as many projects as necessary may be defined.

Neebe and Rao considered a special case of [NR] with integral demands and capacities. While the resulting integer program is not particularly amenable for obtaining optimal solution, it allows the authors to develop a Lagrangian relaxation procedure that is effective for moderate sized problems that are representative of real life applications. It is also interesting to note that the NR model is sparse in data requirements. Furthermore, it is consistent with accounting methods and costs c_{it} can incorporate the cash flows associated with each project. The model can also be modified easily to capture additions in a phased manner.

Multiregion Models. When demand is distributed over a geographical area and transportation costs are substantial, it might be economical to build a large number of small plants—each located close to a different market area. The smaller plants would be unable to take advantage of economies of scale in manufacturing but would be able to serve their local markets at low transportation costs.

The single region models described in the previous section provide a natural starting point and motivated extensions to multiple regions. For example, Erlenkotter (10, 13) examined the two region, infinite horizon problem with linear demand growth and characterized the structure of the optimal expansion plans. Based on these properties, he developed an efficient dynamic programming formulation and algorithms for deriving optimal investment plans. Extensions to the multiple region problem are described in Erlenkotter (13). Fong and Rao (17) considered the two region problem with general demand patterns and used a discrete time, finite horizon model to address the trade-offs between scale economies and transportation costs. Using a network representation of the problem, they present efficient algorithms to obtain optimal solutions. However, these procedures do not extend to the multiple region problem, and the focus of subsequent research shifted to the development of heuristics. In what follows, we describe, in detail, the model and the results due to Fong and Srinivasan (18–20) to illustrate these developments.

Notation

L	The set of time periods
I	The set of producing regions
I'	The same set as I except that it has one more dummy supply region
K	The sets of markets
K'	The same set as K except that it has one more dummy market
x_{ik}^t	Amount shipped from region i to market k during period t
c_{ik}^t	Unit cost of production in region i and shipment to market k in period t
z_i^t	Amount of capacity expansion in region i in period t
$f_i^t(z_i^t)$	Investment cost for establishing a capacity of size z_i^t at region i in period t
q_i^0	Initial capacity at region i
d_k^t	Demand in region k in period t

Model [FS]

$$\text{Minimize} \quad \sum_{t \in L} \sum_{i \in I'} \sum_{k \in K'} c_{ik}^t x_{ik}^t + \sum_{i \in L} \sum_{i \in I'} f_i^t(z_i^t)$$

subject to

$$\sum_{k \in K'} x_{ik}^t = q_i^0 + \sum_{\tau=1}^{t} z_i^\tau \qquad \text{for } i \in I', t \in L , \qquad (8)$$

$$\sum_{i \in I'} x_{ik}^t = d_k^t \qquad \text{for } k \in K', t \in L , \qquad (9)$$

$$x_{ik}^t, z_i^t \geq 0 \qquad \text{for } i \in I', k \in K', t \in L . \qquad (10)$$

In the model above, constraint (8) states that the total shipment (including the shipment to the dummy market) from region *i* at *period t* would be equal to the initial capacity plus the cumulative expanded capacity up to period *t*. Constraint (9) ensures that the supply from all regions (including the dummy supply) satisfies the demand in market *k* at period *t*. The model FS is a multiperiod version of the capacitated warehouse location problem and thus belongs to the class of NP-hard problems. Since the prospect of developing efficient algorithms to obtain optimal solutions for such problems is not very encouraging, the authors focus on heuristics to derive good expansion schedules. In the first stage, an initial solution is obtained by solving a sequence of transportation problems that represent one-period versions of FS. In the second stage, improvements are obtained by considering movements of capacity expansions among regions and/or time periods. Essentially, the improvements are myopic in nature. It may be noted that these heuristics evolved from the earlier work of Hung and Rikkers (27) and Rao and Rutenberg (53).

It is interesting to note that the multiregion model (FS) can be modified to capture some of the issues related to technology selection in a multiproduct environment. This may be done by defining regions to represent products and investments in a region to denote use of dedicated technology for that product. Transportation costs may be used to capture product mix flexibility characteristics and to define conversion costs. While this scheme is attractive in principle, we note that its application is rather limited. This is primarily due to the fact that the heuristics of Fong and Srinivasan do not extend to this general case and additional research is required to exploit these features. For a further elaboration of these aspect, we refer the reader to Li and Tirupati (38).

Models with Technology Improvement. Continuous and, in some cases, dramatic improvements in technology are key characteristics in electronics, semiconductor, computer, and other high-technology industries. Investment in new technologies in these industries has been motivated primarily by technical breakthrough in process technologies rather than economic analysis. This is partly due to the lack of suitable models that address issues related to technology improvement. A limited amount of literature has been reported on this topic. In this section, we provide a brief overview of the modeling literature on the subject.

Hinomoto (26) presented one of the earliest models that explicitly incorporate technology improvements. Hinomoto considers two forms of improvement: (a) reduction in investment costs and (b) reduction in input requirement, which in turn leads to reduction in operating costs. These features are modeled in the following manner:

Investment Cost. Let $K(z)$ denote the cost of adding z units of capacity at time 0. The corresponding cost for addition at time t is then given by $K(z) e^{-kt}$. The factor e^{-kt} captures the effect of technology improvement, and k is the improvement constant.

Operating Cost. The operating cost consists of two components—number of units of input required and the price of the input. The product of the two terms results in operating costs. Hinomoto assumes that unit price of the input is exogenously specified by a dynamic function, $q(t)$. The input requirements depend on the production level (y) and capacity (z) and are specified by a production function $X[y, z]$. The production function is defined with the technology at time 0 as the reference. Reductions in requirements due to technological improvements are modeled in a manner similar to investment costs with an improvement rate h. Thus operating cost at time t due to production level $y(t)$ with a capacity z installed at time t_1 is given by $q(t)X[y(t), z] e^{-ht_1}$.

In Hinomoto's model, there are no explicit restrictions on demand and it is assumed that all production is sold. However, unit price of the product is a function of the total production and thus indirectly models the demand process. It is interesting to note that the model permits a dynamic price function. In the simplest case, termed as one-step model, Hinomoto examines the problem with exactly one opportunity to invest in new capacity. The initial conditions are specified by the age and amount of capacity available at time 0. The objective is to determine (a) amount and time of new investment, (b) the optimal length of the production interval (planning horizon), and (c) the production plan during this plan horizon. It is interesting to note that, with this specification, the planning horizon can be partitioned into two periods (of possibly unequal lengths)—before and after the investment in new technology. The two periods are referred to as Period 0 and Period 1 and within each period capacity remains constant. We now describe the model in detail. Let

n	Number of alternative technologies
t_0	Purchase time of the existing facility ($t_0 < 0$)
t_1	Time the new facility is purchased
z_0	Size of the existing facility
z_1	Size of new facility
$y_0(t)$	Total amount of production at time t in period 0
$y_1(t)$	Total amount of production at time t in period 1
$y_{1,1}(t)$	Amount produced by the new facility at time t
T	Planning horizon
r	Discount factor

Hinomoto developed the profit function for the one-step model as follows:

$$G = \int_0^{t_1} e^{-rt}\{P[y_0(t), t]y_0(t) - q(t)X[y_0(t), z_0] e^{-ht_0}\}\, dt$$

$$+ \int_{t_1}^{T} e^{-rt}[P[y_1(t), t]y_1(t)] - q(t)\{X[y_1(t) - y_{1,1}(t), z_0] e^{-ht_0}$$

$$+ X[y_{1,1}(t), z_1] e^{-yt_1}\}]\, dt - K(z_1) e^{-(r+k)t_1}. \qquad (11)$$

In Eq. (11), the first term represents the net profit from period 0 to t_1, the second term represents the revenue from t_1 to T less operating costs of existing capacity and new facility, and the third term is the investment cost at t_1. Thus G denotes the present value of the discounted profit. For cases in which G can be treated as a continuous function of the decision variables— T, t_1, z_1, $y_0(t)$, $y_1(t)$, $y_{1,1}(t)$—Hinomoto derives first-order conditions that are necessary for optimality. He also provides an economic interpretation for these conditions and develops some interesting characterizations of the relationship between these variables. While it is not easy to obtain closed form expressions for optimal values of the decision variables, it is rather straightforward to solve the necessary conditions and obtain numerical values in specific instances. (Several standard computer programs may be used in these computations.) We note that the necessary conditions are sufficient for optimality for cases in which the profit function G is concave.

In recent years, work related to technology choice in the presence of improvement has been reported in the context of both monopolistic and competitive situations. For example, Klincewicz and Luss (35) examined timing decisions for introduction of new technology facilities. They first examine linearly growing demand and then extended the analysis to non-linear demand. Boyd et al. (3) developed a model of the share of an energy product market served by each technology of a set of competing technologies. The objective is to minimize product cost. Gaimon (23) developed an optimal control model in which the strategic decision concerning the optimal acquisition of flexible technology is examined. Specifically, the model focuses on a profit maximizing firm that optimally derives its prices, its level of output, and its level and composition of productive capacity over time. Gaimon (24) considered competition between two firms and examined its impact on new technology adoption and capacity addition.

Models with Stochastic Demand. The literature on technology evaluation models with stochastic demand is quite sparse and fairly recent. Some of the early work in this area is due to Freidenfelds (21) and Whitt and Luss (59). Freidenfelds showed that when demand process is a time-homogeneous Markov process and capacity expansion cost is time-invariant, the stochastic model can be reformulated as an equivalent deterministic model. Whitt and Luss computed capacity utilization when demand growth follows geometric Brownian motion. The demand process in these models is quite specialized and has limited potential for application. More recently, Cohen and Halperin (6) examined a more general model (CH) that considers technology alternatives as well. In this section, we describe this model in some detail.

In the CH model, the authors use a discrete period, finite horizon model to examine the trade-off between alternative technologies. Each technology is characterized by a dynamic purchase cost and an age-dependent salvage value. The operating costs are comprised of a fixed component and a variable cost that is linear in production volume. In addition, the capacity of each alternative is specified so that capacity decisions are not included in this model. An interesting feature of the CH model is provision for dynamic demands. The demand process is described by the specification of its probability density function for each period. The unit price is assumed to be deterministic but need not be constant over the planning horizon. Cohen and Halperin do not permit carry-over of inventories and/or shortages between periods. Thus shortages due to inadequate production lead to loss of revenue and potential profit. Likewise, excess production gives rise to inventories that cannot be used and incurs unnecessary operating costs.

A key feature of the CH model is the assumption that only one type of technology may be used in each period. Furthermore, as described earlier, technology choice defines the available capacity. Thus change in either technology and/or capacity requires retirement of existing capacity and investment in a new facility. These rather restrictive assumptions make the model unrealistic and limit its application. The objective is to maximize the expected profit over the planning horizon. The following notation is useful in describing the details of the model:

n	Number of alternative technologies
z_i	annual production volume in year i
D_i	Annual demand in year i (a random variable)
$F_i(\cdot)$	Cumulative distribution of demand in year i
$f_i(\cdot)$	Probability density function of demand in year i
r_i	Revenue per unit in year i
A_i	Set of technology alternatives available in year i
K_α	Annual fixed cost associated with technology α
V_α	Unit variable cost for technology α
$S_\alpha(t)$	Salvage value at age t for technology α
C_α^i	Purchase cost of technology α in year i
γ	$= 1/(1 + \rho)$, annual discount factor with discount rate ρ

The evolution of the CH model with uncapacitated technology alternatives may be described in three steps. First, note that a single-period version of the model reduces to n newsboy problems. Thus for a given production technology (say, α) in period i, the optimal production volume is given as

$$z_i^*(\alpha) = F_i^{-1}\left[\frac{(r_i - V_\alpha)}{r_i}\right].$$ (12)

The corresponding profit $\Pi_i(\cdot)$ may be computed as follows:

$$\Pi_i(z_i^*(\alpha), \alpha) = r_i\left\{\int_0^{z_i^*(\alpha)} uf_i(u)\,du + z_i^*(\alpha)[1 - F_i(z_i^*(\alpha))]\right\} - K_\alpha$$
$$- V_\alpha z_i^*(\alpha).$$ (13)

It is interesting to note that once the technology is given, the fixed cost

K_α does not figure in the production volume. However, it is included in the profit function. For the one-period problem, the technology that maximizes $\Pi_i(z_i^*(\alpha), \alpha)$ represents the optimal choice. A negative value for the corresponding $\Pi_i(\cdot)$ implies that the product is not profitable.

Extension of the above model to multiple periods constitutes the second phase. Thus given the technology (α) to be used over a finite number of periods (say, from i through $j - 1$, $j > i$), the objective is to determine the production levels in each of the periods $i, i + 1, \ldots, j - 1$. Again, the goal is to maximize the profits over this time interval. Since the capacity of each alternative is well defined and fixed, the corresponding profit function, $P_{ij}(\alpha)$, is computed rather easily by solving $(j - i)$ one-period problems and can be expressed as follows:

Let $P_{ij}(\alpha)$ denote maximum expected discount profit associated with using technology $\alpha \in A_i$ from periods i through $j - 1$ plus the discounted salvage value for α, which is received in period j when it is of age $j - i$.

$$P_{ij}(\alpha) = \sum_{t=i}^{j-1} \gamma^t \Pi_t(z_t^*(\alpha), \alpha) + \gamma^i S_\alpha(j - i) . \tag{14}$$

Computation of $P_{ij}(\alpha)$ for all feasible i, j combinations for each technology permits a dynamic programming formulation of the problem considered by Cohen and Halperin. (The reader may note that this third phase can also be described as the shortest path problem.) The recursive relation for a backward algorithm can be described as follows:

Let $W(i, \alpha)$ denote the profit realized by following an optimal policy from period i through T, given technology α is used in period i:

$$W(i, \alpha) = \max_{\substack{j > i \\ \beta \in A_j}} \{P_{ij}(\alpha) - \gamma^j C_\beta^j + W(j, \beta)\} \quad \text{with } W(T, \cdot) = 0 . \tag{15}$$

In addition to developing a solution procedure, Cohen and Halperin present an analysis to derive some qualitative insights for the technology choice problem in this restrictive environment. For example, they provide conditions under which it is not economical for a firm to switch to a technology with higher variable costs. Likewise, they also show that switch-over to a technology with lower variables costs depends on the probability distributions of demand, the relationship of selling price to variable cost, the relationship of fixed cost to contribution margin. The authors also extended their model to the case of finite capacity and inventory holding and shortage cost.

10.3.3. Modeling Approaches with Two Products

The presence of multiple products introduces several interesting issues and further complicates the technology selection problem. These relate to the alternative technologies available, product mix characteristics, and demand patterns. To obtain insights into the relevant trade-offs, researchers focused on the two-product problem and the results parallel the single-product case described in the previous section. The multiple-product problem with more than two products (described in Section 10.3.4) is considerably more complex and the contributions relate to the development of heuristics and approximation procedures.

As noted earlier, three types of technology describe the alternatives available for the two-product case: (a) dedicated technology capable of producing one product, (b) flexible technology that can produce both products, and (c) convertible technology that may be converted from one product to another. It is pertinent to note that flexible technology may be interpreted in two ways: (a) technology such as flexible manufacturing systems (FMSs) capable of producing the two products and (b) technology that produces only one product (presumably superior and more expensive) but can be used to meet the demand of the other. We now describe some key results that have been obtained in the context of two products.

Models with Deterministic Demand. As in the single-product case, early work examining capacity planning issues with two products focused on the infinite horizon problem with linearly growing demands. For example, Kalotay (31) considered the problem with two technology alternatives— dedicated for one of the products and flexible technology that can produce both products. Kalotay's analysis does not provide an optimal solution to the problem but presents lower bounds and characteristics of the optimal plan. Subsequently, Erlenkotter (10) and Freidenfelds (22), respectively developed exact (dynamic programming) and heuristic methods to solve the technology selection and capacity planning problem. A variant of this problem with convertible technologies has been examined by Kalotay (32), Merhaut (49), and Wilson and Kalotay (60). In this problem, in addition to investments in new capacities, the key decisions relate to timing and amount of conversion of technology.

In a series of articles, Luss (40, 41) examined the two-product finite horizon problem using a discrete period model. The model, which provides for convertible technologies, is formulated in the following manner.

Notation

i	Index for product families, $i = 1, 2$
d_{it}	Demand i in period t
R_{it}	$\max_{\tau=1,t} d_{i\tau}$
r_{it}	Incremental demand i incurred at period t, $r_{it} = \max\{0, d_{it} - R_{it-1}\}$
x_{it}	Amount of capacity addition of type i in period t (available for use in period t)

y_{it} Amount of capacity of type i converted to the other type in period t

I_{it} Amount of ideal capacity of type i at the beginning of period t (or, equivalently, the idle capacity at the end of period $t-1$, $t = 2, 3, \ldots, T+1$)

$c_{it}(x_{it})$ Investment cost of x_{it}; $c_{it}(x_{it})$ is a concave function

$g_{it}(y_{it})$ Conversion cost of y_{it}; $g_{it}(y_{it})$ is a concave function

$h_{it}(I_{it+1})$ Holding costs of I_{it+1} from t to period $t+1$; $h_{it}(I_{it+1})$ is a concave function

Model [LM]

$$\text{minimize} \quad \left\{ \sum_{t=1}^{T} \sum_{i=1}^{2} [c_{it}(x_{it}) + g_{it}(y_{it}) + h_{it}(I_{it+1})] \right\}$$

subject to

$$I_{it+1} = I_{it} + x_{it} + y_{jt} - y_{it} - r_{it} \quad i = 1, 2, j = 1, 2 \, (j \neq i); t = 1, 2, \ldots, T, \tag{16}$$

$$x_{it}, y_{it}, I_{it} \geq 0 \quad i = 1, 2, j = 1, 2 \, (j \neq i); t = 1, 2, \ldots, T, \tag{17}$$

$$I_{i1} = 0, I_{iT+1} = 0 \quad i = 1, 2, j = 1, 2 \, (j \neq i); t = 1, 2, \ldots, T. \tag{18}$$

The objective in the formulation above is to minimize the total cost incurred, specifically the investment, conversion, and holding costs over the planning horizon. Constraints (16) ensure that product demands are satisfied in each period and define the corresponding unutilized capacity. Luss' major contribution is to show that the model LM is equivalent to a network flow problem. Figure 10.4 describes this network representation of LM. The arcs into each node represent unutilized capacity in the previous period, expansion, and conversion from other capacity in this period; the arcs out of each node represent demand, unutilized capacity, and conversion into the other capacity in the period. While the network representation provides a useful structure, the solution of LM is not trivial due to nonlinearities in the cost functions. Using extreme point properties, Luss developed a dynamic programming approach that provides optimal solution for the planning problem with concave cost functions. As discussed earlier, concave functions are typical for most investments and hence the solution procedures have significant potential for applications. In concluding this section, we note that some recent work has focused on the two-product problem similar to LM with dedicated and flexible technology alternatives. We do not describe these results in this chapter, but refer the reader to Li and Tirupati (38) and references therein.

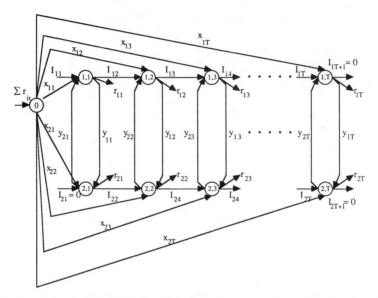

FIGURE 10.4. A network-flow representation for problem LS (Section 3.3).

Models with Stochastic Demand. Uncertainty in demand introduces one more dimension of complexity to the technology selection problem. The literature on the subject is rather sparse and fairly recent. Analytical approaches that provide for stochastic demand usually employ two simplifying assumptions: (a) linear costs and (b) static demand distribution. Thus, in these models, scale economies and demand dynamics are ignored. The key decisions in this environment may be described as a two-stage hierarchical process. In the first stage, technology choice and investment decisions are made, while capacity allocation and production decisions constitute the second phase. In a multiperiod model with static demands, the former is a one-time decision made at the beginning of the planning horizon while the allocation decisions are made in every period and depend on the realizations of product demands. In this section we describe the model examined by Gupta et al. (25) to illustrate the approach. Related work can be found in Hutchinson and Holland (28), Fine and Freund (16), Fine (15), and references therein. We also note that similar work has been reported by Dempster et al. (9) and Lenstra et al. (37) in an entirely different context of machine selection in a stochastic scheduling environment.

Notation

α Flexibility parameter, $0 \le \alpha \le 1$

$A, B, AB(\alpha)$ Index for the two dedicated and the flexible capacities, respectively

D_i	Demand for product $i (i = 1, 2)$ (random variable)
Q_i	Production of type i with flexible capacity $(i = 1, 2)$
Q	Vector of (Q_1, Q_2)
R_i	Net profit per unit of product i $(R_1 > 0, R_1 \geq R_2)$
K_A, K_B	Number of units of dedicated capacities purchased
$K_{AB(\alpha)}$	Number of units of flexible capacity purchased
C_A, C_B	Unit purchase prices of dedicated capacities A, B
$C_{AB(\alpha)}$	Unit purchase price of flexible capacity
P_i	Production of product i with dedicated capacity $(i = 1, 2)$
$f_{12}(\cdot), F_{12}(\cdot)$	Joint probability density and distribution functions of the demands for products 1 and 2
$f_i(\cdot), F_i(\cdot)$	Marginal probability density and distribution functions of the demands for product i, $i = 1, 2$

Modeling of flexibility characteristics is one of the interesting features of the model examined by Gupta et al. They define a flexibility factor—$T(Q; \alpha)$—to capture loss of capacity in flexible technology due to product switch-overs. This factor, which is a function of production levels (Q_1, Q_2) and the flexibility parameter α, is defined as follows:

$$T(Q; \alpha) = \left[\left(\frac{Q_1}{Q_1 + Q_2} \right)^\alpha + \left(\frac{Q_2}{Q_1 + Q_w} \right)^\alpha \right]. \qquad (19)$$

Observe that $T(\cdot)$ can take values between 1 and 2. A value of 1 for the factor implies complete flexibility and no loss of capacity due to product switch-over. A factor of 2 would provide for a 50% loss in capacity. It is interesting to note that if flexible technology is used to produce only one product (either 1 or 2), there is no loss of capacity. The capacity constraint for flexible technology can be described as follows:

$$(Q_1 + Q_2) T(Q; \alpha) \leq K_{AB(\alpha)}. \qquad (20)$$

We now describe the complete model.

Model [GM]

First-Stage Investment Model

$$\text{Max: } \psi(\alpha, K_A, K_B, K_{AB(\alpha)}) = -C_A K_A - C_B K_B - C_{AB(\alpha)} K_{AB(\alpha)}$$
$$+ E\{\phi(\alpha, K_A, K_B, K_{AB(\alpha)}, D_1, D_2)\}, \qquad (21)$$

where $\phi(\alpha, K_A, K_B, K_{AB(\alpha)}, D_1, D_2)$ is defined as the solution of the second-stage problem described below. It should be noted that the objective function $\psi(\cdot)$ is an unconstrained nonlinear function except for nonnegativity requirements on the value of $(\alpha, K_A, K_B, K_{AB(\alpha)})$.

Second-Stage Capacity Allocation Model

$$\phi(\alpha, K_A, K_B, K_{AB(\alpha)}, D_1, D_2) = \text{Max}: \sum_{i=1}^{2} R_i(P_i + Q_i)$$

subject to

$$(P_i + Q_i) \le d_i \qquad\qquad i = 1, 2, \qquad\qquad (22)$$

$$P_1 \le K_A, \qquad\qquad\qquad\qquad (23)$$

$$P_2 \le K_B, \qquad\qquad\qquad\qquad (24)$$

$$(Q_2 + Q_2)T(Q; \alpha) \le K_{AB(\alpha)}, \qquad\qquad (25)$$

$$Q_i, P_i \ge 0 \qquad\qquad\qquad i = 1, 2, \qquad\qquad (26)$$

where d_i denotes realization of product demands.

Observe that capacity allocation decisions are made after product demands have been observed. Also, in this model, the demands need not be completely satisfied and represent upper bounds on the amount that may be sold. Constraints (22) capture this feature of the model. Constraints (23), (24), (25) guarantee that production amounts in dedicated and flexible technologies cannot exceed their available capacities. Even with restrictive assumptions, the model GM is not easy to solve optimally. Based on first-order optimality conditions, Gupta et al. derived several relationships between the decision variables. These represent necessary conditions for optimality of the first-stage decisions. They are necessary and sufficient for production and allocation decisions at the second stage. The authors examine in detail the special case with complexity flexible technology ($\alpha = 1$ and is no longer a decision variable). For this special case, they provide some useful characterizations of the optimal policy. All-flexible capacity and no flexible capacity represent the two extreme policies in this environment. The former policy is optimal if $C_{AB} \le C_A$, $C_{AB} \le C_B$ and in that case the two products may be aggregated and treated as one. The latter policy is optimal if $C_{AB} > C_A + C_B$ and in that case the problem decomposes into two newsboy problems. Gupta et al. present several numerical results describing optimal mix of technology choices for cost parameters within these extremes.

10.3.4. Modeling Approaches with Multiproducts

The literature dealing with technology selection and capacity planning in a multiproduct environment closely parallels the developments in the two-product case. As indicated earlier, the multiproduct models are more complex, and, with few exceptions, the focus has been on the development of heuristic and approximation procedures. Thus it is not surprising that theoretical results characterizing the nature of optimal choices have been few and not particularly insightful. [As explained in detail later in this section, the model due to Fine and Freund (16) is an exception, but it involves several simplifying assumptions.]

Models with Deterministic Demand. The single- and two-product models described earlier provide a natural starting point for extensions to multiple products. Thus most of the work reported on this topic is based on finite horizon, discrete period models. Typically, these models provide for dynamic, general demand patterns and alternative technology choices. For example, the LM model described in Section 3.2.1 extends to multiple products in a straightforward manner. This extension and other variants of the problem with convertible technologies have been explored in Luss (43, 44) and Lee and Luss (36). These developments are based primarily on the network representation of the problem and draw on the extensive literature in the network optimization area. The trade-off between dedicated and flexible technologies in this framework has been examined by Li and Tirupati (39). In what follows, we describe their model and some key results from that paper.

Notation

N	Number of product families
T	Planning horizon
i	Index for product families and technology choices; technology of type i refers to dedicated technology capable of producing items in product i, $i = 0$ refers to flexible technology capable of producing all products
X_{it}	Amount of capacity addition of type i in period t; X_{i0} is the initial capacity of type i
Y_{it}	Amount of flexible capacity allocated to product family i in period t
d_{it}	Demand for product family i in period t
$f_{it}(\cdot)$	Cost function for capacity type i in period t

Model [LT]

$$Z_{LT} = \text{Min} \sum_{i=0}^{N} \sum_{t=1}^{T} f_{it}(X_{it})$$

subject to

$$\sum_{\tau=1}^{t} X_{it} + Y_{it} \geq d_{it} \quad i = 1, 2, \ldots, N; t = 1, 2, \ldots, T, \quad (27)$$

$$\sum_{i=1}^{N} Y_{it} \leq \sum_{t=0}^{t} X_{0\tau} \quad t = 1, 2, \ldots, T, \quad (28)$$

$$X_{it} \geq 0 \quad i = 0, 1, \ldots, N; t = 1, 2, \ldots 3., T, \quad (29)$$

$$Y_{it} \geq 0 \qquad\qquad i = 1, 2, \ldots, N; t = 1, 2, \ldots, T. \qquad (30)$$

In this model, constraints (27) state that dedicated capacity of technology i plus flexible capacity allocated to product family i should be sufficient to satisfy the demand in period t. Constraints (28) specify that the total allocation in flexible capacity cannot exceed the availability. The objective is to minimize the discounted investment cost and to determine the optimal mix of dedicated and flexible capacity. Since the optimal solution involves many complex considerations such as economies of scale, demand patterns, and mix flexibility, the authors developed a two-phased approach and present heuristics to obtain an approximately optimal expansion schedule. Their computational experiments provide many interesting results. For example, their results suggest that substantial investments in flexible technology are economically justified even when it is significantly more expensive than dedicated technologies. Furthermore, these investments occur early in the planning horizon. The heuristics require modest computational effort and thus the model can be used to support managerial decisions in examining choices in technologies together with sizing and timing of capacity additions.

Models with Stochastic Demand. As mentioned earlier, Fine and Freund (16) analyze a simple model to examine the trade-offs between dedicated and flexible technologies in a static environment. Their market model is akin to perfect competition and it is assumed that the firm can sell whatever is produced. The demand uncertainties are captured by the specification of the revenue function for each product. In the simplest case, this is equivalent to specifying product prices. Fine and Freund assume that for each product one of a finite set of discrete alternatives will be realized. Thus the market uncertainties are described by a set of scenarios, each of which is characterized by a probability of realization and a corresponding revenue function for each product. The relevant decisions in this environment may be classified into two categories: (a) strategic decisions dealing with the capacity additions for each type of technology and (b) capacity allocation and production decisions. The latter are made after demand realizations have been observed, and hence it is necessary to specify, for each scenario, allocation of flexible technology to individual products and the production levels. We now describe the model in detail.

Notation

n	Number of products
j	Index for product families and the corresponding dedicated technology
F	Index for flexible technology which can produce all products
K_j	Amount of dedicated technologies purchased, $j = 1, \ldots, n$
K_F	Amount of flexible technology purchased

$r_j(K_j)$	Acquisition costs for dedicated technologies
$r_F(K_F)$	Acquisition costs for flexible technology
k	Number of alternative scenarios
i	Index for scenarios
P_i	Probability of occurrence of scenario i, $\Sigma_{i=1}^{k} P_i = 1$
Y_{ij}	Production of product j with dedicated technologies in scenario i
Z_{ij}	Production of product j using flexible technology in scenario i
$R_{ij}(\cdot)$	Revenue function for product j in scenario i
$C_j(Y_{ij})$	Production cost function with dedicated technologies
$C_F(Z_{ij})$	Production cost function with flexible technology

Model [FF]

$$\text{Maximize} \quad -r_F(K_F) - \sum_j r_j(K_j)$$

$$+ \sum_i P_i \sum_j \{R_{ij}(Y_{ij} + Z_{ij}) - C_j(Y_{ij}) - C_F(Z_{ij})\}$$

subject to

$$Y_{ij} - K_j \leq 0 \qquad i = 1, \ldots, k; \, j = 1, \ldots, n \,, \qquad (31)$$

$$\sum_j Z_{ij} - K_F \leq 0 \qquad i = 1, \ldots, k \,, \qquad (32)$$

$$Y_{ij} \geq 0, \, Z_{ij} \geq 0 \qquad i = 1, \ldots, k; \, j = 1, \ldots, n \,, \qquad (33)$$

$$K_j \geq 0 \qquad j = 1, \ldots, n \,, \qquad (34)$$

$$K_F \geq 0 \,. \qquad (35)$$

The objective in the model is to maximize the total revenue less the investment and production costs. Equations (31) and (32), respectively, represent capacity constraints for dedicated and flexible technologies. With linear cost and revenue function, the model FF reduces to a convex program that is amenable for analysis. Using Kuhn–Tucker conditions, the authors derive several interesting properties and derive results that have policy implications for technology selection. For example, their results help explain the economics of investment in flexible capacity and the sensitivity of profit function and investment levels to changes in acquisition costs. The model also increases the understanding of the advantages of investments in flexible technologies.

10.4. APPLICATIONS IN TECHNOLOGY SELECTION AND CAPACITY EXPANSION

It is interesting to note that several of the models described in the previous section were not mere academic exercises but were motivated by practical considerations. Thus many of the results have been used for generating long-range plans at the corporate level (by large firms) and at the industry level by countries with planned (centralized) economies. In this section we briefly describe some of these applications. The section is organized by industry. For each case, we identify its distinctive characteristics and relate them to the appropriate models and/or results.

10.4.1. Process and Chemical Industry

The process and chemical industry motivated some of the early research on capacity planning. This is a fairly mature sector with homogeneous products and well established process technologies. The stable product mix permits product aggregation and single-product models are usually adequate for capacity planning decisions. Technology is highly capital intensive, with significant scale economies and discrete, well defined alternatives. Thus for a rough cut analysis, the continuous time models of Manne (45, 46) and Srinivasan (54) are adequate. For detailed planning involving specific projects, the discrete time models of Erlenkotter (11, 12) and Neebe and Rao (52) may be used. Applications of these methods in the aluminum, caustic soda, cement, and fertilizer industries in India have been reported by Manne (46). Vietorisz and Manne (57) present a similar study for the synthetic fertilizer industry in Latin America. We note that for cases with spatially distributed demands and significant transportation costs (as in the cement and fertilizer industries), the multiregion model is appropriate.

10.4.2. Communication Industry

In the communication industry some of the models described in the previous section have been used in capacity planning of cable production. The cable market may be described as being made up of multiple products with limited substitution. Each cable type is characterized by the diameter of the wire pairs in the cable and cable size is defined by the number of wire pairs. Cables with larger numbers of wire pairs are more expensive (in terms of investment and operational costs) but can be used to satisfy the demand for cables for fewer wire pairs.

As described in Section 10.3, three types of models have been used for capacity planning in this environment. In the simplest case, Kalotay (31) considered two types of cables and the objective was to develop expansion plans for each cable type. The trade-off is between scale economies in investment and demand patterns that make substitution with more expensive cables (for part of the time) economically viable. Subsequent models, such as those of Luss (40, 41), permit modifications of technology to convert capacity from one type to the other. In this case, in addition to the factors described earlier, we have the trade-off between one-time conversion costs and the cost of meeting demand with more expensive cable. Extension of these models to multiple-product types and other related work can be found in Luss (43, 44) and Lee and Luss (36).

It is interesting but not surprising to note that most of the research in this context had origins at AT&T Bell Laboratories. And, in most cases, the emphasis was on development of implementable algorithms to facilitate ready application.

10.4.3. Automobile, Fabrication, and Discrete Part Manufactures

These industries are characterized by increasing variety in product mix and wide uncertainty in product demands. The technology problem is further complicated by recent developments in modern technologies, which provide substantial operational flexibility and a host of incidental benefits. As pointed out in Section 10.3, models and analytical tools in this area have not kept pace with market and technological developments. Accordingly, there is substantial scope for modeling and development of evaluation methodology. However, in spite of these weaknesses, the importance of technology selection has led to some model-based analysis. For example. Burstein (4) developed a two-product model incorporating productions and choices between flexible and dedicated technology. Hutchinson and Holland (28) developed a simulation model to examine trade-off between flexible manufacturing systems (FMSs) and transfer lines (dedicated technology). Bird (2) examined capacity changes by investing new technologies under uncertain demands. The trade-off is between increasing expected revenue from more facilities and the costs of additional investments.

10.4.4. Water Resources

Water resources are essential for irrigation, flood control, transportation, and recreation. Expansion of water resource systems requires substantial investments in various projects such as dams, reservoirs, and canals. Erlenkotter (11, 12) examined the problem of finding the sequence of a finite set of expansion projects to meet a deterministic demand projection at minimum discounted cost. If each project is considered as one particular type of technology, the water resource problem is equivalent to selecting technologies to satisfy demand. Other related work can be found in Butcher et al. (5), Young et al. (61), Jacoby and Loucks (29), Morin and Esogbue (51), and Erlenkotter (11–14).

10.5. CONCLUSION

In this chapter we have discussed the key factors of the capacity planning and technology selection problem and described some basic analytical models that have been developed to support these strategic decisions. The discussion in this chapter should make it clear to the reader that a considerable amount of methodology has been developed in the literature to address a variety of issues associated with technology choice, equipment selection, and capacity planning. It is interesting to note that much of this work was motivated by practical considerations and several of the results presented in this chapter have successfully been applied for making related strategic choices.

While the literature on capacity planning and technology selection is quite extensive, it is also clear that most of the models do not adequately capture the key characteristics associated with technologies such as CIM and FMSs in the discrete part manufacturing industry. In this section, we briefly describe the weaknesses in the methodologies currently available and point out some avenues for further work. First, since modern technologies are capable of producing a variety of products, an obvious and important direction of extension is to provide for several alternative technologies with varying degrees of flexibility. Second, we note that increased flexibility is consonant with risk avoidance and has potentially more impact (and benefits) in environments characterized by uncertainty. An important issue is the development of capacity expansion models with multiple products and dynamic uncertain demands. Although the case of static uncertain demands has been examined by several researchers, further analysis and a more extensive study of dynamic uncertain demands are needed to support the economic evaluation of flexible technologies. Uncertainty in acquisition costs and complexity of new technologies are other issues that are important in this context. Another interesting problem with strategic implications relates to the integration of technology and expansion choices with marketing decisions. In such scenarios the objective is to choose demand and price levels and appropriate mix of production technologies. Hence extensions that incorporate these factors will result in a comprehensive model that permits simultaneous evaluation of several key elements.

REFERENCES

1. Berliner, C., and Brimson, J. A. (Eds.) (1988). *Cost Management for Today's Advanced Manufacturing*, Harvard Business School Press, Boston.
2. Bird, C. G. (1987). A Stochastic Programming with Resource Approach to Capacity Planning, Working Paper, General Motors, Warren, MI.
3. Boyd, D. W., Phillips, R. L., and Regulinski, S. G. (1982). A Model of Technology Selection by Cost Minimizing Producers, *Management Science*, Vol. 24, No. 4, pp. 418–424.
4. Burstein, M. C. (1986). Finding the Economical Mix of Rigid and Flexible Automation for Manufacturing Systems. In: *Proceedings of the Second ORSA/ TIMS Conference on Flexible Manufacturing Systems: Operations Research Models and Applications*, Elsevier Publishers, Amsterdam.

5. Butcher, W. S., Haimes, Y. Y., and Hall, W. A. (1969). Dynamic Programming for Optimal Sequencing of Water Supply Projects, *Water Resources Research*, vol. 5, pp. 1196–1204.

6. Cohen, M. A., and Halperin, R. M. (1986). Optimal Technology Choice in a Dynamic-Stochastic Environment, *Journal of Operations Management*, Vol. 6, No. 3, pp. 317–331.

7. Cooper, R., and Kaplan, R. S. (1988). How Cost Accounting Systematically Distorts Product Costs, *Management Accounting*, April, pp. 96–105.

8. Cooper, R., and Kaplan, R. S. (1988). Measure Costs Right: Make the Right Decisions, *Harvard Business Review*, Vol. 66, No. 5, pp. 96–105.

9. Dempster, M. A. H., Fisher, M. L., Jansen, L., Lageweg, B. J., Lenstra, J. K., and Rinnooy Kan, A. H. G. (1981). Analytical Evaluations of Hierarchical Planning Systems, *Operations Research*, Vol. 9, pp. 707–716.

10. Erlenkotter, D. (1967). Two Producing Areas—Dynamic Programming Solutions. In: *Investments for Capacity Expansion: Size, Location, and Time-Phasing*, edited by A. S. Manne, MIT Press, Cambridge, MA, pp. 210–227.

11. Erlenkotter, D. (1973). Sequencing of Interdependent Hydroelectric Projects, *Water Resources Research*, Vol. 9, pp. 21–27.

12. Erlenkotter, D. (1973). Sequencing Expansion Projects, *Operations Research*, Vol. 21, pp. 542–553.

13. Erlenkotter, D. (1975). Comments on "Optimal Timing Sequencing and Sizing of Multiple Reservoir Surface Water Supply Facilities" by L. Becker and W. W.-G. Yeh, *Water Resources Research*, Vol. 11, pp. 380–381.

14. Erlenkotter, D. (1976). Coordinating Scale and Sequencing Decisions for Water Resources Projects. In: *Economic Modeling for Water Policy Evaluation*, North-Holland/TIMS Studies in the Management Sciences, Vol. 3, North-Holland, Amsterdam, pp. 97–112.

15. Fine, C. H. (1989). Development in Manufacturing Technology and Economic Evaluation Models, Working Paper, #02139, Sloan School of Management, MIT, Cambridge, MA.

16. Fine, C. H., and Freund, R. M. (1990). Optimal Investment in Product-Flexible Manufacturing Capacity, *Management Science*, Vol. 36, No. 4, pp. 449–466.

17. Fong, C. O., and Rao, M. R. (1975). Capacity Expansion with Two Producing Regions and Concave Costs, *Management Science*, Vol. 22, pp. 331–339.

18. Fong, C. O., and Srinivasan, V. (1981). The Multi-region Dynamic Capacity Expansion Problem, Part I, *Operations Research*, Vol. 29, pp. 787–799.

19. Fong, C. O., and Srinivasan, V. (1981). The Multiregion Dynamic Capacity Problem, Part II, *Operations Research*, Vol. 29, pp. 800–816.

20. Fong, C. O., and Srinivasan, V. (1986). The Multi-region Dynamic Capacity Expansion Problem: An Improved Heuristic, *Management Science*, Vol. 32, pp. 1140–1152.

21. Freidenfelds, J. (1981). Capacity Expansion when Demand Is a Birth–Death Process, *Operations Research*, Vol. 28, pp. 712–721.

22. Freidenfelds, J. (1981). Near Optimal Solution of a Two-Type Capacity Expansion Problem, *Computers Operations Research*, Vol. 8, pp. 221–239.

23. Gaimon, C. (1986). The Strategic Decision to Acquire Flexible Technology, Working Paper, The Ohio State University, Columbus.

24. Gaimon, C. (1989). The Strategic Decision to Acquire Flexible Technology, *Operations Research*, Vol. 37, pp. 410–425.

25. Gupta, D., Buzacott, J. A., and Gerchak, Y. (1988). Economic Analysis of Investment Decisions in Flexible Manufacturing Systems, Working Paper, Department of Management Science, University of Waterloo, Waterloo, Ontario, Canada.

26. Hinomoto, H. (1965). Capacity Expansion with Facilities Under Technological Improvement, *Management Science*, Vol. 11, pp. 581–592.

27. Hung, H. K., and Rikkers, R. F. (1974). A Heuristic Algorithm for the Multi-Period Facility Location Problem, Paper presented at the 45th Joint National Meeting or ORSA/TIMS, Boston.

28. Hutchinson, G. K., and Holland J. R. (1982). The Economic Value of Flexible Automation, *Journal of Manufacturing Systems*, Vol. 1, pp. 215–228.

29. Jacoby, H. D., and Loucks, D. P. (1972). Combined Use of Optimization and Simulation Models in River Basin Planning, *Water Resources Research*, Vol. 8, pp. 1401–1414.

30. Jaikumar, R. (1986). Post-industrial Manufacturing, *Harvard Business Review*, Vol. 64, No. 6, pp. 69–76.

31. Kalotay, A. J. (1973). Capacity Expansion and Specialization, *Management Science*, Vol. 20, pp. 56–64.

32. Kalotay, A. J. (1975). Joint Capacity Expansion Without Rearrangement, *Operations Research Quarterly*, Vol. 26, pp. 649–658.

33. Kaplan, R. S. (1983). Measuring Manufacturing Performance: A New Challenge for Managerial Accounting, *The Accounting Review*, Vol. LVII, No. 4, pp. 686–705.

34. Kaplan, R. S. (1986). Must CIM be Justified by Faith Alone? *Harvard Business Review*, Vol. 64, No. 2, pp. 87–95.

35. Klincewicz, J. G., and Luss, H. (1985). Optimal Timing Decisions for the Introduction of New Technologies, *European Journal of Operations Research*, Vol. 20, pp. 211–220.

36. Lee, S. B., and Luss, H. (1987). Multifacility-type Capacity Expansion Planning: Algorithms and complexities, *Operations Research*, vol. 35, pp. 249–253.

37. Lenstra, J. K., Rinnooy Kan, A. H. G., and Stougie, L. (1984). A Framework for the Probabilistic Analysis of Hierarchical Planning System, *Annals of Operations Research*, vol. 1, pp. 23–42.

38. Li, S., and Tirupati, D. (1990). Technology Choice and Capacity Expansion with Two Product Families: Tradeoffs Between Scale and Scope, Working Paper, Graduate School of Business, University of Texas, Austin.

39. Li, S., and Tirupati, D. (1990). Dynamic Capacity Expansion Problem with Multiple Products: Technology Selection and Timing of Capacity Additions, Working Paper, Graduate School of Business, University of Texas, Austin.

40. Luss, H. (1979). A Capacity Expansion Model for Two Facility Types, *Naval Research Logistics Quarterly*, Vol. 26, pp. 291–303.

41. Luss, H. (1980). A Network Flow Approach for Capacity Expansion Problems with Two Facility Types, *Naval Research Logistics Quarterly*, Vol. 27, pp. 597–608.

42. Luss, H. (1982). Operations Research and Capacity Expansion Problems: A Survey, *Operations Research*, Vol. 30, pp. 907–947.

43. Luss, H. (1983). A Multi-facility Capacity Expansion Model with Joint Expansion Set-Up Costs, *Naval Research Logistics Quarterly*, vol. 30, pp. 97–111.

44. Luss, H. (1986). A Heuristic for Capacity Planning with Multiple Facility Types, *Naval Research Logistics Quarterly*, Vol. 33, pp. 686–701.

45. Manne, A. S. (1961). Capacity Expansion and Probabilistic Growth, *Econometrica*, Vol. 29, pp. 632–649.

46. Manne, A. S. (Ed.) (1967). *Investments for Capacity Expansion: Size, Location and Time-Phasing*, MIT Press, Cambridge, MA.

47. Manne, A. S., and Veinott, A. F. Jr. (1967). Optimal Plant Size with Arbitrary Increasing Time Paths of Demand. In: *Investments for Capacity Expansion: Size, Location and Time-Phasing*, edited by A. S. Manne, MIT Press, Cambridge, MA. pp. 188–190.

48. Meredith, J. R., and Suresh, N. (1986). Justification Techniques for Advanced Manufacturing Technologies, *International Journal of Production Research*, vol. 24, No. 5.

49. Merhaut, J. M. (1975). *A Dynamic Programming Approach to Joint Capacity Expansion Without Rearrangement*, Masters Thesis, Graduate School of Management, University of California at Los Angeles.

50. Meyer, A., Nakane, J., Miller, J. G., and Ferdows, K. (1989). Flexibility: The Next Competitive Battle the Manufacturing Futures Survey, *Strategic Management Journal*, vol. 10, pp. 135–144.

51. Morin, T. L., and Esogbue, A. M. O. (1971). Some Efficient Dynamic Programming Algorithms for the Optimal Sequencing and Scheduling of Water-Supply Projects, *Water Resources Research*, Vol. 7, pp. 479–484.

52. Neebe, A. W., and Rao, M. R. (1983). The Discrete-time Sequencing Expansion, *Operations Research*, vol. 31, No. 3, pp. 546–558.

53. Rao, R. C., and Rutenberg, R. P. (1977). Multilocation Plant Sizing and Timing, *Management Science*, Vol. 23, pp. 1187–1198.

54. Srinivasan, M. N. (1967). Geometric Rate of Growth of Demand. In: *Investments for Capacity Expansion: Size, Location and Time-Phasing*, edited by A. S. Manne, MIT Press, Cambridge, MA.

55. Tombak, M., and Meyer, A. D. (1988). Flexibility and FMS: An Empirical Analysis, *IEEE Transactions on Engineering Management*, Vol. 35, no. 2, pp. 101–107.

56. Vienott, A. F. Jr. (1967). Optimal Plant Size with Arbitrary Increasing Time Paths of Demand. In: *Investments for Capacity Expansion: Size, Location and Time-Phasing*, edited by A. S. Manne, MIT Press, Cambridge, MA.

57. Vietorisz, T., and Manne, A. S. (1963). Chemical Process, Plant Location, and Economics of Scale. In: *Studies in Process Analysis: Economy-Wide Production Capabilities*, edited by A. S. Manne and H. M. Markowits, Wiley, New York, pp. 136–158.

58. Wagner, H. M., and Whitin, T. M. (1958). Dynamic Version of the Economic Lot Size Model, *Management Science*, Vol. 5, pp. 89–96.

59. Whitt, W., and Luss, H. (1981). The Stationary Distribution of a Stochastic Clearing Process, *Operations Research*, Vol. 29, pp. 294–308.

60. Wilson, L. O., and Kalotay, A. J. (1976). Alternating Policies for Nonrearrangeable Networks, INFOR, Vol. 14, pp. 193–211.

61. Young, G. K., Moseley, J. C., and Evenson, D. E. (1970). Time Sequencing of Element Construction in a Multi-Reservoir System, *Water Resource Bulletin*, Vol. 9, pp. 528–541.

■■■■■■ CHAPTER ELEVEN
_____

Group Technology

ANDREW KUSIAK

Intelligent Systems Laboratory, Department of Industrial Engineering,
The University of Iowa, Iowa City, Iowa

11.1. INTRODUCTION

The basic idea of group technology (GT) is to decompose a manufacturing system into subsystems. Introduction of GT in manufacturing results in reduced production lead time, work-in-process, labor, tooling, rework and scrap materials, setup time, delivery time, and paper work.

Group technology can be applied to manufacturing systems in two ways: logical or physical (6). In the logical layout, machines are dedicated to part families but their positions in a factory are not altered. In the physical machine layout, dedicated cells containing different machines are created for part families to exploit flow shop efficiency.

To facilitate the implementation of GT, a number of classification and coding systems have been developed. Kusiak (7) reported a number of these systems. A classification and coding system assigns codes to parts. Based on these codes, parts are grouped into part families. The disadvantage of a classification and coding approach is that a significant effort is required to code parts. Also, the classification and coding process is ambiguous.

An approach that is more practical and easier to implement is based on cluster analysis (13). Cluster analysis is concerned with grouping objects into homogeneous clusters (groups) based on the object features. There are two basic formulations of the clustering model: (a) matrix formulation and (b) integer programming formulation. Unlike classification and coding systems, cluster analysis uses only information that is available in a manufacturing system. In the matrix formulation, a machine and part incidence matrix is directly derivable from process plans. Other information such as the cost of

_____

Intelligent Design and Manufacturing, Edited by Andrew Kusiak.
ISBN 0-471-53473-0 © 1992 John Wiley & Sons, Inc.

machines, part demand, and the cost of intercellular moves required in some integer programming models can easily be collected.

One of the frequently used representations of the GT problem is a machine–part incidence matrix $[a_{ij}]$, which consists of "$*$" and "empty" entries, where an entry $*$ (empty) indicates that machine i is used (not used) to process part j. For example, machines 1 and 3 are used to manufacture part 2 in matrix (1). Typically, when an initial machine–part incidence matrix $[a_{ij}]$ is constructed, clusters of machines and parts are not visible. A clustering algorithm allows one to transform an initial incidence matrix into a structured (possibly block diagonal) form. To illustrate application of the clustering concept to GT consider the machine–part incidence matrix (1):

$$
[a_{ij}] = \begin{array}{c} \\ 1 \\ 2 \\ 3 \\ 4 \end{array}
\begin{array}{c} \text{Part number} \\ \begin{array}{ccccc} 1 & 2 & 3 & 4 & 5 \end{array} \\
\left[\begin{array}{ccccc}
 & * & & * & * \\
* & & * & & \\
 & * & & * & \\
* & & * & &
\end{array} \right]
\end{array}
\begin{array}{l} \\ \text{Machine} \\ \text{number}. \end{array}
\qquad (1)
$$

Rearranging rows and columns in matrix (1) results in matrix (2):

$$
\begin{array}{cc}
 & \begin{array}{cc} \text{PF-1} & \text{PF-2} \end{array} \\
 & \begin{array}{ccccc} 1 & 3 & 2 & 4 & 5 \end{array} \\
\text{MC-1} \left\{ \begin{array}{c} 2 \\ 4 \end{array} \right. & \left[\begin{array}{ccccc}
* & * & & & \\
* & * & & & \\
 & & * & * & * \\
 & & * & * &
\end{array} \right] \\
\text{MC-2} \left\{ \begin{array}{c} 1 \\ 3 \end{array} \right. &
\end{array}
\qquad (2)
$$

Two machine cells (clusters), MC-1 = $\{2, 4\}$ and MC-2 = $\{1, 3\}$, and two corresponding part families, PF-1 = $\{1, 3\}$ and PF-2 = $\{2, 4, 5\}$, are visible in matrix (2).

Clustering of a binary incidence matrix may result in the following two categories of clusters:

1. Mutually separable clusters.
2. Partially separable clusters.

The mutually separable clusters are shown in matrix (2), while the partially separable clusters are presented in matrix (3):

$$
\begin{array}{cc}
 & \begin{array}{ccccc} 1 & 2 & 3 & 4 & 5 \end{array} \\
\text{MC-1} \left\{ \begin{array}{c} 1 \\ 2 \end{array} \right. & \left[\begin{array}{ccccc}
* & * & & & * \\
* & * & & & \\
 & & * & * & * \\
 & & * & * &
\end{array} \right] \\
\text{MC-2} \left\{ \begin{array}{c} 3 \\ 4 \end{array} \right. &
\end{array}
\qquad (3)
$$

Matrix (3) cannot be separated into two disjoint clusters because of part 5, which is to be machined in two cells, MC-1 and MC-2.

Removing part 5 from matrix (3) results in the decomposition of matrix (3) into two separable machine cells, MC-1 = $\{1, 2\}$ and MC-2 = $\{3, 4\}$, and two part families, PF-1 = $\{1, 2\}$ and PF-2 = $\{3, 4\}$. The two clusters are called partially separable clusters and the overlapping part is called a bottleneck part. To deal with the bottleneck part 5, one of the following three actions can be taken.

1. It can be machined in one machine cell and transferred to the other machine cell by a material handling carrier.
2. It can be machined in a functional facility.
3. It can be subcontracted.

Analogous to the bottleneck part, a bottleneck machine is defined. A bottleneck machine is a machine that does not allow decomposition of a machine–part incidence matrix into disjoint submatrices. For example, machine 3 in matrix (4) does not permit decomposition of that matrix into two machine cells and two part families.

$$
\begin{array}{c}
\\
\text{Machine}\\
\text{number}
\end{array}
\begin{array}{c}
\\ \\
\begin{array}{c}1\\2\\3\\4\\5\end{array}
\end{array}
\overset{\begin{array}{c}\text{Part number}\\ 1\ \ 2\ \ 3\ \ 4\ \ 5\ \ 6\end{array}}{
\left[
\begin{array}{cccccc}
* & * &   &   &   &   \\
* & * &   &   &   &   \\
* & * & * &   & * & * \\
  &   & * & * & * & * \\
  &   & * &   & * & *
\end{array}
\right]}.
\qquad (4)
$$

A way to decompose matrix (4) into two disjoint submatrices is to use an additional copy of machine 3. The latter leads to the transformation of matrix (4) into matrix (5):

$$
\begin{array}{c}
\text{MC-1}\left\{\begin{array}{c}1\\2\\3(1)\end{array}\right.\\
\text{MC-2}\left\{\begin{array}{c}3(2)\\4\\5\end{array}\right.
\end{array}
\overset{\begin{array}{c}\text{PF-1}\qquad\ \text{PF-2}\\ 1\ \ 2\ \ 3\ \ 4\ \ 5\ \ 6\end{array}}{
\left[
\begin{array}{cc|cccc}
* & * &   &   &   &   \\
* & * &   &   &   &   \\
* & * &   &   &   &   \\ \hline
  &   & * &   & * & * \\
  &   & * & * & * & * \\
  &   & * &   & * & *
\end{array}
\right]}.
\qquad (5)
$$

Two machine cells, MC-1 = $\{1, 2, 3(1)\}$ and MC-2 = $\{3(2), 4, 5\}$, and two corresponding part families, PF-1 = $\{1, 2\}$ and PF-2 = $\{3, 4, 5, 6\}$, are shown in matrix (5).

To solve the matrix formulation of the GT problem, the following heuristic approaches have been developed:

- Similarity coefficient methods (11, 14)
- Sorting based algorithms (3, 5)
- Bond energy algorithms (2, 12, 15)
- Cost based methods (1, 9)
- Extended cluster identification algorithm (7).

In this chapter, a branching algorithm is developed for solving the GT problem with bottleneck parts and bottleneck machines. The algorithm is based on the cluster identification algorithm presented in Kusiak and Chow (9), which is briefly summarized in the next section.

11.2. CLUSTER IDENTIFICATION ALGORITHM

The cluster identification (CI) algorithm solves a special case of the GT problem, namely, one where the mutually separable clusters exist in a binary machine–part incidence matrix. The greatest advantage of the CI algorithm is its efficiency and simplicity.

CI Algorithm

Step 0. Set iteration number $k = 1$.

Step 1. Select any row i of incidence matrix $[a_{ij}]^{(k)}$ and draw horizontal line h_i through it ($[a_{ij}]^{(k)}$ is read: matrix a_{ij} at iteration k).

Step 2. For each entry of $*$ crossed by the horizontal line h_i, draw a vertical line v_j.

Step 3. For each entry of $*$ crossed-once by a vertical line v_j, draw a horizontal line h_k.

Step 4. Repeat steps 2 and 3 until there are no more crossed-once entries $*$ in $[a_{ij}]^{(k)}$. All crossed-twice entries $*$ in $[a_{ij}]^{(k)}$ form machine cell MC-k and part family PF-k.

Step 5. Transform the incidence matrix $[a_{ij}]^{(k)}$ into $[a_{ij}]^{(k+1)}$ by removing rows and columns corresponding to all the horizontal and vertical lines drawn in steps 1 through 4.

Step 6. If matrix $[a_{ij}]^{(k+1)} = \mathbf{0}$ (where $\mathbf{0}$ denotes a matrix with all empty elements), stop; otherwise set $k = k + 1$ and go to step 1.

Note that the cluster identification algorithm scans each element of matrix $[a_{ij}]$ two times. Since there are mn elements in matrix $[a_{ij}]$, its computational time complexity is $O(mn)$.

Application of the CI algorithm for solving the GT problem is illustrated in Example 1.

Example 1. Consider machine–part incidence matrix (6):

$$
[a_{ij}] =
\begin{array}{c}
 \\
1 \\ 2 \\ 3 \\ 4 \\ 5 \\ 6 \\ 7
\end{array}
\begin{array}{c}
\text{Part number} \\
\begin{array}{cccccccc}
1 & 2 & 3 & 4 & 5 & 6 & 7 & 8 \\
 & * & * & & * & & & \\
* & & & & & * & & \\
 & & & * & & & * & \\
* & & & & & * & & \\
 & & * & & * & & & * \\
 & & * & & & & & \\
 & * & * & & * & & & *
\end{array}
\end{array}
\quad \text{Machine number .}
\tag{6}
$$

Step 0. Set iteration number $k = 1$.

Step 1. Row 1 of matrix (6) is selected and horizontal line h_1 is drawn. The result of steps 1 and 2 are presented in matrix (7).

Step 2. Three vertical lines v_2, v_3, and v_5 are drawn.

$$
[a_{ij}]^{(1)} = 4
\tag{7}
$$

As a result of drawing these three vertical lines, five new crossed-once entries $*$ are created in matrix (7), that is, entries $(5,3)$, $(5,5)$, $(7,2)$, $(7,3)$, and $(7,5)$.

Step 3. Two horizontal lines h_5 and h_7 are drawn through all the crossed-once nonempty entries of matrix (7), as shown in matrix (8).

$$
[a_{ij}]^{(1)} = 4
\tag{8}
$$

Step 4. Since the entries $(5, 8)$ and $(7, 8)$ of matrix (8) are crossed once, the vertical line v_8 is drawn, as shown in matrix (9).

$$[a_{ij}]^{(1)} = \begin{array}{c} \\ 1 \\ 2 \\ 3 \\ 4 \\ 5 \\ 6 \\ 7 \end{array} \begin{array}{c} \begin{array}{cccccccc} 1 & 2 & 3 & 4 & 5 & 6 & 7 & 8 \end{array} \\ \left[\begin{array}{cccccccc} --- & * & -- & * & ----- & * & ------- & - \\ * & & & & & * & & \\ & & & * & & & * & \\ * & & & & & * & & \\ ---- & -- & * & ----- & * & ----- & * & - \\ & & & * & & & & \\ --- & * & -- & * & ----- & * & ------- & * & - \end{array} \right] \end{array} \begin{array}{c} -h_1 \\ \\ \\ \\ -h_5 \\ \\ -h_7 \end{array} \quad (9)$$

$$\begin{array}{cccc} \;\; v_2 & v_3 & v_5 & v_8 \end{array}$$

Since there are no more crossed-once nonempty entries, all the crossed-twice entries $*$ of matrix (9) form machine cell $MC\text{-}1 = \{1, 5, 7\}$ and part family $PF\text{-}1 = \{2, 3, 5, 8\}$.

Step 5. Matrix (9) is transformed into matrix (10).

$$[a_{ij}]^{(2)} = \begin{array}{c} 2 \\ 3 \\ 4 \\ 6 \end{array} \begin{array}{c} \begin{array}{cccc} 1 & 4 & 6 & 7 \end{array} \\ \left[\begin{array}{cccc} * & & * & \\ & * & & * \\ * & & * & \\ & * & & \end{array} \right] \end{array} . \quad (10)$$

In the second iteration $(k = 2)$, steps 1 through 4 are performed on matrix (10). This iteration results in incidence matrix (11):

$$[a_{ij}]^{(2)} = \begin{array}{c} 2 \\ 3 \\ 4 \\ 6 \end{array} \begin{array}{c} \begin{array}{cccc} 1 & 4 & 6 & 7 \end{array} \\ \left[\begin{array}{cccc} * & ---- & * & ---- \\ & * & & * \\ -* & ---- & * & ----- \\ & * & & \end{array} \right] \end{array} \begin{array}{c} -h_2 \\ \\ -h_4 \\ \\ \end{array} .$$

$$\begin{array}{cc} v_1 & v_6 \end{array}$$

Also, machine cell $MC\text{-}2 = \{2, 4\}$ and part family $PF\text{-}2 = \{1, 6\}$ are obtained. In the third iteration $(k = 3)$, matrix (12) is generated:

$$[a_{ij}]^{(3)} = \begin{array}{c} 3 \\ 6 \end{array} \begin{array}{c} \begin{array}{cc} 4 & 7 \end{array} \\ \left[\begin{array}{cc} -* & -- * - \\ -* & --- \end{array} \right] \end{array} \begin{array}{c} -h_3 \\ -h_6 \end{array} . \quad (12)$$

$$\begin{array}{cc} v_4 & v_7 \end{array}$$

From this matrix $MC\text{-}3 = \{3, 6\}$ and $PF\text{-}3 = \{4, 7\}$ are obtained. The final clustering result is illustrated in matrix (13):

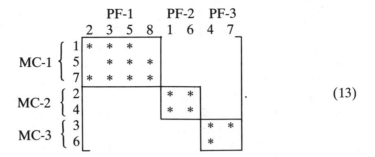

$$\tag{13}$$

11.3. BRANCHING ALGORITHM

The implicit enumeration algorithm presented in this section is using the CI algorithm at each node to solve a GT problem.

Algorithm

Step 0. (Initialization): Begin with the incidence matrix $[a_{ij}]$ at level 0. Solve the GT problem presented with $[a_{ij}]$ using the CI algorithm.

Step 1. (Branching): Using the breadth-first search strategy, select one of the active nodes (not fathomed) and solve the corresponding GT problem with the CI algorithm.

Step 2. (Fathoming): Exclude a new node from further consideration if:

Test 1: Cluster size is not satisfactory.

Test 2: Cluster structure is not satisfactory.

Step 3. (Backtracking): Return to an active node.

Step 4. (Stopping rule): Stop when there are no active nodes remaining; the current incumbent solution is optimal; otherwise go to step 1.

Branching. Branching is performed in one of the following two ways:

1. The CI algorithm when the incidence matrix partitions into mutually separable submatrices.
2. Removing one column at a time from the corresponding incidence matrix when the matrix does not partition into mutually separable submatrices.

Bounding. Note that the bounding scheme has not been included in the branching algorithm. As a possible bound the measure of effectiveness (ME) introduced by McCormick et al. (12) has been considered. For a nonnegative 0–1 matrix $[a_{ij}]_{m \times n}$ the ME that indicates the quality of clusters is defined as follows:

$$\mathrm{ME} = \tfrac{1}{2} \sum_{i=1}^{m} \sum_{j=1}^{n} a_{ij}(a_{i,j-1} + a_{i,j+1} + a_{i-1,j} + a_{i+1,j})$$

where $a_{0,j} = a_{m+1,j} = a_{i,0} = a_{i,n+1} = 0$.

To illustrate the ME consider the two matrices shown in Fig. 11.1. Matrix (b) in Fig. 11.1 has been obtained from matrix (a) with the CI algorithm. Of course, the objective of the GT problem considered in this chapter is to maximize the value of the measure of effectiveness ME.

Another possible bound is the total distance D between any two pairs of rows and columns of the 0–1 incidence matrix:

$$D = \sum_{p=1}^{m-1} \sum_{q=p}^{m} d_{pq} + \sum_{p=1}^{n-1} \sum_{q=p}^{n} d'_{pq},$$

where

$$d_{pq} = \sum_{k=1}^{n} \delta(a_{pk}, a_{qk}) \quad \text{and} \quad d'_{pq} = \sum_{k=1}^{m} \delta(a_{kp}, a_{kq})$$

for

$$\delta(a_{pk}, a_{qk}) = \begin{cases} 1 & \text{for } a_{pk} = a_{qk} = 1 \\ 0 & \text{otherwise} \end{cases}$$

$$\text{and} \quad \delta(a_{kp}, a_{kq}) = \begin{cases} 1 & \text{for } a_{kp} = a_{kq} = 1 \\ 0 & \text{otherwise}. \end{cases}$$

Since in the bounding scheme, columns of the incidence matrix are being removed, none of the two measures (ME or D) can provide a proper upper bound for the GT problem. Due to the nature of the CI algorithm, which does not change the sequence of the rows and columns in the incidence matrix, finding a proper upper or lower bound is difficult. Such a bound should not only be tight but also easy to compute. The computational time complexity of calculating the distance D is $O(m^2 n + n^2 m)$, while the complexity of the CI algorithm is only $O(mn)$. Due to the low computational

Part number
1 2 3 4

Machine 2
number 3

1 [* *]
2 [* *]
3 [* *]
4 [* *]

ME = 2
(a)

Part number
1 3 2 4

Machine 4
number 2

1 [* *]
4 [* *]
2 [* *]
3 [* *]

ME = 8
(b)

FIGURE 11.1. Matrices with different values of the measure of effectiveness.

time complexity of the CI algorithm, the proposed branching algorithm is efficient too. In fact, the branching algorithm might be the first optimal technique published for solving the GT problem.

For an efficient bounding scheme see Kusiak and Cheng (8).

Fathoming. The fathoming is based on the following tests:

Test 1: Testing the size of the incidence matrix
Test 2: Testing the structure of the incidence matrix

One of the parameters of grouping might be an upper limit on the number of machines or parts in a cluster. This requirement is checked in test 1. Fathoming might be performed based on the structure of the incidence matrix. The ideal structure of an incidence matrix is a rectangle filled with 1's. In order to effectively apply test 2, one can define the following two quality measures for the incidence matrix $[a_{ij}]_{m \times n}$:

$$QS_M = \frac{ME}{mn} \quad \text{or} \quad QS_D = \frac{D}{mn},$$

where ME is the measure of effectiveness and D is the total sum of distances.

As an example, consider the two matrices in Fig. 11.1. Values of quality measures are $QS_M = \frac{1}{8}$ and $QS_M = \frac{1}{2}$, respectively, while the $QS_D = 1/2$ for the two matrices. The designer could set a lower limit on the value of the quality measure QS_M or QS_D for each cluster.

The developed branching algorithm is illustrated in Example 2.

Example 2. Solve the GT problem represented with the machine–part incidence matrix (14):

$$
\begin{array}{c}
\text{Part number} \\
\begin{array}{ccc}
 &
\begin{array}{c}
\text{Machine} \\
\text{number}
\end{array}
\begin{array}{c}
1 \\ 2 \\ 3 \\ 4 \\ 5 \\ 6 \\ 7 \\ 8
\end{array}
&
\begin{array}{ccccccc}
1 & 2 & 3 & 4 & 5 & 6 & 7 \\
* & & * & & * & & \\
& * & & & & * & \\
& & & * & & & \\
& & & * & & & * \\
& & & * & & & \\
& * & & & & & \\
* & & & * & & & * \\
& & * & * & & &
\end{array}
\end{array}
\end{array}
\quad (14)
$$

Assume that the size of each machine cell is to be not smaller than 2 and not greater than 3.

Applying the CI algorithm, one observes that matrix (14) partitions (step 1: branching) into two submatrices presented in (15):

$$
\begin{array}{c}

\end{array}
\begin{array}{c}
2\\6\\1\\3\\4\\5\\7\\8
\end{array}
\left[
\begin{array}{ccccccc}
* & * & & & & & \\
* & & & & & & \\
 & & * & * & & * & \\
 & & & & & * & \\
 & & & * & & & * \\
 & & & * & & & \\
 & & * & & * & & * \\
 & & & * & & * &
\end{array}
\right]
\qquad (15)
$$

with the starting column headers: 2 6 1 3 4 5 7

with the corresponding machine cells MC-1 = {2, 6} and MC'-2 = {1, 3, 4, 5, 7, 8} and part families PF-1 = {2, 6} and PF'-2 = {1, 3, 4, 5, 7}. Since none of the columns has been removed, a value 0 is assigned to each branch (see Fig. 11.2).

Also, since the size of the matrix representing MC-1 is satisfactory (step 2: fathoming) the corresponding matrix becomes an incumbent solution.

Further branching (step 1) is performed on an active matrix. The resulting submatrices are presented in Fig. 11.2. The five matrices at level 2

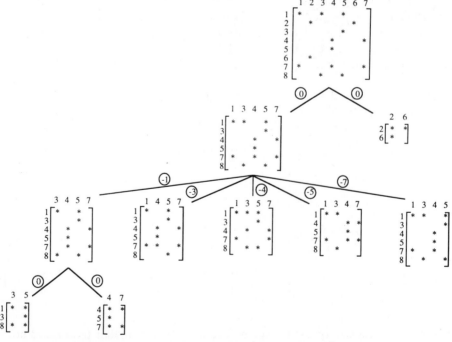

FIGURE 11.2. An enumeration tree for the branching algorithm.

in Fig. 11.2 are generated by removing one column at a time from the matrix at level 1. Only the matrix with the column (part) 1 removed has partitioned into two submatrices (level 3 in Fig. 11.2). Each of the two submatrices satisfies test 1. The final solution of the GT problem is shown in matrix (16):

$$
\begin{array}{c}
\begin{array}{ccccccccc}
 & \text{PF-1} & & \text{PF-2} & & & \text{PF-3} & & \\
 & 2 & 6 & 3 & 5 & & 4 & 7 & 1
\end{array} \\
\begin{array}{c}
\text{MC-1} \left\{ \begin{array}{c} 2 \\ 6 \end{array} \right. \\[2pt]
\text{MC-2} \left\{ \begin{array}{c} 1 \\ 3 \\ 8 \end{array} \right. \\[2pt]
\text{MC-3} \left\{ \begin{array}{c} 4 \\ 5 \\ 7 \end{array} \right.
\end{array}
\begin{bmatrix}
* & * & & & & & & \\
* & & & & & & & \\
& & * & * & & & & * \\
& & & * & & & & \\
& & * & * & & & & \\
& & & & & * & * & \\
& & & & & * & & \\
& & & & & * & * & *
\end{bmatrix}
\end{array}
\qquad (16)
$$

Tree machine cells, three part families, and the bottleneck part 1 are visible in matrix (16).

11.4. SOLVING THE GT PROBLEM WITH BOTTLENECK MACHINES

The branching algorithm can be used to solve the GT problem with bottleneck machines provided that its step 1 (branching) is modified. Rather than removing columns, rows representing multiple machines are added. The branching scheme is discussed next.

Branching Scheme. Consider the group technology problem with multiple machines that is represented in matrix (17):

$$
\begin{array}{c}
\begin{array}{cc}
 & \text{Part number} \\
 & \begin{array}{ccccc} 1 & 2 & 3 & 4 & 5 \end{array}
\end{array} \\
\begin{array}{cc}
\text{Machine} \\ \text{number}
\end{array}
\begin{array}{c} 1 \\ 2 \\ 3 \\ 4 \end{array}
\begin{bmatrix}
 & * & * & & * \\
* & & & * & \\
 & * & * & & * \\
* & & * & * &
\end{bmatrix}
\end{array}
\qquad (17)
$$

Since each of the four machines in matrix (17) could potentially be a bottleneck machine, the maximum number of multiple copies that one may consider for each machine is equal to the number of parts processed on that machine.

Assuming that machine 1 is the bottleneck machine, matrix (17) can be transformed into matrix (18), where each copy of machine 1 corresponds one part:

$$
\begin{array}{c}
\begin{array}{ccccc} 1 & 2 & 3 & 4 & 5 \end{array} \\
\begin{array}{c} 1(1) \\ 1(2) \\ 1(3) \\ 2 \\ 3 \\ 4 \end{array}
\left[
\begin{array}{ccccc}
  & * &   &   &   \\
  &   & * &   &   \\
  &   &   &   & * \\
* &   &   & * &   \\
  & * & * &   & * \\
* &   & * & * &
\end{array}
\right].
\end{array}
\qquad (18)
$$

Applying the CI algorithm to matrix (18) does not lead to the decomposition of the matrix. This proves that machine 1 is not the bottleneck machine. The same is true for machines 2 and 3.

Considering machine 4 as the bottleneck machine results in matrix (19):

$$
\begin{array}{c}
\begin{array}{ccccc} 1 & 2 & 3 & 4 & 5 \end{array} \\
\begin{array}{c} 1 \\ 2 \\ 3 \\ 4(1) \\ 4(2) \\ 4(3) \end{array}
\left[
\begin{array}{ccccc}
  & * & * &   & * \\
* &   &   & * &   \\
  & * & * &   & * \\
* &   &   &   &   \\
  & * &   &   &   \\
  &   &   & * &
\end{array}
\right].
\end{array}
\qquad (19)
$$

In the first iteration of the CI algorithm, matrix (20) is obtained:

$$
\begin{array}{c}
\begin{array}{ccccc} 1 & 2 & 3 & 4 & 5 \end{array} \\
\begin{array}{c} 1 \\ 2 \\ 3 \\ 4(1) \\ 4(2) \\ 4(3) \end{array}
\left[
\begin{array}{ccccc}
  & * & * &   & * \\
* &   &   & * &   \\
  & * & * &   & * \\
* &   &   &   &   \\
  & * &   &   &   \\
  &   &   & * &
\end{array}
\right]
\begin{array}{l} \\ -h_2 \\ \\ -h_{4(1)} \\ \\ -h_{4(3)} \end{array}
\end{array}
\qquad (20)
$$
$$
\quad v_1 \qquad v_4
$$

Two copies 4(1) and 4(3) of machine 4 are involved in the first machine cell. However, it is enough to consider only one (the first copy), as illustrated in matrix (21):

$$
\begin{array}{c}
\begin{array}{ccccc} 1 & 4 & 2 & 3 & 5 \end{array} \\
\begin{array}{c} 2 \\ 4(1) \\ 4(2) \\ 1 \\ 3 \end{array}
\left[
\begin{array}{ccccc}
* & * &   &   &   \\
* & * &   &   &   \\
  &   & * &   &   \\
  &   & * & * & * \\
  &   & * & * & *
\end{array}
\right].
\end{array}
\qquad (21)
$$

In the case considered, only two out of three copies of machine 4 are needed to decompose incidence matrix (19). Introducing additional multiple machines would result in further decomposition of the two submatrices in (21).

11.5. ARTIFICIAL INTELLIGENCE AND GT PROBLEM

Attempts have been made to solve the GT problem considered in this chapter with various artificial intelligence approaches. Kusiak et al. (7) presented an A* algorithm that did not perform better in terms of CPU time than a heuristic proposed in the same paper. Harhalakis et al. (4) applied the simulated annealing algorithm, which appears to be sensitive to the values of the initial parameters. Based on the author's experience, the AI approach—in particular, a knowledge-based system linked with heuristics—is suitable for solving GT problems that involve qualitative and quantitative constraints, that is, problems that are much more general than the formulations considered in this chapter. One of the most recently developed knowledge-based systems for GT, named KBGT, is presented in Kusiak (7).

11.6. CONCLUSION

In this chapter, an implicit enumeration algorithm for solving the group technology (GT) problem with bottleneck parts and bottleneck machines was developed. The branching scheme for the GT problem with bottleneck parts is based on removing parts. The branching scheme for the GT problem with bottleneck machines utilizes multiple machines. The algorithm presented is perhaps the first optimal algorithm developed for the GT problems discussed. The efficiency of the branching algorithm is derived from the efficiency of the cluster identification algorithm that is used at each node of the algorithm.

REFERENCES

1. Askin, R., and Subramanian, S. (1987). A Cost-Based Heuristic for Group Technology Configuration, *International Journal of Production Research*, Vol. 25, No. 1, pp. 101–114.
2. Bhat, M. V., and Haupt, A. (1976). An Efficient Clustering Algorithm, *IEEE Transaction on Systems, Man, and Cybernetics*, Vol. SMC-6, No. 1, pp. 61–64.
3. Chan, H. M., and Milner, D. A. (1982). Direct Clustering Algorithm for Group Formation in Cellular Manufacturing, *Journal of Manufacturing Systems*, Vol. 1, No. 1, pp. 65–74.
4. Harhalakis, G., Proth, J. M., and Xie, X. L. (1990). Manufacturing Cell Design Using Simulated Annealing: An Industrial Application, *Journal of Intelligent Manufacturing*, Vol. 1, No. 3, pp. 185–191.
5. King, J. R. (1980). Machine–Component Group Formation in Production Flow Analysis: An Approach Using a Rank Order Clustering Algorithm, *International Journal of Production Research*, Vol. 18, No. 2, pp. 213–232.
6. Kusiak, A. (1985). The Part Families Problem in Flexible Manufacturing System, *Annals of Operations Research*, Vol. 3, pp. 279–300.

7. Kusiak, A. (1990). *Intelligent Manufacturing Systems*, Prentice-Hall, Englewood Cliffs, NJ.

8. Kusiak, A., and Cheng, C.-H. (1990). A Branch-and-Bound Algorithm for Solving the Group Technology Problem, *Annals of Operations Research*, Vol. 26, pp. 415–431.

9. Kusiak, A., and Chow, W. S. (1987). Efficient Solving of the Group Technology Problem, *Journal of Manufacturing Systems*, Vol. 6, No. 2, pp. 117–124.

10. Kusiak, A., Boe, W. J., and Cheng, C.-H. (1990). Solving the Group Technology Problem, Working Paper No. 90-21, Department of Industrial Engineering, The University of Iowa, Iowa City, Iowa.

11. McAuley, J. (1972). Machine Grouping for Efficient Production, *The Production Engineer*, February, pp. 53–57.

12. McCormick, W. T., Schweitzer, P. J., and White, T. W. (1972). Problem Decomposition and Data Reorganization by Clustering Technique, *Operations Research*, Vol. 20, pp. 992–1009.

13. Propen, M. (1990). Grappling with Group Technology, *Manufacturing Engineering*, July, pp. 80–82.

14. Seifoddini, H., and Wolfe, P. M. (1986). Application of the Similarity Coefficient Method in Group Technology, *IIE Transactions*, Vol. 18, No. 3, pp. 271–277.

15. Slagle, J. L., Chang, C. L., and Heller, S. R. (1975). A Clustering and Data Reorganization Algorithm, *IEEE Transactions on Systems, Man, and Cybernetics*, Vol. SMC-5, pp. 125–128.

CHAPTER TWELVE

Design of Machining Systems

HORST TEMPELMEIER

Department of Production Management, Technical University of Braunschweig,
Braunschweig, West Germany

12.1. INTRODUCTION

A flexible machining system (FMS) is a production system consisting of a set
of computer numerically controlled machines (CNC machines) connected by
an automatic material handling system (MHS), all under common computer
control. A FMS is capable of simultaneously processing a large variety of
different workpieces (within a part family) in a random order without
significant utilization losses due to setups. This is possible because the CNC
machines have automatic tool change capabilities, which in connection with
a local tool magazine (containing 20–200 tools) allow a switch from one
workpiece to another virtually within seconds. Workpieces are introduced
into the FMS via load and unload (L/UL) stations, where they are mounted
on pallets—usually by human operators—and circulate in the FMS until
completion. The correct positioning of the pieces on the pallets is achieved
by means of clamping devices or fixtures. Flexible machining systems and
flexible assembly systems are special types of flexible manufacturing systems. However, many papers written about flexible manufacturing systems
implicitly assume the machining type.

A detailed classification of various types of FMS is given by Kusiak (19),
who distinguishes between flexible manufacturing module (FMM), flexible
manufacturing cell (FMC), flexible manufacturing group (FMG), flexible
machining system (FMS), and flexible production system (FPS), based on
the type and number of different processes and types of machines included
in the system. In the following classification, the term FMS is used to
subsume all variants of systems including more than one machine.

Intelligent Design and Manufacturing, Edited by Andrew Kusiak.
ISBN 0-471-53473-0 © 1992 John Wiley & Sons, Inc.

12.1.1. Flexible Machining Module

The basic component of a FMS is a *flexible machining module* (FMM; machining center) (see Fig. 12.1).

A FMM is a CNC machine augmented by a number of auxiliary devices such as a local tool magazine, an automatic tool changer, and an automatic pallet changer (19). Often a small local part buffer able to store up to two workpieces is attached to the FMM. The main difference between a FMM and a conventional NC machine is that it is able to perform a number of different consecutive operations using different tools without any human interaction, given the workpieces are available on the machining table or in the local buffer. The operations are described by a NC program, defining which types and order of operations are to be performed and which tools are needed. If the FMM is part of a FMS, the NC program is downloaded from the central FMS computer immediately before beginning the processing of a workpiece. In general, the processing time per machine load is longer than it would be on a conventional NC machine (20) because due to the flexibility of a machining center several "conventional" operations are aggregated into one "new" operation, including all necessary tool changes. This point may affect the design of the machining processes. Upon completion of processing in a FMM, the workpiece is unloaded from the machining table, put into the local output buffer of the machine (if any), and replaced by another workpiece already waiting in the local input buffer of the machine. As shown in Fig. 12.1, the tool magazine may be segmented, allowing the exchange of tool cassettes without interference with the machining operation. Tools may also be stored in chains, drums, turrets, or disks (20).

12.1.2. Flexible Machining Cell

A FMM augmented by a stand-alone part buffer and a L/UL station is called a *flexible machining cell* (FMC). In Fig. 12.2 instead of the standard

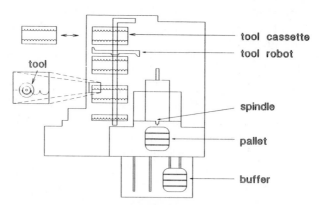

FIGURE 12.1. Flexible machining module.

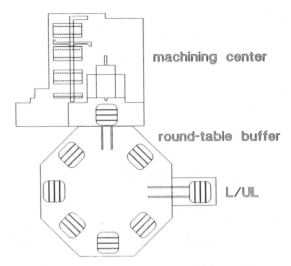

FIGURE 12.2. Flexible machining cell.

local input–output buffer a round-table pallet buffer is attached to the FMM. Workpieces are introduced into the round-table pallet buffer via a L/UL station. The MHS functions are integrated into the round-table pallet buffer. A different FMC design concept often found in industrial practice includes a linear pallet buffer connected to the machining center by a rail-guided vehicle.

Compared to a FMM, a FMC has the advantage of being able to run in a stand-alone mode during a third shift or even during part of the weekend, due to the availability of unprocessed workpieces stored in the attached buffer, with the result of a much higher machine utilization.

In the literature, the term FMC sometimes is used when *several* FMMs are combined with a stand-alone buffer and a L/UL station. In the current classification, this configuration would be called a flexible machining *system* (FMS).

12.1.3. Flexible Machining System

Several FMMs, stand-alone workpiece buffers, and L/UL stations may be interconnected by an automatic MHS [usually an automatic wire-guided vehicle (AGV) or a rail-guided vehicle (RGV) system], constituting a *flexible machining system* (FMS; flexible manufacturing network; flexible machining group). A FMS can be built according to *process type*. Then it consists of several *identical* FMMs (parallel machines). Or it can be built according to *product type*. Then several nonidentical machines are combined that are able to perform several consecutive operations of different types in the FMS (L/UL, lathe, drill, wash, debur). In Fig. 12.3 a schematic layout

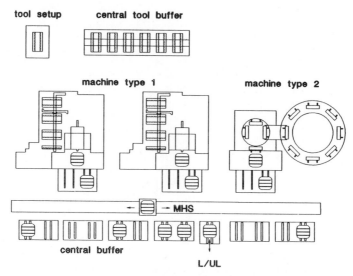

FIGURE 12.3. Flexible machining system.

of a FMS is shown, including three FMMs (e.g., two lathes and one drilling machine) interconnected by a rail-guided vehicle (RGV), a stand-alone workpiece buffer, a L/UL station, a tool setup station, and a central tool buffer, all integrated under common computer control (not depicted). The tool cassettes may be transported by a forklift cart.

Several layout examples of FMS currently under operation in industrial practice are depicted in Carrie (9).

12.1.4. Functional Subsystems of a Flexible Machining System

A typical FMS consists of several interconnected subsystems, namely, the technical system, the information system, and the operator system (18) (Fig. 12.4).

The *technical system* may be considered to consist of the *processing system*, the *workpiece supply system*, the *tool supply system*, and several auxiliary subsystems, namely, the *energy supply system*, the *coolant system*, and the *waste disposal system*.

Although the automation of production processes in part has substituted machines for human operators, several people are still needed to run a FMS. The *human operator system* comprises all people directly involved in operating the FMS. Among these are shop floor operators who mount workpieces on pallets, set up tools, and sometimes perform tool transports, as well as repair people who perform maintainance operations. In addition, due to the limited automatic planning and scheduling capabilities of the current available FMS software systems, supervisors and dispatchers are part of the human operator system (4).

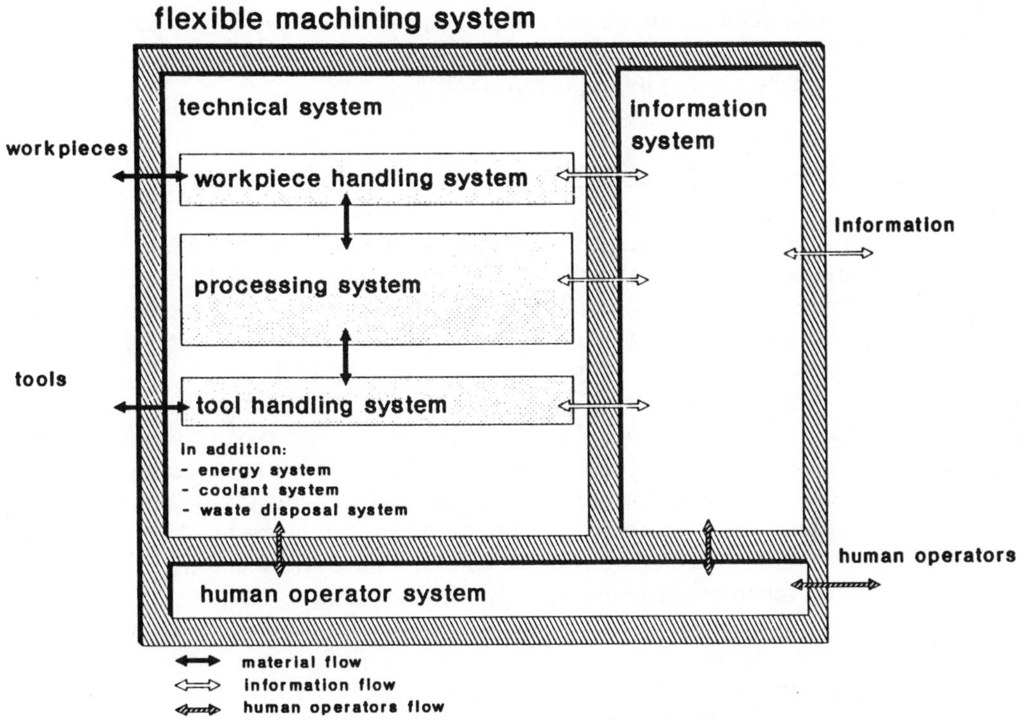

flexible machining system

FIGURE 12.4. Functional subsystems of a FMS.

The *information system* includes all computer hardware and software to monitor and control the FMS.

12.1.5. Expected Benefits of Flexible Machining Systems

In 1988 there were about 1000 flexible manufacturing (machining and assembly) systems in operation worldwide, with 800 being compact systems having two to five CNC machines and investment costs up to 4 million U.S. dollars (23).

A FMS is designed to simultaneously achieve both the high productivity of a (product-oriented) automatic transfer line and the flexibility of a (process-oriented) conventional job shop (17). It aims at the resolution of the classical conflict between (high) utilization of machines and (short) throughput times of workpieces usually found in low-volume, small-batch production environments. FMSs are considered to be one building block of the computer integrated (CIM) factory of the future.

There are several benefits usually associated with the installation of a FMS. Among these are labor and capital saving, increased utilization of machines, decreased delivery times, increased flexibility and the ability to adapt to market changes, and improved quality (17, 23).

In order to achieve the benefits expected, however, great care has to be taken during the design process of the FMS. If not designed properly, a FMS may fail to realize its opportunities.

12.2. THE DESIGN PROCESS

The high financial investment involved (investment volumes of up to $100 million are not uncommon) and the long-term consequences of the investment decision necessitate a careful analysis of FMS design alternatives and a systematical structure of the design process of a FMS. The design process may be divided into three phases, as shown in Fig. 12.5 (5).

In the *planning phase*, an analysis of the workpiece spectrum is performed. Based on technological criteria (part geometry, type of processing, etc.), workpieces with similar processing requirements are grouped together. Taking into consideration the annual planned production volumes and expected batch sizes, those parts that should be produced by the planned FMS concept are selected. Group technology planning techniques (workpiece coding schemes, machine–part matrix clustering techniques, etc.) may be used to solve the problems of simultaneous part grouping and machine type selection (20, 41).

The planning phase ends up with a specification of the workload (part types, described by processing sequences and annual production quantities) for which the FMS will be designed and the specification of the type of equipment (machinery and MHS) needed. Often the conventional job shop processing plans need restructuring due to the increased flexibility of the new technology under consideration.

Having defined the workpieces to be produced in the new FMS and given the type of machinery most suitable to process them, in the *configuration design phase* the capacity of each equipment type to be installed in the FMS is determined. The number of machine tools, the number of pallets, the number of MHS carts, and the number of central buffers are determined. Typically, a great number of design alternatives must be evaluated numerically with respect to the design criteria (production rate per part type, throughput time, utilization, etc.). The design stage ends up with a specification of the "optimal" FMS configuration alternative.

After the decision has been made regarding the FMS configuration to be installed, in the *installation planning phase* the FMS hardware is constructed and the software needed to run the FMS is developed. Main activities of this phase are concerned with the preparation of the building, the manufacturing, installation, and testing of the equipment, and the installation and testing of the computer hardware and software (10).

From the beginning of the design process the problem of economic evaluation of a FMS exists. A first economic evaluation must be performed on a strategic decision level, taking into account expected chances as well as

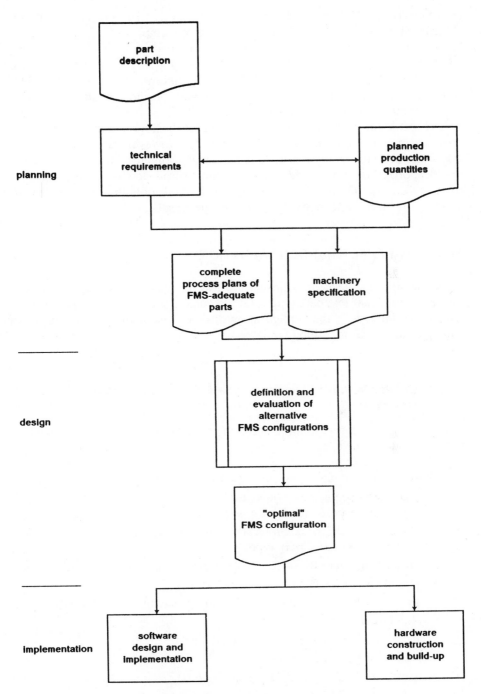

FIGURE 12.5. FMS design process.

expected risks. However, the identification and quantification of these changes and risks are often very difficult, if not impossible. Nevertheless, based on the actual state of knowledge, an economic evaluation must accompany the total design process, thus offering at any stage the chance to cancel the FMS concept and switch to a different, perhaps nonflexible production system, which may be a conventional job shop organized according to group technology principles.

The problem of economic justification of flexible manufacturing has attracted much attention by researchers (30). However, up to now many questions are still unanswered, leaving much room for human judgment-based approaches. Only a few steps of the overall design process may be supported by quantitative performance evaluation tools that are needed to quantify the input and output of a design proposal.

12.3. FMS PERFORMANCE EVALUATION

Once the strategic decision to invest in FMS technology has been made and the set of FMS-suitable part types and the corresponding types of machinery have been defined, in the configuration design phase many alternative FMS configurations are identified and evaluated with respect to the relevant performance criteria.

12.3.1. Design Variables, Performance Criteria, and Design Restrictions

A FMS configuration is defined through the specification of several design variables. These include

- Subset of *part types* to be processed in the FMS
    Processing sequences, for example, the number and order of operations to be performed on a workpiece
    Routing mix, for example, the mix of alternative processing sequences used for a part type
    Production quantities (part type mix) to be produced in the FMS
- *Components of the FMS*
    Type of CNC machines and number of machines of each type
    Number of L/UL stations
    Type of MHS and number of MHS carts
    Number of central buffer places
    Number of pallets circulating in the FMS
- *Layout* of the FMS
- Type of computer system and structure of the *planning system*
- Number and skills of *human operators*

A FMS configuration alternative may be characterized by a specific combination of the above-mentioned design variables. Many of these variables are highly interrelated. For example, the planned production quantities and the processing sequences of the part types assigned to the FMS determine the type and capacity of the system components (e.g., machines needed).

In general, the number of FMS configuration alternatives is combinatorial in the design variables. During the evaluation phase, the problem of ranking the feasible configuration alternatives according to the relevant performance criteria arises.

Each design alternative is evaluated with respect to the design criteria, some of which are defined such that they act as restrictions. In the design phase, perhaps the most important performance criterion is connected with the question, whether a FMS configuration alternative under consideration will be able to produce the desired production quantities of the part types assigned to the FMS (throughput per time period). If a lower limit for the desired throughput is specified, then all design alternatives not reaching that level are considered infeasible and excluded from the set of alternatives. Another restriction may arise from an upper limit of the investment volume. If the FMS is to be built up in an existing machining hall, then often layout restrictions must be considered, specifying an upper limit for the number of machines or central buffer places.

In order to identify the optimal design proposal among the configurations satisfying the restrictions imposed, a multicriterion decision process is necessary that often involves several persons. In the design process several alternatives are compared with respect to several nonquantifiable and quantifiable goals. Unfortunately, many of the effects are hard to quantify, leaving much room for human judgment. In the next subsection, special attention is given to the determination of those goals that can be evaluated by numerical methods.

12.3.2. Performance Evaluation Methods

As stated earlier, many of the design variables are strongly interrelated. The numerical quantification of the dynamic interactions and their influence on the overall FMS performance is a very difficult problem. However, it is needed as a prerequisite for the economic justification of a FMS. Considering the number of design alternatives to be evaluated numerically, a performance evaluation tool is needed that is fast, accurate, and available to the system planner on his/her desktop.

Figure 12.6 depicts the performance evaluation tools that have been proposed to aid in the configuration design phase of a FMS (31).

Static Allocation Models. Static allocation models compare the total processing time needed to produce a given workload to the time the equipment is available. The data provided by the routing sheets of the parts are

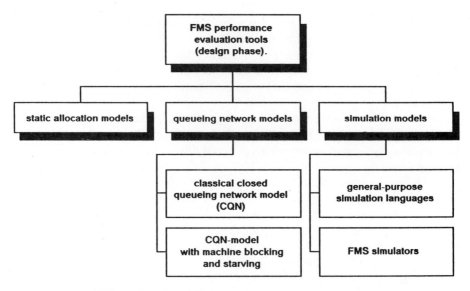

FIGURE 12.6. FMS performance evaluation tools.

aggregated to station-dependent mean processing times per server (machine). The station with the longest mean processing time is identified as the bottleneck (with a utilization of 100%). To estimate the utilization of the other stations, their mean processing times per server are divided by the mean processing time per server of the bottleneck station.

Static allocation models totally neglect the dynamic queueing effects occurring in a FMS and the influence of a limited number of pallets in the FMS. However, they are easy to understand and to use, implementable by means of microcomputer spreadsheet programs, and therefore are frequently used in industrial practice as a cheap evaluation tool that is easily adjustable to user-specific needs. Their accuracy is, in general, not very good, but they can be used in a first-step screening process to determine a lower bound for the bottleneck equipment with respect to a planned production volume.

Queueing Network Models. Queueing network models use results from queueing theory to model the stochastic dynamic interaction effects between workpieces and resources in a FMS.

The Classical Closed Queueing Network Model. In the classical closed queueing network (CQN) model, each equipment type is modeled as a single-stage queueing system. A CQN consists of M stations (with one or several parallel servers), each with an associated queue of unlimited size. Station 1 is the central server station (material handling system). Service times are assumed to be exponentially distributed with mean B_i. There is a fixed number of N general purpose (universal) pallets circulating in the CQN. A pallet after being processed at station i proceeds to the next station j with probability r_{ij}. These probabilities, which can easily be extracted from the routing sheets of the part types, are assumed to be independent of the actual state of the system. The state of the system is described by an M-tuple $\mathbf{n} = (n_1, n_2, \ldots, n_M)$, where n_i denotes the number of workpieces (pallets) in station i (waiting or in service). Gordon and Newell (16) showed that the joint queue length distribution has the following product form:

$$\text{Prob}\{\mathbf{n}\} = \frac{1}{G(N, M)} \prod_{i-1}^{M} \frac{(r_i B_i)^{n_i}}{A_i(n_i)} \,,$$

where

$n_i =$ number of pieces in station M_i (waiting or in service)

$r_i =$ stationary arrival probability of a workpiece at station i

$r_{ij} =$ transition probability between station i and j

$i = 1 =$ index of the MHS

$i > 1 =$ index of a processing station (machines; L/UL)

$M =$ number of stations (including MHS)

$N =$ number of universal pallets circulating in the FMS

$G(M, N)$ is the normalization constant over the state space and

$$A_i(0) = 1 \qquad\qquad i = 1, 2, \ldots, M \,,$$

$$A_i(k) = \prod_{j=1}^{k} a_i(j) \qquad k > 0; i = 1, 2, \ldots, M \,,$$

$$a_i(j) = \begin{cases} j & \text{if } j < S_i \\ S_i & \text{if } j \geq S_i \end{cases} \qquad i = 1, 2, \ldots, M \,.$$

The probability that the next station visited is station i, r_i, is the solution to the system of M linear flow balance equations:

$$r_i = \sum_{j=1}^{M} r_j r_{ij} \qquad i = 1, 2, \ldots, M \,.$$

It can be shown that the relevant performance criteria are functions of the normalization constant, $G(N, M)$, which can be computed with the convolution algorithm developed by Buzen (6, 7). For example, the throughput of the FMS, $X(N)$, is given by

$$X(N) = p_1 G(M, N - 1)/G(M, N) \,,$$

where p_1 specifies the mean number of operations (visits to a processing station) of a typical part. According to Little's law, the throughput is inversely related to the mean flow time D of parts in the FMS. If there are N pallets circulating in a FMS, the following relation holds: $X = N/D$. An auxiliary performance criterion that helps in the design phase to determine which type of equipment must be enlarged is the utilization of equipment type m, U_m. The utilization is the product of the throughput X_m and the mean processing time, B_m: $U_m = B_m X_m$.

To evaluate the performance of a FMS under consideration by a CQN model, the part type descriptions are aggregated into station-specific visiting frequencies and mean processing times (13). Together with the design variables (number of pallets, number of servers at the stations) these data are fed into the CQN algorithm to compute the performance criteria needed (Fig. 12.7).

Consider an example, where a FMS with two machine types with one machine each and one L/UL station producing four part types with estimated production ratios 20%, 20%, 40%, and 20%, respectively. The MHS is a rail-guided vehicle with a mean transportation time of 5.9 minutes. In addition to the load and unload operations, the L/UL station is used to refixture the parts on the pallets between two processing stages. The processing sequences of the part types are shown in Table 12.1.

Table 12.2 shows the results of the CQN model when four pallets circulate in the FMS. In the last two columns the throughputs and utilizations computed by the static allocation model are shown.

Since the numerical evaluation of the design alternative using the CQN model only takes about 2 seconds on a standard PC, it is easy to perform a sensitivity analysis with respect to the number of pallets circulating in the FMS. For the example considered, the results are shown in Figs. 12.8 and 12.9. With increasing number of pallets, the FMS throughput converges asymptotically to an upper bound, dictated by the processing time per server of the bottleneck station. In the current example, pallet numbers larger than

TABLE 12.1. Part Type Descriptions

Part type number	Operation Sequences (Processing Times in Minutes)												
1 (20%)	L/UL	M1	L/UL	M1	L/UL	M1	L/UL	M1	L/UL	M1	L/UL	M2	L/UL
	60	428	90	57	90	57	90	57	90	351	90	229	30
2 (20%)	L/UL	M1	L/UL	M2	L/UL								
	60	470	90	639	30								
3 (40%)	L/UL	M1	L/UL	M1	L/UL								
	60	185	90	130	30								
4 (20%)	L/UL	M1	L/UL	M1	L/UL	M2	L/UL						
	60	47	90	223	90	19	30						

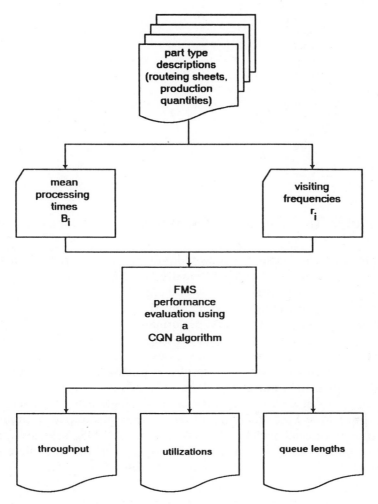

FIGURE 12.7. CQN model data flow.

TABLE 12.2. Throughputs and Utilizations (Given $N = 4$ Pallets)

	CQN Model		Static Allocation Model	
Equipment Type	Throughput (per Shift)	Utilization	Throughput	Utilization
L/UL	2.82	52.93%	3.10	58.19%
M2	0.56	34.77%	0.62	38.23%
M1	2.26	90.95%	2.48	100%
MHS	5.65	6.94%	6.18	7.6%

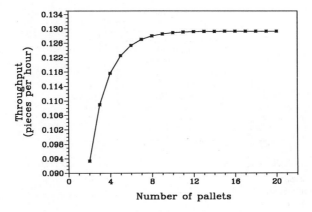

FIGURE 12.8. Throughput of the FMS versus number of pallets.

8 would be senseless because they would only increase the queue lengths of the bottleneck station (M1) almost linearly (see Fig. 12.9).

The CQN model was first used by Solberg (28) to model FMS and was implemented in a computer program called CAN-Q. It has been widely applied to evaluate the stationary performance of several real-life FMS. Despite the restrictiveness of the underlying assumption (infinite buffers, exponential processing times) in many cases, the approximations obtained coincided very well with simulation results. A discussion of the robustness of the model is presented in Co and Wysk (12).

The CQN model described so far assumes a single pallet type (universal pallets). In real-life FMS, there are often several types of pallets circulating in the system (special pallets). For these type of FMS, several heuristic approximation models have been developed (26, 29, 33) and successfully used to predict the behavior of existing FMS (34, 35).

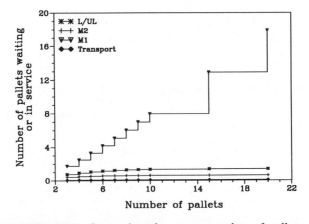

FIGURE 12.9. Queue lengths versus number of pallets.

Floss and Talavage (14) have integrated the CQN model approach (as implemented in the program CAN-Q) in a two-stage FMS design methodology, which in the first stage applies the fast evaluation capabilities of the queueing network model and in the second stage uses simulation to evaluate the detailed design of the FMS under realistic assumptions. Both stages are integrated via an artificial intelligence approach. An expert system interprets the results from the CQN model and provides recommendations for the modification of the design proposal that should be undertaken in order to enhance the performance of the FMS design alternative. If the recommendation is accepted by the designer, the design modification is chosen and a new CQN model run is performed. This process continues until no further meliorations are possible. At this point the second phase of the approach starts where a detailed simulation is constructed that is used to study the dynamic behavior of the best FMS design alternative found so far in detail. The procedure is supported by a knowledge base that contains design rules. In this concept the expert system is used to substitute the experienced systems planner.

The classical CQN model has also been used in conjunction with design optimization approaches, where several of the design variables mentioned earlier are optimized with respect to some criterion. For a discussion of these approaches see Tetzlaff (39).

The Closed Queueing Network Model for FMS with limited local buffers. The CQN model as described is classical in the sense that several assumptions, which may not hold in industrial FMS practice, are necessary to compute the stationary probability distribution of the system states. Whereas the assumption of exponentially distributed processing times does not pose severe problems, as many simulation tests with deterministic processing times have shown (34), the assumption of *unlimited local buffers* does. Real-life flexible manufacturing systems (FMSs) often suffer to a considerable extent under utilization losses, which are due to limited local and/or central buffer space and the limited velocity of the material handling system (MHS). In particular, two forms of performance deterioration must be taken into account: *blocking* and *starving*.

A station giving service to a workpiece is *blocked* if it cannot dispose of a finished piece because there is no buffer space. *Blocking* has been studied by several researchers (1, 2, 15, 22, 32, 42, 43, 44) with special emphasis on *MHS blocking*, where a cart arriving at a workstation finds its destination full and either must wait (block-and-wait mechanism) or put the workpiece into the central buffer (block-and-recirculate mechanism). The main type of blocking observed in industrial practice, however, is *machine blocking*, when a machine cannot dispose of a finished workpiece because there is no buffer space available—neither local nor central. Tempelmeier et al. (37) have made a proposal to estimate the influence of machine blocking on the performance of an FMS using standard closed queueing network algorithms.

They use a two-step approximation procedure, where in the first step the classical CQN model is employed to compute the stationary probability distribution of system states. This probability distribution is used to quantify the blocking conditions and to compute modified mean processing times, B_i. ($i = 2, \ldots, M$). These modified processing times are then iteratively fed into the CQN algorithm until a stable solution has been found.

The procedure has been tested against detailed simulation models written in the SIMAN IV language for a large number of different FMS configurations. As a representative example, in Fig. 12.10 the "classical" CQN results (CQN) are compared to the results computed with the proposed procedure (CQNBLK) and results of a SIMAN-simulation model (SIM). The FMS considered consists of five stations, two of which have a local buffer with room limited for two pallets. In the FMS there is no central buffer, giving rise to the possibility of blocking. The results presented in Fig. 12.10 show that in case of blocking the classical CQN model overestimates the performance of a FMS.

In industrial practice, machine blocking often does not seem to be an insurmountable problem during the design phase of a FMS, because in many cases it is possible to allow for enough central buffer space at comparatively moderate costs to prevent machine blocking. However, even if the central buffer space is large enough, the limited velocity of the material handling system (MHS) in combination with limited local buffer space may lead to a form of reduction of machine utilization that is known as *starving*.

If the local buffer space at a particular downstream machine is limited, the situation may arise that a workpiece to be processed on this machine next must wait in the local buffer of the upstream station (where it has just been processed) or is stored in the central buffer until the downstream machine is ready to take over the piece. Starving begins at the point in time when the downstream machine is ready to process the piece *and* the piece is

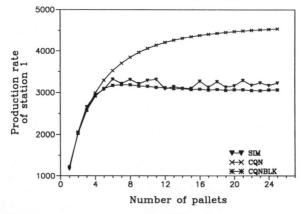

FIGURE 12.10. Results for a FMS with machine blocking.

still located elsewhere in the system, for example, waiting or in transit to the machine.

To complicate things, in real-life FMS *two additional factors* must be considered that have a direct influence on the workload of the MHS and thus indirectly upon the extent of starving. These are the *trips to and from the central buffer* and the *empty trips* of the MHS carts. Both types of trips constitute a form of "workload" for the MHS that is *not known at all in classical closed queueing network models* as described in the preceding paragraph.

Tempelmeier et al. (38) have proposed a two-stage procedure to account for starving in a FMS. They proceed as follows. They first estimate *idle trip times* and the *disposal trip times* to the central buffer and modify the mean transportation time accordingly, achieving a hopefully good approximation of the "real" MHS workload. Then station-specific mean starving times are estimated and added to the mean processing times at the stations. With the transportation and processing times modified in this way, a classical CQN algorithm is employed to compute the performance measures of a specific FMS design proposal under consideration.

The approximation scheme has been tested by means of numerous simulation runs with respect to several different FMS configurations. Figure 12.11 shows the production rate of a particular machine in a FMS including four machine types, some of which have only one local buffer space. As observed with the blocking example shown above, the approximation is very good, whereas the classical CQN approach overestimates the throughput of the machine and consequently of the FMS. The procedure has been transferred to the special pallet case as well and has shown good results (38).

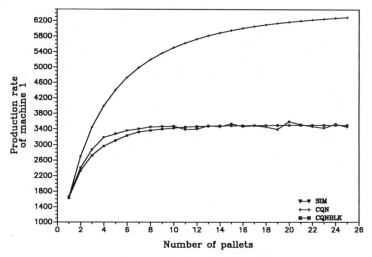

FIGURE 12.11. Results for a FMS with starving.

Simulation Models. Simulation models are certainly the most widely used computer-based tools to evaluate the performance of a FMS. This is in part due to the rapidly growing availability of fast microcomputers and high-quality simulation software. Simulation is the representation of a system by a numerical model (typically a computer model) to study the dynamic behavior of the model over time. Simulation models are typically used to evaluate a detailed design proposal including several aspects that cannot be modeled with CQN models in an adequate way (nonstationary behavior, tool availability, loading and scheduling strategies).

There are a great number of simulation tools available to simulate FMS (8). These may be classified as *special-purpose FMS simulators* (e.g., MAST, GRAFSIM), *general-purpose simulation languages* (e.g., GPSS, SLAM II, SIMAN IV, SIMSCRIPT II), and *code generators* that, based on a specific application domain, can be used to generate a FMS simulation model written in a general purpose simulation language.

A *FMS simulator* is a software system able to simulate a generic FMS. The specific structure of the FMS to be simulated is specified by the user via a set of parameters. The rapid development of computer graphics had led to the provision of graphics-based user interfaces for the definition of the FMS simulation model (27). Whereas special-purpose FMS simulators are the fastest tool to perform a FMS simulation, requiring only very little input, they are limited to those types of FMS that the software developers had in mind when they programmed the simulator. Simulation experience shows that simulation users are greedy in the sense that they want to include more and more details into a simulation model (e.g., a particular scheduling rule not provided by the simulator), once the first simulation runs have been evaluated. In this case, a user often reaches the capability limits of the simulator being employed and is forced to switch to a general-purpose simulation language, for example, SIMAN IV. However, because considerably more time is requested to write a FMS simulation program from scratch in a general-purpose simulation language, *code generators* have recently emerged, offering the potential to generate application-specific FMS simulation code. The output of a FMS code generator is a FMS simulation program written in a general-purpose simulation language, adaptable by the user to particular needs. See Schroer and Tseng (24), who uses artificial intelligence techniques to generate manufacturing simulation code (11, 25). Like simulators, the facilities of the code generators are limited to the application domain that the developer of the code generator kept in mind. However, in many cases, the generated simulation program may serve as a skeleton from which a more detailed program may evolve by manual programming.

Unlike conventional code generators that are bound to a clearly defined application domain, the SIMAN Module processor (SMP) (36) offers a platform to generate program code based on user-defined modules. These modules are stored in a module library and can be used to generate a simulation model for the application domain for which the module library was developed. The module library developed for FMS comprises modules depicting L/UL stations, central buffer places, machining stations with and without local buffers, and so on. With this open concept, the type of simulation model that can be generated automatically may evolve under the control of the user parallel to the development of FMS technology.

Simulation models are widely used to evaluate FMS proposals. Mellichamp and Wahab (21) describe an expert system for the design of a FMS that uses a simulation model to evaluate the performance of a FMS proposal generated by the expert system.

Integrated Application of Performance Evaluation Tools. The modeling approaches described may be combined in an efficient way to solve the performance evaluation problem during the FMS design phase. Typically, in the first iterations of the design phase there is a need for a fast screening and elimination of nonpromising alternatives. Starting with the (usually large) set of design alternatives, the classical CQN model may be applied to get a first idea of the fundamental functional interactions between the several components of the FMS design alternative. In this phase, one would perform a *sensitivity analysis* with regard to the number of pallets in the system, for example. Based on the insights gained, more detailed CQN models including the *blocking* and *starving* effects of limited local buffers could be used to analyze the reduced set of design alternatives, leaving usually a few alternatives that would be evaluated by use of a detailed simulation model. The successive reduction of the set of FMS design alternatives is shown in Fig. 12.12.

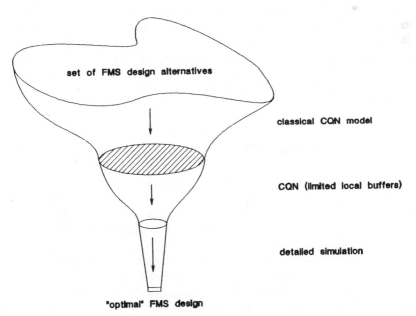

FIGURE 12.12. Successive reduction of the set of FMS design alternatives.

Taken together, the process outlined is a systematical search process with increasing modeling detail. For each step in the process, there are suitable software tools available. In the first step, CAN-Q (28), MVAQ (33), MULTIQ (29), FFS-EVAL (35), or any general-purpose CQN model evaluation software [PNET (6), QUANTRUM (3)] may be applied. The CQN model with limited local buffers may be evaluated with FFS-EVAL. To perform a simulation analysis of the FMS design proposal, the SIMAN Module Processor combined with a module library for FMS or any code generator for FMS may be used.

12.3.3. *Economic Evaluation of FMS*

In the design process described so far, the quantitative, nonmonetary aspects of the FMS performance using criteria like production volume and utilization are considered. However, it must be emphasized that cost and financial investment considerations play an important role in each phase of the design process. Classical methods of investment appraisal (e.g., discounted cash flows and risk analysis) may be applied to gain insight into the financial consequences of the investment decision.

As with other long-term investment decisions, forecasts of future developments in the marketplace and in the technological field are needed. The nature of a FMS, its modularity, and the ability to exchange components of the machines according to new technological developments, as well as the ease of assigning new product types to the FMS in the future, make it very difficult to financially evaluate a FMS proposal. Troxler and Blank (40) give an overview of current research efforts trying to aggregate the tangible and nontangible effects describing the value of a FMS in an overall system value. It must be stated that the problem of economic justification of FMS has not yet been solved.

12.4. CONCLUDING REMARKS

The overall process of designing a flexible machining system has been described. It has been shown that closed queueing network models are widely applicable tools for performance evaluation of a FMS design alternative. These methods can be systematically integrated into the process of searching for the best FMS proposal according to quantitative performance criteria. Once the best proposal has been identified, a financial evaluation can be performed and the other nontangible effects neglected so far can be taken into account. In a multicriterion decision-making process, the decision to invest in FMS technology can be taken.

REFERENCES

1. Akyildiz, I. F. (1988). Mean Value Analysis for Blocking Queueing Networks, *IEEE Transactions on Software Engineering*, Vol. 14, No. 4, pp. 418–428.
2. Akyildiz, I. F. (1988). On the Exact and Approximate Throughput Analysis of Closed Queueing Networks with Blocking, *IEEE Transactions on Software Engineering*, Vol. 14, No. 1, pp. 62–70.
3. Barack, R., Lemoine, A., and Wenocur, M. (1990). *QUANTRUM—Queueing Analytic Network Transaction Resource Utilization Model*, Introductory Guide, Ford Aerospace, Western Development Laboratories Division.
4. Barfield, W., Hwang, S.-L., and Chang, T.-C. (1986). Human-Computer Supervisory Performance in the Operation and Control of Flexible Manufacturing Systems. In: *Flexible Manufacturing Systems: Methods and Studies*, edited by A. Kusiak, North-Holland, Amsterdam.
5. Browne, J., Chan, W. W., and Rathmill, K. (1985). An Integrated FMS Design Procedure, *Annals of Operations Research*, Vol. 3, pp. 207–237.
6. Bruell, S. C., and Balbo, G. (1980). *Computational Algorithms for Closed Queueing Networks*, North-Holland, New York.
7. Buzen, J. (1973). Computational Algorithms for Closed Queueing Networks with Exponential Servers, *Communications of the ACM*, Vol. 16, pp. 527–531.
8. Carrie, A. S. (1986). The Role of Simulation in FMS. In: *Flexible Manufacturing Systems: Methods and Studies*, edited by A. Kusiak, North-Holland, Amsterdam.
9. Carrie, A. (1988). *Simulation of Manufacturing Systems*, Wiley, Chichester.
10. Carrie, A. S., Adham, E., Stephens, A., and Murdoch, I. C. (1984). Introducing a Flexible Manufacturing System, *International Journal of Production Research*, Vol. 23, No. 6, pp. 907–916.
11. Co, H. C., and Chen, S. K. (1988). Design of a Model Generator for Simulation in SLAM, *Engineering Costs and Production Economics*, Vol. 14, pp. 189–198.
12. Co. H. C., and Wysk, R. A. (1986). The Robustness of CAN-Q in Modelling Automated Manufacturing Systems, *International Journal of Production Research*, Vol. 24, No. 6, pp. 1485–1503.
13. Dallery, Y. (1986). On Modelling Flexible Manufacturing Systems Using Closed Queueing Networks, *Large Scale Systems*, Vol. 11, pp. 109–119.
14. Floss, P., and Talavage, J. (1990). A Knowledge-Based Design Assistant for Intelligent Manufacturing Systems, *Journal of Manufacturing Systems*, Vol. 9, No. 2, pp. 87–102.
15. Gershwin, S. B., and Schick, I. C. (1983). Modeling and Analysis of Three-Stage Transfer Lines with Unreliable Machines and Finite Buffers, *Operations Research*, Vol. 31, pp. 345–380.
16. Gordon, W. J., and Newell, G. F. (1967). Closed Queueing Networks with Exponential Servers, *Operations Research*, Vol. 15, pp. 252–267.
17. Huang, P. Y., and Chen, C.-S. (1986). Flexible Manufacturing Systems: An Overview and Bibliography, *Production and Inventory Management*, Vol. 27, No. 3, pp. 80–90.

18. Kuhn, H. (1990). *Einlastungsplanung von flexiblen Fertigungssystemen*, Physica, Heidelberg (in German).

19. Kusiak, A. (1986). Flexible Manufacturing Systems: A Structural Approach, *International Journal of Production Research*, Vol. 23, No. 6, pp. 1057–1073.

20. Kusiak, A. (1990). *Intelligent Manufacturing Systems*, Prentice-Hall, Englewood Cliffs, NJ.

21. Mellichamp, J. M., and Wahab, A. F. A. (1987). An Expert System for FMS Design, *Simulation*, Vol. 48, No. 5, pp. 201–208.

22. Perros, H. G. (1984). Queueing Networks with Blocking: A Bibliography, *Performance Evaluation Review*, Vol. 12, No. 2, pp. 8–12.

23. Ranta, J., and Tchijov, I. (1990). Economics and Success Factors of Flexible Manufacturing Systems: The Conventional Explanation Revisited, *International Journal of Flexible Manufacturing Systems*, Vol. 2, pp. 169–190.

24. Schroer, B. J., and Tseng, F. T. (1989). An Intelligent Assistant for Manufacturing System Simulation, *International Journal of Production Research*, Vol. 27, No. 10, pp. 1665–1683.

25. Schroer, B. J., Tseng, F. T., Zhang, S. X., and Wolfsberger, J. W. (1988). Automatic Programming of Manufacturing Simulation Models. In: *Proceedings of the 1988 Summer Simulation Conference*, Society for Computer Simulation, San Diego.

26. Shalev-Oren, S., Seidmann, A., and Schweitzer, P. J. (1985). Analysis of Flexible Manufacturing Systems with Priority Scheduling: PMVA, *Annals of Operations Research*, Vol. 3, pp. 115–139.

27. Siemens AG (1988). *GRAFSIM—Simulationssystem für Fertigungseinrichtungen*, Systembeschreibung (in German).

28. Solberg, J. J. (1977). A Mathematical Model of Computerized Manufacturing Systems, *Proceedings of the Fourth International Conference on Production Research*, Tokyo, Japan.

29. Solot, P., and Bastos, J. M. (1988). MULTIQ: A Queueing Model for FMSs with Several Pallet Types, *Journal of the Operational Research Society*, Vol. 39, No. 9, pp. 811–821.

30. Suresh, N. C., and Sarkis, J. (1989). A MIP Formulation for the Phased Implementation of FMS Modules, In: *Proceedings of the Third ORSA/TIMS Conference on Flexible Manufacturing Systems: Operations Research Models and Applications*, edited by K. E. Stecke and R. Suri, Elsevier, Amsterdam.

31. Suri, R. (1985). An Overview of Evaluative Models for Flexible Manufacturing Systems, *Annals of Operations Research*, Vol. 3, pp. 13–21.

32. Suri, R., and Diehl, G. W. (1986). A Variable Buffer-Size Model and Its Use in Analyzing Closed Queueing Networks with Blocking, *Management Science*, Vol. 32, No. 2, pp. 206–224.

33. Suri, R., and Hildebrant, R. R. (1984). Modelling Flexible Manufacturing Systems Using Mean-Value Analysis, *Journal of Manufacturing Systems*, Vol. 3, No. 1, pp. 27–38.

34. Tempelmeier, H. (1988). Kapazitätsplanung für Flexible Fertigungssysteme, *Zeitschrift für Betriebswirtschaft*, Vol. 58, No. 9, pp. 963–980 (in German).

35. Tempelmeier, H. (1989). Konfigurierung flexibler Fertigungssysteme auf dem

Personal Computer, *ZwF/CIM Zeitschrift für wirtschaftliche Fertigung und Automatisierung*, Vol. 84, No. 8, pp. 448–450 (in German).

36. Tempelmeier, H. (1990). Template-Bases Automatic Generation of SIMAN IV Simulation Models. In: *Modelling and Simulation, Proceedings of the 1990 European Simulation Multiconference*, edited by B. Schmidt, Society for Computer Simulation, San Diego.

37. Tempelmeier, H., Kuhn, H., and Tetzlaff, U. (1989). Performance Evaluation of Flexible Manufacturing Systems with Blocking, *International Journal of Production Research*, Vol. 27, No. 11, pp. 1963–1979.

38. Tempelmeier, H., Kuhn, H., and Tetzlaff, U. (1991). Performance Evaluation of Flexible Manufacturing Systems with Starving. *Proceedings of the International Conference on Modern Production Concepts*, Hagen, August 20–24 1990, edited by G. Fandel and G. Zäpfel, Springer, Berlin.

39. Tetzlaff, U. (1990). *Optimal Design of Flexible Manufacturing Systems*, Physica, Heidelberg.

40. Troxler, J. W., and Blank, L. (1989). A Comprehensive Methodology for Manufacturing System Evaluation and Comparison, *Journal of Manufacturing Systems*, Vol. 8, No. 3, pp. 175–183.

41. Wemmerlöv, U., and Hyer, N. L. (1987). Research Issues in Cellular Manufacturing, *International Journal of Production Research*, Vol. 25, No. 3, pp. 413–431.

42. Yao, D. D., and Buzacott, J. A. (1985). Modeling a Class of State-Dependent Routing in Flexible Manufacturing Systems, *Annals of Operations Research*, Vol. 3, pp. 153–167.

43. Yao, D. D., and Buzacott, J. A. (1986). Models of Flexible Manufacturing Systems with Limited Local Buffers, *International Journal of Production Research*, Vol. 24, No. 1, pp. 107–118.

44. Zhuang, L., and Hindi, K. S. (1990). Mean Value Analysis for Multiclass Closed Queueing Network Models of Flexible Manufacturing Systems with Limited Buffers, *European Journal of Operational Research*, Vol. 46, pp. 366–379.

CHAPTER THIRTEEN

Design of Assembly Systems

JERRY L. SANDERS

Department of Industrial Engineering, University of Wisconsin—Madison

13.1. INTRODUCTION

Assembly of fabricated parts into a final product has become an increasingly important part of modern manufacturing for a number of reasons. In many industries, the labor costs in assembly amount to over 50% of the total labor cost in the entire product (2). In addition, since assembly has proved to be relatively resistant to automation in comparison to fabrication, this labor cost has been rising as a percentage of total labor cost in the product in many industries. This set of difficulties has led many firms in the 1980s to send labor intensive assembly operations overseas in search of cheap labor or to "outsource" the entire production of the assembly to other firms. These attempts to reduce the cost of assembly that led many companies to overseas assemblers and outsourcing caused the parent firm to lose manufacturing capability and capacity, often created new competition in the parent firm's own market niches, made component and product quality very difficult to control and inserted long delays and high in-process inventories in the product production cycle. Consequently, many firms have made a major effort to redesign the product for easier and less expensive assembly and/or to move to assembly automation.

The organization of assembly operations can generally be grouped into two major groups: (a) project-oriented assembly and (b) assembly flow lines. (Functional or job shop organization is relatively rare in assembly operations.) Project-oriented assembly operations include building construction, ship building, aircraft assembly, and other assembly operations where fabrication and/or delivery of components to the assembly area occur in very small lots (often size 1) and/or where the final product does not move

Intelligent Design and Manufacturing, Edited by Andrew Kusiak.
ISBN 0-471-53473-0 © 1992 John Wiley & Sons, Inc.

during the assembly process. By way of contrast, the assembly flow line moves the product from one assembly station to another and at each station a substantial inventory of components is used to add components to the passing products in a predefined series of assembly operations. Component parts may be fabricated at the assembly station but more frequently they are fabricated "off-line" and moved to the assembly flow line in moderate sized batches. The literature and analysis tools for project-oriented assembly have grown up separately from that of the assembly flow line. Some of the first modern analysis tools for this set of applications included the development of critical path and PERT approaches to project planning, design, and management. More recently, Wilhelm et al. (9) developed new tools for analyzing manufacturing systems of this type. We do not consider this class of system further in this development.

Since the introduction of assembly line concepts into automobile assembly in the early 1900s, the organization of the assembly flow shop has become increasingly automated and sophisticated. In the earliest forms of assembly systems there was little automation except a tow line that pulled the undercarriage of the item to be assembled (the automobile) along at a constant rate and forced the manual assembly tasks to be executed at a predetermined rate. By the 1940s, one saw the emergence of more automated forms of assembly including totally automated assembly machines for the assembly of such items as artillery shell fuses. These systems were mechanical marvels but were totally specialized in the sense that they could produce only one type of assembly per machine albeit at very high rate. After the second World War, this class of automation was put to work making such items as electrical plugs and switches and a variety of other small household and industrial assemblies. More recently, one sees increasing sophistication in both the variety and technology of assembly systems concepts.

Boothroyd et al. (2) classify assembly systems or machines into two primary classes. These include synchronous and asynchronous systems. By their nature, the synchronous systems are usually totally automated and often are "hard-automated," that is, capable of assembling only a single product type. These systems are currently used in both mechanical and electrical assembly work with outstanding success in tasks that involve components with low mass and high production volume requirements. In mechanical assembly applications, such as assembly of electrical fuses, fountain pens, and similar products, where the number of components is typically less than 10, these systems are capable of production rates of up to 7200 assemblies per hour in certain applications and rates of 1800–3600 per hour are not unusual. It would be inaccurate to characterize these systems as exclusively dedicated to low mass assembly operations because auto body framing lines with hard automated or robotic welding would also fall in this category.

13.2. PROBLEMS IN ASSEMBLY SYSTEMS DESIGN

In this section we examine formal statements of the analysis and design questions that an engineer can be expected to encounter in the systems design of an assembly system. We use the term "systems design" advisedly because there are an enormous number of engineering design problems that we cannot possibly address here that are related to hardware problems involved in the design of parts feeding systems, fastening methods, material handling methods, design of transport systems, and so on. (For a good introduction to these hardware issues see ref. 6.) We are forced here to consider only a limited number of design issues that occur at a relatively macroscopic view of the assembly system. These issues include those that involve choices that have a substantial effect on the overall production rate of the system. Variables in this class of design parameters include grouping of assembly tasks into stations, choices of technologies (from among a well defined set) for station formation, choice of the number of workpieces to be loaded on an assembly system, choice of the number of workpiece buffer units to be placed between each pair of stations, and choice of the number of human operators required for maximum productivity of a large auto-mated system.

Any informed approach to system design depends on the existence of a set of analysis tools and in the sections that follow we present a series of analysis and (rarely) optimization tools and attempt to explore the implica-tions of these tools for the problems of systems design outlined previously. The reader will discover that the tools developed are all analytical and mathematical in nature as opposed to approaches that use discrete event computer simulation as the base analysis methodology. The power of simulation is well understood and problems of great complexity can often be modeled using these methods, but in modern assembly systems design these tools encounter two classes of problems. Both problems stem from the fact that simulation results take a good deal of time to produce and accurately interpret. Since design, by its very nature, requires the examination of large numbers of alternatives and since simulation does not offer much qualitative insight into the nature of the performance of complex systems, it has become an unwieldly tool for assembly systems design and analysis. Having said this, it must be recognized that it is still the "court of last resort" in the analysis of complex systems and often is the baseline against which we validate our analytic models.

13.3. DESIGN OF MANUAL ASSEMBLY FLOW LINES

We mention only briefly the design issues in the design of manual assembly flow lines. Since the turn of the century, the principal design problem for these lines has been to accomplish a given production rate for the assembly of a single product at a total labor cost that is a minimum, or to design the line in such a way that worker idle time is minimized, or to design the line so that a maximum throughput is achieved with a given number of workers. These problems can all be seen as variants of a problem known as "assembly line balancing."

Analysis and design methods in this arena have concentrated primarily on methods for estimating assembly cycle times, methods to reduce cycle times, and methods to attempt to balance the line by making the station cycle times as equal as possible. (See ref. 1 for a recent review of the literature of line balancing.)

The principal historical approach to manual assembly flow line design centered around the design of single-product manual assembly operations and the major technical problem was the solution of one of the variants of the assembly line balancing problem. The required production rate was specified as well as a set of timed assembly tasks that were required. In addition, a specific technological sequence for the execution of the tasks was assumed given (e.g., a bolt would have to be inserted into a transmission cover before it was torqued into place—bolt insertion precedes the torqueing operation). With these data the line balancing problem attempted to find the minimum number of stations to accomplish the tasks at the specified production rate or an attempt was made to find the station design that minimized the total worker idle time summed over all stations. This design formulation makes sense under the assumption of single-product manual assembly because minimizing the number of stations would minimize the number of workers required. The same can be said of attempting to minimize the total idle time since idle time costs the same as productive time and minimizing idle time minimizes the cost of assembly for the specified production volume.

13.4. MULTIPRODUCT ASSEMBLY SYSTEM DESIGN (EQUIPMENT SELECTION AND TASK ASSIGNMENT)

On the other hand, when one considers flow line design using assembly automation and the requirement for the assembly of several distinct products on the same equipment, then minimizing the idle time or the number of stations makes very little sense. We assume that production volumes for each of the product types are specified and that assembly task times and their sequences have been predetermined. In addition, we assume that for each assembly task a number of assembly technology and tooling alternatives are available to accomplish the task. The problem now becomes one of selecting the technology and tooling for each task and grouping that technology into assembly stations to be located in an assembly flow line that will minimize the total cost of production at the required volumes for all the products to be assembled. An important aspect of this formulation is that we assume that all cycle times are deterministic and known and that no other stochastic influences are present to disturb the predictable performance of the system. This is useful as a first approximation but it is unrealistic in practice. All the models of system performance beyond this section explicitly incorporate random events in the models.

In attempting to formulate and solve this set of problems we follow Graves and Redfield (3). To introduce the methodology, we consider an example where there are two models of an assembly to be manufactured. Model A requires assembly tasks 1, 2, 3, 5, 6, 8, 9, 10, and 12. Model B requires the execution of tasks 1, 2, 4, 5, 6, 7, 10, 11, and 12.

The tasks must be executed in the sequence given and a partially ordered graph showing the task sequence is shown in Fig. 13.1.

We assume that two different types of assembly automation are available and the task capability, task times, and special tooling requirements are shown in Table 13.1.

We assume one 8-hour shift production 240 days/year (1920 hours of production). We assume a required production of 216,000 per year for each product and we split the production time in equal halves or 960 hours each per product. As a consequence, the line must be capable of turning out either product once every 16 seconds; that is, the cycle time must be a maximum of 16 seconds. We assume that the transport delay is 2 seconds per move, leaving us a 14-second interval in which to execute the assembly operations. Therefore any assembly station on the final line design may use at most 14 seconds to accomplish one or more of the assembly tasks.

The next step is to generate a set of "candidate workstations." A candidate workstation is simply a grouping of "adjacent" assembly tasks from the graph of Fig. 13.2.

Note that a candidate workstation may not be a feasible workstation because in order to be a feasible workstation the set of tasks must be able to be accomplished by at least one of the alternative automation modes within the cycle time required—that is, 12 seconds in this case. We proceed as follows. We first generate all possible candidate workstations and then

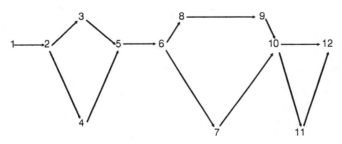

FIGURE 13.1.

TABLE 13.1.

	Type 1			Type 2		
Task	Operation Time (seconds)	Tool #	Tool Cost (%)	Operation Time (seconds)	Tool #	Tool Cost (%)
1	5.6	100	11,000	—	—	—
2	3.6	120	8,000	2.4	220	8,000
3	3.6	121	3,000	2.4	221	3,000
4	1.8	121	3,000	1.8	221	3,000
5	1.8	121	3,000	1.8	221	3,000
6	3.6	131	8,000	2.4	231	3,000
7	3.6	141	7,000	3.0	241	7,000
8	2.0	142	2,000	2.0	242	2,000
9	4.0	150	7,000	3.6	250	7,000
10	7.2	160	4,000	7.2	260	4,000
11	5.4	170	10,000	—	—	—
12	5.4	170	10,000	—	—	—
Annualized fixed cost			$40,000			$50,000
Variable cost/hour			$ 4.8			$ 5.3
Tool change time (seconds)			$ 2.0			2.0

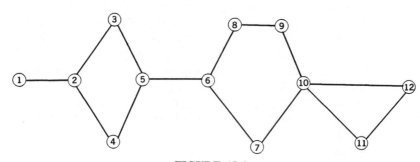

FIGURE 13.2.

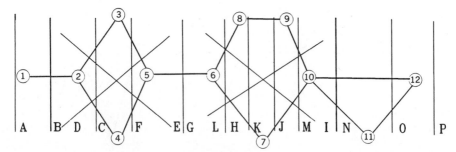

FIGURE 13.3.

examine each candidate for feasibility. The set of all possible workstations can be generated by generating all the cut sets of the graph of Fig. 13.2 and then considering the set of all cut sets and their differences. See Fig. 13.3.

This final set is the set of all candidate workstations. [Graves and Redfield (3) discuss an algorithm for developing the set of candidate workstations.] We now proceed to test feasibility of each workstation. Consider the feasibility of a candidate workstation composed of tasks 4, 5, 6, and 7. With mode 1 technology, we have the following time requirements for this station:

MODE 1

Product	Operation time	Tool Change Time	Total Station Time
1	5.4	2.0	7.4
2	10.8	4.0	14.8

We can see immediately that mode 1 automation violates the station cycle time requirement on product 2 (12 seconds); hence the station is infeasible using automation mode 1. Actually, this station is closer to feasible than it would appear because the total cycle time requirement is 14 seconds including transport time and of the 14.8 seconds of total station time the transport (2 seconds) could overlap with tool change operations and as a consequence the station misses feasibility by only 0.8 seconds (rather than 2.8 seconds). With mode 2 technology we have the following:

MODE 2

Product	Operation time	Tool Change Time	Total Station Time
1	4.2	2.0	6.2
2	10.0	4.0	14.0

The workstation is now feasible with mode 2 automation.

If more than one automation mode is feasible for a candidate workstation, the least cost alternative is chosen by calculating the annualized fixed costs associated with original equipment acquisition and tooling and that is added to the annual variable costs, which are calculated based on the number of hours of operation. At the end of this process, a lowest cost technology has been selected for every possible feasible station. We now need to find a sequence of feasible least cost stations that will accomplish the entire set of assembly tasks. Graves and Redfield (3) proceed as follows.

A network representation of the set of feasible workstations is developed, as shown in Fig. 13.4, where each node in the diagram represents a cut in the original task network. Each arc in Fig. 13.4 connects a pair of cuts in Fig. 13.3 and represents a candidate workstation. (Actually, only arcs that represent feasible workstations need be entered.)

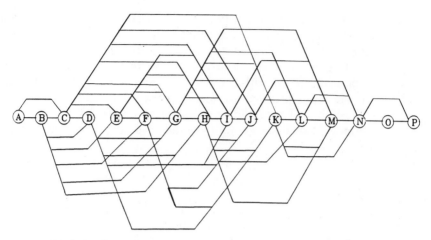

FIGURE 13.4. Network representation of feasible workstations.

To find the optimal (least cost) system, we label each arc with the annual cost of the least cost technology for that feasible workstation and we find the least cost path from node A to node P in the diagram. The algorithm to accomplish this task is known in operations research as the shortest path algorithm and is easily computerized. The results for this example are shown below.

LEAST COST SYSTEM

Workstation	Automation Mode	Task Numbers	Tool Numbers	Station Cost
1	1	1, 2	100, 120	$68216
2	2	3, 4, 5, 6, 7, 8	221, 231, 241, 242	$75167
3	1	9, 10	150, 160	$60216
4	1	11, 12	170	$59216

Total System Cost = $262,824

WORKSTATION SUMMARY BY PRODUCT

	Product 1		Product 2	
Workstation	Station Time	Tasks	Station Time	Tasks
1	11.2	1, 2	11.2	1, 2
2	12.6	3, 4, 5, 6, 7, 8	13.0	4, 5, 6, 7
3	13.2	9, 10	7.2	10
4	5.4	11, 12	10.8	11, 12

This analysis is predicated on an equal production volume for each product of 216,000 per year. We could now run the analysis several times for various production volumes of each product type and examine how the design of the line and the total system cost change as the product volume and mix change. One modification that is required as product volumes increase is the introduction of stations involving two (or more) pieces of assembly hardware in order to cut the cycle times to feasible levels. For example, if we have an automation mode (say mode 1) that has a cycle time of 12 seconds on a single task but the production volume requires a cycle time of 10 seconds, then we would introduce a new automation mode that would consist of two units of mode 1 (mode 1(2)) operating at a combined cycle time of 6 seconds (one-half of the old time of 12 seconds) and we would run the algorithm with this new option. Actually, nothing requires that the station technologies be automatic. We could include manual mode stations as well, and often for small production volumes manual stations will prove optimal.

The advantage of this algorithm is that it is quite simple and permits a good deal of exploration of the impact of various parameters on the final system design. Graves and Redfield (3) provide a more extensive example of the assembly of an automotive steering column with three models and three types of assembly technology and they examine six different levels of production. As one might expect, at low volumes manual assembly is generally the most cost effective and as volume increases fixed automation comes into the picture more and more. Also, as volume increases, the number of tasks done per station declines and doubling of resources in some stations is required. However, the results of the sensitivity analysis are not all that straightforward because at the lowest product volume a hard-automated station is the optimal solution, but at the next higher level of production examined the hard-automated station drops out of the solution to be replaced by a manual station.

13.4.1. Performance Analysis of Synchronous Assembly Systems

As is the case for every automated manufacturing station, the assembly stations on a synchronous assembly machine are subject to parts misfeeds, insertion jams, or joining system failure. (See Figs. 13.5 and 13.6. Figure 13.5 shows a synchronous rotary dial machine. Figure 13.6 shows a synchronous carousel system.) The machine is usually designed with sensors to determine when these events occur but the restoration of the station to operating conditions must be done by a human operator.

One of the major weaknesses of the synchronous system is that when a single station experiences a misfeed or jam the entire system stops because the transfer system stops. What one would like to find is the dependence of the expected production rate of the machine as a function of the rates of malfunction at each station and as a function of the number of human operators available to restore the system to an operating conditions. In addition, the time required for the operators to achieve the restoration must

FIGURE 13.5. Synchronous rotary dial machine. (Courtesy of Gilman Engineering and Manufacturing Company.)

FIGURE 13.6. Synchronous carousel system. (Courtesy of Gilman Engineering and Manufacturing Company.

also be accounted for in the model. Once a performance model is developed, design decisions can be made that will indicate the practical limits on the number of stations that can be considered in a single machine and the number of operators necessary to achieve reasonable production rates. It is important to note that the type of machine failures considered here are the operation or cycle-dependent failures (CDF). Standard machine breakdowns are usually considered to be time-dependent failures (TDF) and must be modeled separately. We examine both mechanisms in detail in the section that follows.

We assume that CDF failures at each station are mutually independent of failures at other stations. More than one workstation can experience a CDF in the same machine cycle and a single operator is all that is required to clear a failure at a single station. The mean clear time is usually fairly short and is on the order of a low multiple of the machine cycle time. We assume that a failure is discovered at the end of the cycle in which it occurs and subsequently the machine will be shut down until all failed workstations are cleared. If the assembly involved in the CDF has to be scrapped, then it is either removed from the machine or in some cases the machine "knows" the assembly is bad and at subsequent stations the assembly is passed through without further work. A station receiving an empty fixture or a part marked as bad is not subject to CDFs during that cycle.

In the subsequent development we follow Leung and Kamath (7).

We introduce the following notation:

c (Deterministic) cycle time of the machine if no failures occur; $c = c_t + \max\{cp_i: i = 1, \ldots, M\}$, where c_t is the station to station transport time and cp_i is the assembly cycle time at station i

γ_i $P\{$workstation i experiences a CDF during a cycle$\}$

q_i $P\{$workstation i produces a good part during a cycle$\}$

P $P\{$the machine experiences a CDF (one or more) during a cycle$\}$

D Random duration of a line stoppage due to CDFs

λ_i' Mean production rate of good parts at workstation i, in parts per unit time

ρ_i' Mean proportion of time that workstation i is up and working (not idle, jammed, or forced down)

h Efficiency of the machine = (mean actual production in a given time period)/(production in the same period if no workstations have failed)

W Mean time a good part spends in the machine from entry to exit at the last station

θ_i Mean working time between failures for station i

Steady-State Performance. For $j = 1, \ldots, M$,

$\alpha_i = P($workstation i experiences a CDF in a cycle given that it received a part in that cycle$)$

$\beta_i = P$(a part is scrapped at workstation i given that the workstation encountered a CDF while processing the part)

Hence

$$P\{\text{workstation } j \text{ gets a part in a given cycle}\} = \prod_{k=1}^{j-1} (1 - \alpha_k \beta_k)$$

so

$$\gamma_j = \alpha_j P\{\text{workstation } j \text{ gets a part}\} = \alpha_j \prod_{k=1}^{j-1} (1 - \alpha_k \beta_k).$$

The machine will be working at the end of any cycle if no station experiences a jam during the cycle, so if p is the probability that the machine experiences at least one jam in a cycle then

$$1 - p = \prod_{i=1}^{j-1} (1 - \gamma_i).$$

The mean cycle time of the machine can be calculated as

$$(1 - p)c + p(c + E(D)) = c + pE(D)$$

and the mean throughput rate at station j is $\lambda'_j = q_j/(c + pE(D))$, where

$$q_j = \prod_{k=1}^{j} (1 - \alpha_k \beta_k).$$

The utilization at station j (ignoring the transfer time ct) is

$$\rho'_j = \left[cp_j \prod_{k=1}^{j-1} (1 - \alpha_k \beta_k) \right] / [c + pE(D)].$$

All the major performance parameters can now be calculated if we can find a way to obtain $E(D)$.

In the case where there is a single operator on the machine, the computation of $E(D)$ is quite simple and is not dependent on the distributional form of the pdf of D.

Let Z_i be an indicator random variable defined at the end of the machine cycle that is one if workstation i has jammed and zero otherwise, given that the machine is down. Let

$$Z = \sum_{i=1}^{M} Z_i,$$

$$E(Z) = \sum_{i=1}^{M} EZ_i$$

$$= \sum_{i=1}^{M} P\{\text{workstation } i \text{ jammed} \mid \text{the machine is down}\}$$

$$= \sum_{i=1}^{M} \frac{\gamma_i}{p}$$

Consequently

$$E(D) = \mu E(Z) = \frac{\mu}{p} \sum_{i=1}^{M} \gamma_i \, .$$

Another useful statistic is

$$P1 = P\{\text{more than one station has jammed} \mid \text{the machine is down}\}$$

$$= 1 - P\{\text{exactly one station is jammed} \mid \text{the machine is down}\}$$

$$= 1 - \frac{1}{p} \sum_{i=1}^{M} \gamma_j \prod_{k=1}^{M} (1 - \gamma_k) \quad (\text{where } k \neq j) \, .$$

$P1$ is useful because if it is near zero then it is an indication that a single operator is enough to handle the jams on the machine.

In the event that more than one operator is required to keep the production rate of the machine close to its maximum achievable performance, then the models become more complex. Leung and Kamath (7) derive exact results for the case where the number of operators L is larger than 1 but where D is assumed to have an exponential distribution. They also derive upper and lower performance bounds for the case where D has a general distribution and where L is larger than 1.

In general, for high-speed synchronous assembly machines it is assumed that M is on the order of 10 or less. As a consequence, since to get reasonable production rates the maximum value of c_k will often be less than 0.02, we expect $P1$ to be small and a single operator is almost always the most that can be justified on economic or practical grounds.

Time-Dependent Failures. Now let us add time-dependent failures to the picture. The availability of a series machine with repair is a, where

$$\frac{1}{a} = \left[1 + \sum_{i=1}^{M+1} \frac{(\text{mean down time of workstation } i)}{(\text{mean up time of workstation } i)} \right]$$

$$= \left[1 + \sum_{i=1}^{M+1} \frac{ct_i}{\theta_i(c + pE(D))} \right] .$$

Now the throughput of station i corrected for time-dependent failures is

$$\lambda_i = a^*\lambda_i' + (1 - a)^*0$$

and λ_M is the production rate (in good parts) of the machine. Similarly, $\rho_i = a\rho_i'$ and $\eta = \lambda_M c$, where λ_M is the throughput of the last station in the machine.

$$W = M[c + pE(D)] \quad \text{without TDFs}$$

$$W = M[c + pE(D)]/a \quad \text{with TDFs}.$$

We now have all the necessary equations to complete a performance analysis of a synchronous system. Tables 13.2 through 13.5 show the results of the models applied to two different synchronous systems. Table 13.2 shows the input data for a 10-station system and Table 13.3 shows the results of the analysis. Table 13.4 provides the data for a five-station system and Table 13.5 in turn shows the results of the analysis.

TABLE 13.2. System Data for 10-Workstation SPL (Time Unit–Second)

Global Data

M	c_t	μ
10	1	30

Workstation Data

Workstation i	c_{pi}	α_i	β_i	τ_i	θ_i
1	10	0.03	0.0	3600	115,200
2	10	0.05	0.0	3600	115,200
3	10	0.03	0.0	3600	115,200
4	10	0.05	0.1	3600	115,200
5	10	0.05	0.1	3600	115,200
6	10	0.03	0.1	3600	115,200
7	10	0.05	0.1	3600	115,200
8	10	0.05	0.1	3600	115,200
9	10	0.05	0.1	3600	115,200
10	10	0.03	0.1	3600	115,200
11[a]	—	—	—	7200	230,400

[a]Transfer mechanism.

TABLE 13.3. Model Results for 10-Workstation SPL (Time Unit = Second)

Global Results

L	1	2	3
$E(Z)$	1.201	1.201	1.201
$P1$	0.179	0.179	0.179
η	0.391	0.407	0.407
W	272.6	262.2	261.8

Workstation Results

$L_i = (L_i)(16)(3600) =$ mean throughput per 16-hour day

	1		2		3	
L_i	ρ_i	λ_i	ρ_i	λ_i	ρ_i	λ_i
1	.367	2113	.381	2197	.378	2200
2	.367	2113	.381	2197	.378	2200
3	.367	2113	.381	2197	.378	2200
4	.367	2102	.381	2186	.378	2189
5	.365	2092	.380	2175	.376	2178
6	.363	2086	.378	2169	.374	2172
7	.362	2075	.376	2158	.373	2161
8	.360	2065	.375	2147	.371	2150
9	.358	2055	.373	2136	.369	2139
10	.357	2048	.371	2130	.367	2133

TABLE 13.4. System Data for 5-workstation SPL (Time Unit = Second)

Global Data

M	c_t	μ
5	1	30

Workstation Data

Workstation i	c_{pi}	α_i	β_i	τ_i	θ_i
1	6	0.005	0.5	3600	115,200
2	6	0.02	0.5	3600	115,200
3	6	0.005	0.5	3600	115,200
4	6	0.01	0.5	3600	115,200
5	6	0.005	0.5	3600	115,200
6^a	—	—	—	7200	230,400

[a]Transfer Mechanism.

TABLE 13.5. Model Results for 5-Workstation SPL (Time Unit–Second)

Global Results

L	1	2
$E(Z)$	1.016	1.016
$P1$	0.016	0.016
h	0.709	0.710
W	48.26	48.21

Workstation Results

$\lambda_i = (\lambda_i)(16)(3600) =$ mean throughput per 16-hour day

	1		2	
L_i	ρ_i	λ_i	ρ_i	λ_i
1	.622	5953	.622	5960
2	.620	5893	.621	5900
3	.614	5879	.615	5885
4	.612	5849	.613	5856
5	.609	5835	.610	5841

13.4.2. Performance Analysis of Asynchronous Assembly

In an asynchronous system the transfer mechanism is of the free transfer type, where the assembly moves from one station to the next as soon as it has completed service at the previous station. In conveyor-based systems the assembly circulates through the system on a fixtured pallet and the pallet rides or "floats" on the conveyor or transfer chain. In other systems the transfer system may use AGVs (automatic guided vehicles) or independent monotractors to move the workpieces from station to station. This class of system permits queuing of workpieces at stations and a number of phenomena can occur in asynchronous assembly systems. These systems are almost always "closed" in the sense that the number of assembly workpieces in the system is constant at all times and the carriers (either pallets or AGVs or monotractors) recirculate through the system from start to finish to start and the assembly base is loaded at the start and the completed assembly is removed at the finish station (see Fig. 13.7). Figure 13.8 shows a large asynchronous system built by Gilman Engineering and Manufacturing Company.

These systems often have fairly short station cycle times (on the order of 6–10 seconds per operation) but in some applications cycle times can be as long as 40–60 seconds. In fully automatic systems the cycle times at each

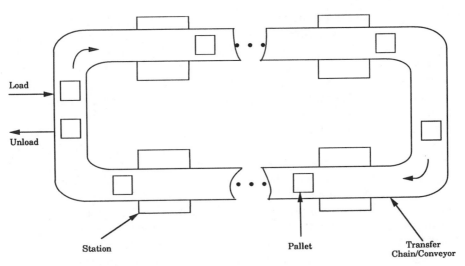

Load →

Unload ←

Station

Pallet

Transfer
Chain/Conveyor

FIGURE 13.7. Asynchronous automatic assembly system.

FIGURE 13.8. Asynchronous system built by Gilman Engineering and Manufacturing Company.

station are of fixed duration and are usually the same at each station. Hybrid systems, where some stations are automatic and some are manual, are also quite common. The asynchronous nature of the system is the mechanism that permits the mixing of manual and automatic stations. In the automatic portion of the system we have the same problems with parts jams with required manual clearing and resetting of the system by a human operator as in the synchronous case. We usually assume that data are available for this phenomenon in the form of a percent jam rate, which specifies how many parts per 100 on average will experience a failed assembly operation. In addition, we assume that we know the mean and variance of the time required to clear the station once the jam has occurred. In addition to the jam and clear phenomena that occur in these systems, the queuing effects can cause station starvation, where a station may be fully operational but has no workpieces due to delays upstream on the machine. In addition, because the space between stations is limited, the number of workpieces that can queue up between stations is limited and when that capacity is exhaused the station upstream of the queue can be forced to stop work due to blocking because the station has nowhere to discharge completed workpieces. Finally, in addition to these effects, all of which affect the productivity of the machine, we need to account for the effects of transport delays between stations because a workpiece can have been released from an upstream station and the downstream station can be empty and ready for the workpiece but an amount of time elapses before the transport system can bring the workpiece to the receiving station for processing. All these phenomena must be reflected in a complete performance model in addition to the impact of the number of operators that are used to monitor the machine. In these systems, unlike the synchronous case, the number of stations can be quite large. Systems with 60–100 stations are not unusual and consequently more than one operator is often justified in terms of increased system performance.

Among the system parameters that a design engineer would need to consider in addition to the mechanical and control design decisions are the following:

1. How many workpieces should be loaded on the system in order to achieve optimal performance?
2. How many operators are needed to get near maximal throughput?
3. How many buffer spaces should be permitted between each pair of stations on the line?
4. How will the variability of manual assembly operations affect the throughput of the system?
5. Does the system performance fall off rapidly as the number of stations in the system grows?

The answer to all these questions can be found by using one or more of the performance models developed below. The approach to performance modeling of these systems has emerged only in the last few years due to advances in queuing theory. These systems are treated as closed networks of GI/G/1 queues and the relevant theory is used to obtain approximations for production rates, station utilization, expected lead time, and other relevant performance measures. Closed queuing networks have been treated in the literature since the early 1970s, but the theory has until recently been confined to stations that satisfy the M/M/1 or M/M/k assumptions. This, of course, requires that the service times be exponentially distributed random variables. This is a wildly inaccurate assumption for nearly all asynchronous assembly systems. The more recently theoretical results permit the development of performance models that are quite accurate over a wide range of system parameters and whose results can be obtained with only modest computational requirements.

We now discuss three different classes of models for asynchronous systems:

Model 1. Large buffer approximations (local approximations, no simultaneous resource requirements)

Model 2. Operator models (local approximations, simultaneous resource possession)

Model 3. Blocking models (nonlocal phenomena, no simultaneous resource possession)

If the number of buffer spaces between each pair of stations is "large" (usually at least twice the average number of workpieces waiting for service) and we are not concerned with examining the possibility of two or more machine operators, then station blocking can be ignored and the class of phenomena known as operator interference can be ignored and the models that result are quite simple and easy to compute. In this development we follow Kamath and Sanders (4) and Kamath et al. (5).

If the probability that more than one station is down at a time is substantial, one would usually want to be able to investigate how long a "down" station is required to wait for the services of an operator. The use of more than one operator could substantially improve the productivity of the system. The simultaneous modeling of the productivity of the basic machine and the utilization of multiple operators requires the modeling of simultaneous resource possession [a "down" station requires a pallet (customer) and the presence of an operator to return to an operational state]. This class of problems requires the simultaneous solution of two sets of stochastic network equations, but the numerical requirements are not excessive and results are quite accurate for a large range of system sizes and the numbers of operators. We briefly review the work of Kamath and Sanders (4) for this class of models.

Finally, when buffer sizes are small, the number of workpieces is large, jam rates are large, or mean clear times are large (or any significant combination of the above), then station blocking cannot be ignored. This gives rise to a very difficult class of problems because blocking at one station can quickly propagate upstream in the system and one can find that a station is blocked by a problem several stations downstream. The "nonlocal" phenomenon has proved to be very difficult to model even approximately. A recent approximate method (8) known as "synchronous model interpolation" is discussed, which provides reasonable approximations for many asynchronous assembly systems.

Model 1. The Large Buffer Approximations. We introduce the following notation:

T	Total time in station; $T = D + XR$, where D is a constant for automatic stations and a random variable (processing time) for manual stations
X	Indicator random variable ($X = 0$ or $X = 1$) depending on whether the cycle (for automatic stations) is jam-free or whether a jam occurs; X can be assumed to be identically 0 for manual stations or equivalently $E(X) = 0$
R	Random variable that specifies the time required to clear the jam if it occurs
ET	$ED + EX\ ER = t + am$
Var T	Var D for manual stations
Var T	$\alpha\{v = m^2(1 - \alpha)\}$, where $v = $ Var R
τ	ET by definition
$c^2{}_s$	Var $\{T\}/(ET)^2$
M	Number of assembly stations in the system
N	Number of pallets (workpieces) in the system
ρ	Average station utilization
λ	System production rate; $\lambda = \rho/\tau$

In fact, all the notation above, which describes the characteristics of individual stations, should be subscripted by the station index; but if we assume that every station in the system has exactly the same parameters, we can obtain a particularly simple expression for the system production rate. In this case,

$$\lambda = \rho/\tau = \{[(M + N)^2 + 4MN(c^2{}_s - 1)]^{-1/2} - (M + N)\}/\{2M\tau(1 - c^2{}_s)\}.$$

If $c^2{}_s = 1$, then $\lambda = N/[\tau(M + N)]$.

Table 13.6 shows the accuracy of these approximations for a large (100)

TABLE 13.6. 50-Station Balanced AAS Example with Geometrically Distributed Clear Times

Mean Clear Time	% Defective	Number of Pallets	Squared Coefficient of Variation of Service Times	Simulation Estimates		Analytic Estimates					
						Product Form Analysis[a]			Renewal Approximations		
				Station Utilization	Throughput Rate	Station Utilization	Throughput Rate	% Error in Throughput	Station Utilization	Throughput Rate	% Error in Throughput
6	0.5	50	0.0091	0.918 ±0.004	0.1522 ±0.0005	0.505	0.0838	−45.94	0.913	0.1514	−0.53
6	0.5	100	0.0091	0.990 ±0.002	0.1641 ±0.0002	0.671	0.1113	−32.18	0.991	0.1644	0.18
6	3.0	50	0.0510	0.817 ±0.004	0.1322 ±0.0005	0.505	0.0817	−38.20	0.816	0.1320	−0.15
6	3.0	100	0.0510	0.953 ±0.004	0.1542 ±0.0003	0.671	0.1086	−29.57	0.955	0.1546	0.26
36	0.5	50	0.3338	0.649 ±0.011	0.1049 ±0.0016	0.505	0.0817	−22.12	0.634	0.1026	−2.19
36	0.5	100	0.3338	0.811 ±0.012	0.1312 ±0.0017	0.671	0.1086	−17.23	0.814	0.1317	0.38
36	3.0	50	1.5065	0.452 ±0.009	0.0637 ±0.0007	0.505	0.0713	11.93	0.449	0.0634	−0.47
36	3.0	100	1.5065	0.601 ±0.013	0.0848 ±0.0009	0.671	0.0948	11.79	0.605	0.0854	0.71

[a]Exact solution for the symmetric cyclic exponential network.

station system with a variety of system parameters whose clear times are geometrically distributed. The results are compared to simulation results (10 runs and 100,000 time units with the first 10% of each run eliminated to remove transient effects). The column headed "Product Form Analysis" provides the classical queuing theory results, which require the assumption of exponential distribution of time in station. The column headed "Renewal Approximations" contains the results based on the formulas shown above. We see from the table that, if we ignore blocking, operator interference, and transport delays, these methods produce very accurate results for large systems with a trivial amount of computational requirements.

If we examine the production rate equation given above, we see that a number of the questions posed earlier can be answered in a qualitative fashion immediately.

We can see that for $c_s^2 = 1$, λ approaches $1/\tau$ as M becomes large if we let N grow as some large multiple of the number of stations M. (Note that $1/\tau$ is the maximum possible t production rate.) Hence, if we are willing to let the number of workpieces grow and the buffer sizes grow accordingly, the performance of the system need not fall off as the number of stations is allowed to grow. This is in strong contrast to the situation in the synchronous case.

If c_s^2 is small, then again λ approaches $1/\tau$ but this time the result is independent of M and N, as we would expect since each station has (identical) deterministic processing times and no starvation occurs.

Finally, if c_s^2 is large, the situation is rather complex but the performance of the system falls off (very roughly) proportional to $1/c_s$. We would expect something of this sort since, clearly, variability in station cycle times takes a toll on system performance through the production of station starvation.

Production rate (in these approximations) can be seen to be a monotone increasing function of N (the number of workpieces) for fixed M and c_s^2. While this is true under the assumptions we have made, it requires that buffer sizes grow to keep blocking insignificant. If the buffer sizes are fixed, then as the value of N grows the throughput grows and then declines as N approaches the total number of buffer spaces plus the number of stations in the machine. At the limit where $N = B + M$, where B is the total number of buffer spaces in the machine, the system becomes a synchronous machine. This observation is the basis for a set of blocking approximations that will be presented as Model 3.

When N is small compared to $B + M$, it is easy to see that one can starve the machine by providing too few pallets. The optimal number of pallets is near $(B + M)/2$, but it usually is somewhat larger than this value.

We have come quite a distance with a simple approximation formula, but obviously real machines do not have identical parameters at each station even if the other conditions of the model are met. Kamath et al. (5) provide the extension of the present approximations to the general case and improve the original approximations so that the results are valid down to systems with as few as 10 stations. The computations are essentially identical to the ones required here but the arithmetic is more involved since very station is permitted to have different parameter values.

Model 2. Operator Interference. In attempting to understand what effect the number of operators will have on the productivity of an asynchronous assembly system, we can formulate a station model in much the same way as we did for the single operator case previously.

We define (for a typical station) $T = D + XR$ as in the single station case, where T is the total time in station, D is the assembly cycle time, X is an indicator random variable that signals the occurrence of jams, and R is total time from the onset of the jam to the return of the station to an operational condition. The difference here is that R must be written as $R = C + I$, where C is the actual time required for the operator to clear the jam once he/she reaches the machine and I is the waiting time that the station spends waiting for an operator to arrive. I is called the interference time because it results from the fact that other stations may command the attention of the operator or operators, causing the station in question to wait for the onset of service. Since our models of assembly system performance all require only the mean and variance of the (total) station service time, if we had the mean and variance of I we could apply the models of the previous section to calculate all the standard performance criteria.

A little reflection will tell us that this may not be a simple task, because we need these calculations as a function of the number of operators available to service the "down" stations. In addition, even for a fixed number of operators, the interference time I depends on the number of stations that are down and the rate at which stations go down depends on the production rate of the machine. However, the production rate of the machine depends on the interference time, so we have come full circle. In fact, we have two interdependent queuing systems (see Fig. 13.9). The first is the network of circular queues that includes the assembly machines and the second system is the set of (parallel) servers that make up the service operators. We have to solve for the throughput rates for both of these systems simultaneously. Kamath and Sanders (4) present a convergent algorithm that solves for $E(I)$, $\text{Var}(I)$, and the relevant performance statistics of the assembly machine.

Table 13.7 demonstrates how accurately the algorithm models the essential features of the multiple operator-assisted assembly machine. It is easy to see from these examples that unlike most synchronous assembly machines more than one operator often makes a substantial difference in the net production rate of the machine. Since the more complex of these systems can easily cost from 1 to 10 million dollars and they often produce product valued at $10,000 per day and up, it is relatively easy to justify the addition of extra operators to achieve significantly higher production rates.

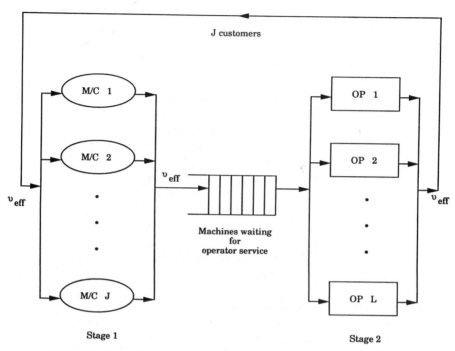

FIGURE 13.9. Two-stage machine–operator cyclic queuing model.

Transport System Delays and Station Blocking. In all the analysis discussed previously we have assumed that both blocking and transport delays had negligible effect on the system throughput when compared to station down times due to the time required to reset the station after a parts jam or assembly cycle error. The results of these assumptions permit one to obtain models that give a good deal of qualitative insight into the performance of this class of systems, but in certain design contexts we simply must have a way of understanding the impact of transportation delays and especially the impact of station blocking.

In a detailed simulation analysis of these systems, when transport delays and blocking are taken into account, the production rate of the system rises as the number of workpieces is increased from one per station; but as the number approaches the total number of buffer spaces plus the number of stations in the system, the production rate falls. This reduction in production rate is caused by station blocking that occurs with increasing frequency as the system becomes saturated with pallets. In short, if we want to investigate problems of buffer sizing or optimal pallet loading, we have to find analysis methods that incorporate blocking effects. On the reverse side, if the system includes large interstation distances, then transport delays can have a significant impact on the performance of the system especially when the pallet loading is low relative to the total available space on the machine.

TABLE 13.7. 30-Station Balanced AAS Example with Exponential Clear Times, 3% Jam Rate, and 45 Pallets

	Simulation Estimates					Analytical Results						
Operators	Mean Interference	Operator Efficiency	Mean Time in Station	Station Utilization	Production Rate (per second)	Mean Interference	Operator Efficiency	Mean Time in Station	SCV of Service Time	Station Utilization	Production Rate (per second)	% Error in Production Rate
1	115.88 (±3.74)	0.99 (±0.001)	10.38 (±0.20)	0.381 (±0.008)	0.0367 (±0.004)	112.16	0.994	10.26	5.891	0.378	0.0368	0.27
2	29.2 (±1.37)	0.898 (±0.002)	7.79 (±0.07)	0.519 (±0.006)	0.0666 (±0.0006)	23.95	0.912	7.62	2.029	0.515	0.0676	1.50
3	7.75 (±0.68)	0.725 (±0.004)	7.14 (±0.05)	0.576 (±0.004)	0.0807 (±0.0006)	6.05	0.726	7.08	1.329	0.571	0.0807	0.00
4	1.95 (±0.23)	0.571 (±0.004)	6.97 (±0.04)	0.591 (±0.001)	0.0847 (±0.0007)	1.7	0.57	6.95	1.174	0.587	0.0844	−0.35
5	0.4 (±0.11)	0.462 (±0.003)	6.92 (±0.04)	0.593 (±0.005)	0.0856 (±0.0007)	0.47	0.463	6.91	1.132	0.593	0.0857	0.12
6	0.06 (±0.03)	0.386 (±0.003)	6.91 (±0.04)	0.593 (±0.005)	0.0858 (±0.0007)	0.12	0.386	6.90	1.121	0.593	0.0859	0.12
30	0 (±0.00)	0.077 (±0.001)	6.91 (±0.04)	0.593 (±0.005)	0.0858 (±0.0007)	0	0.077	6.90	1.117	0.593	0.0859	0.12

Cycle time = 6 seconds; mean clear time = 30 seconds.

The development of analysis methods for systems with significant amounts of blocking has proved to be very difficult and only recently have reasonably accurate methods become available for closed asynchronous systems. We follow Venkateswar and Sanders (8) in this analysis. The development presented here is approximate but is applicable to general station service time distributions.

In general terms, the blocking and transport delay analysis proceeds as follows. The mean throughput of the line satisfies the result due to Little's law:

$$\lambda \sum_{i=1}^{M} [EW_i + ES_i + ED_i] = N ,$$

where λ is the mean throughput rate, M the number of stations, ES_i the expected time in station (including blocked time), EW_i the mean time spent waiting for entry to service at station i, and ED_i the expected transport time delay between station $i - 1$ and station i.

From this equation the algorithm proceeds as follows:

1. Start with an initial estimate of the station utilizations r_i.
2. Calculate ES_i, EW_i, and ED_i for all i.
3. Use the equation above to calculate throughput λ.
4. Recalculate the utilizations: new $\rho_i = \lambda\, ES_i$ for all $i = 1, \ldots, M$.
5. If $|$new $\rho_i -$ old $\rho_i| < e$, then stop; else return to step 2. (e is a small constant set to control accuracy of convergence.)

Clearly, we need methods to calculate ES_i, EW_i, and ED_i for all i.

Station Service Times. We write $S_i = c_i + t + X_i C_i + Z_i B_i$, where c_i is the deterministic cycle time at station i; t is the unit transport time (time to transit a single buffer space); $X_i = 1$ if a jam occurs at station i, $X_i = 0$ otherwise; C_i is the time needed to clear a jam once it occurs at station i; $Z_i = 1$ if blocking occurs at station i, $Z_i = 0$ otherwise; and B_i is the time required for a blocked condition to be cleared.

Expected Transport Delay Time. For this condition,

$$ED_i = t[b_i + 1 - ELq] ,$$

where ELq is the expected number of pallets in the queue at station i when the next pallet arrives.

Expected Waiting Times. To calculate EW_i we begin by using results from the theory of GI/G/1 queues, where

$$EW_i = \nu_a [\rho^2(c_a^2 + c_s^2) - 2\rho g c_a c_s]/2(1 - \rho) ,$$

where ν_a is the mean time between arrivals at station i, c_a is the coefficient of variation of the interarrival times, and c_s is the coefficient of variation of the service time at station i. ρ is the utilization of station i and g is the interarrival time correlation of successive workpiece arrivals to station i.

From Little's law applied to station i we can calculate the expected number of customers at station i from

$$EL_i = \lambda(EW_i + ES_i) .$$

Approximation Requirements. If we examine the data required in step 2 of the algorithm, we see that while we have formulas for each of the three components, several problems remain. In the calculation of ES_i the basic data from the machine are sufficient to specify everything except EZ_i and EB_i. EZ_i is the probability of blocking occurring during a randomly chosen cycle and EB_i is the expected time the station is blocked after blocking begins.

In ED_i we need the expected number of customers waiting at station i when an arrival appears. When we examine EW_i we see that a fundamental problem has arisen because EW_i as given is for open systems and if we use it for closed systems we find that it will often imply (through Little's law) a total number of customers in the system that is not equal to the number of actual pallets loaded, or it may imply a number of customers (pallets) that is larger than the number of available pallets. In addition, the interarrival correlation g is nearly impossible to calculate for a closed system with blocking.

In order to overcome these problems, the following approximations or heuristics are introduced:

1. We substitute ρ for g since ρ, like g, increases as the traffic intensity increases. This effect (in the correlation) is due to increased blocking. Both parameters lie between 0 and 1 (at least for this class of problems).

2. After calculating the EL_i for each station from EW_i, the resulting EL_i values are reduced in such a way that EL_i is less than or equal to the total buffer capacity at that station and also the sum of the EL_i values (adjusted) is equal to N.

3. The adjusted L_i values are used to calculate ELq_i for the ED_i calculation.

4. In ES_i the blocking probability EZ_i is obtained by calculating the blocking probability for a system with one pallet for each available station and buffer space on the machine (a totally synchronous system) using, for example, the method of Leung and Kamath (7) cited for synchronous machines above. The blocking probability for the asynchronous machine is

obtained by calculating the fraction of spaces on the asynchronous machine that are occupied and using the following interpolation formula:

$$EZ_i = (c_i + t + EX_iEC_i)p_i/[(1-p_i)EB_i],$$

where p_i is obtained from

$$p_i = (\gamma p_i)^{b(i+1)+1}(ES_{i+1}/ES_i)/\max(ES_{j+1}/ES_j) - (fb_i),$$

where fb_i is the blocking probability from the asynchronous machine station i and γ is the saturation fraction of the asynchronous machine.

5. To calculate EB_i and Var B_i we use the equilibrium residual clear time for station $i + 1$:

$$EB_i = EC_{i+1}/2 + \text{Var } C_{i+1}/2EC_{i+1} \quad \text{and} \quad \text{Var } B_i = \text{Var } C_i.$$

The results of the application are shown in the form of a factorial experiment in Tables 13.8 and 13.9. The conditions of the experiment are as follows:

1. The machine is a 30-station machine and the transport time t is one time unit per buffer space. Station cycle times (c_i) were chosen at random from a uniform distribution from 1 to 12 time units.

TABLE 13.8. Different Sets Designed to Test the Performance of Algorithm

Set Number	Cycle Times	Buffers	Saturation	Mean Clear Times	Jam Rates
1	+	−	−	−	−
2	+	−	−	−	+
3	+	−	−	+	−
4	+	−	−	+	+
5	+	−	+	−	−
6	+	−	+	−	+
7	+	−	+	+	−
8	+	−	+	+	+
9	+	+	−	−	−
10	+	+	−	−	+
11	+	+	−	+	−
12	+	+	−	+	+
13	+	+	+	−	−
14	+	+	+	−	+
15	+	+	+	+	−
16	+	+	+	+	+

TABLE 13.9. Comparison of the Results of Algorithm with Those from Simulation

| Set Number | Throughput | | Error (%) |
	From Simulation	From Analytical Model	
1	0.0756	0.0768	−1.59
2	0.0742	0.0745	−0.40
3	0.0726	0.0729	−0.41
4	0.0687	0.0747	−8.73
5	0.0756	0.0750	0.79
6	0.0742	0.0714	2.16
7	0.0726	0.0720	0.83
8	0.0685	0.0725	−5.84
9	0.0817	0.0816	0.12
10	0.0729	0.0755	−3.57
11	0.0737	0.0744	−0.95
12	0.0751	0.0775	−3.20
13	0.0817	0.0788	3.55
14	0.0731	0.0729	0.27
15	0.0737	0.0730	0.95
16	0.0747	0.0745	0.27

2. The high (+) and low (−) conditions for each factor were determined as follows:

Buffer Spaces
   (+) Uniform random variable from 1 to 7
   (−) Four buffers per station

Pallet Saturation
   (+) Pallets equal $M + 0.8$ times total number of buffers
   (−) Pallets equal $M + 0.2$ times total number of buffers

Mean Clear Times
   (+) Values uniform between 10 and 60 time units
   (−) Values constant at 30 time units

Jam Rates
   (+) 20 of the stations with values from a uniform distribution between 0 and 3% (remainder 0%)
   (−) 9 stations with values from a uniform distribution between 0 and 3% (remainder 0%)

For each example, simulations were run for a total of 10 independent runs for each case with a total of 10,000 products completed during each run. The 95% confidence intervals were established for the throughput rate; comparisons are made with the mean of this interval.

By way of time comparison, the simulation results took 1800 seconds each on a specially optimized Pascal simulation program running on a 16-MHz IBM PS/2 Model 80. As a consequence, each confidence interval requires 5 hours of computer time. (Standard manufacturing simulation languages typically require roughly five to six times this amount of time.) The analytic model took from 5 to 20 seconds depending on conditions.

REFERENCES

1. Baybars, I. (1986). A Survey of Exact Algorithms for the Simple Assembly Line Balancing Problems, 1986, *Management Science*, Vol. 32, No. 8, pp. 909–932.
2. Boothroyd, G., Poli, C., and Murch, L. (1984). *Automatic Assembly*, Marcel Dekker, New York.
3. Graves, S., and Redfield, C. (1988). Equipment Selection and Task Assignment for Multiproduct Assembly System Design, *International Journal of Flexible Manufacturing Systems*, Vol. 1, pp. 11–50.
4. Kamath, M., and Sanders, J. L. (1990). Modeling Operator/Workstation Interference in Asynchronous Automatic Assembly Systems, *Journal of Discrete Event Systems*, Vol. 1, No. 1, pp. 93–124.
5. Kamath, M., Suri, R., and Sanders, J. L. (1989). Analytical Performance Models for Closed Loop Flexible Assembly Systems, *International Journal of Flexible Manufacturing Systems*, Vol. 1, pp. 51–84.
6. Lentz, Kendrick W. Jr. (1985). *Design of Automatic Machinery*, Van Nostrand Reinhold, New York.
7. Leung, Y. T., and Kamath, M. (1990). Performance Analysis of Synchronous Production Lines, *IEEE Journal of Robotics and Automation*, Vol. 7, No. 1, pp. 1–8.
8. Venkateswar, K., and Sanders, J. L. (1990). Optimization of Assembly Systems, *Technical Report 90-7*, Department of Industrial Engineering, University of Wisconsin–Madison.
9. Wilhelm, W., Saboo, S. and Wang, L. (1989). Recursion Models for Describing and Managing Transient Flow of Materials in Generalized Flowshops, *Management Science*, Vol. 35, No. 6, June, pp. 722–743.

■■■■■■ **CHAPTER FOURTEEN**

Design of Manufacturing Control Systems

U. NEGRETTO

Institute for Real-Time Computer Systems and Robotics, University of Karlsruhe, Karlsruhe, West Germany

14.1. INTRODUCTION

The design of a flexible manufacturing system (FMS) involves choosing suitable machines to manufacture or assemble certain types of products, defining the needed operations and elaborating an appropriate control system. For discrete event dynamic systems like FMSs, a well defined theory for the control design is not yet available. This, however, is the case for linear, continuous, or synchronous systems.

The problem is to integrate the resources into an efficiently working control structure taking into account concurrent and asynchronous processes (9, 12, 14). Essential for the control design is consideration of important FMS properties like absence of overflows and deadlocks and the capability of reconfiguration of the system. The latter property addresses the reachability of the initial system state or another working state from any current (i.e., also exceptional) state. The reconfigurability plays an important role in the context of exception handling and has to be supported by the control model (6). Flexibility in the sense used here means that the system is not restricted to only one product or product family, but also enables the manufacturing of various products.

This chapter illustrates a method for the design of FMS control based on the representation of the system in augmented predicate transition nets, a class of high-level Petri nets. The focus is put on the representation of the static and dynamic functionality of the real system and on the verification of the control algorithm through simulation of the planned process. The

Intelligent Design and Manufacturing, Edited by Andrew Kusiak.
ISBN 0-471-53473-0 © 1992 John Wiley & Sons, Inc.

affinity between predicate transition nets and the rule-based language Prolog enabled the symbiosis of both, thus resulting in a gain in modeling power and complexity, easier rapid prototyping, and application of artificial intelligence (AI) methods.

In this context, planning and programming of a robot-based assembly cell discussed are currently under development at the Institute for Real-Time Computer Systems and Robotics at the University of Karlsruhe.

The purpose of this work is a task-independent representation of the functionality of the real FMS, modeling of the control flow, programming of the FMS, and a simulation of the process based on this specification. In Section 2, a problem statement is given and the requirements for methods of modeling of FMSs are discussed. Based on these requirements the selected predicate transition nets are introduced. In Section 3, the methodology is presented with the discussed assembly cell. Different specification levels dependent on the degree of abstraction are presented, using a top–down approach. In Section 4, the combination of the net representation and the rule-based language (Prolog) is discussed. The net is transformed and implemented in a Prolog notation. Exceptions that require rescheduling of the process are analyzed. Different rescheduling strategies are discussed. The simulation of the process is shown and the achieved results are discussed. In Section 5, the implemented prototype simulation system is described. The simulation of the process is shown and the achieved results are discussed.

14.1.1. System Configuration: The Assembly Cell

Planning and programming of robot-based assembly cells with interacting components are the main research areas of the institute. The most important resource within the examined assembly cell is the Karlsruhe Autonomous Mobile RObot (KAMRO). All other components have a stationary layout that allows the KAMRO to reach them and perform operations with them; see Fig. 14.1. Each operation of the mobile robot is preceded by a definite transport and docking operation at the desired device to achieve a defined reference position. The components of the cell are:

1. A conveyor to transport parts between the cell and its environment. The function of the conveyor is to feed the cell with single parts and to shift the completed assemblies.
2. A stationary robot (Puma 600) to pick up the parts from the conveyor and to put them into a magazine in defined positions.
3. The magazine is used as a cell storage device. It serves as a material buffer.
4. An assembly station (AS), which consists of an assembly table with fixtures for pallets or parts.

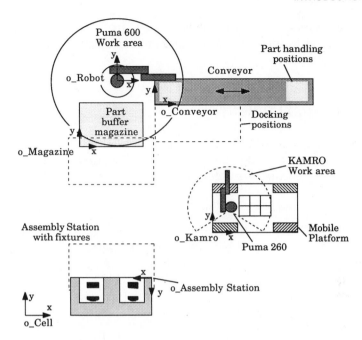

FIGURE 14.1. The flexible assembly cell.

5. The autonomous mobile robot, which performs a variety of tasks. First, it performs transportation tasks within the cell; that is, it transports parts from the magazine to the assembly station and assemblies from the AS to the conveyor. Second, it performs the handling operations to assemble the parts; that is, it gets parts from the magazine and performs the assembly task at the AS. The KAMRO consists of a mobile omnidirectional carrier and two robots (Puma 260) in portal configuration hanging upside down on gauges fixed on the carrier. It is also equipped with a part buffer device for transportation of several parts. Traveling, docking, and manipulation are supported by tactile sensors, as well as infrared, sonar, and vision systems.

For specification of the cell program and its simulation, the following basic data are required:

- Geometric data (layout, size of cell devices)
- Kinematic data (e.g., joint types, relation between joints, kinematic chains)
- Technological data (e.g., storage capacity, material, load capacity, required accuracy)

Furthermore, part-specific workplans, which are related to the desired product, are used as input. A technical database can be used to access the required information.

14.2. PROBLEM STATEMENT

Given the specification of a flexible assembly system (FAS) or cell, model the system such that it can be controlled, programmed, and simulated realistically. The specification refers to:

- The layout of the FAS (list of resources)
- Resource descriptions (functionality, capacity)
- Task description (assembly graphs, material data)
- Operation capabilities of active devices
- System start and goal state

14.2.1. Objectives

The correct and complete representation of the system and its functionality requires:

- A task-independent model of the topology of the flexible assembly system
- The static representation of layout, component properties, and relations between components
- The dynamic representation of concurrency and synchronization of resources
- A workplan description of the task coherent to the functional FMS model

14.2.2. Model Requirements

To execute the assembly task, the cell components have to cooperate. For this reason, a cell level control system is required to synchronize and monitor the processes in the cell. This cell controller needs, in addition to the given workplans, a task-independent representation of the target system, that is, the real work cell. For this purpose, local component states and, as a combination of these, global cell states have to be represented. In addition to those states, allowed transitions between the different states, that is, the dynamic behavior of the system, have to be defined in the model.

To model the functionality of a FMS, various methods can be taken into consideration that fulfill the following requirements.

1. Operating with various independent components, flexible systems are able to execute time parallel operations. Therefore the model has to support the representation of the existing parallel processes running on different devices and synchronization between them.

2. The development of such complex systems is facilitated by dividing the system into subsystems. In this case, the specification method has to allow hierarchical structuring of the system into subcomponents. Furthermore, the specification of the relations between the sub-components has to be supported.

3. Depending on the local capabilities of the single device control, that is, the degree of autonomously executable operations, the level of detail represented in the model has to vary accordingly. The method can fulfill this requirement if it allows the specification of the components' functionality on various levels of abstraction, so that the specific model reflects only the necessary information.

4. The actual cell state has to be retrievable at all times, because checking of all possible states would require a large search space. Moreover, it would be useful to know the next reachable states to reduce the number of conditions to be monitored before the next step.

5. The processes in flexible manufacturing systems are event driven. In the model, the discrete behavior has to be represented to simulate the processes. In addition to the simulation, which is the basis for improving the operation flow in the cell, the model has to ensure that the programs and the control system for the real plant can be applied.

6. During simulation of a FMS, various problems can arise due to unforeseen conflict situations and deadlocks. The model should be accessible to mathematical analysis and verification methods for dead-lock and conflict detection at an early stage.

7. In real processes, exceptions like loss of a part or positioning failures can occur. First the exceptions have to be detected and classified and then the planned recovery operations have to be executed. The selected method has to be accessible to dedicated exception handling procedures or systems (i.e., expert systems). Furthermore, it has to allow a restart from the interrupted system state without the need of restarting from the beginning.

8. Finally, the specification method should be applicable in each project phase. The information representation has to be the communication link between the involved experts in the different phases.

To meet the above-mentioned requirements, we choose special Petri nets with extensions proposed in recent publications (8, 13). For an introduction in the field of Petri nets, the literature in Brauer et al. (2), Peterson (9), and Reisig (10) is recommended. In the following section the applied method is shown.

14.2.3. Predicate Transition Nets

Petri nets with their various extensions have already been used to model the dynamic properties of discrete systems like distributed systems, distributed databases, FMSs, communication protocols, and others (1, 5, 11). To reduce the size of the representation and thus to provide a more compact and efficient model, the concept of predicate transition nets (Pr/Ts), which is an extension of Petri nets, is introduced.

Predicate transition nets are directed graphs with two different types of nodes, predicates and transitions. The relations between the nodes are represented by arcs. An arc always connects a predicate with a transition or vice versa. Pr/Ts have a structure to define predicates, operators, and a set of individual tokens. To each transition and each arc in the net, an inscription is associated specifying either variables (to arcs) or logic equations (to transitions). A marking, that is, the system state, is defined through the individual token tuples in each predicate at a definite time (also called partial states). A capacity relation is associated to each predicate, defining the maximal amount of tokens for each individual type. A firing rule is associated to each transition, defining the change of partial states between two or more predicates.

Because predicate transition nets are abstract models, it is possible to give an interpretation to the elements of the net, assigning them to elements of the physical system. This assignment is more or less arbitrary and thus very important for an exact representation of the system's behavior.

A robot can be assigned to a token and the execution of a task can be represented through the partial state "a token ⟨identifier⟩ is on the predicate ⟨identifier⟩". Transitions represent events and cause the state changes of the real system. The interpretation also defines the firing rules of the transitions. A transition that is assigned to an event fires if and only if it is enabled and if a predefined value specified in the inscription (i.e., a sensor value) is reached. A transition is enabled if the tokens marking the input predicates are members of the individual token set assigned to the arc inscription. When a transition fires, the tokens associated to the input arcs are removed from the input predicates. According to the logic equations specified in the transition inscriptions, the information related to the tokens may change. The sets of the output predicates are checked and the predicates are marked with the appropriate tokens.

14.3. METHODOLOGY

The first step of the method consists in the task-independent representation of the flexible assembly system. In a second phase, the task has to be scheduled using the available resources in the best possible way.

14.3.1. Modeling of the Application

The chosen interpretation follows the assignment rules stated above. According to these rules, the functional representation of the cell in the Pr/T net is shown in Fig. 14.2.

Each possible state of the components is assigned to a predicate. The states of the components on the highest level of abstraction are shown in Table 14.1; the states on a more detailed level are not shown.

For each component type a subnet is modeled that reflects its physical states. The allowed changes between the states are represented by transitions. The transitions are also used for the synchronization of the components.

The robot and the conveyor are synchronized through the transition, which starts the procedure: "The robot gets part ⟨name⟩ of the conveyor ⟨part a, pos_on_conveyor a⟩". The transition triggers the execution of the operation only if the conveyor is inactive and has a part available. The real movement of the robot and the actual transport operation of the conveyor have to be further detailed on adequate levels.

The five subnets, one for each component, make up the cell net. The net is a task-independent model of a cell. Simulation and control of processes in a FMS with such nets require the fulfillment of two additional conditions:

1. The informal representation has to be detailed to a formal level, where the model can be directly associated to processes and signals.

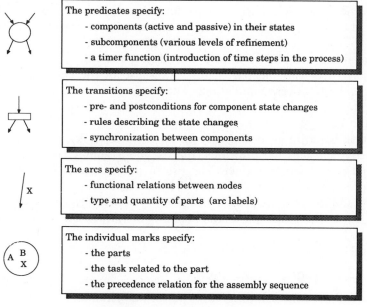

The predicates specify:
- components (active and passive) in their states
- subcomponents (various levels of refinement)
- a timer function (introduction of time steps in the process)

The transitions specify:
- pre- and postconditions for component state changes
- rules describing the state changes
- synchronization between components

The arcs specify:
- functional relations between nodes
- type and quantity of parts (arc labels)

The individual marks specify:
- the parts
- the task related to the part
- the precedence relation for the assembly sequence

FIGURE 14.2. Modeling rules.

TABLE 14.1. Cell Components and Their States on the Highest Level of Abstraction

Cell Components	States
KAMRO	Is docked at ⟨comp_name⟩, robots are inactive Is docked at ⟨comp_name⟩, robots perform operation ⟨op_name, parameters⟩ Is moving from Pos_p_1 to Pos_p_2
Conveyor	Carries ⟨parts⟩ into the cell Carries ⟨parts⟩ out of the cell Is inactive
Robot	Performs operations ⟨op_name, parameters⟩ Is inactive
Magazine	Is filled with ⟨(part a, pos mag_a), (part b, pos mag_b), ...⟩
Assembly station	Is filled with ⟨(part a, pos ass_a), (part b, pos ass_b), ...⟩

2. The execution of cell tasks requires the representation of the task workplan on the cell model (cell programming).

In regard to condition 1, as mentioned previously, the predicate transition net (Pr/T) formalism provides methods to refine or coarsen nodes of the net. In this approach, a top–down method to gradually refine structures is used. The inscriptions become more formal with each step. The nodes can be refined to linear predicate transition sequences or be expanded to subnets (4).

In regard to condition 2, the introduced individual tokens and the inscriptions with variable parameters allow use of the cell model for the task-dependent process simulation. The variables in the inscriptions are assigned to task-specific constants, which direct the flow of individual tokens within the net.

The task description (in this case for an assembly task) has to be given in a form that specifies the priorities, that is, the assembly constraints of the individual parts of the product. The use of precedence graphs is a widely used representation for this purpose. The nodes represent the parts of the assembly and the arcs the precedence relation between two nodes (or parts); see Fig. 14.3. The graph allows the planning of various assembly sequences.

14.3.2. Task Program Generation

The generation of the cell program is performed automatically. The planner takes as input the functional model of the cell and the description of the

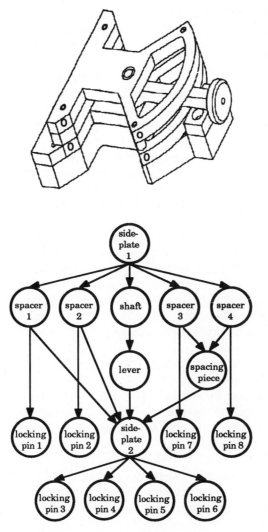

FIGURE 14.3. The Cranfield benchmark and its precedence graph.

assembly task. An assembly consists of a number of subtasks (abstract operators), which have a dependency relation given by the application. The dependency relations indicate the necessary order in which two tasks have to be carried out. A cell program is generated by transforming a global task description into a sequence of subtasks. This process is repeated until the subtasks are detailed enough to be directly expressed in the operating primitives of the cell components. These operating primitives are called elementary operations (EO). These elementary operations serve as the standard projection level of all planning and programming modules. The task decomposition is done analogously for all active components of the cell. As output of this phase for each active component a sequence of subtasks is planned, which is expanded to elementary operation sequences during runtime (see Fig. 14.4).

With this description and the cell model, the cell controller is able to monitor and execute the process. Details will be discussed in the following.

Because of the discrete event-driven strategy of the system, the consideration of time is not explicitly necessary. The transitions are assumed to fire immediately if they are enabled and if the conditions hold.

Time and time intervals, however, are used in the system as monitoring conditions. Since in the first phase, no feedback mechanism (i.e., sensor) is connected to the simulation system, the duration of the operations is used to trigger the transitions. This allows simulation of beginning and end of the processes.

To model the time, an additional predicate and two transitions are used. The first transition starts the process and the time (clock, watchdog), the predicate represents the active process, and the second transition specifies the end of the process and of the time interval. After the integration of sensors, this concept is used as an additional monitoring condition.

Separating both cell model and cell programming, the system can be analyzed for different properties: cell specific and task dependent. In the first case, known algorithms are applicable to detect conflicts in the net, where strategies have to be brought in to resolve those conflicts. In the simplest case, a random selection algorithm can be used or else priorities can be specified in the arc inscriptions to solve the conflicts. The analysis of the net in dependence on the task can be performed to detect deadlocks due to specified system parameters, for example, a buffer with insufficient capacity. The performance of the components can also be observed in the simulation runs.

The simulation of the process is executed by a "player" program, which has a general, task-independent form. This program monitors the token flow in the net and executes the operations specified in the inscriptions in cycles. A cycle consists of the following steps:

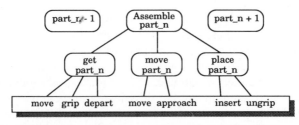

FIGURE 14.4. Elementary operations and task knowledge for an assembly operation.

1. Determine enabled transitions through actual marking.
2. Check the preconditions of enabled transitions if individual tokens of the input predicates match the parameters of the arc inscriptions.
3. Firing of the enabled transitions depending on the fulfilled conditions, change of the actual marking.
4. Execute the operations connected with the new marked predicates.

In conclusion, this method provides an efficient model for a FMS, which is used to simulate the processes in these systems. The fixed structure of the predicate transition net model allows the application of the method to an a priori well known system. To enlarge the application range of the method to systems with degrees of uncertainty, which have to be considered during runtime, the following extension was made.

14.4. COMBINING PREDICATE TRANSITION NETS AND THE RULE-BASED LANGUAGE PROLOG

The combination of Pr/Ts and Prolog is reasonable because of the affinity between the two. This affinity is given by the correspondence between different elements of predicate transition nets and elements of Prolog (3, 11):

1. Transitions can be represented by rules, if the input marking and the preconditions of the transitions are associated to the condition part of the rule and if the operation part of the rule causes the new marking that would result after firing.
2. Facts can be associated to the markings or to instantiated variables.
3. The inference engine of Prolog corresponds to the player program of the Pr/T formalism because it checks the rules if they succeed and performs the related operations. These procedures create new facts or delete old ones and generate the new fact base.

A general description of predicate transition nets was implemented in Prolog and adapted to the needs of the system. In the transformation the net properties were saved. Therefore rules could also be defined to detect deadlocks or to check the reachability of a marking. The elements of the real system are represented in a Prolog notation in analogy to the previous interpretation.

To simulate the overall task process, the precedence graph was also transformed into a Prolog notation. The part-specific task was assigned to each part of the assembly and stepwise refined to subtasks and elementary operations.

Out of this description a module generates component-specific elementary operation sequences for each active components, that is, KAMRO, conveyor, and stationary robot. In these sequences, the priority information of the precedence graph is still kept. This information is necessary for the later application of rescheduling strategies in case of exceptions in the process. Because of the general representation of the cell in the Pr/T net model, it is not necessary to explicitly specify task dependent parallel procedures and synchronization points. The cell model and the component-specific EO sequences are combined and the process is simulated, driven by the inference engine of Prolog. The combination is effected with the instantiation of the variables of the cell model. The part-specific parameters are matched with the variables during the simulation. In Fig. 14.5, the model levels of the flexible assembly cell are shown.

Level I of Fig. 14.5 shows the topology of an example cell configuration. This topology can either be specified interactively by a user with the system

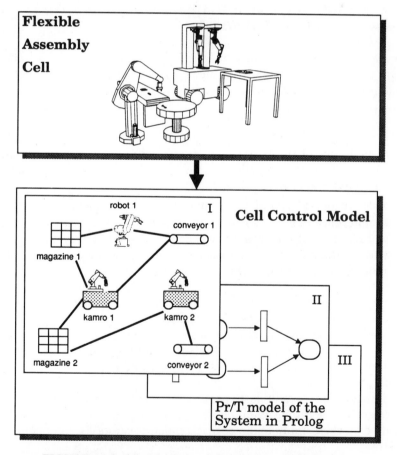

FIGURE 14.5. Model levels of the flexible assembly cell.

described in Section 14.5 or directly derived by a knowledge-based representation such as described by Camarinha-Matos et al. (3). The system offers a library of predefined component types, which, as mentioned previously, are specified by type-specific subnets with default values. Selection of a component type instantiates the component in the database of Prolog. Now the user is allowed to change the default parameters to fit the real cell. The links between the components define the possible relations and the synchronization strategies between the different pairs of interacting components.

Level II is the augmented Pr/T net representation of the cell automatically derived out of the specified topology following the rules of the previous section.

Level III is the representation of the cell net and of the Pr/T net formalism in Prolog.

One of the problems that arises when transforming Pr/T into Prolog is the loss of the parallelism specified in the net. In general, Prolog follows a sequential top-down processing method. The use of existing parallel Prolog versions like Parlog was not taken into consideration because the process time of the used sequential Prolog (IF/Prolog running on a SUN 3) needed to complete the cycle "matching–conflict resolution–action" was smaller by some order of magnitude than the operation time. Therefore a quasiparallel behavior of the cell process simulation could be achieved. Quasiparallelism means that, although processing is sequential, no difference to real parallelism was measured for the modeled cell. This could become a problem in the case of a much larger and more complex system.

14.4.1. Exception Handling at Cell Level

The declarative definition of objects and their relations in Prolog considerably facilitates the systematic modeling of the system. In contrast to the strict structure of the predicate transition nets, Prolog offers methods to change the model during execution of the process. Two different types of changes due to exceptions can be necessary: The first concerns the cell-specific part and the second the task-specific part. The former change is necessary if the exception is caused by a wrong representation of the system in the model. This case requires a completely new specification. Because of the clearly structured and small sized system, we provided a correct model. The latter change is caused by exceptions due to errors in the performance of operations by the components. In this case, the cell model is not affected; only the task-related elements like tokens or instantiated arc and transition inscriptions are taken into account.

The sequence of EOs is the executable procedure for the task related to one part. Each elementary operation also contains the information about which subtask and task it is a part of. This information is necessary to use the variety of possible assembly sequences defined in the precedence graph.

Before combining these EO sequences with the cell model as explained earlier, the part-specific sequences are reordered to component-specific sequences. The active component sequences are necessary to create executable programs for single devices. The synchronization of the overall process is performed by the cell controller. Different kinds of exceptions can occur due to the hierarchical structuring of the task in subtasks and elementary operations. The EOs are planned with an offset or boundary to enable a certain degree of autonomy in correcting small deviations. This offset can be a collision-free 3D pipe around the path, a force threshold for compliance operations, or a time interval in which an operation is supposed to succeed. If an exception arises where the measured value is not as expected but is within the given boundaries, a control mechanism specific to the EO is able to recover (i.e., to plan and execute recovery actions) without consequences for the process. For a more detailed description we refer the reader to Meijer and Negretto (7).

If the measured value exceeds the EO boundaries, the recovery, if possible, affects the process. The analysis of the exception and the replanning is based on the application of the given rules on the status knowledge before and after the exception and on the planned goal state. The difference of this AI method to the explicit declaration of exception handling mechanisms in Pr/Ts lies in the implicitly generated conclusions. They result from different combinations of rules to which a probability factor can be assigned. These criteria are used to choose the adequate recovery actions. The replanning can require some time, but because of the concurrency representation in the model, the operational continuation of the components not involved in the failure is guaranteed. However, the continuation of the process is enabled only until a synchronization point is reached, which requires the interaction with the component involved in the failure.

The handling of an exception can therefore be split up into the analysis and recovery of the failed operation and the rescheduling of the operation sequences. The analysis of the failure is the key to the decision whether or not a rescheduling is necessary. Obviously the precondition is to have redundant functionalities in at least one of the remaining components. The

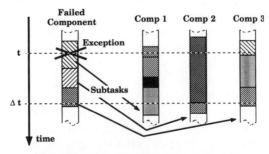

FIGURE 14.6. Rescheduling scheme.

rescheduling will then assume the cell status at the end of the last running subtask without interrupting the current process; that is, current time + a time interval Δt. Starting from this state, production planning will reschedule the remaining subtasks for the available components; see Fig. 14.6. This solution is not optimal due to transitory idle times of components, but it ensures satisfactory process continuation and termination.

14.5. CELL CONTROL SIMULATION

A cell control simulator (CECOSI) was implemented in Prolog, based on the presented method and offering functions for modeling the control and for analyzing and validating the planned cell program (see Fig. 14.7). The interaction is performed through a user interface in a mouse-based window environment on a SUN (Sunviews) in Unix. For the emulation of the specified control algorithm, the system is connected with a graphic simulation system developed at the University of Karlsruhe. The results improve the quality of the previously generated program in respect to performance and time. The system also detects deadlocks due to unsolvable conflict situations or for material flow reasons. The user is able to generate interrupts in the execution simulation to verify the system state or to change a set of parameters and check the effect of the change after restart of the simulation.

The user interface presents the status of each component together with the actual active operation and a preview of the next operation to be performed. The trace flow of the execution is prompted to the user and is updated at each start of a new operation of one or more (i.e., parallel)

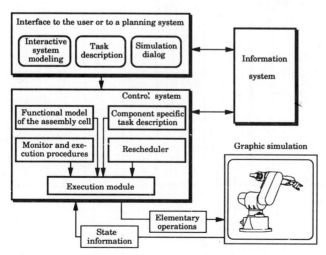

FIGURE 14.7. Structure of the cell control simulator (CECOSI).

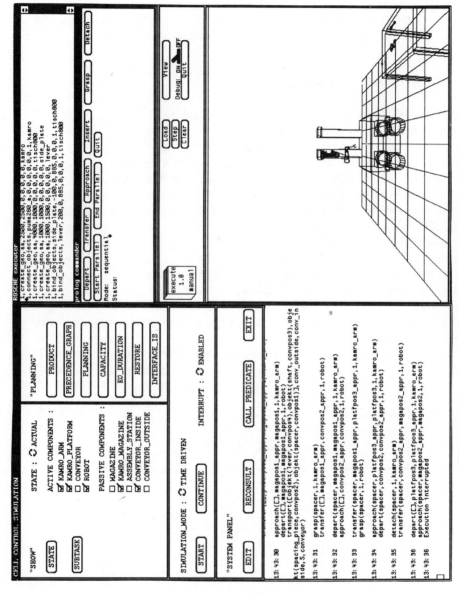

FIGURE 14.8. User interface of the cell control simulator.

components. The user interface has been realized as a prototype implementation. It offers the above-mentioned system interactions in a mouse-based window environment implemented in Prolog. The integration with a 3D graphical robot simulation system is partly realized and will be further extended; see Fig. 14.8.

14.6. CONCLUSION

A prototype version of the cell controller was implemented and is running. It was tested in simulation of the assembly cell described previously. The program consists of three parts: the predicate transition representation in Prolog to start up and run any arbitrary Pr/T net, the task-independent cell model (i.e., the rules and predicates to specify the cell functionality), and the task-dependent representation (i.e., the rules associated to the arc and transition inscriptions). The cell model was analyzed with regard to deadlocks, loops, and reachable markings, and it was found to be deadlock free. For the assembly procedure, this means that the generated elementary operation sequences can be performed by the components without conflicts in the process flow. Verification of the model altered after certain exceptions is part of future work.

In conclusion, this method allows direct transition from a functional specification of a FMS to the control and monitoring of such a system. A major advantage lies in the simple model representation of synchronization and concurrency of processes.

Future work will concentrate on the generalization of strategies dedicated to rescheduling, on the verification of the new generated model, and on the integration of sensor components with dedicated elementary operations.

ACKNOWLEDGMENT

This work is based on two EEC sponsored ESPRIT projects (Project No. 623: Operational Control of Robot System Integration into CIM and Project No. 2202: CIM Planning Toolbox). The work was performed at the Institute of Real-Time Computer Systems and Robotics (Prof. Dr. Ing. U. Rembold and Prof. Dr. Ing. R. Dillmann) at the University of Karlsruhe, Germany.

REFERENCES

1. Alanche, P., et al. (1984). PSI: A Petri Net Based Simulator for FMS, 5th European Workshop on Applications and Theory of Petri Nets, AARHUS.
2. Brauer, W., Reisig, W., and Rozenberg, G. (September 1986). Petri Nets:

Central Models and Their Properties, Parts I and II, Proceedings of an Advanced Course, Bad Honnef, Germany.

3. Camarinha-Matos, L. M., Negretto, U., Meijer, G. R., Moura-Pires, J., and Rabelo, R. (1989). *Information Integration for Assembly Cell Programming and Monitoring in CIM*, ISATA, Wiesbaden, Germany.

4. Jensen, K. (1983). High Level Petri Nets. In: *Applications and Theory of Petri Nets*, edited by A. Pagoni and G. Rozenberg, Informatik Fachberichte, Vol. 66, Springer Verlag, Berlin.

5. Kasturia, E., DiCesare, F., and Desrochers, A. (1988). Real Time Control of Multilevel Manufacturing Systems Using Colored Petri Nets, *IEEE Robotics and Automation*, Vol. 2, pp. 1114–1119.

6. Meijer, G. R., and Hertzberger, L. O., (1988). Exception Handling for Robot Manufacturing Process Control, *CIM Europe Conference*, IFS Publications, Madrid, Spain.

7. Meijer, G. R., and Negretto, U. (July 1990). Exception Handling and Task Level Programming of Flexible Manufacturing Systems, ISRAM, Vancouver, Canada.

8. Negretto, U. (July 1990). Process Control for Flexible Assembly Systems, ISRAM, Vancouver, Canada.

9. Peterson, J. L. (1981). *Petri Nets Theory and Modelling of Systems*, Prentice-Hall, Englewood Cliffs, NJ.

10. Reisig, W. (1986). *Petrinetze–Eine Einführung*, Springer Verlag, Berlin.

11. Sahraoui, A., Atabakhche, M., Courvoisier, M., and Valette, R. (1987). Joining Petri Nets and Knowledge Based Systems for Monitoring Purposes, *Proceedings of 1987 IEEE International Conference on Robotics and Automation*, Raleigh, NC.

12. Stienen, H., and van der Weerd, P. R. (1988). Object-Oriented Design of a Flexible Manufacturing System, IFIP Working Conference, Galway, Ireland.

13. Valette, R. (1986). Nets in Production System, GMD Advanced Course on Petri Nets, Bad Honnef, Germany.

14. Zhou, M., DiCesare, F., and Desrochers, A. (1989). Top Down Approach to Systematic Synthesis of Petri Net Models for Manufacturing Systems, *IEEE Robotics and Automation*, Vol. 1, pp. 534–539.

PART III

MANUFACTURING OPERATIONS

CHAPTER FIFTEEN

Knowledge-Based Systems for Process Planning

H. J. WARNECKE and H. MUTHSAM

Fraunhofer Institute for Manufacturing Engineering and Automation (IPA), Stuttgart, West Germany

15.1. INTRODUCTION

The efficiency of the production area in a company, particularly that of the production equipment, has risen considerably in recent years. The pre-manufacturing areas, on the other hand, have been neglected by this development, although many firms admit they are responsible for a large percentage of production costs. Measures must be taken to solve this problem, which is detrimental to the increasing efforts to rationalize and automate production, for the purpose of optimizing the production process as a whole.

Only with the implementation of computer integrated manufacturing is appropriate emphasis put on process planning and other premanufacturing areas. Regarding preparation of process plans, this may originate from the fact that process planning connects the technical information flow from the CAD system to the NC programming and the planning and scheduling area using the MRP system.

Until now, the use of computer aided process planning systems was limited to workpieces, with geometry that could easily be described, such as turning parts or sheet metal components. For prismatic workpieces, only a limited amount of assistance could be given for special machining processes or individual planning functions. No general system existed that could be used for all planning functions.

Intelligent Design and Manufacturing, Edited by Andrew Kusiak.
ISBN 0-471-53473-0 © 1992 John Wiley & Sons, Inc.

15.2. PROCESS PLANS AS A BASIS FOR PLANNING AND OPERATION OF FLEXIBLE MANUFACTURING SYSTEMS

Considering first the individual planning and introduction phases of flexible manufacturing systems, it is evident that both the manufacturer and the system user have a very high one-time cost responsibility in the planning stage. The costs occurring during system operation are in fact smaller by comparison; however, when accumulated over the total service life, they represent a considerable cost saving potential (6). From the point in time where the planning phase is completed, the task of economically operating the flexible manufacturing system is up to its user and operator. In general, this can be divided into four phases (4).

Since essential tasks in the conceptual as well as in the operative phase of the manufacturing system are assigned to operations scheduling, considerable interactions are to be recorded, particularly between process planning and the economical introduction of flexible manufacturing systems. Thus operations scheduling determines the investment cost for both processing and material flow facilities, we well as the manufacturing system operating cost (3).

With the planning and layout of a flexible manufacturing system, process plans are used for analysis of the actual production, for formation of part families in the workpiece spectrum according to their processing requirements, and as the basis for new concepts of manufacturing cycles (Fig. 15.1).

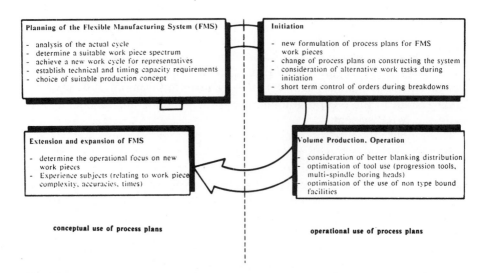

FIGURE 15.1. Use of process plans in the different phases of a flexible manufacturing system.

In addition to being used in the planning stage, the process plans for all workpieces to be processed in the flexible manufacturing system must be prepared in its initiation phase. Thus a multitude of special requirements must be considered with respect to the manageable complexity of workpieces or the process reliability of machine tools operated at low levels of staffing.

It is necessary to take these requirements into consideration on the part spectrum not only in the planning stage of a flexible manufacturing system. In the setting-up phase and in volume operation, the workpieces for the manufacturing system must be checked as well for economical production when preparing the process plan. Consequently, large-scale production on flexible manufacturing systems requires further process plan modifications. In consideration of process reliability at each point in time, an optimization of cutting parameters and of machine utilization is sought. In addition, in running flexible manufacturing systems, further technological developments of operating resources such as tools or fixtures are to be considered.

The experiences from initiation phase and production operation are significant not only for planning expansion measures but also for workpieces manufactured conventionally up to this point. A comparatively higher expense is required to prepare the process plans for these workpieces, which, with regard to the operations to be carried out and demands on the planner, are essentially based on different focal points and courses of action.

Various influences on the manufacturing technology and organizational boundary conditions of a company require a continuous process plan revision, with respect to planning new processing and material handling facilities (Fig. 15.2).

FIGURE 15.2. Causes for the obsolescence of process plans.

This is based on the fact that, due to market demands for shorter and shorter lead times and the general aim for reduction of capital tied up in the manufacturing area (work-in-process), an increased tendency for smaller batch sizes exists.

Therefore it is obvious that process plans prepared for a certain batch size range but applied to manufacturing orders outside these batch sizes increasingly counteract economical production.

Using a plan set up for a certain batch size for another order, with an essentially smaller batch size, leads to less planning expense. However, only in a few cases has this practice resulted in an economical production of the parts.

An actualization of process plans is in fact often carried out in the course of an adaptation of setups or process times to new manufacturing concepts with higher production. Nevertheless, a fundamental consideration of changed chucking situations or of changed tool requirements occurs only in a few cases.

Considering the importance of process plans for the economical production of workpieces, one is even forced to retain the detailed knowledge about the manufacturing capabilities of a FMS obtained in the course of planning activities, not only with the objective to facilitate further planning but also to secure the foundations for a computer aided system for process plan preparation.

15.3. COMPUTER AIDED PREPARATION OF PROCESS PLANS

15.3.1. Basic Tasks, Methods, and Organizational Principles of Process Planning

Within the production development process, operations scheduling represents the connection between product definition in design and product realization in manufacturing. The tasks of operations scheduling generally consist of planning, scheduling, and controlling the manufacturing processes.

Within process planning, the tasks can be classified according to the time scale of the planning activities:

Short term (i.e., processing parts lists, preliminary calculation, process plan preparation, NC programming)

Medium term (i.e., accounting, profitability considerations, manufacturing methods planning)

Long term (i.e., procedure planning, investment planning, preparing operation standards, time study matters)

Because of the different objectives of these individual task areas, consideration of all demands on a planning aid is not within the practical bounds in a single computer aided system. However, individual modules of a single process planning system should also be used for medium- and long-term planning tasks. This requirement results in an essential demand on the concept of CAPP systems, which is difficult to achieve with current state-of-the-art developments.

Of equal importance is the planning method used for preparing process plans, since it considerably influences the concept.

Here a classification into four basic types is presented, which, having different prerequisites, result in different courses of action, which in turn permit different degrees of computer aid.

1. *Repeat Planning*. As the simplest type of process planning, only slight or formal changes on already existing plans are made. However, no new planning data are generated.

2. *Variant Planning*. Already existing standard process plans are used as a basis for planning of workpieces within corresponding part families. Results depend on geometry or quantity variations.

3. *Similar Planning*. By means of the plan of a similar workpiece with respect to geometry and manufacturing technology, the planning takes place by changing and adapting individual processing operations. Finding a similar part presents an essential problem using this method.

4. *Generative Planning*. This method is applied to preparation of process plans "from scratch," whereby no existing plans whatsoever are taken as a basis.

As a constituent of administration and MRP systems, edit and text processing functions supporting the repeat planning can be offered from a computer to modify existing plans. By variation of parameters of a parametered standard process plan, variant planning permits the development of plans up to partly automatic with computer aid. This applies only to a strictly limited workpiece spectrum.

With similar planning, the search for an already planned, geometrically classified workpiece and the change of the most similar process plan are possible, to a limited extent, with computer aid. Since generative planning is not based on the principle of comparison and applies to the whole workpiece spectrum, it is the most complex task for the planner. Because of the geometric multiplicity generally present in a production program, the universal planning process is not automated. However, part functions can be implemented with computer aid.

The organizational principle of process planning represents a further significant parameter for a computer aided system. Here two different principles have been established in operational practice. In object-oriented process planning, a single planner prepares the complete process plan of a workpiece throughout all planning functions. Therefore the planner has to possess a comprehensive and detailed knowledge about procedure and machines. Function-orientated planning, on the other hand, requires detailing of the individual process operations by specialists, after rough planning of the workpiece has previously been carried out in a central location.

Furthermore, this principle of organization can be differentiated into procedure-specific (turning planner, milling planner) or technology-specific (NC planner, planner for flexible manufacturing system) subgroups.

15.3.2. Integration of Process Planning into the Intercompany Information Flow Concept of a Company

From endeavors to optimize the entire information flow system, it was found that the CAPP system has to be completely integrated into the interorganizational information flow (Fig. 15.3). Apart from an extensively redundancy free data storage, this achieves a maximum relevance of data and information.

Advances in the design area have resulted from the introduction of software systems (CAD, calculation program, databases of subject characteristics), but current process planning, as a tie between CAD and production planning and control, receives little computer aid for planning activities.

For example, it would be meaningful for the design department to be able to fall back on a module of a CAPP system to generate a cost estimate of design alternatives. This task is difficult to realize, because even slight geometric modifications of a workpiece, such as reduction in wall thickness between two pockets, can generate considerable changes in work cycles. Therefore, in such cases, completely new process plans have to be prepared. In contrast, a basic feasibility check of the design drawings represents a solvable problem for a software system. On the part of the production

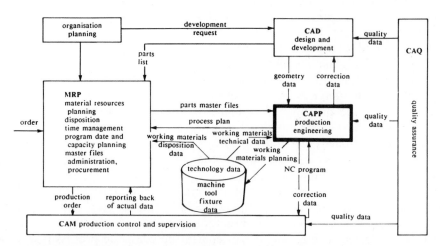

FIGURE 15.3. Integration of CAPP in the operational information flow.

planning and control system in which the order data are administrated, the requirement exists to have process plans automatically available for timing and capacity planning. Part master files are transferred a priori to the CAPP system in order to create the process plans.

In this context, particularly with reference to a short-term preparation of process plans, it is desirable to be able to react to breakdowns on the shop floor, adopting the most economical solution for the situation even in exceptional cases. Hence the requirement arises that one be able to introduce single components from CAPP systems. In practice, when a machining center fails and a deadline is approaching, a change is made to universal boring and milling machines without considering available facilities and tools for reduction of chucking.

The pursued goal, however unachievable at this point in time, is to couple the planning system to all industrial management and technical areas of a company, making process plans or the information contained therein available to the entire company.

15.3.3. Demands on Systems for Computer Aided Preparation of Process Plans

The special demands imposed on a system for computer aided process planning must be represented by various views. Thus, for example, expectations of the company management are substantially of an economical and organizational form, whereas process planners as system users require a simplification of plan preparation with adequate independence and possibilities for intervention.

The general and economical requirements can be listed as follows:

- Time and cost effective process plan preparation
- Shorter throughput time of process plans by operations scheduling
- Improved planning results
- Better utilization of available resources (i.e., machines, tools) by considering more influencing variables
- Immediate adaptation of process planning to rationalization measures in the manufacturing area
- Available documents right up to date (timetables, tool catalogs)
- Formal uniform process plans
- Complete integration of the interorganizational information flow between CAD, MRP, and NC programming

Requirements on the part of the process planner include the following:

- Support of the planning process through best possible provision of information

- Applicability to as many planning functions as possible, at least for the time intensive tasks and constantly repeated activities
- Consideration of all relevant manufacturing procedures
- Planning capability for the entire workpiece spectrum
- Simple system management
- Low input expense
- Good clarity of results

These requirements are to be viewed as universally valid. In a particular firm planning case, they have to be supplemented with the company's specific expectations, which can vary according to planning methods and organizational principles. However, they represent the basis for a requirement profile on systems for computer aided process planning, irrespective of those being realized by conventional programs, decision table systems, or expert systems.

15.4. CONVENTIONAL AND KNOWLEDGE-BASED SYSTEMS FOR COMPUTER AIDED PROCESS PLANNING

Depending on company-specific boundary conditions and influencing variables in the field of process planning, different degrees of automation, realizable with various software systems, may be necessary for CAPP systems. The ideal case, which for reasons of complexity and extent of the process planning task is not yet achievable, would be the fully automatic preparation of process plans from geometry data, with subsequent automatic generation of the associated NC programs.

For this reason, it is indispensible to balance and to limit requirements on the CAPP system. The evaluation of functions according to various criteria represents an aid for choosing a suitable software system as well as the basis for implementation of the process planning system. For the separate process planning steps, different forms of software system can be considered suitable, depending on the planning methods to be portrayed (5).

15.4.1. Conventional Software Systems

Special systems for computer aided process planning are those implemented by means of a conventional programming language. As a rule, these systems are company specific and are prepared with considerable programming expense for solving problems of limited scope. Therefore they are not transferable to other planning tasks having different boundary conditions.

In contrast, user-oriented systems, which support detailed planning of special types of workpieces such as turned or sheet metal parts, are more general. The systems offer dialogue controlled support for individual functions, such as the search for similar workpieces, the input of text for process operations, or the evaluation of cutting parameters and process times as well as the calculations.

15.4.2. Decision Table Systems

The decision table systems represent a preliminary stage in the direction of automatic process planning systems. As an aid, they offer procedures from manual dialogue-oriented plan preparation, via partly automatic functions, up to fully automatic process plan generation. By means of decision table technology, intelligent dialogue can be achieved, resulting in reduced input expense. Above all, however, projection of the process planning logic contributes to the determination of raw material operations, machines, and means of production, as well as to time studies. Furthermore, the required hierarchical architecture of the system contributes considerably to the systemization of planning activities. With the aid of decision table generating and interpreting programs, the tables can be generated and modified relatively easily. Decision tables are to be considered as a form of knowledge-based system, in which company-specific proceedings for problem solving can be represented relatively simply (compared to conventional programs) in a rule-oriented manner (1). In the ideal case, building of these systems should be carried out by the user with the available internal system aids provided. Reality shows, however, that the manufacturer's planning experience is required, particularly in the difficult stage of structuring the knowledge.

15.4.3. Expert Systems

In comparison to conventional systems and as an extension of decision table systems, expert systems make it possible to process the planning knowledge in the form of rules and facts. Knowledge is no longer provided in firmly programmed or tabular form but is given to the knowledge base without a compulsory structure. The processing of this planning knowledge takes place through a so-called inference component, which deduces results from the user's inputs for a specific application.

In recent years, process plan preparation has been tackled intensively with the help of expert systems, particularly by university institutions. Today just a few realizable formulations are known—in the area of rotational workpieces and sheet metal components—which can be considered promising.

Because of the success of expert systems in different diagnosis applications, at present there is a big potential for their introduction especially in planning tasks. This is because the planner, even with partial problems, at first considers numerous alternative solutions and only in the course of his/her activities determines the solution to be pursued further.

The combinatorial multiplicity that arises when one considers all the alternatives makes the universal use of conventional software aids impossible and requires knowledge-based problem solving strategies.

15.5. EXPLAN: AN EXPERT SYSTEM FOR GENERATING PROCESS PLANS

15.5.1. Project Framework and Objective

The EXPLAN project (Expert System Application to Computer Aided Generation of Process Plans) is subproject of the research project on "The Factory of the Future," as part of the European research initiative EUREKA.

The aim of the project is to remove the burden of routine tasks from the process planner by using an expert system. The process of preparing a process plan should unfold in a way that, although automated to a large extent, is yet so transparent that the process planner can correct every important intermediate result in dialogue mode.

The feasibility of a universal planning cycle was tested and demonstrated using a first prototype implementation. The main emphasis here was put on the planning of milling with its complex sequence of operations. The design specification for a second prototype is currently being prepared as part of the project.

15.5.2. Model of Process Planning

Of basic importance for the structure, but predominantly for the manner, of operation of knowledge-based systems is the precise model of the fields of its application. On the one hand, a model can lead to the development of the mechanisms of knowledge representation required; on the other hand, it forms the basis for the derivation of application-specific inference systems. Initial systematizing of the world of process planning allows us to break it down into three models: workpiece geometry, machining, and planning.

These geometrical data are used by the process planner as the basis both for the interpretation of the existing workpiece drawing and for almost all activities in the course of planning. Thus, in any case, the data have to be represented and processed in a form acceptable to the planner within the expert system. In the *workpiece model*, prismatic workpieces must be reproduced in the machining-oriented mode familiar to the process planner instead of the geometry-oriented mode of the design. Thus it is necessary to break a workpiece down into individual coherent objects, the processing elements (BEs) (Fig. 15.4). The purpose of using this form of representation is to create a definition of geometrical elements that approximate a way of interpretation and a linguistic usage familiar to the process planner. In

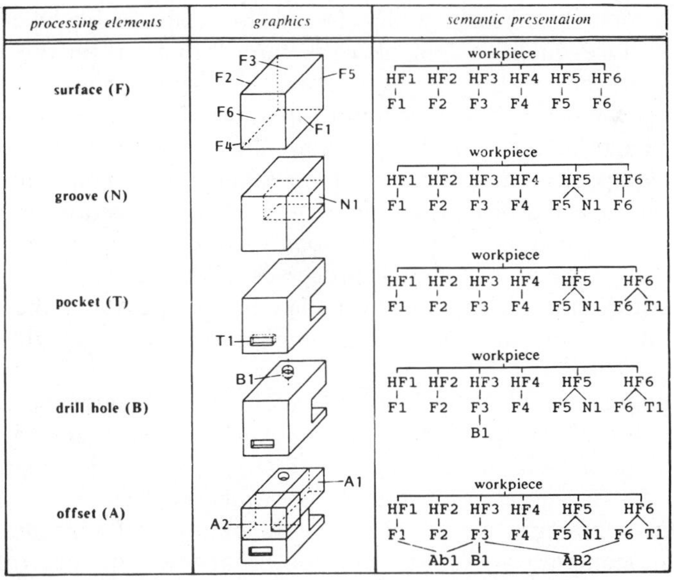

processing elements	graphics	semantic presentation
surface (F)	F3, F2, F5, F6, F1, F4	workpiece HF1 HF2 HF3 HF4 HF5 HF6 F1 F2 F3 F4 F5 F6
groove (N)	N1	workpiece HF1 HF2 HF3 HF4 HF5 HF6 F1 F2 F3 F4 F5 N1 F6
pocket (T)	T1	workpiece HF1 HF2 HF3 HF4 HF5 HF6 F1 F2 F3 F4 F5 N1 F6 T1
drill hole (B)	B1	workpiece HF1 HF2 HF3 HF4 HF5 HF6 F1 F2 F3 F4 F5 N1 F6 T1 B1
offset (A)	A1, A2	workpiece HF1 HF2 HF3 HF4 HF5 HF6 F1 F2 F3 F4 F5 N1 F6 T1 Ab1 B1 AB2

FIGURE 15.4. Development of a semantic network for workpiece representation.

addition, this definition can be enhanced by attributes relevant to production engineering, such as surface or quality requirements.

The *machining model* comprises various types of planning knowledge, as can be seen in the schematic representation of a process plan (Fig. 15.5). This is just process-specific knowledge concerning the changes in geometry and properties of the workpiece, which can be generated by individual

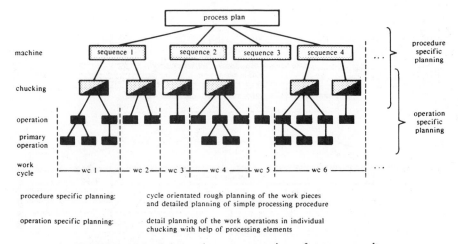

procedure specific planning: cycle orientated rough planning of the work pieces and detailed planning of simple processing procedure

operation specific planning: detail planning of the work operations in individual chucking with help of processing elements

FIGURE 15.5. Schematic representation of a process plan.

production processes. This includes, for instance, obtainable levels of precision and surface qualities, but also the burr formation on the workpiece resulting from the metal cutting operation. The operation-specific planning requires knowledge of the machining operations that can be carried out using particular setups. In the planning the influence of the machines, clamping equipment, and tools must be taken into account to the same extent as the specific processing elements and their position on the workpiece. These yield important conclusions on the obligatory or sequential relationships between machining processes. The central object-oriented categories of the machining model are elementary operations that represent the smallest possible unit of a machining process (e.g., rough and finish milling of a surface).

In the *planning model*, the processing elements, and later the elementary operations, are allocated to the individual clamping positions and included in the plan on the basis of adherence to all sequential restrictions applicable. The aim of planning is to maximize the work content of the individual setups, thus obtaining a reduction in the number of machine changes, changes of clamping positions, and tool changes. The overriding goal, however, is to enable the economical machining of the workpiece involved in the planning, with all the influencing variables being taken into account.

15.5.3. System Structure

The structure of the expert system—in the first stage composed of a knowledge base, dialogue, and inference components—is characterized by being connected to the CAD system and to a knowledge base distributed over various modules (Fig. 15.6). The dialogue component not only repre-

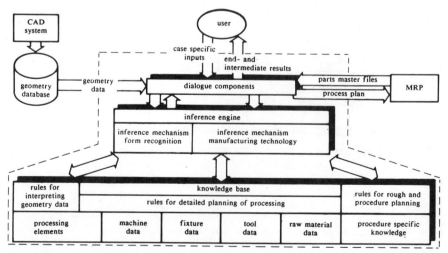

FIGURE 15.6. Structure of the expert system EXPLAN for process planning.

sents the inference to the user during the actual course of planning but also assists the user in structure recognition. The process of structure recognition in the knowledge base is to be considered dissociated from the other planning knowledge, since it is concluded with the transfer of the knowledge-based workpiece description prior to planning. In this stage, the results to the rough-controlled dialogue and detailed planning are available. The resource data are filed directly into the knowledge base. Links to external data files can be considered at a later date, as can connecting up the expert system with an MRP system.

15.6. MILLING AND BORING PLANNING MODULE

As the most important part in the EXPLAN system, the milling and boring planning module has the complex task of creating the link to the elementary operations and their sequence, on the basis of the workpiece geometry and the relationships of the processing elements to one another (2). The procedure can be divided into four individual steps (Fig. 15.7): structure recognition, planning preparation, rough planning, and detailed planning.

15.6.1. Structure Recognition

In practical applications, a computer aided production planning system for geometrically complex workpieces cannot be used on an economically meaningful basis unless the workpiece geometry is known to the system in a form that is relevant to process planning. Since CAD systems are increasingly used in design today, the most important basis for the electronic transfer

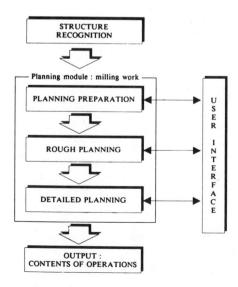

FIGURE 15.7. Functional scope of a prototype for knowledge-based generation of process plans for milling operations.

of a workpiece drawing has already been established. However, the complete technological data required are not obtainable from the CAD system. Furthermore, the available data are stored in a form that is initially unsuitable for planning work. It is thus necessary to prepare the CAD drawing in a way that corresponds to the requirements of production planning for structure recognition (Fig. 15.8).

The starting point for structure recognition is the 3D CAD system used by the designer to prepare a drawing. The drawing must be available to the structure recognition module on a file that corresponds to the IGES standard. The transfer module uses this standard to compose the drawing for the transfer into the IAOGraph format. The IAOGraph (InterActive Object-oriented Graphics tool) has an object structure that is distinguished by the fact that an object consists of several surfaces: a surface of a closed line of straight edges (loop) and an edge of two points, which in turn consist of their coordinates. Any surface and its position in space can be identified using this procedure. For the recognition of solid bodies, for example, 3D objects composed of geometrical surfaces, it is necessary to determine the relative position of individual surfaces to one another. For this purpose, the edges connecting the individual surfaces have to be examined. To establish the edge orientation, the following definition has been laid down:

If the angle between two external surfaces of the workpiece is smaller (greater) than 180°, then it is an internal (external) edge.

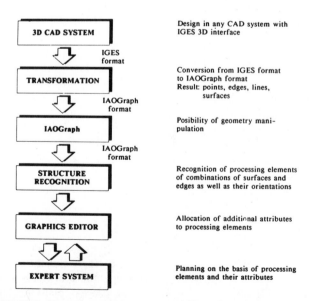

FIGURE 15.8. Process structure recognition of workpiece geometries.

The orientation of an edge thus determined is allocated to the latter as an attribute. Complete processing elements can be recognized by scanning the workpiece for preset patterns of external and internal edges for the various processing elements.

The pattern for the simplest processing element "surface" can be, for example, that it consists of one or more geometrical surfaces that (a) all lie in the same plane and (b) are associated loops made up of edges with an "external" orientation only.

As a result of structure recognition, all processing elements recognized on the basis of the geometry/topology of the workpiece are retained. If the workpiece cannot be fully recognized by the system—that is, edges or surfaces still exist that cannot be allotted to any processing element—the user has to identify them manually.

The next step in structure recognition specifies the relationships between processing elements. Here it is of particular importance to construct the relationships governing the position of processing elements to one another, such as "adjacent to," "penetrates," and "lies in." These relationships supply essential instructions for later sequence planning. For example, a relationship describing the position of a "groove" in another processing element could be:

Relationship of "groove lies in": A groove lies in another processing element *if* a side face of the groove has an edge in common with a face of this processing element *and* this edge is not adjacent to the edge of the side face which displays the "internal" orientation.

The workpiece geometry and the workpiece representation can be manipulated in the IAOGraph. The processing elements are transferred into the internal data structure (frames) of the expert system, with all relationships concerning them, for further analysis by the expert system. Additional attributes, such as tolerances, specifications for surfaces, and dimensional references, can be allocated to the individual processing elements by the user in the course of planning through an intermediately located graphics-assisted editor.

15.6.2. Planning Preparation

The aim of planning preparation is to prepare the initial data for the subsequent planning algorithms, taking into account the limiting conditions specific to machining. Here planning preparation is to be carried out in terms of a sequence of independent jobs (Figs. 15.9 and 15.10).

It is necessary to determine *dimensional references*, since all processing elements can be dimensioned with respect to the six external surfaces (dimensional reference planes) on the workpiece. At least three main dimensional reference planes must be given for the workpiece. If necessary,

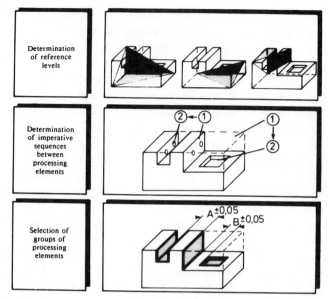

FIGURE 15.9. Function of the "planning preparation" module as part of the prototype (I).

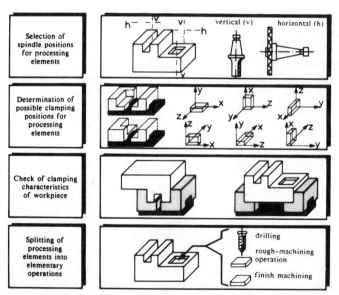

FIGURE 15.10. Function of the "planning preparation" module as part of the prototype (II).

the other dimensional references are to be transformed into one of eight possible combinations of dimensional reference planes.

In the *determination of obligatory sequences*, obligatory sequential relationships are derived from the position of the processing elements (BE) on the workpiece (e.g., BE-1 "penetrates" BE-2, BE-1 "lies in" BE-2) and from general machining rules (e.g., the dimensional reference planes of a BE must be created before the BE can be machined). The obligatory relationships between processing elements can be broken down into various classes, which relate either to finish machining or to rough machining. For reasons involving specific combinations of processing elements or close tolerances between processing elements, it can happen that these must be manufactured either in one setup or, if necessary, even with the same tool. In *determination* involving *processing element groups*, the relationships between processing elements are checked and, if necessary, processing elements are combined into groups.

Another module allocates the *spindle positions* with which the processing element can be rough and finish machined to all processing elements recognized during structure recognition. The spindle positions are given in relation to the relevant processing element (parallel or vertical) and in relation to the workpiece (machining from above, below, right, left, front, and rear). Restrictions that relate to specific spindle positions are considered in such a way that, through geometrical or production engineering restrictions, spindle positions are either completely eliminated or restricted.

With the dimensional reference planes taken into account, all processing element groups and processing elements are allocated to the *clamping positions* on the machine at which the processing element group or the processing element can be machined. This allocation can only be done if the available clamping equipment is taken into account. The determination of possible clamping positions is carried out by establishing either that all the dimensional reference planes required for the machining of the processing element lie against the bearing surfaces of the clamping equipment, or that the processing element can be machined when one of its dimensional reference planes is being machined. For every processing element, a check must be made as to whether there are clamping positions in which the workpiece can no longer be fixed after the current processing element has been machined.

Since the volume of metal removed by cutting tool and the geometry of a processing element play a part, among other factors, in determining the *clamping possibilities*, there is no differentiation between rough and finish machining.

Splitting the processing element up into *elementary operations* (EOs) depends on the spindle position in which a processing element can be machined. Because of tolerances, surface requirements, and the volume of metal removed by cutting tools, rough machining may be required, as well as the finish machining elementary operation, using the characteristic types of tools for the individual machining operations.

15.6.3. Rough Planning

Rough planning covers the establishment of the setups that are absolutely necessary for the production of the workpiece. The basis for this statement is the idea that it is a common goal of all process planners to prepare a plan with the minimum number of setups. Rough planning is carried out only with regard to the finish machining of the processing elements. The necessary types of machines, spindle positions, and clamping positions required for the production of the workpiece are passed over to detailed planning.

In the two-stage planning algorithm, all processing elements are first divided into possible combinations of clamping position and spindle position based on the results of the planning preparation. These combinations are obtained by processing on any type of milling machine with three to five axes, and thus represent the comprehensive spectrum of machining capability.

The machining elements are next included in the plans, starting with the processing element groups and the processing elements occurring least frequently. This is performed by means of a search, in which the sequence restrictions have to be considered as constraints. The number of setups and machine changes is entered into the analysis function. The interface to the detailed planning stage is a list with specified setups, in which all processing elements of a workpiece can be manufactured redundantly.

15.6.4. Detailed Planning

Considering the sequence restrictions between processing elements, two steps emerge for the detailed planning: the determination of the sequences between the setups and the definite allocation of elementary operations to the processing elements.

In the determination of the sequences, setups weighted according to the type and number of limiting conditions are selected and entered into the plan. Simultaneously, a check is made as to whether the sequence conditions are still being met for setups already planned or for their respective processing elements. If they are not being met, a backtracking takes place, and the planning is continued using the next weighted setup. In the case of an unsolvable conflict, various strategies are available.

Definite allocation of elementary operations to the individual setups is based on economical criteria (minimizing tools changes and machining times) and also takes into consideration the existing obligatory conditions for the finish machining of processing elements.

15.7. EXPERIENCES

Since the availability of experts is usually limited, the development of a suitable procedure for knowledge acquisition should not be underestimated

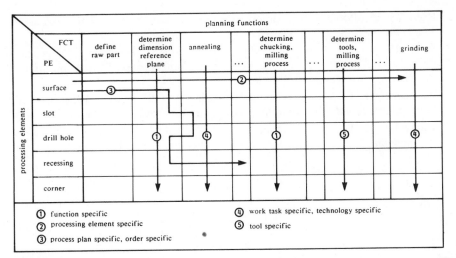

FIGURE 15.11. Strategies of knowledge acquisition to build up computer aided systems for process planning.

(Fig. 15.11). When a project starts, it is recommended that working sessions should be held with the process planner on the basis of simple workpieces from actual orders, in order to develop a common understanding of the difficulties of planning and the company-specific term̰s.

The effect of consistent model formation, especially with regard to highly complex problems, is that individual modules are obtained that are easily comprehensible in terms of function and data. These are considerably easier to reproduce with reference to representation or the inference mechanism. In the CAPP area, the effectiveness of computer assistance depends on the existence of "technology databases" that give a comprehensive and up-to-date picture of all the information/specifications (e.g., tools and resources) and their function-specific parameters.

Drawings or data files, which have already been set up in the design stage on the basis of production engineering objects, would make a substantial contribution, not only in terms of simplifying production preparation but also of pushing forward the functional integration of design and production planning and thus also encouraging efforts to achieve the organizational goals of the CIM concept.

REFERENCES

1. Hüllenkremer, M. (1987). Arbeitsplanerstellung: Schnittstelle zwischen CAD und PPS. *CAD-CAM Report*, Vol. 6, No. 3, pp. 102–106.
2. Muthsam, H., and Mayer, C. (1990). An Expert System for Process Planning of Prismatic Workpieces, *Proceedings of the 1st Conference on Artificial Intelligence and Expert Systems in Manufacturing*, London, UK, pp. 211–220.

3. Roth, H.-P. (1988). Rationalisierungspotential Arbeitspläne, *VDI-Fachtagung: Praxis der FFS*, Stuttgart, pp. 135–148.

4. Roth, H.-P., Zeh, K.-P., and Muthsam, H. (1988). Artificial Intelligence in Future Manufacturing, *Proceedings of the 7th Conference on Flexible Manufacturing Systems*, Stuttgart, pp. 249–263.

5. Warnecke, H. J., Mayer, C., and Muthsam, H. (1989). New Tools in CAPP, *CIRP International Workshop on CAPP*, Hanover, West Germany, pp. 169–184.

6. Warnecke, H. J., and Steinhilper, R. (1985). *Flexible Manufacturing Systems: International Trends in Manufacturing Technology*, IFS Publications, London.

CHAPTER SIXTEEN

Process Planning for Electronic Components

JOHN W. PRIEST

Department of Industrial Engineering, The University of Texas at Arlington

JOSE M. SANCHEZ

Instituto Tecnologio y de Estudios Superiores de Monterrey, Mexico

KEVIN PARÉ

Tandy Instruments, Fort Worth, Texas

16.1. INTRODUCTION

Computer aided process planning (CAPP) is the use of computer technology for developing optimum manufacturing plans that translate product design requirements into the sequence of processes needed to fabricate a product. In the last decade, computer assistance for process planning has commonly been applied in the realm of metal machined components (24, 33–36). Unfortunately, only a reduced number of computerized process planning systems exist today which are applicable in the electronic field. The use of artificial intelligence (AI) methodologies will, however, be a critical component in future computerized process planning systems especially in concurrent engineering environments.

There are basically two approaches for process planning that are recognized as being of general practice: variant and generative (28). The first approach creates new process plans by the retrieval of an existing plan for a "similar" product and then making the needed modifications to accommodate the desired characteristics of the new product. A variant method is based on the existence of a standard manufacturing plan that is identified by a predefined set of decisions rules. This plan is a permanently established ordered sequence of production steps for a specific category of product.

Intelligent Design and Manufacturing, Edited by Andrew Kusiak.
ISBN 0-471-53473-0 © 1992 John Wiley & Sons, Inc.

Computer assists for the variant approach generally consists of providing an efficient system for the management, retrieval, editing, and transmission of process planning data.

The advantage of variant systems is attributed to the productivity increase they provide for the expert process planner. These systems provide the company with the benefits of standardization, reducing the variance between plans for similar products or even for plans of the same products. Variant systems are also attractive because they require a minimal investment on the part of the company and a minimal requirement for involvement of the expert planners during development. A known commercial variant process planning system, which supports the development of manual assembly of wiring harness (1), is called the PICAPP system. The plans produced by this system are a sequence of assembly processes for assembly workers. A disadvantage of the variant method, however, is that it still requires the participation of the expert process planner for the creation of the standard process plan. Another disadvantage, which is as serious, is that with the variant method it is not easy to introduce new processes or processing approaches into the system. Likewise, variant methods are only applicable where there is a large number of similar components in the company's parts base. Significant variations in the form or processing of the electronic components in an organization generally will rule out the variant methods and dictate a generative approach as a more feasible alternative.

Intelligent generative process planning systems employ a completely computerized approach using decision logic, processing formulas, and geometric models to determine the required operation steps. Work to date on completely generative systems has shown that a truly universal system is probably not feasible. However, in restrictive domains the feasibility of generative systems has been demonstrated. In general, a computer-based generative process planning system for electronic components has the functional architecture shown in Fig. 16.1.

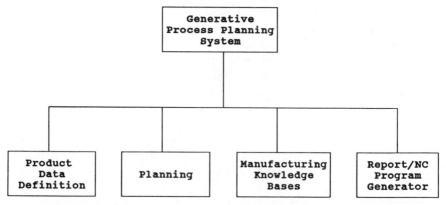

FIGURE 16.1. Generative process planning functional architecture.

The first module serves the function of acquisition of the product's descriptive information. The most proven technique for this data acquisition is the use of group technology (GT) classification and a coding system augmented with interactive prompts for unclassified components. Another method is automated capturing of information from a computer-aided-design (CAD) database. In a systems development effort at Contact Systems, Inc., conventional CAD data are being directly interpreted for the purpose of generating semiautomatic assembly plans for printed circuit boards (8). The second module is the planning or reasoning component that actually generates the required operation steps. This module normally functions in a two-pass manner. The first pass determines the required operations and the second pass sequences the operations. One successful system in this area covers the domain of electronic subassembly and sheet metal fabrication for simple forms (18). The third module of the computerized planning system is the manufacturing knowledge base, which includes the manufacturing logic and data used for the generation of the plan sequences. The fourth component is the plan generator module, which organizes the plan data into the appropriate format, adds in "special notes" information and produces a hard copy or a numerical control (NC) program for the final plan. The disadvantages of a generative method center around the lack of generality, the high investment required to build the decision rule knowledge base, and the general reliance on manual coding and classification practices.

16.2. BACKGROUND OF PROCESS PLANNING FOR ELECTRONIC ASSEMBLY

The assembly process plan is a series of steps that describes the process required to make an assembly. The plan serves as the communication link between design and manufacturing.

The goals of assembly process planning for electronic assembly are as follows:

1. Define process and assembly steps necessary to produce the product.
2. Indicate necessary resources.
3. Optimize sequence of inserted components.
4. Assign components to assembly machines.
5. Provide work estimates for scheduling.
6. Preparation of assembly machine program files.

There are several factors that should be considered before developing a process plan for electronic assembly. These factors can be grouped into five major categories (31):

1. Circuit board attributes.
2. Component attributes.
3. Handling attributes.
4. Placement attributes.
5. Manufacturing capabilities and assembly method.

16.2.1. Circuit Board Attributes

The attributes of the printed circuit board (PCB) itself will affect the development of the process plan. Board size, for example, can affect which machines can handle a particular board as well as process parameters such as solder and reflow temperatures. Board attributes that affect process planning are dimensions, shape, and printed circuit board type.

Printed circuit boards have been categorized into four types (31). The classification is based on the type of components as well as the side on which components are mounted. Type 1 is considered to be the conventional board with all components inserted on one side. Boards that have surface-mounted components on one side and insert components on the other are classified as Type 2. Type 3 boards are boards that have a mix of inserted of surface-mounted components on one or both sides of the board. Boards that have all surface-mounted components on one or both sides are classified as Type 4.

16.2.2. Component Attributes

Component attributes describe the type of component as well as its physical shape and size. Some components are available only in certain types of package or case formats. They determine the mounting method as well as the type of machine that will assemble that particular component. Components can be grouped into two major categories: "through hole" or inserted components and surface-mounted type components. Within these categories are several types of formats. Figure 16.2 illustrates various types of inserted and surface-mounted components.

16.2.3. Handling Attributes

Some components require special care so that they are not damaged during the manufacturing process. The format in which the components are packaged also has an effect on how the process plan will be devised. The three major handling attributes for electronic manufacturing are static sensitivity, heat sensitivity, and packaging.

Static-sensitive devices are a great concern in electronic assembly. Assembly workers in electronic manufacturing facilities are required to wear foot straps or other devices that "ground" the worker to prevent static

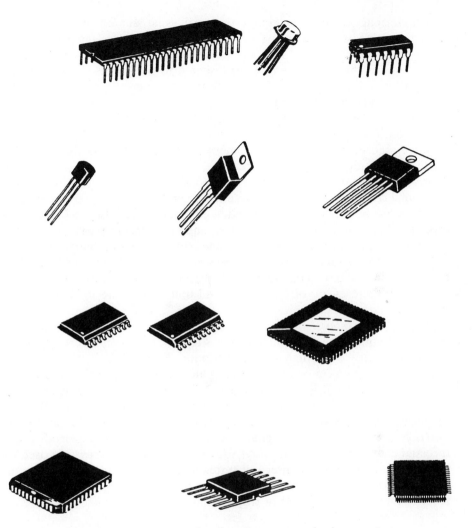

FIGURE 16.2. Inserted and surface-mounted components.

discharge. Precautions such as this must be built into the process plan to prevent damage to static-sensitive devices. Components that are sensitive to electrostatic discharge should be avoided by the designer.

Most soldering processes involve reflowing of solder through the direct application of heat or the passing of leads through molten solder. Some components are not capable of withstanding the heat stress caused by these soldering processes. Many components must be protected from extreme heat sources or attached after the solder reflow process has been performed. Heat-sensitive components, therefore, require special operations that affect the process plan.

The format in which a component arrives at an assembly station is determined by its packaging. The format may not always integrate with a chosen assembly method. The automation of component insertion/placement requires that components are presented to the process in a particular orientation and packaging media. The four main categories of packaging are bulk, reel, magazine, and tray.

Bulk packages, as the name implies, are loose components packaged in a bin or bag. Bulk packing is used most often in manual assembly but can also be integrated into automated equipment through the use of vibratory bowl feeders or by repackaging the parts.

Reel packages are used extensively with autoinsertion machines for axial and radial components. The components are embedded into a paper tape by its leads or body. Surface-mounted components are also found in reel format, where the entire component is embedded in the packaging media. The reel format usually has holes, much like computer paper, used for indexing by insertion equipment. Tape can be found rolled onto a reel or fan-folded into a box. This format is not recommended for manual assembly, as components are often difficult and time consuming to remove from the paper by hand.

Magazines or *plastic tubes* are used to present various components into automated equipment. The magazine has a track that is formed to handle a component with a particular case attribute. Plastic tubes are a type of magazine feeder where components are placed end-to-end inside a rectangular plastic tube. Dual in-line packaged (DIP) and surface-mounted components are often integrated into autoinsertion equipment with this method. Magazines and tubes can also be used with manual and robotic assembly.

Trays are a type of packaging where the components is placed into a recessed cavity, much like the trays used in boxed candy. Trays are used by manual assembly, robotic insertion systems, and surface-mount equipment. Odd-shaped components, in-house kitted parts, and large surface-mount components are often found in tray format.

16.2.4. Placement Attributes

Placement attributes determine how a component will be assembled on a board. In order to effectively plan the assembly process, knowledge of the mounting method, mounting location, and component orientation must be acquired.

There are generally two methods to mount a component: through-hole and surface mounting. Each method requires different process steps. Through-hole components are attached to the circuit board by inserting the leads of the components through holes in the board or into sockets that have been attached to the board. Surface-mounted components, as the name implies, are mounted directly on the surface of the board. The components are held in place with a solder paste or epoxy.

The mounting location on a circuit board is the X-Y position where the component is to be located. Some electronic assemblies such as cables, connectors, and systems require a three-dimensional position. Although electronic components appear to be symmetrical, the polarity of some components necessitates an exact orientation. With autoinsertion/placement equipment, the proper orientation of the components must be maintained throughout the placement process. Most components are physically marked to indicate the orientation of the part. This mark is used by manual assembly workers to assure proper placement.

16.2.5. Manufacturing Capabilities and Assembly Method

Process planning requires an in-depth knowledge of the manufacturing capabilities of the company. The problems that would occur are obvious if the process plan called for the autoinsertion of certain components when the shop floor is not capable or the parts cannot be provided in the proper package. Manufacturing capabilities can be broken down into three categories: manual, semiautomatic, and automatic.

Manual assembly is still used extensively for the insertion of components and other aspects of electronic assembly. Odd-shaped components, those that cannot be obtained in the appropriate packaging, cables, and connectors lend themselves to manual assembly. Often, the economics of low volume production require that components be attached manually even though they could have been autoinserted or inserted by robot. The flexibility of the human worker when compared to the automated method makes manual assembly an important option in the process plan. Repair tasks, because of the complexities of electronic rework and component insertion after solder of heat-sensitive components, are almost exclusively accomplished manually.

Semiautomated work cells integrate manual workers with some form of automated equipment. The possible combinations of systems are endless. A particular example of such a workstation uses an automated parts carousel and a light beam to assist the manual assembly worker. The parts carousel presents the proper part to be inserted while a thin light beam indicates the proper location for the part on a fixtured circuit board. After the part is in place, the worker pushes a button to index the system to the next part. An intelligent process planning system for this type of operation is discussed later.

Automation is becoming extensively used in the electronic industry due to improved quality, production rates, and producibility. Some assembly lines are fully automated, while others exist as "islands of automation." Automated assembly has been estimated to place more than 70% of all components (21). There are two types of automated equipment: autoinsertion/placement machines and robotic assembly stations.

The use of automated insertion/placement machines has become the preferred method for mounting "standard configured components." These machines are capable of reliable assembly at very high placement rates. An insertion machine can mount 2000–4000 dual in-line packages (DIP) per hour, 7000–30,000 axials per hour, 7000 radials per hour, and up to 100,000 surface-mounted components per hour (19). These machines are found as stand-alone processes or integrated with other processes through complex material handling systems.

Another type of automated assembly system is the robot workstation. Robots are often used to assemble odd-shaped components to the board. Robotics is typically an expensive option, because the end-effectors (end of arm tooling) and material handling are often custom designed and built. The speed of the robotics is considerably slower than the insertion/placement equipment at rate of 1000–2000 parts per hour (19). Robotics, however, has the distinct advantage of flexibility. The robot offers flexibility to the system with capabilities of tool changing, multiple packaging formats, and board handling.

16.3. PROCESS PLANNING PROBLEMS

A computerized process planning function traditionally falls short of expectations. The cause of this can usually be attributed to lack of integration and communication between company functional units. The problems with the process plan can be grouped into three major categories: design, scheduling, and manufacturing.

16.3.1. Design Problems

More often than not, the design of an assembly is developed without sufficient concern as to how it will be manufactured. Some of the factors that should be considered when developing a plan for electronic components are as follows:

1. Does the design provide sufficient accessibility for component placement?
2. Is the part autoinsertable by autoinsertion or robotic equipment?
3. Is the part available in package format compatible with existing equipment?
4. Does the part require special handling or processing such as heat sensitivity and static sensitivity?
5. Is the board density too high for automated placement?
6. Are all components on one side of the board?
7. Are all components oriented in well ordered rows?

Improved communication between the design and production disciplines will answer many of these questions.

16.3.2. Scheduling Problems

Just as the process plan is bound by the design, scheduling is bound by the process plan. The process plan does not consider all factors, such as equipment availability, that are necessary to produce a workable schedule. Some of the shortfalls of the process plan as suggested by Khoshnevis (16) have been:

1. Process plans are considered to be fixed and infallible.
2. The process plan assumes unlimited resources of the shop floor. Resources, however, are not always available as a result of equipment breakdowns, employee absentees, and maintenance.
3. Process plans do not always suggest alternate methods when a desired process is not available.

As a result, process plans are not always followed on the shop floor. For example, supervisors and managers will change the plan to improve machine utilization balances. Scheduling considerations should be considered in the construction of an intelligent process planning system.

16.3.3. Manufacturing Problems

A characteristic that is unique to the electronic assembly process plan is that many of the process steps are often strictly sequenced, that is, ordered. For example, leads must be formed prior to insertion and components must be inserted before soldering. This characteristic greatly simplifies the development of an intelligent assembly process planning system, because without some rigidly sequenced process steps the number of different possible plans that would be developed could increase exponentially.

As a minimum, the process plan should be optimized to:

1. Maximize number of components that can be autoinserted/placed.
2. Minimize assembly workstation operation time.

In order to minimize the operation time of workstations, three rules must be followed. First, maximize the use of autoinsertion/placement equipment since these machines are considerably faster than manual assemblers. Second, reduce the distance to be traveled between successive placements. Third, minimize downtime when changing product mix.

16.4. PROCESS PLANNING TOOLS AND METHODOLOGIES

The preparation of the electronic process plan has traditionally been performed by an experienced individual who has an in-depth knowledge of the shop floor. This person would complete an operations sheet, which would provide the assembly steps in process order. The plan would also include:

1. The components and parts to be placed/inserted, including part number and location.
2. Order in which components are to be placed/inserted.
3. Method to be used and identification of the workstation in which work will occur.
4. Schedule for when installation will occur.

In today's printed circuit board market of high-density boards, frequent engineering design changes, and short product life, it is becoming increasingly difficult to generate manual process plans in a timely manner. However, the expense of computer-based systems, the simplicity of some designs, and the perceived risk in new technology make it economically feasible for some companies to continue with manual methods. This section describes some of the tools used by the process planner when developing a process plan. These tools provide a foundation of information for the development of an intelligent system. These tools are:

1. Process planning guidelines.
2. Classification and coding.
3. Formula and algorithms.
4. Computer based systems.

16.4.1. Process Planning Guidelines

Printed guidelines are the simplest of the process planning tools. Guidelines are basically a set of rules or a checklist for the planner to assure that he/she has met certain criteria and goals. When published in the form of a series of questions, they can encourage process planners to check their work for possible simplifications and improvements.

As mentioned in the previous section, the basic steps for developing a workable but nonoptimal process plan for electronic assembly are fairly straightforward. Developing an optimal process plan, however, is much more difficult. Board assembly becomes even more complex when designs call for both surface-mount technology and through-hole insertion on both sides of a board. In this case, guidelines can help to assure that the process planner has considered all necessary process steps. In addition, these guidelines can be helpful in the future when developing rules for an intelligent planning system.

16.4.2. Classification and Coding

Another tool used by a process planner is classification and coding. Classification is the grouping of components or assemblies according to their similarities. Coding is the assignment of a numbering system or symbols by which the classification system can be referenced.

These techniques are the basis for group technology. Group technology is "the realization that many problems are similar, and that, by grouping similar problems, a single solution can be found to a set of problems; thus saving time and effort" (14). By grouping parts into part families based on design characteristics, it is possible to take advantage of similarities during process planning.

A classification system should group parts according to their design and manufacturing characteristics. Two broad areas that should be investigated when developing part families are board characteristics and component characteristics. For example, printed circuit boards (PCBs) can be classified by several characteristics including dimensions, shape, and board type.

By classifying these boards into part families with similar attributes, process planning can be simplified. The process planner can refer to process plans of similar assemblies rather than starting from scratch. Similar process plans would indicate such things as machines capable of handling board size, solder height for wave soldering, and material handling techniques.

Components to be placed on the board can also be classified and coded (3, 4) as follows:

1. Axials.
2. Radials.
3. Dual in-line packages (DIPs).
4. Pin and array.
5. Surface-mounted devices.

Components are typically classified by their case and package attributes not their functions. The coding system aims at capturing manufacturing characteristics not design characteristics. When components are properly classified and coded, they can easily be referenced to find such manufacturing attributes as preferred mounting method (insertion/placement/manual), alternate mounting method, format of packaging, and machines that are capable of handling that component. The classified and coded components provide a direct link to the design engineers and can greatly simplify the development of an intelligent process planning system in the concurrent engineering environment.

16.4.3. Formulas and Algorithms

Over the years, various academic and industry researchers have developed methods for solving the more complex problems of process planning. The techniques of operations research have been especially helpful in this effort. These methods are often in the form of algorithms and formulas. Two algorithms that can be computerized to solve specific process planning problems are to be discussed.

One method is to use Venn diagrams to solve sequence and grouping problems (9). When a company assembles many different circuit boards on a single production line, process planners attempt to group similar boards together in order to minimize any disruption in production due to changeover and to maximize productivity. Boards are often grouped together based on how many components they have in common in order to minimize the number of changes required. When a production line is changed from making one board to the next, the more components that are different, the more involved the changeover will be. For example, in a particular placement machine has enough tape feeders for 20 different components and if three boards are to be manufactured, each with 15 components and where the component mix of each board differs, then a Venn diagram will assist the process planner in developing groups and even sequence the order in which the boards are to be manufactured (9). An example of a Venn diagram is shown in Fig. 16.3.

Assuming that the current sequence of jobs is A, B, C, it can be seen by the Venn diagram that the changeover between board A and B will involve changing 9 tape reels. This is the number of parts on board B, but not on A. To change from B to C would involve changing 10 feeders. The total amount of changes made during the manufacture of the three boards in an A–B–C sequence is 19. Being that these reel changes result in downtime of machinery, it is important to minimize the number of tape reel changes. By using the Venn diagram to change the order in which these boards are produced, the number of reel changes can be reduced. Using this method,

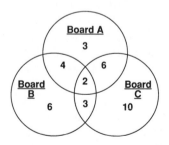

Number of Reel Changes	
A to B	9
B to C	10
A to C	7
C to B	10

FIGURE 16.3. Venn diagram for scheduling circuit boards.

Optimal Sequence is A-C-B

we find that the number of reel changes is reduced when the sequence is changed to A–C–B. This provides a resultant reduction in the amount of downtime. This method can also be used to group boards by grouping those with similar components.

A second approach is the use of branch and bound method to solve component insertion sequence problems. The sequence in which individual components are inserted into the board is probably the most complex task of the electronics process planner. Factors that must be considered are board design, distance between successive placements, component height, insertion gripper clearance, and component dimensions. For the manual planner, making an optimal sequence considering all these factors would be extremely time consuming. So the manual planner will concentrate only on those factors that have been determined to be of greater importance. When one considers a simple board with only 30 components, there are 29 factorial or 8.84×10^{30} possible sequences. Even the fastest of computers would take considerable time to analyze all possible sequence combinations, and manual methods would take a lifetime!

The purpose of the branch and bound method is to compare time or cost values associated with a given action and compare this value to alternative actions. For example, imagine we are required to determine the insertion sequence for a given group of components. The time or cost values for a given sequence are determined by the amount of time to index the positioning table to the proper location, amount of time for the machine to retrieve the part from the part feeder, or the time required to clear the feeding head.

In order to make this a more manageable problem, the branch and bound algorithm is implemented by dividing the components into groups of manageable size and then scheduling the groups and components within the group. By dividing the components into smaller groups, the problem is made simpler. Let's divide the 30 components into groups of 7, 7, 7, and 9. The number of possible combinations is reduced to $7! + 7! + 7! + 9! = 5040 + 5040 + 5040 + 362,800 = 378,000$. This number, although still large, is considerably more manageable. As other sequence factors are considered, which add constraints to the problem, further reductions in the possible combinations are made. Groups of components can then be scheduled according to the various sequence factors. One approach is to schedule all the components with smaller heights so that they are not susceptible to collisions from the mounting head. Another approach is to insert one entire geographical region of the board at a time.

16.4.4. Computer-Based Systems

Although manual methods are still widely used, this method of process planning is quickly becoming inadequate. Since the electronic field is a dynamic industry, it is important for companies to be able to get their products from design to the customers in the shortest time possible and to optimize productivity. With boards of up to several hundred components and the typical factory producing hundreds of different boards, it becomes evident that the ability of manual process planners to keep up with this pace will be difficult.

With the aid of computers, process planners can be given a valuable tool to produce and improve process plans. In fact, computer aided process planning (CAPP) can be the direct link between CAD and computer aided manufacturing (CAM), where process plans are generated directly from CAD drawings. By implementing a CAPP system, a company can take advantage of the following benefits:

1. Reduce process planning time and skill requirements of process planners.
2. Reduce manufacturing costs by developing optimum process plans.
3. Create more consistent and accurate plans.
4. Increase productivity.

There are basically two fundamental concepts on which the CAPP system is based. The concepts can be compared to the two types of process planning, variant and generative, the functions and benefits of which were discussed in the beginning of the chapter.

16.5. ARTIFICIAL INTELLIGENCE SOLUTIONS

In order to address process planning problems in electronics, such as its complexity, lack of ability to interface with other systems to automatically generate NC programs, and the lack of capability to extract product information directly from product CAD representations, most researchers in the area of process planning have turned to the methods and tools of artificial intelligence (AI). The hope is that the technology would provide the needed leverage. Although a great deal of progress has been made for the process planning of metal parts, limited success has been found in electronic assembly. The published results to date have limited use of knowledge representation or new knowledge-based programming tools. A predominant approach seems to be one of recording IF–THEN rule structures generally with an attendant loss of efficiency. Another popular approach has been to program the planning systems into an AI language, such as Lisp or Prolog.

The HICLASS system at Hughes Aircraft Company (18) represents one of the early successful applications of AI techniques to the process planning problem. Initially written in Pascal, the current system has been reprogrammed in C language for performance and portability issues. The HICLASS system owes some of the success to the product family chosen for planning (electronic subassemblies and simple sheet metal parts). Given this application domain, the product descriptions required for planning decisions are nearly identical to those required for the engineering definition of the product. The result is a simpler translation from the engineering definition to the form that the rule-based logic of the planner can manipulate. One of the points that the system demonstrates is the potential that such automatic translation provides. While HICLASS was originally envisioned as an assembly sequence step generator, it has been extended into an entire system that interfaces from engineering to the shop floor. Based on the schedule of the production facility, HICLASS retrieves the necessary engineering product definitions and develops the assembly plan sequences. HICLASS then downloads information to an operator terminal, where the system supports the interactive display and manipulation of planning data as the operator performs the assembly instructions.

The main difference between an expert system and generative CAPP is the way the knowledge is managed. Generative CAPP captures its decision logic and algorithms in rigidly programmed code, requiring major recoding of logic should a change in the process occur. An expert system knowledge base has the capability of being updated and modified without changing the underlying logic rules used to generate the process plan.

A successful application of expert system techniques is the expert system for circuit board assembly (CBA) developed for the Navy by CIM Systems (26). The CBA system goal is to enable users to quickly and efficiently define and optimize the placement and insertion sequence of electronic components for a printed wiring assembly using kitted parts. By reviewing PWA layout information, parts list, component engineering data, and available manufacturing capabilities and mimicking the way a human expert would go about the component insertion sequencing task, the system generates an optimal component insertion sequence. With the available design information as well as the built-in operator considerations embedded in the reasoning system, CBA helps to minimize the human errors leading to significant savings at the design and programming stages, thus resulting in increased assembly productivity.

The conceptual design of the CBA system provides flexibility for configuration suitable for various sized manufacturing operations. It is organized in a modular architecture, providing flexibility to be modified for different assembly environments. User access to the system is supported by menu-driven prompting, on-line help aids, and on-line procedural instructions. The system is interactive in that it requires some intervention by designers in the initial establishment of assembly requirements. These are interpreted from CAD generated blueprints along with decision support checklists. An abstraction program is then used to produce the inference conditions and symbolic representations required to drive the system's reasoning process. Figure 16.4 shows a schematic overview of the CBA system operation.

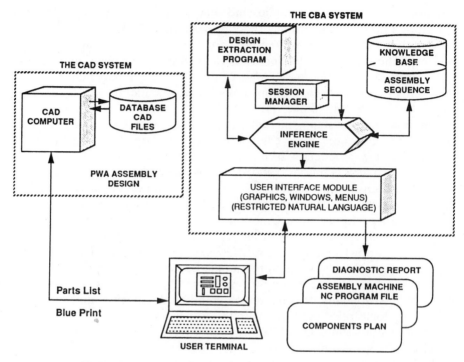

FIGURE 16.4. CBA schematic working overview (26).

16.6. EXPERT SYSTEM FOR SEMIAUTOMATED ASSEMBLY

The following discussion provides details in the development of the CBA expert system. The system takes into account the following issues:

1. Insertion planning criteria
2. Insertion sequencing rules
3. Component sequencing methodology
4. Machine table tour planning

16.6.1. Insertion Planning Criteria

The first step in developing component insertion plans requires that all the assembly planning criteria be identified and classified. All decision rules needed to satisfy these criteria must then be developed. In the PCB assembly domain, the assembly planning criteria are the point of reference by which an assembly plan's effectiveness will be judged. These criteria depend on management policies and are influenced by the assembly environment. Some of them may refer to the method of assembly, whereas others may refer to assembly equipment characteristics. Although the number of assembly planning criteria can be extensive, a process planning approach is described based on the following two criteria (26):

1. Simplify the operator's role in the semiautomatic component insertion process.
2. Minimize the machine bed tour's traveling distance.

The simplification of the semiautomatic insertion process aims at decreasing the level of component insertion difficulty for the manual assembly tasks while improving overall quality. Taking the operator as the first order of concern optimizes the overall process throughput by reducing employee fatigue and downstream rework. Similarly, the second criterion aims at minimizing the movement of the bed that holds and positions the PCB, thus maximizing the specific assembly equipment throughput.

16.6.2. Insertion Sequencing Rules

Modern PCB assembly requires a new systems approach, to automate the entire planning process including manual operations. That means the ability to call upon an on-line program that produces the optimum component insertion sequence for a new PCB assembly in a way that minimizes the overall manufacturing cycle time. In order to accomplish this objective, the component insertion methodology described in this example is based on a series of sequencing decision rules that meet the criteria previously described (26).

Rule 1. Insert smaller components before the larger components.
Reasons:

1. Avoids component "valleying." The formation of valleys during the component insertion process creates difficulties for subsequent component insertion, orientation, and lead guiding through the insertion holes because the access space is limited by the previously inserted components.
2. Avoids the blocking of the light guiding system. Because the movable light mechanism shines on the PCB from the side and/or rear, it is possible for tall components to block the light guiding system from shining on lead holes. Therefore a component that is near and/or below a tall component should be inserted before the tall component.

Rule 2. Insert, in one pass, all the components of the same type and value.
Reason:

1. Reduces the number of wrong value components inserted by the operator.

2. Promotes the separation of identical appearing components by alternating the kitting sequence.

Rule 3. Kit components with identical size and shape but different electrical value with other components of nonsimilar size and shape.
Reason:

1. Reduces the risk of mixing look-alike parts in the bin trays when retrieving parts from or returning extra components to the rotary bin trays.

Rule 4. Select the insertion sequence that results in the shortest machine bed movement.
Reasons:

1. Minimizes the PCB assembly time by eliminating erratic movements.
2. Improves accuracy, since long bed movements tend to reduce the position accuracy of the insertion bed and cut and clinch mechanisms.

The adequate application of the previous rules requires the definition of the component insertion priority shown in Table 16.1. This table represents the consensus of several PCB design and manufacturing experts from companies in the public and private sector as well as experts working in the academic field. The table is organized based on the component packaging (i.e., DIP, axial, radial, can, box, etc.) and component prep configuration (i.e., laydown and stand-up). These features are then used as the ranking priority criterion. Components with priority number 1 must be inserted first followed by the ones with priority number 2 and so on.

TABLE 16.1. Component Insertion Priority

Priority	Order	Packaging and Configuration
1	Ascending pin count	DIP and SIP components: network resistors, capacitors, and "U" type ICs
2	Ascending diameter	Laydown axials: resistor, capacitors, and diodes
3	Ascending height, then footprint	Transistors and "can" (AR) ICs
4	Ascending height, then footprint	Box-type: variable resistors, tombstone capacitors, and other two-leaded radial components
5	Ascending height	Stand-up axials: resistors, capacitors, and diodes

See ref. 26.

Once the component insertion priority has been established, the success-ful use of the insertion sequencing rules is facilitated by organizing or partitioning the component data into several primary but independent categories. For example, by dividing the sequencing decision rules by component type, it is possible to eliminate a group of components not included in the methodology or components not being considered at a particular time during an analysis. This heuristic approach facilitates a search of a smaller knowledge space that is most likely to have the answer to a question being posted. Heuristic searches thus incorporate some informa-tion about the nature and structure of the problem domain and limit the search space. Following this line of thought, a component insertion tax-onomy (classification of similar items into a system) is defined. The goal of this taxonomy is to organize the rules and criteria that will determine the relative order of insertion of components of the same type but with different configuration.

The previous partition has divided the PCB component types into five different priority groups. Each group enumerates the different types of component configuration included in the methodology and its associated order of insertion. The taxonomy is later used to guide the creation of the methodology decision tree structures.

16.6.3. Component Sequencing Methodology

In order to detail the component sequencing methodology, a series of decision trees based on the previous taxonomy was developed. That ap-proach was taken because of the forward chaining nature of the PCB assembly planning process. In such as environment, a knowledge search normally proceeds in a stepwise fashion through a knowledge base with a new branch to be decided at each step. A decision tree is a grouping of questions that are related to a list of facts and follow a specific consulting path.

The path may take several directions and explore several possibilities for those facts, providing different results, but when the end of a path is reached, the methodology would have come to a decision about these facts and, as such, reached a goal. Figure 16.5 depicts the resistor decision tree as developed for the CBA system.

The previous decision trees can then be used to derive some of the decision rules of the methodology's knowledge base. This can be done by tracing all paths leading to each terminal node of each single decision tree. A terminal node is one represented by a square box and does not have any branches leaving. For example, in Fig. 16.5 the node marked number 7, "resistor is a stand-up component," is a terminal node. The path leading to this node is via the nodes 1, 3, and 5. A decision-making rule can then be written for arriving to this node as follows:

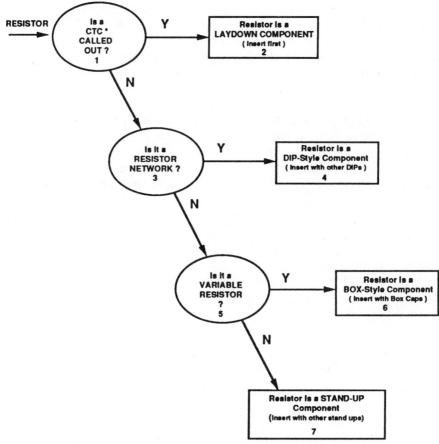

* CTC = Center-to-Center lead spacing

FIGURE 16.5. Resistor sequencing decision tree (26).

```
IF     (the CTC is not called out) AND
       (the resistor is not a resistor network) AND
       (the resistor is not a variable resistor)

THEN   (the resistor is a stand-up component) AND
       (insert first)
```

16.6.4. Machine Table Tour Planning

The last issue to be addressed in the fulfillment of the second insertion planning criterion, which is to minimize insertion bed tour. To that end, the sequencing methodology applies the guidelines detailed by Rule 4. This rule, which applies to the insertion of components of the same size, shape, and electrical value, aims at selecting the component insertion sequence that results in the shortest machine table movement.

This problem has historically been treated as a traveling salesman problem (23). The traveling salesman problem can be stated as follows: Given a group of cities to be visited with known distance between any two of them, find the shortest route to visit each city once and return to the starting point. It has been proved that for a problem including n cities, the number of possible rounds trips is $(n-1)!$ (29). If $n = 50$, there would be 6.083×10^{62} possible tours. Because such an exhaustive search is generally unfeasible for most practical purposes, heuristic methods have been applied to find solutions to this kind of problem that could be considered as acceptable if not optimal. There are several solution methodologies to this problem, such as the branch and bound methods (30), clustering methods (17), and the nearest-neighbor algorithms (11). Among them, the nearest-neighbor method is the most widely used heuristic in industry and therefore the one that was applied in this project. The adaptation of this algorithm is summarized as follows:

Step 1. Select a component value for which the insertion location (node) is closest to the current machine bed position as the beginning of the component insertion sequence.

Step 2. Find a location closest to the last insertion location. If not previously visited, add this to the insertion sequence.

Step 3. Repeat step 2 until all the components with the same value are included in the insertion sequence.

16.7. SUMMARY

Computer aided process planning (CAPP) is the use of computer technology for developing optimum manufacturing plans that translate product design requirements into the sequence of processes needed to fabricate a product. Unfortunately, only a reduced number of computerized process planning systems exist today which are applicable in the electronic field. The use of artificial intelligence methodologies will, however, be a critical component in future computerized process planning systems, especially in concurrent engineering environments. This chapter has reviewed process planning for electronics and an example of how artificial intelligence can be implemented into an automated planning process.

REFERENCES

1. Assembly Engineering (1981). CAPP Speeds Electronic Assembly, *Assembly Engineering*, June, pp. 46–49.
2. Arpino, F., and Groppetti, R. (1988). Assyt: A Consultation System for the Integration of Product and Assembly System Design, *Developments in Assembly Automation*, March, pp. 167–180.

3. Bao, H. P. (1988). Group Technology Classification and Coding for Electronic Components. In: *Productivity and Quality Improvement in Electronics Assembly*, edited by J. Edosomwam and A. Ballakur, McGraw-Hill, New York, pp. 785–836.

4. Bao, H. P. (1990). Classification and Coding of SMT Devices: Part I—Part Family, *IE News Electronics Industry*, Vol. 25, No. 1, pp. 1–2.

5. Berenji, R., and Khosnevis, B. (1986). Use of Artificial Intelligence in Automated Process Planning, *Journal of Computers in Mechanical Engineering*, pp. 47–55.

6. Chang, C. M., Smith, M. L., and Blair, E. L. (1988). Analysis of Assembly Planning Problem for a Robotized Printed Circuit Board Assembly Center. In: *Recent Developments in Production Research*, edited by A. Mital, Elsevier Science Publishers, Amsterdam, pp. 472–484.

7. Chang, T. (1983). *Advances in Computer Aided Process Planning*, Bureau of Standards Report, NBS-GCR-83-441.

8. Contact Systems, Inc. (1988). *CS-400 Component Locator Programming Manual*, Danbury, CT.

9. Cunningham, P., and Brown, J. (1986). A LISP-Based Heuristic Scheduler for Automatic Insertion in Electronics Assembly, *International Journal of Production Research*, Vol. 24, No. 6, pp. 1395–1408.

10. Freedman, H. S., and Frail, R. P. (1986). OPGEN: The Evolution of an Expert System for Process Planning, *The AI Magazine*, Winter, pp. 58–70.

11. Gavet, J. N. (1965). Three Heuristic Rules for Sequencing Jobs to Single Production Facility, *Management Science*, Vol. 11, No. 8, pp. 166–176.

12. Gavish, B., and Seidmann, A. (1988). Printed Circuit Boards Assembly Automation—Formulations and Algorithms. In: *Recent Developments in Production Research*, edited by A. Mital, Elsevier Science Publishers, Amsterdam, pp. 624–635.

13. Gupta, T. (1990). An Expert System Approach in Process Planning: Current Development and Its Future, *Computers in Industrial Engineering*, Vol. 18, No. 1, pp. 69–80.

14. Han, C., Li, J., and Ham, I. (1987). Development of an In-House Computer Automated Process Planning System Based on Group Technology Concept, *Proceedings of 15th North American Research Conference*, May, pp. 41–116.

15. Hird, G., Swift, K. G., Bassler, R., Seidel, U. A., and Richter, M. (1988). Possibilities for Integrated Design and Assembly Planning, *Developments in Assembly Automation*, pp. 15–166.

16. Khoshnevis, B. (1990). Integrated Process Planning, *Proceedings of Manufacturing International 90*, Atlanta, GA, Vol. 1, pp. 243–248.

17. Litke, I. D. (1982). Minimizing PWB NC Drilling. In: *Proceedings of the 10th Design Automation Conference*, IEEE, New York, pp. 444–447.

18. Liu, D. (1984). Utilization of Artificial Intelligence in Manufacturing. In: *Proceedings of Autofact 6*, Society of Manufacturing Engineers, Dearborn, MI.

19. Mangin, C., and Salvatore, D. (1984). Printed Circuit Board Assembly Picks Up Automation, *Electronics*, January, pp. 171–174.

20. Mason, A.K., and Young, A. (1987). *Computer Aided Process Planning for Printed Circuit Board Assembly*, Technical Report MS87-722, Society of Manufacturing Engineers, Dearborn, MI.

21. Nash, T. F. (1985). Automation for Surface Mount and Thro-hole Technology, Proceedings of The Technical Program of NEPCON EAST: National Electronic Packaging and Production Conference, Boston, MA, June.

22. Nau, D. S. (1987). Automated Process Planning Using Hierarchical Abstraction, *Texas Instruments Technical Journal*, Winter, pp. 39–46.

23. Parker, R. G., and Rardin, R. L. (1983). The Traveling Salesman Problem: An Update of Research, *Naval Research Logistic Quarterly*, Vol. 30, pp. 78–107.

24. Phillips, R. H., Zhou, X. D., and Mouleeswaran, C. B. (1984). An Artificial Intelligence Approach to Integration CAD/CAM Through Generative Process Planning, *Proceedings of ASME International Computers in Engineering*, Las Vegas, NV.

25. Randhawa, S. U., McDowell, E. D., and Faruqui, S. D. (1985). An Integer Programming Application to Solve Sequencer Mix Problems in Printed Circuit Board Production, *International Journal of Production Research*, Vol. 23, No. 3, pp. 543–552.

26. Sanchez, J. M., and Priest, J. W. (1991). Optimal Component Insertion Sequencing Methodology for the Semi-automatic Assembly of Printed Circuit Boards, *Journal of Intelligent Manufacturing*, Vol. 2, pp. 177–188.

27. Srihari, K., and Greene, T. J. (1990). Expert Process Planning for Flexible Manufacturing, *CIM Review: The Journal of Computer-Integrated Manufacturing Management*, Vol. 6, No. 2, pp. 43–50.

28. Steudel, H. J. (1984). Computing Aided Process Planning: Past, Present and Future, *International Journal of Production Research*, Vol. 22, No. 2, pp. 253–266.

29. Taha, H. A. (1986). *Operations Research: An Introduction*, Macmillan, New York.

30. Tarjan, R. E. (1972). Depth First Search and Linear Graph Algorithms, *SIAM Journal on Computing*, Vol. 1, No. 2, pp. 146–160.

31. Terwilliger, J. P. (1985). Process Planning for Electronic Assembly, unpublished Master's thesis, Purdue University, West Lafayette, IN.

32. Tulkoff, J. (1988). *CAPP from Design to Production*, Society of Manufacturing Engineers, Dearborn, MI.

33. Vogel, S. A., and Adlard, E. J. (1981). The AUTOPLAN Process Planning System, *Proceedings of the 18th Numerical Society Annual Meeting and Technical Conference*, Dallas, TX, pp. 729–742.

34. Wang, H. P. (1985). Microcomputer-Based Process Planning Systems, Unpublished Master's thesis, Pennsylvania State University, University Park.

35. Wang, H. P., and Wysk, R. A. (1987). Turbo-CAPP: A Knowledge-Based Computer Aided Process Planning System, *19th CIRP Seminar on Manufacturing Systems*, pp. 161–167.

36. Wysk, R. A. (1977). Automated Process Planning and Selection Program: APPAS, unpublished Ph.D. thesis, Purdue University, West Lafayette, IN.

CHAPTER SEVENTEEN

Intelligent Scheduling of Automated Machining Systems

ANDREW KUSIAK and JAEKYOUNG AHN

Intelligent Systems Laboratory, Department of Industrial Engineering,
The University of Iowa, Iowa City, Iowa

17.1. INTRODUCTION

Automated machining systems may involve sophisticated information systems to control automated equipment. The equipment typically includes (13):

- Automated machine tools to process parts
- Automated assembly machines
- Industrial robots
- Automated material handling and storage systems
- Computer hardware for planning, data collection, and decision-making to support manufacturing activities

The benefits offered by automated manufacturing systems are as follows (10):

- Fast response to market demands
- Better product quality
- Reduced cost
- Enhanced performance
- Better resource utilization
- Shorter lead times

Intelligent Design and Manufacturing, Edited by Andrew Kusiak.
ISBN 0-471-53473-0 © 1992 John Wiley & Sons, Inc.

- Reduced work in process
- Flexibility

With the development of automation technology, its supporting systems—planning, scheduling, and control—have gained importance. Production planning involves establishing production levels for a known length of time. It determines production parameters, such as product mix, production levels, resource availability, and due dates. With the specified production parameters, the goal of scheduling is to make efficient use of resources to complete tasks in a timely manner (24).

There have been extensive studies on scheduling manufacturing systems. These studies can be divided into three basic approaches:

- Operations research (OR) approach
- Artificial intelligence (AI)-based approach
- Combination of OR and AI-based approaches

The literature on scheduling manufacturing systems using operations research techniques is rather extensive. Panwalkar and Iskander (27) divided these studies into the following two categories:

- Theoretical research dealing with optimization procedures
- Experimental research dealing with dispatching rules

The theoretical research has focused on the development of mathematical models and optimal or suboptimal algorithms (4, 6, 11). The theoretical results have not been widely used in industry due to high computational complexity of the scheduling problem. However, the theoretical approach has its own merits, mainly in capturing the problem structure. It allows an analyst to construct a model according to the characteristics of the scheduling environment.

The experimental research has primarily been concerned with dispatching rules and heuristics that efficiently solve the scheduling problem. This approach has appealed to both researchers and practitioners. To date, over 100 dispatching rules have been developed. The research on the dispatching rules has been presented in a number of publications, for example, Gere (12), Panwalkar and Iskander (27), Blackstone et al. (7), Alexander (3), Koulamas and Smith (17), Schultz (32), and Kusiak and Ahn (20).

Cohen and Feigenbaum (9) categorized expert systems in manufacturing as hierarchical, nonhierarchical, script-based (skeleton), opportunistic, and constraint-directed expert systems. The most common characteristics in the expert systems category are as follows (33):

- On-line decision support
- Dynamic scheduling of operations

- Coordination of manufacturing resources
- Synchronization of processes for different jobs
- Monitoring of the execution of plans

Several surveys of expert systems for manufacturing applications have been published in the literature (15, 21, 22, 34).

The operations research-based approach usually focuses on finding the "best" schedule under the deterministic constraints, while a number of artificial intelligence approaches focus on finding a "feasible" schedule subject to probabilistic constraints. As pointed out in Phelps (29), there are some similarities between the two approaches:

- Face similar problems
- Use models for problem solving
- Use heuristics when optimal methods are not suitable
- Use mathematics
- Use computers for their implementations
- Employ interdisciplinary analysts and designers

O'Keefe et al. (26) presented a view that expert systems and operations research methods are complementary instances of a broad range of decision-making tools. Kanet and Adelsberger (16) suggested that the expert scheduling systems of the future will have the reformulative ability (by expert system techniques) along with the best available algorithmic scheduling knowledge (by operations research techniques). Jaumard et al. (15) identified operations research tools that can be useful in intelligent problem solving.

Perhaps the most promising architecture that is able to incorporate operations research and artificial intelligence techniques in scheduling manufacturing systems is the tandem architecture (see Fig. 17.1) suggested by Kusiak (18). The tandem system has been designed so that a knowledge-based system interacts with algorithms. The algorithm deals mainly with quantitative and deterministic components of the scheduling problem, guaranteeing rigorous generation of schedules. At the same time, the knowledge-based system deals mainly with qualitative and probabilistic elements of the scheduling problem. Incorporation of the two approaches is possible through the communication channel. In subsequent sections, a rule-based scheduling system implemented in the tandem architecture is described.

A tandem expert system architecture considered in the chapter has the following characteristics (21):

- Capability of solving difficult problems
- Flexibility in solving problems of various types

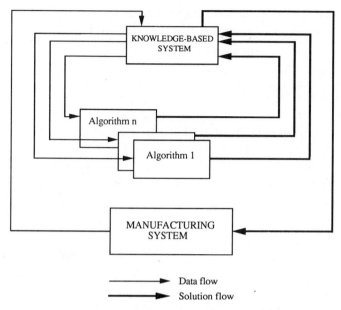

FIGURE 17.1. Multimodel tandem architecture.

- Modularized development and implementation
- Sharing of intellectual resources with other control and management systems
- Increased role of communication between subsystems

In this chapter, a multimodel tandem architecture embedded in an expert system shell NEXPERT OBJECT is used to develop an intelligent scheduling system for automated manufacturing.

17.2. THE SCHEDULING FRAMEWORK

Scheduling an automated machining system involves concurrent use of manufacturing resources, alternative process plans, and flexible routings. Kusiak and Ahn (20) developed a dispatching rule (MDR rule) designed to maximize the utilization of resources in a resource-constrained machining system. Ahn and Kusiak (1) analyzed the performance of a number of dispatching rules for various scheduling scenarios. The analysis was done under the assumption that the required data are complete and certain. Likewise, the objectives were assumed to be specified in advance. In many practical applications, however, scheduling under the assumption that data are complete is not practical due to unpredictable and changing manufacturing conditions. For the schedules to be flexible so that they could be updated and modified in response to the changes in the scheduling environment, algorithms should be integrated with a rule-based system.

17.2.1. Scheduling Algorithm

In this section, a dispatching rule and a scheduling algorithm (1) are briefly described.

The MDR Dispatching Rule. The most dissimilar resources (MDR) dispatching rule has been developed for efficient scheduling of operations in an automated machining system where the maximization of the utilization rate of manufacturing resources is a major concern because of the following:

1. Considerable capital investment is needed to install the system.
2. Production is lost when manufacturing resources are idle.
3. Scheduling with the minimum makespan and minimum tardiness criteria tends implicitly to maximize resource utilization over the scheduling horizon (30).

For further use in this chapter, a schedulable operation is defined. An operation o_k is schedulable at time t if (1):

1. No other operation that belongs to the same part is being processed at time t.
2. All operations preceding operation o_k have been completed before time t.
3. All resources required by the process plan to perform operation o_k are available at time t.
4. The part with operation k has been moved to a specified machine by the material handling system.

The MDR dispatching rule selects the largest set of schedulable operations that satisfy the following conditions:

1. Each operation belongs to a distinct part; that is, more than one operation in a part cannot be processed at the same time.
2. All the operations in the set can be processed at the same time; that is, there is no resource conflict.

The MDR dispatching rule can also be represented with an operation–resource graph, G. Let $G(V, E, \psi)$ be a graph, where:

V A set of schedulable operations

E A set of incidence relations among the schedulable operations

ψ $\psi(e) = v_k v_j$, if $\Sigma_{q=1}^{z} m_{kq} m_{jq} = 0$, and v_k and v_j do not belong to the same part, where $m_{kq} = 1(0)$ if operation v_k uses (not uses) resource q.

For n parts, the operation–resource graph becomes an n-partite graph and is not necessarily complete. In the operation–resource graph, each edge indicates the nodes (operations) that can be processed simultaneously, that is, the nodes that do not share the same resources. Note that the MDR dispatching rule selects the largest set of nodes such that each node in the set belongs to a distinct part and there is no resource conflict between the nodes in the set.

A clique of a simple graph, G, is defined as a subset S of V, such that $G[S]$ is complete (8). Thus a clique in the operation–resource graph satisfies the conditions under which the MDR dispatching rule selects the schedulable operations from the operation–resource incidence matrix, and the maximum clique in the graph corresponds to the largest set of schedulable operations in the incidence matrix. Therefore selecting operations with the MDR priority rule is equivalent to the problem of finding the maximum clique in the operation–resource graph, G.

Since the MDR priority rule is to be used in real-time scheduling, it is important that the maximum clique problem is solved almost instantaneously. A general integer programming code, such as LINDO (31), is not likely to meet the latter requirement for the problem sizes encountered in industry. A more efficient way to solve the maximum clique problem is to use a specialized algorithm.

To date, a number of exact and heuristic algorithms (5, 14, 28) have been developed for solving the maximum clique problem. A simple pairwise counting heuristic of computational time complexity $O(n^2)$ is presented below.

Before the algorithm is presented, the following notation is introduced:

P A set of operations to be dispatched
$D(k)$ Distance index of operation k, defined as the number of parts in which the number of 0's in the kth column in the distance matrix is greater than 1

The algorithm constructs a set of operations to be dispatched, P, from a set of schedulable operations, S.

Dispatching Algorithm

Step 0. Initialize the set of schedulable operations, S, and the set of operations to be dispatched, $P = \varnothing$.

Step 1. If $S \neq \varnothing$, then go to step 2, otherwise, STOP.

Step 2. From the set of schedulable operations, S, construct an operation–resource incidence matrix $[m_{kj}]$.

Step 3. From the operation–resource incidence matrix $[m_{kj}]$, construct the distance matrix $[d_{kj}]$.

Step 4. For each operation $k \in S$, compute the distance index $D(k)$ from the operation–resource incidence matrix $[m_{kj}]$.

Step 5. Find maximum $D(k^*)$. If a tie occurs, break it, arbitrarily.

Step 6. Move operation k^* to P.

Step 7. Update the set of schedulable operations:

(a) Delete operation k^* from S.

(b) Delete from S the operations that belong to the same part as the operation k^*.

(c) Delete from S the operations that share the same resources with the operation k^*.

Step 8. Go to step 1.

The MDR dispatching rule is illustrated in Example 1.

Example 1. Assume that the resource–operation matrix (1) is given.

		Machine				Tool				Fixture				
		m_1	m_2	m_3	m_4	t_1	t_2	t_3	t_4	f_1	f_2	f_3	f_4	
Part 1	o_1	1	0	0	0	0	1	0	0	0	0	1	0	O
	o_2	0	0	0	1	0	0	0	1	0	0	0	1	p e r a t i o n
Part 2	o_3	0	1	0	0	1	0	0	0	0	1	0	0	
	o_4	1	0	0	0	0	0	1	0	0	1	0	0	
	o_5	0	0	1	0	0	1	0	0	1	0	0	0	
Part 3	o_6	0	0	1	0	0	0	1	0	0	0	0	1	

$$[m_{kj}] = \quad (1)$$

For the matrix (1), the distance matrix (2) is computed in step 3.

		Operation						
		o_1	o_2	o_3	o_4	o_5	o_6	
Part 1	o_1	M	M	0	1	1	0	O
	o_2	M	M	0	0	0	1	p e r a t i o n
Part 2	o_3	0	0	M	M	M	0	
	o_4	1	0	M	M	M	1	
	o_5	1	0	M	M	M	1	
Part 3	o_6	0	1	0	1	1	M	

$$[d_{kj}] = \quad (2)$$

In step 4, the distance index, $D(k)$, is computed:

$$D(1) = 2, \ D(2) = 1 ,$$
$$D(3) = 2, \ D(4) = 1 ,$$
$$D(5) = 1, \ D(6) = 2 .$$

The maximum value of the distance index $D(k)$ is attained for operations 1, 3, and 6. In step 5, operation 1 is selected. The set of operations to be dispatched, P, is updated, $P = \{1\}$ in step 6. In step 7, operations 1, 2, 4, and 5 are deleted, and then go to step 1. After two more iterations, the resulting set of operations to be dispatched, $P = \{1, 3, 6\}$, is obtained.

In a real-time scheduling environment, schedules are usually determined by dispatching an operation whenever a decision is required. For instance, the MDR rule dispatches an operation that can be processed with as many other operations as possible at the same time while COVERT rule dispatches an operation that has a potential of being late. In a sense, dispatching an operation affects not only the schedulable operations that can be processed simultaneously at a specific time, but also the set of operations to be dispatched in the subsequent scheduling horizon. In this chapter, the MDR dispatching rule (myopic) is combined with promising look-ahead dispatching rules. Since the MDR dispatching rule quite frequently provides a tie among operations, the myopic information and look-ahead information are combined.

The dispatching rules considered in this chapter are shown in Tables 17.1 and 17.2. In the two tables, the following notation is used:

t	Current time
t_{ik}	Processing time of operation k of part i
G_i	Set of pairs of operations $[k, l]$ of part i, where k precedes l
U_{ik}	Set of operations succeeding operation k in G_i
V_{ik}	Set of unprocessed operations of part i except of operation k
d_i	Due date of part i
sl_k	Slack of operation k; $sl_k = d_i - t - \sum_{l \in V_{ik}} t_{il}$
$c_k(\rho_1, \rho_2)$	Expected delay penalty for operation k,

$$c_k(\rho_1, \rho_2) = \begin{cases} 1, & \text{if } sl_k < 0 \\ 0, & \text{if } sl_k \geq \sum_{l \in V_{ik}} t_{il} \\ \dfrac{\sum_{l \in V_{ik}} \rho_1 \cdot \rho_2 t_{il} - sl_k}{\sum_{l \in V_{ik}} \rho_1 \cdot \rho_2 t_{il}}, & \text{otherwise} \end{cases}$$

TABLE 17.1. Non-Due-Date-Related Dispatching Rules and Priority Indexes for Operation _k_ of Part _i_

Rule Number	Rule Name	Max/Min	Priority Index
R_1	Most dissimilar resource	Max	$D(k)$
R_2	Most subsequent work remaining	Max	$\sum_{l \in U_{ik}} t_{il}$
R_3	Most subsequent operations remaining	Max	$|U_{ik}|$
R_4	Most work remaining	Max	$\sum_{l \in V_{ik}} t_{il}$
R_5	Most operations remaining	Max	$|V_{ik}|$
R_6	Shortest processing time	Min	t_{ik}
R_7	Longest processing time	Max	t_{ik}
R_8	Least work remaining	Min	$\sum_{l \in V_{ik}} t_{il}$

TABLE 17.2. Due-Date-Related Dispatching Rules and Priority Indexes for Operation _k_ of Part _i_

Rule Number	Rule Name	Max/Min	Priority Index
R_9	Minimum slack remaining	Min	sl_k
R_{10}	Remaining allowance per operation	Min	$\dfrac{d_i - t}{|V_{ik}|}$
R_{11}	Slack per operation	Min	$\dfrac{d_i - t - t_{ik}}{|V_{ik}|}$
R_{12}	Slack per remaining work	Min	$\dfrac{d_i - t - t_{ik}}{\sum_{l \in V_{ik}} t_{il}}$
R_{13}	COVERT1	Max	$\dfrac{c_k(2,2)}{t_{ik}}$
R_{14}	COVERT2	Max	$\dfrac{c_k(3,2)}{t_{ik}}$

The combination of the MDR rule with other dispatching rules are listed later in this chapter as $MDR + R_i$. Thus the number of dispatching rules considered is 27 including the single and the combined ones (for details, see ref. 20). The scheduling algorithm has been implemented so that static as well as dynamic part arrivals are handled. Before the algorithm is presented, the following notation and definitions are introduced:

f_{ik} 　　　　Completion time of operation k

rt_k 　　　　Remaining processing time of operation k

s_k 　　　　Status of operation k;

$$s_k = \begin{cases} 1, & \text{if operation } k \text{ is schedulable} \\ 2, & \text{if operation } k \text{ is nonschedulable} \\ 3, & \text{if operation } k \text{ is being processed} \\ 4, & \text{if operation } k \text{ has been completed} \\ 5, & \text{if operation } k \text{ satisfies the first two conditions} \\ & \text{in the definition of schedulability} \end{cases}$$

sr_{c^r} 　　　Status of resource r of type c;

$$sr_{c^r} = \begin{cases} 1, & \text{if resource } c^r \text{ is available} \\ 0, & \text{otherwise} \end{cases}$$

S_j 　　　　Set of operations with $s_k = j$, $j = 1, \ldots, 5$

C_j 　　　　Temporal set of operations

A_{ik} 　　　Set of alternatives of operation k of part i

Q_i 　　　　Set of ordered pairs of operations $[k, l]$ of part i, where k and l can be performed in any order

N_{cr} 　　　Set of operations to be performed by resource r of type c

R_k 　　　　Set of resources used by operation k

$\tau[m, m']$ 　Travel time between machines m and m' (or the corresponding operations)

Scheduling Algorithm

Step 0. Initialize the variables.

(a) Set current time $t = 0$.

(b) Set resource status, $sr_{c^r} = 1$, for all r and c.

(c) Operation status and sets of operations with the operation status are initialized as follows: (i) Set $S_1 = S_2 = S_3 = S_4 = \varnothing$. (ii) For each part, if operation k has no predecessor operations then $s_k = 1$; otherwise, $s_k = 2$. (iii) Construct S_1 and S_2 for operations in (ii).

Step 1. If all the operations have been completed, STOP; otherwise go to step 2.

Step 2. If $S_1 = \varnothing$, go to step 5; otherwise, go to step 3.

Step 3. Select an operation $k_{q^\#}^*$ in the set S_1, based on a dispatching rule provided (part i^* corresponding to operation $k_{q^\#}^*$ is automatically selected).

Step 4. Set:

(a) Remaining processing time of operation k^*, $rt_{k^*} = t_{ik_{q^\#}^*}$.

(b) Resource status $sr_{c^r} = 0$ for $c^r \in R_{k_{q^\#}^*}$.

Construct:

(a) $C_1 = \{q \mid q \in A_{i^*k^*} - q^\#\}$.

(b) $C_2 = \{k_q \mid k_q \neq k_{q^\#}^*, [k_{q^\#}^*, k_q] \in Q_i^*, k_q \in S_1\}$.

(c) $C_3 = \{k_q \mid \{k_{q^\#}^*, k_q\} \in N_{cr} \text{ for all } c \text{ and } r, k_q \in S_1\}$.

Update:

(a) Set of schedulable operations $S_1 = S_1 - \{k_{q^\#}^*\} - \{C_1 \cup C_2 \cup C_3\}$.

(b) Set of nonschedulable operations $S_2 = S_2 \cup C_2 \cup C_3$.

(c) Set of processing operations $S_3 = S_3 \cup \{k_{q^\#}^*\}$. If $S_1 \neq \emptyset$, go to step 3; otherwise, go to step 5.

Step 5. Set:

(a) Completion time $f_k = rt_k + t, k \in S_3$.

(b) Current time $t = t + 1$.

(c) Remaining time $rt_k = rt_k - 1$. If $rt_{k^*} = 0$ in $k \in S_3$, then resource status $sr_{c'} = 1$ for $c' \in R_{k_{q^\#}^*}$.

Step 6. In set S_2, construct:

(a) $C_1 = \{k_q \mid \text{all the preceding operations of operation } k_q \text{ have been completed}\}$.

(b) $C_2 = \{k_q \mid sr_{c'} = 1, c' \in R_{k_q}\}$.

(c) $C_3 = \{k_q \mid \text{current time} - \text{completion time of the immediate preceding operation of operation } k_q > \tau[\text{immediately preceding operation of operation } k_q, k_q]\}$.

Update:

(a) Set of schedulable operations $S_1 = S_1 \cup \{C_1 \cap C_2 \cap C_3\}$.

(b) Set of nonschedulable operations $S_2 = S_2 - \{C_1 \cap C_2 \cap C_3\}$.

(c) Set of processing operations $S_3 = S_3 - \{k^\#\}$.

(d) Set of completed operations $S_4 = S_4 \cup \{k^\#\}$.

Step 7. Go to step 1.

17.2.2. Rule-Based Scheduling System

In the algorithm presented in the preceding section, once a dispatching rule is selected, all parts are scheduled regardless of manufacturing conditions. In some cases, the schedules may be infeasible due to unpredictable and changing manufacturing conditions (e.g., blocking or machine breakdowns). Moreover, it is evident that the "blind" selection of a dispatching rule might result in an inefficient schedule.

In the chapter, the manufacturing conditions are divided into the following categories:

1. Exogenous manufacturing conditions.
    (a) Scheduling objectives (e.g., maximization of resource utilization, minimization of the number of tardy parts).

    (b) System load levels.

    (c) Resource constrainedness (the ratio of the number of constrained resources to the number of unconstrained resources).

    (d) Due date assignment (e.g., constant, slack-based, total work-based).

    (e) Due data tightness.

2. Endogenous manufacturing conditions

    (a) Changing shop status (e.g., inventory status, queuing status, bottleneck resource status, machine breakdown).

    (b) Preference constraints (priority of temporal scheduling objectives).

Based on the two categories of manufacturing conditions, appropriate dispatching rules and scheduling algorithms are selected (exogenous manufacturing condition) and/or modified (endogenous) by a rule-based scheduling system. The need to construct such a system arises from the fact that:

1. An automated manufacturing system requires a dynamic and accurate scheduling.
2. There is no evidence in the literature that there exists a dispatching rule that performs best under all manufacturing conditions.
3. The existing scheduling studies present the performance of dispatching rules only for narrow domains.
4. The computational results indicate that combining the MDR and other dispatching rules may depend on the manufacturing conditions.

A rule-based system developed in this chapter consists of four components: algorithm selector, rule selector, process reactor, and rule base.

Algorithm Selector. Algorithm selector determines a scheduling algorithm to be used in solving a problem considered. It is comprised of a set of production rules or, alternatively, a user may specify the name of the model desired. One of the main advantages of the tandem architecture is that it handles multiple models. In this chapter, only one algorithm is introduced while there are many other scheduling algorithms available in the literature (e.g., see ref. 19).

Rule Selector. Rule selector provides a global dispatching rule for the scheduling algorithm, based on the exogenous manufacturing conditions. A global dispatching rule selected is fired whenever a dispatching decision is made (see step 3 of the algorithm), unless there is a significant change in the machining shop status during the scheduling horizon. Once an inadmissible change is detected on the shop floor, the process reactor is activated to minimize the schedule disruption. The process reactor fires a temporary dispatching rule. If the equilibrium on the shop floor is restored, then the global rule begins dispatching operations.

Process Reactor. Process reactor communicates on-line with the manufacturing facility in order to respond to the endogenous shop conditions. It modifies the process of selection of dispatching rules in the "warning" state and imposes selection of operations in the "urgent" state. A warning state is issued when the system is likely to generate infeasible schedules; for example, the number of waiting parts in front of a machine exceeds 75% of its capacity. In the warning state, the process reactor consults with the rule base, and it assigns a temporary dispatching rule that may help in attaining a normal condition. The selected temporary dispatching rule is used by the rule selector. An urgent state is issued when the schedule obtained is infeasible or it might become infeasible in the next scheduling time horizon due to, for example, blocking or preventive machine maintenance. If the above immediate situation calls for action extraneous to the dispatching rule, then the process reactor takes exception to the rule (see steps 4 and 6 of the algorithm). This is also possible through communication with the rule base.

Rule Base. Rule base plays an important role in the entire scheduling processes. It is implemented using the NEXPERT shell. All the production rules in the rule base are divided into the following classes:

Class 1. Selects an appropriate algorithm to solve the problem (model selector).

Class 2. Selects an appropriate dispatching rule to solve the problem (rule selector).

Class 3. Modifies the selected dispatching rule to solve the problem (process reactor).

Class 4. Selects an appropriate operation to solve the problem (process reactor).

Several sample production rules are presented next.

Class 1

RULE1_1. IF the machining system has more than three machines
AND the number of operations in all the parts being scheduled exceeds 20

AND the scheduling problem considered has alternative process plans
AND traveling times are imposed
THEN solve it using the scheduling algorithm (presented in this chapter).

Class 2

RULE2_1. IF the machining system has more than three machines
AND the number of operations in all the parts being scheduled exceeds 20
AND the scheduling problem considered is static
AND the scheduling objective is to minimize the makespan
AND the resource constrainedness is high ($RC > RC_2$)
THEN use the MDR/MSWR dispatching rule.

RULE2_2. IF the scheduling problem is dynamic
AND the scheduling objective is to minimize the number of tardy parts
AND the resource constrainedness is medium ($RC_1 < RC < RC_2$)
AND due dates are assigned with MWR method
AND the system is light-loaded
THEN use the MDR/COVERT1 dispatching rule.

RULE2_3. IF the scheduling problem is dynamic
AND parts are produced for safety stock
AND the resource constrainedness is medium ($RC_1 < RC < RC_2$)
AND the system is heavy-loaded
THEN use the LWR dispatching rule.

Class 3

RULE3_1. IF the machining system has more than three machines
AND the number of operations in all the parts being scheduled exceeds 20
AND the scheduling problem considered is static

```
AND the work-in-process, W > W₀
AND the LWR dispatching rule is not used
THEN replace the current dispatching rule
with the LWR rule.
```

Class 4

```
RULE4_1. IF the number of parts waiting in front of a
         bottleneck machine, Qₘ > Qₘ.
         THEN override the current dispatching rule
         AND find a feasible schedule.

RULE4_2. IF a machine is down
         THEN set the machine status to unavailable
         during that interval.
```

17.2.3. Expert System Shell

An intelligent system for scheduling automated machining facilities is being developed using an expert system shell NEXPERT (23). NEXPERT allows developers to embed portions of its inference engine in their own code, combining the advantages of a knowledge-based system and algorithms. NEXPERT offers four inferencing methods: backward chaining, forward chaining, semantic gates, and context links. It also allows developers to edit rules and objects as well as build control structures, with the rules and object structures available at all times through a dynamic, graphic browsing mechanism (for details, see ref. 2). The intelligent scheduling system is implemented on an Apollo workstation.

17.2.4. Knowledge Acquisition

An intelligent system should have learning ability. A system that learns is able to improve its own problem solving ability. In this section a discrete simulation-assisted knowledge acquisition process is described. The simulation is used for knowledge acquisition due to the following (25, 35).

- No domain experts are available.
- It is possible to build models that can predict the effects of input parameters on output measures.
- Practical rules of thumb and experience are needed to use simulation as an effective tool.

Computer simulation is a problem solving process of predicting the future state of a real system by studying an idealized computer model of the real system. Simulation experiments are usually performed to obtain predictive information that would be costly or impractical to obtain with real devices (36). In our simulation, two normalized performance measures for parts and five measures for schedules (runs) are considered:

- Normalized average waiting (AW) time

$$AW = \frac{r_{\text{machine}}}{n \cdot m_{\text{avg}}} \sum_{j=1}^{R} \frac{AW_j}{R} , \tag{3}$$

where r_{machine} is the number of machines, m_{avg} is the average number of operations in a part, n is the number of parts

$$AW_j = \sum_{i=1}^{n} \frac{1}{n} \left(\frac{\text{sum of processing time}}{\text{work-in-system time}} \right)_i ,$$

and R is the number of runs.

- Normalized average flow (AF) time

$$AF = \frac{r_{\text{machine}}}{n \cdot m_{\text{avg}}} \sum_{j=1}^{R} \frac{AF_j}{R} , \tag{4}$$

where $AF_j = \sum_{i=1}^{n} \frac{1}{n} (\text{flow time})_i$.

- Normalized average makespan (AM)

$$AM = \frac{r_{\text{machine}}}{n \cdot m_{\text{avg}}} \sum_{j=1}^{R} \frac{T_j}{R} , \tag{5}$$

where T_j is the makespan in the jth run.

- Normalized average percent tardiness (PT)

$$PT = \frac{r_{\text{machine}}}{n \cdot m_{\text{avg}}} \sum_{j=1}^{R} \frac{PT_j}{R} , \tag{6}$$

where PT_j is the percentage of tardy parts in the jth run.

- Normalized average maximum tardiness (MT)

$$MT = \frac{r_{\text{machine}}}{n \cdot m_{\text{avg}}} \sum_{j=1}^{R} \frac{MT_j}{R} , \tag{7}$$

where MT_j is the maximum tardiness in the jth run.

- Normalized average tardiness (AT)

TABLE 17.3. Dispatching Rules Generating the Best and the Second-Best Solutions in the Static Machining System for Various Performance Measures

Measure of Performance	Solutions	Resource Constrainedness (RC)		
		High	Medium	Low
AW(3)	Best solution	LWR	LWR	LWR
	Second-best solution	MDR + LWR	MDR + LWR	MDR + LWR
AF(4)	Best solution	MDR + SPT	MDR + SPT	SPT
	Second-best solution	SPT	SPT	MDR + SPT
AM(5)	Best solution	MDR + MSWR	MDR + MSOR	MSOR
	Second-best solution	MSWR	MDR + MSWR	MDR + MSOR
PT(6)	Best solution	MDR + COVERT2	MDR + MSR	MSR
	Second-best solution	MDR + COVERT1	MSR	MDR + MSR
MT(7)	Best solution	MSR	MSR	MDR + MSR
	Second-best solution	MDR + MSR	MDR + MSR	MSR
AT(8)	Best solution	MDR + MSR	MSR	MDR + MSR
	Second-best solution	MSR	MDR + MSR	MSR
CAT(9)	Best solution	MDR + MSR	MDR + MSR	MSR
	Second-best solution	MSR	MSR	MDR + MSR

TABLE 17.4. Dispatching Rules Generating the Best and the Second-Best Solution in the Medium RC Dynamic Machining Systems for Various Performance Measures

Measure of Performance	Solutions	Load		
		Light Load	Normal Load	Heavy Load
AW(3)	Best solution	MDR + LWR	LWR	LWR
	Second-best solution	LWR	MDR + LWR	MDR + LWR
PT(6)	Best solution	MDR + COVERT1	COVERT1	MDR + COVERT2
	Second-best solution	MDR + COVERT2	MDR + COVERT2	MDR + COVERT1
MT(7)	Best solution	MDR + COVERT1	COVERT1	MDR + COVERT1
	Second-best solution	COVERT1	MDR + COVERT1	COVERT1
AT(8)	Best solution	MDR + COVERT1	COVERT1	MDR + COVERT1
	Second-best solution	COVERT1	MDR + COVERT1	COVERT1
CAT(9)	Best solution	MDR + COVERT1	COVERT1	MDR + COVERT1
	Second-best solution	COVERT1	MDR + COVERT1	COVERT1

$$AT = \frac{r_{machine}}{n \cdot m_{avg}} \sum_{j=1}^{R} \frac{AT_j}{R}, \tag{8}$$

where AT_j is the average tardiness in the jth run.
- Normalized conditional average tardiness (CAT)

$$CAT = \frac{r_{machine}}{n \cdot m_{avg}} \sum_{j=1}^{R} \frac{CAT_j}{R}, \tag{9}$$

where CAT_j is the conditional average tardiness in the jth run.

Tables 17.3 and 17.4 illustrate partial knowledge obtained by simulation. For the details of simulation and output analysis, see Kusiak and Ahn (20).

17.3. ILLUSTRATIVE EXAMPLE

The proposed intelligent scheduling system is illustrated with a numerical example. For simplicity, the scheduling problem is static with the objective of minimizing the makespan. The layout of a machining system is shown in Fig. 17.2.

Each of the two machining cells includes machine(s) that is (are) served by AGVs. AGVs also move parts between the machining cells and the input/output (I/O) buffer. Assume that the buffer capacity for each of the two machine cells is 3. Machine loading and unloading times for each operation are included in the processing time. The intracellular handling (traveling) times are assumed to be negligible, and the intercellular traveling times are shown in Eq. (10).

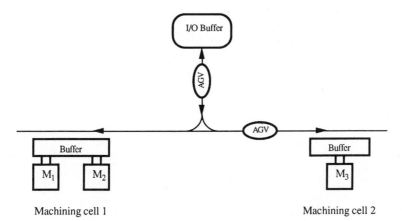

FIGURE 17.2. Manufacturing cell layout.

$$
[\tau_{rr'}] = \begin{array}{c} \\ \text{I/O} \\ \text{MC-1} \\ \text{MC-2} \end{array} \begin{array}{ccc} \text{I/O} & \text{MC-1} & \text{MC-2} \\ \left[\begin{array}{ccc} 0 & 1 & 1 \\ 1 & 0 & 2 \\ 1 & 2 & 0 \end{array} \right] \end{array}. \tag{10}
$$

Eighteen operations belonging to six parts are to be machined in the system (Fig. 17.3).

The resources required by each operation and the corresponding processing times are shown in Table 17.5.

The preventive maintenance period for machine 3 in MC-2 is scheduled during time interval [20, 24].

The intelligent scheduling system selects an appropriate algorithm first (in this case, the algorithm presented in this chapter). Since the level of resource constrainedness is 1:2:3.33, the rule selector recommends the MDR + MSWR dispatching rule as a global dispatching rule (see Rule2_1). Having selected the scheduling algorithm and dispatching rule, the scheduling algorithm system is activated.

Scheduling of operations for machining begins at scheduling time 1 after all the parts have been prepared for manufacturing in the I/O buffer. At scheduling time 1, the MDR + MSWR rule selects the set of operations to

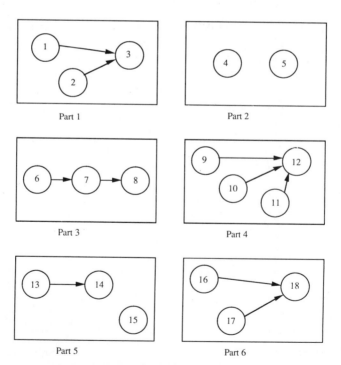

FIGURE 17.3. Six parts with 18 operations.

TABLE 17.5. Process Plans for Six Parts

Part Number	Operation Number	Process Plan	Processing Time	Machine	Tool	Fixture
1	1	1	3	m_1	t_2	f_1
		2	4	m_2	t_3	f_2
	2	1	3	m_3	t_5	f_3
		2	5	m_3	t_6	f_4
	3	1	8	m_2	t_1	f_1
2	4	1	6	m_2	t_3	f_1
	5	1	6	m_1	t_2	f_2
		2	7	m_1	t_2	f_3
3	6	1	3	m_3	t_4	f_4
		2	4	m_3	t_5	f_4
	7	1	3	m_1	t_2	f_2
		2	4	m_1	t_1	f_1
	8	1	4	m_2	t_2	f_1
4	9	1	4	m_1	t_3	f_2
	10	1	3	m_3	t_4	f_3
	11	1	3	m_2	t_1	f_1
		2	4	m_1	t_2	f_1
	12	1	4	m_2	t_3	f_1
		2	5	m_3	t_6	f_4
5	13	1	4	m_3	t_6	f_3
		2	5	m_3	t_4	f_4
	14	1	3	m_2	t_2	f_2
		2	5	m_3	t_4	f_4
	15	1	3	m_1	t_1	f_2
6	16	1	5	m_2	t_1	f_2
	17	1	4	m_1	t_2	f_2
	18	1	8	m_3	t_6	f_4

TABLE 17.6. Machining Cell Status

Time	MC-1		MC-2		AGV	I/O
	Machine	Buffer	Machine	Buffer		
t = 0						P1, P2, P3, P4, P5, P6
					P1, P3, P6	P2, P4, P5
t = 1	P1, P6	P1, P6	P3	P3		P2, P4, P5
t = 4	P6	P1, P4	P5	P3, P5	P4, P5	P2
	P6, P4	P1	P5	P3		P2
t = 6	P4	P1, P6	P5	P3		P2
	P4	P6		P3	P1, P2	P2
t = 8		P2, P4, P6		P1, P3, P5		
t = 11	P2, P6	P4	P1	P5	P3	
	P2, P6	P4		P1, P5	P3	
	P2, P6	P4	P5	P1		
t = 12	P2	P3, P4, P6	P5	P1		
	P2, P3	P4, P6	P5	P1		
t = 14	P3	P2, P4, P6*	P5	P1		
	P3	P2, P6	P5		P1, P4	

$t = 15$		P2, P3, P6	P5		P1, P4	
	P2	P3, P6	P5		P1, P4	
$t = 16$	P2,	P1, P3, P6		P4, P5		
	P1, P2	P3	P4	P5	P6	
$t = 19$	P1, P2	P3		P4, P5, P6*		
	P1, P2	P3	[redacted]	P5, P6	P4	
$t = 21$	P1	P2, P3, P4	[redacted]	P5, P6		
	P1, P4	P3	[redacted]	P5, P6	P2	
$t = 24$	P4	P1, P3	[redacted]	P6	P5	P2
	P3, P4		P6		P1, P5	P2
$t = 25$	P3	P4, P5	P6			P1, P2
	P3, P5	P4	P6			P1, P2
$t = 28$		P3, P4, P5	P6			P1, P2
	P4		P6		P3, P5	P1, P2
$t = 32$	P4			P6		P1, P2, P3, P5
	P4				P6	P1, P2, P3, P5
$t = 33$		P4				P1, P2, P3, P5, P6
$t = 34$						P1, P2, P3, P4, P5, P6

be dispatched, $P = \{1, 16, 6\}$. At scheduling time 4, operations 1 and 6 are completed, and operations $11_{(1)}$ and 13 are selected by the global dispatching rule. In order to machine operations $11_{(1)}$ and 13 at this time, parts 4 and 5 must be waiting in the corresponding machining cell buffer, so that they can be loaded. The schedule generated by the intelligent scheduling system is shown in Fig. 17.4.

Status of the machining cells during the entire scheduling horizon is shown in Table 17.6. For example, at scheduling time 1, parts 1 and 6 are in the MC-1 buffer, part 3 is in the MC-2 buffer, while all the other parts are in the I/O buffer. In this table, * represents "urgent" status of a machining cell buffer, and the dark area shows the downtime of machine 3 due to preventive maintenance.

At scheduling time 8 (when both machine cell buffers are full), urgent state is not issued since part 2 and part 6 have already been scheduled for the next operation in MC-1, and part 1 has been scheduled for MC-2. At scheduling time 14, the buffer of MC-1 is full, and there is no operation in the buffer that has been scheduled for machining. Thus urgent state has been issued at this time. Even is operation 3 is schedulable (since all the required resources are available), it cannot be dispatched (see Rule4_1). Instead, operation 10 of part 4 is scheduled at time 16 in MC-2.

At scheduling time 19, operation 18 is selected by the dispatching rule; however, machine 3 has already been scheduled for preventive maintenance. Thus the operation waits until the maintenance period ends (see Rule4_2). At scheduling time 34, all the parts have been machined, and the scheduling system generates the schedule shown in Fig. 17.4.

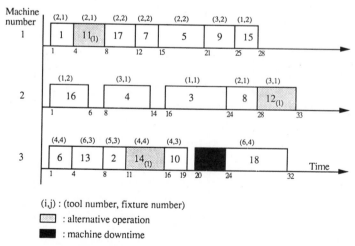

(i,j) : (tool number, fixture number)

☐ : alternative operation

■ : machine downtime

FIGURE 17.4. Gantt chart of the final schedule.

17.4. SUMMARY

In this chapter, an intelligent approach for scheduling automated machining systems was presented. By incorporating operations research and artificial intelligence techniques (tandem architecture), the proposed system has a high potential to provide efficient schedules reflecting details of the automated machining environment and accomplishing the objectives of the system.

The current implementation of the system is being improved. More elaborate knowledge acquisition methods are being sought. Other scheduling algorithms can easily be incorpor; 'ed into the existing system.

REFERENCES

1. Ahn, J., and Kusiak, A. (1990). Scheduling with Alternative Process Plans, *Working Paper #90–18*, Department of Industrial Engineering, The University of Iowa, Iowa City.

2. Aiken, M. W., and Sheng, O. R. L. (1990). NEXPERT OBJECT, *Expert Systems*, Vol. 7, No. 1, pp. 54–57.

3. Alexander, S. M. (1987). An Expert System for the Selection of Scheduling Rules in a Job Shop, *Computers and Industrial Engineering*, Vol. 12, No. 3, pp. 167–171.

4. Baker, K. R. (1974). *Introduction to Sequencing and Scheduling*, Wiley, New York.

5. Balas, E., and Yu, C. S. (1986). Finding a Maximum Clique in an Arbitrary Graph, *SIAM Journal on Computing*, Vol. 15, No. 4, pp. 1054–1068.

6. Bellman, R. E., Esogbue, A. O., and Nabeshima, I. (1982). *Mathematical Aspects of Scheduling and Applications*, Pergamon Press, Oxford.

7. Blackstone, J. H. Jr., Phillips, D. T., and Hogg, G. L. (1982). A State-of-the-Art Survey of Dispatching Rules for Manufacturing Job Shop Operations, *International Journal of Production Research*, Vol. 20, No. 1, pp. 27–45.

8. Bondy, J. A., and Murty, U. S. R. (1976). *Graph Theory with Applications*, Elsevier, New York.

9. Cohen, P., and Feigenbaum, E. A. (1982). *The Handbook of Artificial Intelligence*, William Kaufmann, Los Altos, CA.

10. Cowan, D. A. (1985). Is CIM Achievable?, *Proceedings of the 3rd European Conference on Automated Manufacturing*, Birmingham, UK, May 15–17, pp. 7–16.

11. French, S. (1982). *Sequencing and Scheduling: An Introduction to the Mathematics of the Job-Shop*, Wiley, New York.

12. Gere, W. S. Jr. (1966). Heuristics in Job Shop Scheduling, *Management Science*, Vol. 13, No. 3, pp. 167–190.

13. Groover, M. P. (1987). *Automation, Production Systems, and Computer Integrated Manufacturing*, Prentice-Hall, Englewood Cliffs, NJ.

14. Horowitz, E., and Sahni, S. (1978). *Fundamental of Computer Algorithms*, Computer Science Press, Potomac, MD.

15. Jaumard, B., Ow, P. S., and Simeone, B. (1988). A Selected Artificial Intelligence Bibliography for Operations Researchers, *Annals of Operations Research*, Vol. 12, pp. 1–50.

16. Kanet, J. J., and Adelsberger, H. H. (1987). Expert Systems in Production Scheduling, *European Journal of Operational Research*, Vol. 29, pp. 51–59.

17. Koulamas, C. P., and Smith, M. L. (1988). Look-Ahead Scheduling for Minimizing Machine Interference, *International Journal of Production Research*, Vol. 26, No. 9, pp. 1523–1533.

18. Kusiak, A. (1987). Artificial Intelligence and Operations Research in Flexible Manufacturing Systems, *Information Processing and Operational Research (INFOR)*, Vol. 25, No. 1, pp. 2–12.

19. Kusiak, A. (1990). *Intelligent Manufacturing Systems*, Prentice-Hall, Englewood Cliffs, NJ.

20. Kusiak, A., and Ahn, J. (1990). A Resource-Constrained Job Shop Scheduling Problem with General Precedence Constraints, *Working Paper #90-03*, Department of Industrial Engineering, The University of Iowa, Iowa City.

21. Kusiak, A., and Chen, M. (1988). Expert Systems for Planning and Scheduling Manufacturing Systems, *European Journal of Operational Research*, Vol. 34, pp. 113–130.

22. Marucheck, A. S. (1989). Integrating Expert Systems and Operations Research: A Review, *Working Paper*, Graduate School of Business Administration, University of North Carolina, Chapel Hill.

23. NEXPERT OBJECT (1987). Neuron Data Inc., Palo Alto, CA.

24. Newman, P. A., and Kempf, K. G. (1985). Opportunistic Scheduling for Robotic Machine Tending, *The Second Conference on Artificial Intelligence Applications*, Miami Beach, FL, December 11–13, pp. 168–173.

25. O'Keefe, R. (1986). Simulation and Expert Systems—Taxonomy and Some Examples, *Simulation*, Vol. 46, No. 1, pp. 10–16.

26. O'Keefe, R. M., Belton, V., and Ball, T. (1986). Experiences with Using Expert Systems in O.R., *Journal of Operational Research Society*, Vol. 37, No. 7, pp. 657–668.

27. Panwalkar, S. S., and Iskander, W. (1977). A Survey of Scheduling Rules, *Operations Research*, Vol. 25, No. 1, pp. 45–61.

28. Papadimitriou, C. H., and Steiglitz, K. (1982). *Combinatorial Optimization: Algorithms and Complexity*, Prentice-Hall, Englewood Cliffs, NJ.

29. Phelps, R. I. (1986). Artificial Intelligence—An Overview of Similarities with O.R., *Journal of Operational Research Society*, Vol. 37, No. 1, pp. 13–20.

30. Rodammer, F. A., and White, K. D. Jr. (1988). A Recent Survey of Production Scheduling, *IEEE Transactions on Systems, Man, and Cybernetics*, Vol. 18, No. 6, pp. 841–851.

31. Schrage, L. (1984). *Linear, Integer, and Quadratic Programming with LINDO*, Scientific Press, Palo Alto, CA.

32. Schultz, C. R. (1989). An Expediting Heuristic for the Shortest Processing Time Dispatching Rule, *International Journal of Production Research*, Vol. 27, No. 1, pp. 41–51.

33. Shaw, M. J. P., and Whinston, A. B. (1986). Application of Artificial Intelligence to Planning and Scheduling in Flexible Manufacturing. In: *Flexible Manufacturing Systems: Methods and Studies*, edited by A. Kusiak, North-Holland, Amsterdam, pp. 223–242.

34. Steffen, M. (1986). A Survey of Artificial Intelligence-Based Scheduling Systems, *Proceedings of Fall Industrial Engineering Conference*, December 7–10, Boston, MA, pp. 395–405.

35. Thesen, A., and Lei, L. (1986). An Expert System for Scheduling Robots in a Flexible Electroplating System with Dynamically Changing Workloads. In: *Proceedings of Second ORSA/TIMS Conference on Flexible Manufacturing Systems: Operations Research Models and Applications*, edited by K. E. Stecke and R. Suri, Elsevier Science, Amsterdam.

36. Widman, L. E., and Loparo, K. A. (1989). Artificial Intelligence, Simulation, and Modeling: A Critical Survey. In: *Artificial Intelligence, Simulation, and Modeling*, edited by L. E. Widman, K. A. Loparo, and N. R. Nielsen, Wiley, New York, pp. 1–44.

CHAPTER EIGHTEEN

Scheduling of Flexible Assembly Systems

MICHAEL PINEDO

Department of Industrial Engineering and Operations Research,
Columbia University, New York

18.1. INTRODUCTION

One of the results of the increased emphasis on customer satisfaction is that many manufacturing companies have found that their product mixes have become increasingly complex. For many consumer items the variety of different models and the number of different options have never been greater. Yet pressures to reduce operating costs have forced manufacturers to assemble large numbers of different models, possibly from different product families, at the same facility or even on the same assembly line. This is the case in many assembly environments, including automobile assembly, television assembly, and the assembly of components such as printed circuit boards.

Now that a large number of different items have to be produced at a single facility makes the scheduling function a very important element of the production process. Investment in equipment, such as robots and fixtures, is usually very high. Thus there is an incentive to keep the equipment, as much as possible, fully utilized. Keeping the utilization high implies that setup times have to be kept to a minimum and workloads have to be well balanced. Note that these setup times usually are sequence dependent; they may be large when there is a change of family and they may be negligible when consecutive items belong to the same family.

Mixed model assembly in modern factories is now done with so-called flexible assembly systems (FASs). FASs constitute a subcategory of the more general class of flexible manufacturing systems (FMSs) (6). The class

Intelligent Design and Manufacturing, Edited by Andrew Kusiak.
ISBN 0-471-53473-0 © 1992 John Wiley & Sons, Inc.

of FMSs also includes the subcategory of general flexible machining systems (GFMSs), which are systems designed to manufacture an almost unlimited variety of parts in an almost random order. GFMSs are somewhat similar to conventional job shops. FASs form a somewhat more special subcategory of FMSs. FASs are usually designed to assemble a limited number of different models and each one of these in a reasonably high volume. The scheduling process in a FAS is more stable than the scheduling process in a GFMS and tends to focus on a longer time horizon. The routing in a FAS is usually unidirectional, but it does at times allow for recirculation. This is in contrast with the routing in a GFMS, which may be considerably more complicated.

The scheduling process in a FAS involves the detailed sequencing and scheduling of the different items that are to be processed by the FAS. The FAS schedule must attempt to achieve the requirements of the production planning process, which does the medium- and long-term planning for the system, while at the same time not violate any of the current machine conditions as reported by the shop floor control system. Events that happen on the shop floor do have an impact on the schedule; for example, the breakdown of a particular robot may necessitate changes in the product mix.

This chapter is organized as follows. In Section 18.2 various aspects of FASs are discussed, namely, machine layouts, measures of flexibility, and scheduling objectives. In Section 18.3 various scheduling procedures are described. In Section 18.4 several scheduling systems are described that have been developed and implemented in industry. The last section discusses the directions in which the field is going.

The following terminology and notation is used throughout the chapter. There are m machines or workstations in the system (the terms machines and work stations are used interchangeably). The subscript i refers to station i, while the subscript j refers to job j. A job may refer to a single item, such as an automobile or TV set, or a batch (or lot) of identical items, such as a batch of printed circuit boards (PCBs). Throughout the remainder of the chapter it will be clear from the context whether a job is a single item or a batch of identical items. The following notation is common in job shop scheduling and is used in FAS scheduling as well.

p_{ij} Processing time of job j on machine i.
d_j Due date of job j
w_j Weight or importance factor of job j
C_j Completion time of job j, which, of course, depends on the schedule
T_j $\max(C_j - d_j, 0)$, which is the tardiness of job j

18.2. FAS MODELS AND SCHEDULING OBJECTIVES

This section describes three basic machine environments and the most common scheduling objectives. Any one of the environments could represent a simple assembly system, but, more commonly, such an environment is part of a more complex production process with various types of sub-assemblies merging at specific stations. Similarly, in any real system the overall objective is a mixture of both these basic objectives as well as other objectives, which depend specifically on the system.

18.2.1. Single Stations in Series with Limited Buffers and No Bypass

This system consists simply of a number of stations (manual or robotic) in series with a limited buffer in-between any two successive stations (see Fig. 18.1). When the buffer in-between two stations is full, the station that feeds the buffer cannot release an item for which processing has completed. This phenomenon is called *blocking*. This type of environment is common in the assembly of items that are physically large, such as television sets, copiers, and so on. The fact that the items are large makes it difficult to keep an arbitrary number waiting in front of a workstation. Also, no items are allowed to bypass any others waiting in queue (i.e., while residing in a buffer).

The processing requirements of one item may be significantly different from those of another, due to the fact that different items have to be equipped with different option packages. The timing of the releases from the stations can be determined through various disciplines; the release mechanism at a station may be a function of the queue at the station downstream as well as of the queue at the station itself.

Note that this machine environment can be modeled only with stations; buffers between stations are not really needed. The reason for this is the following: A buffer can simply be modeled by a station at which the processing times of all the items are equal to zero. Based on this observation, a fair amount of theoretical research has been done on stations in series with zero buffers in-between.

18.2.2. Stages in Series with Stations in Parallel at Each Stage

This system consists of a number of stages in series, each stage consisting of a number of stations in parallel. An item, which in this environment often is

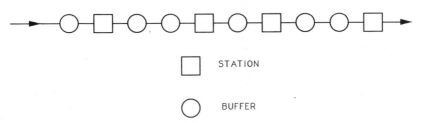

FIGURE 18.1. Single stations in series with limited buffers and no bypass.

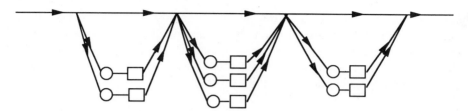

FIGURE 18.2. Stages in series with stations in parallel at each stage.

equivalent to a batch of relatively small identical items such as PCBs, needs to be processed on only one of the stations at every stage. Usually, any one of the stations can do, but at times it may be the case that not all stations at a stage are exactly identical and that a given item has to be processed at a specific station. It also may occur that an item does not need to be processed at a stage at all; the material handling system that moves the items from one stage to the next usually allows an item then to bypass a stage and all the items residing at that stage (see Fig. 18.2). In this environment the buffers at each stage may have limited capacity. If this is the case and the buffer is full, then either the material handling system has to come to a stand still or, in case there is the option to recirculate, the item bypasses the stage and recirculates.

There are many other FAS models that also allow bypass. For example, consider a system with two lanes in parallel. One lane is the assembly lane, which contains all the stations. The second lane is the so-called bypass lane. At set intervals (after a fixed number of stations) there are shuttles that enable an item to go from the assembly lane to the bypass lane and back (see Fig. 18.3). This type of environment has been used for the assembly of

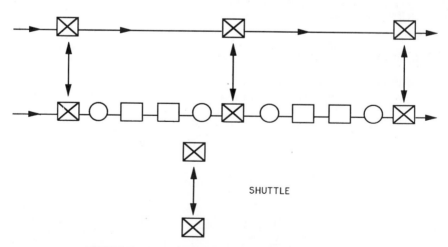

FIGURE 18.3. Stations in series with bypass lane.

copiers. The various items to be assembled are from the same family but may be different because of the various options packages (e.g., automatic document feeder).

18.2.3. Paced Assembly Lines

In a paced assembly system, the material handling system conveys the items with a constant speed from one workstation to the next. The items are moving on the line with equal distances from one another. If the assembly is done manually, the workers at a particular station walk back toward the point in the line which corresponds to the beginning of their station (section). The time it takes to walk back is negligible in comparison with the time it takes to perform the operation. As the tasks to be performed at different stations are different and have different processing times, the amount of space along the line reserved for a particular task is in proportion to the amount of time needed. So for this type of assembly line a station is basically equivalent to a designated section of the assembly line. Clearly, in such an environment bypass is not possible.

This type of assembly system is very common in the automobile industry. Assembly systems in the automobile industry have, of course, a large number of other properties and special characteristics, which are not included in the model described above. One of these features is, for example, the point in the assembly line where the engine and the body come together. Another important feature of automobile assembly is the paint shop. The setup costs for different colors of paint are significantly larger than any of the other setup costs incurred on the line. Therefore some lines allow resequencing, either before or after the paint shop.

18.2.4. Scheduling Objectives

The scheduling process usually has to deal with several competing objectives. Typical among these are throughput maximization, work-in-process (WIP) minimization, and meeting due dates.

Throughput is defined as the long-term average number of items produced per unit time. Throughput maximization may, to a certain extent, actually be equivalent to the maximization of machine utilization (especially of the bottleneck machines) and the minimization of the sum of the setup times (also especially on the bottleneck machines).

The minimization of the work-in-process has led to the just-in-time philosophy, which attempts to keep all inventories of subassemblies and so on at a level that is as low as possible. There are several reasons for minimizing the WIP. If a station is not functioning properly and a quality control station at the end of the line discovers improprieties due to this mal-functioning, then the number of faulty units in need of repair are limited. If a sudden change in the production plans requires a switch over to a different model, the switch over process can be done faster when the WIP is low. If buffer capacities are limited, a low WIP reduces the occurrences of blocking as well.

A third objective is to meet the due dates, that is, the committed shipping dates, or to minimize tardiness. If the items have different weights, the objective to be minimized is the sum of the weighted tardinesses, that is $\Sigma\, w_j T_j$.

The objectives described here are somewhat different from those usually considered in job shop scheduling. In a job shop there are usually only a finite number of jobs. One objective often to be minimized is the makespan, which is the time the last job leaves the system. This objective is to a certain degree equivalent to the maximization of machine utilization, the minimization of setup times, and the balancing of the workload over the various machines. The heuristics used in the algorithms for minimizing the makespan can at times be adapted and used for the maximization of throughput in FASs. Another objective that is popular in job shop scheduling is the minimization of the sum of completion times (the so-called flow time). The heuristics used in this objective can at times be adapted to minimize the WIP in FASs.

18.3. SCHEDULING METHODS

Sequences and schedules used in FASs are often periodic or cyclic. That is, a number of different items are scheduled in a certain way and this schedule is repeated over and over again. Of course, cyclic schedules are a special class of schedules and it is not necessarily true that the optimal schedule is cyclic. More often than not, it is an acyclic schedule that is optimal. However, cyclic schedules have a natural advantage due to their simplicity; they are easy to keep track of and they reduce the amount of confusion in the system. In practice, there is often an underlying basic cyclic schedule from which minor deviations are allowed, dependent on current orders.

Before elaborating further on this class of schedules, it is useful to define the so-called minimum part set (MPS). Let R_j denote the number of items of type j to be assembled in the production target. Suppose there are l different types of items. If k is the greatest common divisor of the integers $R_1, \cdots, R_l$, then the vector

$$R^* = (R_1/k, \ldots, R_l/k)$$

represents the smallest set having the same proportions as the production target. This set is usually referred to as the minimum part set. Cyclic schedules are specified by the sequence in which the MPS is ordered. Cyclic schedules of MPSs are, of course, not practical when there are significant sequence-dependent setup times between two different items. When this is the case it is preferable to have long runs of the same type.

In what follows three heuristics are described that have proved to be useful in several machine environments, including the ones described in the previous section.

18.3.1. The Profile Fitting Heuristic

The profile fitting heuristic (PFH) is a heuristic well suited for the single-lane machine environment described in Section 18.2.1 (4, 5). The PFH attempts to minimize the so-called cycle time in steady state. The cycle time is the time between the first items of two consecutive MPSs entering the system. Minimization of cycle time is basically equivalent to maximization of throughput. In order to illustrate the concept of cycle time, consider an assembly line with four stations and no buffers between stations. Consider a MPS with three different types and a production target

$$R^* = (1, 1, 1) \,.$$

Let the associated processing requirements of the three different types be the following:

$$p_{11} = 0, \ p_{21} = 0, \ p_{31} = 1, \ p_{41} = 1 \,,$$
$$p_{12} = 1, \ p_{22} = 0, \ p_{32} = 0, \ p_{42} = 1 \,,$$
$$p_{13} = 0, \ p_{23} = 0, \ p_{33} = 1, \ p_{43} = 0 \,.$$

It is clear that the second "station" (i.e., the second column of processing times), with zero processing times for all three types, basically functions as a buffer between the first station and the third station. The Gantt charts for this example under the two different sequences are shown in Fig. 18.4. Under both sequences the steady state is reached during the second cycle. Under sequence 1, 2, 3 the cycle time is three, while under sequence 1, 3, 2 the cycle time is two.

The PFH works as follows: One item is selected to go first. The selection of the item to go first may be done arbitrarily or according to some scheme. This first item generates a *profile*. For the time being, it is assumed that the item does not encounter any blocking and proceeds smoothly from one station to the next (in steady state the first item of a MPS may be blocked by the last item of the previous MPS). The profile is determined by the departure time of this first item, denoted as item (1), of type j from station i; that is,

$$D_{i(1)} = \sum_{k=1}^{i} p_{k(1)} = \sum_{k=1}^{i} p_{kj} \,.$$

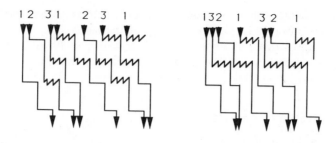

FIGURE 18.4. Profile fitting.

In order to determine which is the most appropriate item to go second, every remaining item in the MPS has to be tried. For each candidate item a computation is carried out to determine the amount of time stations are idle and the amount of time the item is blocked at a station. The departure epochs of a candidate item, denoted as item (2), from station i can be computed recursively:

$$D_{1(2)} = \max(D_{1(1)} + p_{1(2)}, D_{2(1)}),$$

$$D_{i(2)} = \max(D_{i-1(2)} + p_{i(2)}, D_{i+1(1)}), \qquad i = 2, \dots, m-1,$$

$$D_{m(2)} = \max(D_{m-1(2)}, D_{m(1)}) + p_{m(2)}.$$

The time wasted at station i, either being idle or being blocked, is then $D_{i(2)} - D_{i(2)} - p_{i(2)}$. The sum of these idle times and times blocked over all m machines are then computed. The candidate with the smallest total is selected to go next. Observe that in Fig. 18.4 after item 1, item 3 would be selected to go next, as this would cause only one unit of blocking time (on station 2) and no idle times. If item 2 would have been selected to go next, one unit of idle time would be incurred at station 2 and one unit of blocking time on station 3, resulting in two units of time wasted.

After the best fitting item is selected to go next and added to the partial sequence, the new profile, that is, the departure times of this last item from all the stations, is computed and the procedure repeats itself. From the remaining items in the MPS, the best fitting one is again selected and so on.

Experiments have shown that the PFH results in schedules that are close to optimal. However, the heuristic can be refined and made to perform even better. In the description presented above the goodness of fit of a particular item was measured by summing all the times wasted on the m stations. Each station was considered equally important. It is intuitive that lost time on a bottleneck station is worse than lost time on a station that on the average does not have much processing to do. When measuring the total amount of lost time, it may be appropriate to multiply each one of these inactive time periods with a factor that is proportional to the degree of congestion at a particular station. The higher the degree of congestion at a particular station, the larger the weight. One measure for the degree of congestion of a station is easy to calculate; simply determine the total amount of processing to be done on all items in a MPS at the station in question. In the numerical example presented above the third and fourth stations are more heavily used than the first and second stations (the second station was not used at all and basically functioned as a buffer). Time wasted on the third and fourth stations is therefore less desirable than time wasted on the first and second stations. Experiments have shown that such a weighted version of the PFH works exceptionally well.

18.3.2. The Flexible Flow Line Loading Algorithm

The flexible flow line loading (FFLL) algorithm was specifically designed at IBM for the machine environment described in Section 18.2.2 (7). The algorithm was originally conceived for an assembly system that is used for the insertion of components in PCBs. The two main objectives of the algorithm are (a) the maximization of throughput and (b) the minimization of WIP. Instead of maximizing the throughput, an attempt is actually made to minimize the makespan of a whole day's mix. As the amount of buffer space is limited, it is recommended to minimize the WIP in order to reduce blocking probabilities.

The FFLL algorithm consists of three segments:

1. The station allocation segment.
2. The sequencing segment.
3. The release timing segment.

The station allocation segment determines on which particular station in each bank any given item of a MPS is processed. Station allocation has to be done before sequencing and timing, as these last two segments have to know what workload is assigned to each station. The lowest conceivable maximum workload for a bank would be obtained if all the stations in a bank were given an equal workload. In order to obtain nearly balanced workloads over the station at a bank, the longest processing time first (LPT) heuristic is used. According to this heuristic the items are, for the time being, all assumed to be available at the same time and are allocated to a bank one at a time on the next available station in decreasing order of their processing time. After the allocation is determined in this way, the items assigned to a station may actually be resequenced. Resequencing a station, of course, does not alter the workload balance over the stations at a given bank. The output of this segment is merely the allocation of items and not the sequencing of the items or the timing of the processing.

The sequencing segment determines the order in which the items of the MPS are released into the FAS. The chosen order will have a strong effect on the cycle time. The FFLL algorithm uses the *dynamic balancing* heuristic to sequence a MPS. This heuristic is based on the following intuitive observation. Items tend to queue up in the buffer of a station if a large workload is sent to that station during a short timespan. This occurs when there is an interval in the loading sequence, which contains many items with large processing times going to the same station. The following notation is needed. Let n be the number of items in a MPS and m the number of stations in the entire system. Let p_{ij} denote the processing time of item j at station i. So $p_{ij} = 0$ for all but one station in a bank. Let

$$W_i = \sum_{j=1}^{n} p_{ij} ;$$

that is, W_i represents the workload in a MPS destined to station i. Let

$$W = \sum_{i=1}^{m} W_i ;$$

that is, W represents the entire workload in a MPS. For a given sequence let S_j denote the set of items loaded into the system up to and including item j. Let

$$\alpha_{ij} = \sum_{k \in S_j} \frac{p_{ik}}{W_i} ;$$

that is, α_{ij} represents the fraction of the total workload of station i which has entered the system by the time item j is loaded. Clearly, $0 \le \alpha_{ij} \le 1$. The dynamic balancing routine attempts to keep the $\alpha_{1j}, \alpha_{2j}, \ldots, \alpha_{mj}$ as close to one another as possible, that is, as close to an ideal target α_j^*, which is defined as follows:

$$\alpha_j^* = \sum_{k \in S_j} \sum_{i=1}^{m} p_{ik} \bigg/ \sum_{k=1}^{n} \sum_{i=1}^{m} p_{ik}$$

$$= \sum_{k \in S_j} p_k / W ,$$

where

$$p_k = \sum_{i=1}^{m} p_{ik}$$

and p_k represents the workload on the entire system due to item k. So α_j^* equals the total workload put into the system up to and including item j, divided by the system workload due to the entire MPS. The cumulative workload on station i, $\sum_{k \in S_j} p_{ik}$, should be close as possible to the target $\alpha_j^* W_i$. Now let o_{ij} denote a measure of *overload* (or *underload*) at station i due to item j entering the system:

$$o_{ij} = p_{ij} - p_j W_i / W . \tag{1}$$

Clearly, o_{ij} may be negative as well. Now, let

$$O_{ij} = \sum_{k \in S_j} o_{ik} = \left(\sum_{k \in S_j} p_{ik} \right) - \alpha_j^* W_i . \tag{2}$$

That is, O_{ij} denotes the cumulative overload (or underload) on station i due to the items in the sequence up to and including item j. To be exactly on target means that station i is neither overloaded nor underloaded when item j enters the system; that is, $O_{ij} = 0$. The dynamic balancing heuristic now attempts to minimize

$$\sum_{i=1}^{m} \sum_{j=1}^{n} \max(O_{ij}, 0) . \tag{3}$$

The heuristic is basically a greedy heuristic. It selects among the remaining items in the MPS the one that minimizes the objective at that point in the sequence.

The release timing segment now works as follows. From the allocation segment the MPS workloads at each station are known. The station with the greatest MPS workload is the bottleneck. The period of the schedule cannot be smaller than the MPS workload at the bottleneck station. it is easy to determine a timing mechanism that results in a minimum period schedule. First, let all items in the MPS enter the system as rapidly as possible. Consider now the stations one at a time. At each station the items are processed in the order in which they arrive and the processing starts as soon as the station is available. After considering all items in the MPS, determine by how much the workspan at the station exceeds its workload. Delay processing the first item by this difference. Delaying this release is crucial to the timing mechanism; delaying the first item also delays all subsequent items, but the delay ultimately cancels out with the idle time.

This three-step algorithm attempts to find the cyclic schedule with minimum cycle time in steady state. If the system starts out empty at some point in time, it may take a few MPSs to reach steady state. Usually, this transient period is very short. The algorithm tends to achieve short cycle times during the transient period as well.

Extensive experiments with the FFLL algorithm indicate that the method is a valuable tool for the scheduling of flexible flow lines.

18.3.3. The Grouping and Balancing Heuristic

The grouping and balancing heuristic (GBH) has been specifically designed for the paced assembly lines that are common in the automobile industry (2). In this industry the number of different models produced in a single facility is usually very large and there are many options that can be included in an item (e.g., automobile).

For any particular operation there may be a number of configurations. If two consecutive items have different configurations with regard to a particular operation (e.g., automatic transmission versus manual), there is a setup cost involved in the changeover. If the two consecutive items have the same configuration, no setup cost is incurred. This setup cost can be measured in monetary units (the cost of the labor involved). Different operations have different numbers of configurations and different setup costs when a changeover is required between two consecutive items.

Besides setup costs, a second aspect of equal importance has to be dealt with as well; certain operations may have to be done only on a subset of the items (e.g., installation of optional equipment such as a sunroof, application of a vinyl roof cover). A given percentage of the items may have to undergo such a particular operation. Often, such an operation takes more time than those operations that have to be done on all items; the section of the line assigned to such an operation may then have to be longer than the sections of the line assigned to other operations. As the number of workers assigned to this operation is in proportion with the *average* workload, it is important that the arrivals of the items that need to undergo this operation are somewhat balanced over time.

For example, suppose the installation of a sunroof lasts five times longer than another operation, which has to be done on all units. The section of the line corresponding to the long operation has to be at least five times as long as the section of the line assigned to the short operation. If the section of the line corresponding to the short operation contains on the average a single item, then the section of the line corresponding to the long operation contains on the average five items. Suppose only 10% of the items need to undergo the long operation. A perfectly balanced sequence implies that every tenth item on the line has to undergo the long operation. If two consecutive items would need to undergo the long operation, the workers may not be able to complete the work on the second item in time. The reason is clear: they complete the work on the first item when it is about to leave the section and then turn toward the second item and start working on it. However, this second item is already relatively close to the end of the section and it may not be possible to complete the work on this item before it leaves the section. In this case there would not have been any contention if the items under consideration were spaced at least five units apart.

Such operations, which only have to be done on a subset of the items coming through, are therefore capacity constrained; they require a careful balancing of the schedule. Proper balancing impacts the quality of the items coming off the line. This requires certain bounds, in the form of upper limits with regard to the maximum frequency with which items end up in positions in which such operations cannot be completed properly (these are the so-called capacity violations constraints).

The GBH works as follows. The total number of items to be sequenced is determined and a so-called final group size is determined. This final group size is determined by considering capacity violation goals. It is within this final group that items are more or less evenly spaced with regard to operations that have capacity constraints. Setup costs are not taken into consideration when determining the sequence within a final group. If there are many operations with capacity constraints, then this final group size actually has to be fairly large in order to provide sufficient freedom for proper spacing with regard to the capacity constrained operations involved. Each final group constitutes a number of consecutive items, which all have the same configurations with respect to a given number of operations that are *not* capacity constrained (actually, with respect to a number of operations that are subject to high setup costs). So any sequence within the final group has the same setup costs with respect to those particular operations that have high setup costs but are not capacity constrained. This allows the scheduler to select the best sequence with respect to the capacity constrained operations.

Consider the following example. Suppose the total number of items to be sequenced is 4000 and the capacity constrained operations require the final group size to be 40. This implies that there are 100 groups. Now operations have to be identified that have several configurations and setup costs. These setup costs have to be estimated and a rank order of setup costs for all operations has to be established. The items are first grouped according to the operation with the highest setup cost. For example, if the operation with the highest setup cost allows four different configurations and, on average, the demands for the four configurations are equal, then four groups with 1000 items each are formed. The operation with the second highest setup cost is considered next. Assume there are five configurations here, the five having equal demands. Assume also, for the time being, that the demands for these five configurations are in no way correlated with the demands for the four configurations of the first operation. This makes it possible to subdivide each group of 1000 into five subgroups of 200 each. So each group of 200 is characterized by a specific configuration for the first operation and a specific configuration for the second operation. If a third operation with five configurations is included in the grouping procedure, the final group size becomes 40. Now no more operations can be included in the grouping procedure, because the spacing in each one of these final groups has to be done in such a way that capacity violation goals are met.

Spacing algorithms suitable for implementation are significantly more complex than the simple procedure outlined above. For one reason, the demands for the various configurations of an operation are not necessarily equal. Moreover, the demand for the configurations of one operation are usually strongly correlated with the demand for the configurations of other operations. These demands are also correlated with whether an item needs to undergo an operation that is subject to capacity constraints.

18.4. SCHEDULING SYSTEMS

In the preceding section a presentation was given of some of the heuristic ideas and procedures used in FAS scheduling. In this section design and implementation aspects of scheduling systems are discussed.

The models and algorithms described in the previous section are highly stylized and do not capture the many subtleties that occur in practice. For example, mathematical models usually assume a stationary product mix. However, product mixes do change continuously and these changes may require significant rescheduling efforts. Unexpected machine breakdowns are also a reality that has to be faced. Such random events may make it necessary to build a certain amount of slack in the schedules.

18.4.1. General Structure of Scheduling Systems

In recent years a number of scheduling systems have been developed that control scheduling operations in FASs. For a number of reasons the *implementation* of such systems has turned out to be at least as difficult as the *development*. Computer-based scheduling systems often consist of three modules (see Fig. 18.5):

1. A database management module.
2. A schedule generation module.
3. A user interface module.

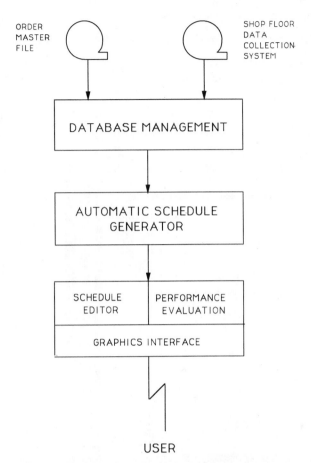

FIGURE 18.5. Possible configuration of scheduling system.

All three modules play a crucial role in the functionality of the system. In practice, a large effort is usually required to make a factory's database suitable for input to the system. The database management module may also have capabilities of manipulating the data, to perform various statistical analyses and so on. The schedule generation module controls model formulation, the objective function selections, and suitable solution heuristics and algorithms. The user interface module is a critical component, especially with regard to the implementation process. The user interface often takes the form of an electronic Gantt chart with corresponding tables and graphs, which enable the scheduler to edit the schedule generated by the system and take last minute information into account (see Fig. 18.6). The database management and user interface aspects of scheduling systems are beyond the scope of this chapter. In what follows, only the schedule generation module is discussed in detail.

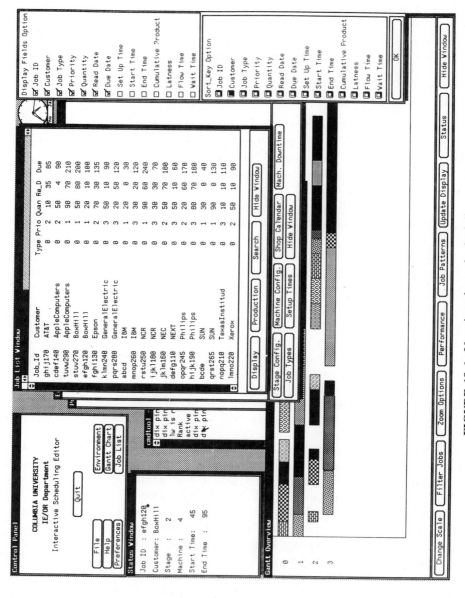

FIGURE 18.6. User interface of scheduling system.

Schedule generation is now done through a combination of different approaches. One approach could be called the *algorithmic* approach. This approach originated with industrial engineers and operations researchers. The other approach is usually called the *knowledge-based* approach. This approach originated with computer scientists and artificial intelligence experts. Today, scheduling systems tend to use concepts from both approaches; these are usually referred to as hybrid systems.

The algorithmic approach usually requires a fairly precise mathematical formulation, including objective functions and constraints. This approach attempts to achieve schedules that are as close as possible to an optimal schedule.

The second approach requires less rigorous problem formulation. The problem is basically formulated in the form of a set of rules. Under this approach the system attempts to find a *feasible* schedule that satisfies all the rules. The more modern programming languages have so-called object-oriented extensions—for example, C++ and LISP—which promote a modular programming style and facilitate coding of these rules.

The remainder of this section contains an overview of two FAS scheduling systems that have been described in the literature. Both systems appear to have been designed with the electronic component industry in mind. However, the authors of the descriptions do not elaborate on any details with regard to the application domain.

18.4.2. A System for an Assembly Line with Bypass

A scheduling system for an assembly line with bypass was developed at the Politecnico di Torino around 1985 (1). It is specifically designed for a machine environment that is a special case of the one described in Section 18.2.2. It assumes that each stage consists of a single station instead of a number of stations in parallel. It also assumes that there are limited buffers not only before but also after each station (see Fig. 18.7). The system consists of two basic modules. One module is concerned with the actual problem solving, that is, with the generation of feasible schedules; the other module checks capacity constraints.

The approach used by the problem solving module is mostly rule based. The rules are designed in order to satisfy the following three types of constraints: production constraints, resource constraints, and capacity constraints. According to the production constraints, a batch of items cannot be worked on before its release time or after its due date. The resource

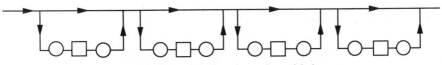

FIGURE 18.7. Stations in series with bypass.

constraints have to be taken care of when several batches use the same fixtures and thus cannot be produced at the same time. The capacity constraints are there to make sure that the stations are not overutilized. Because of the limited buffer capacities, queue lengths are not allowed to exceed predefined limits; otherwise the material handling system is not able to function properly. The schedule generation module is not entirely rule based. A priority rule is used that sets the priority of each batch as the total remaining processing time divided by the time remaining until the due date. A schedule can then be generated by ordering the batches according to their priorities and by releasing them into the system as long as production and resource constraints are satisfied. It then checks the capacity constraints by estimating the FAS performance if the batch were released.

The capacity constraint checking module uses a closed queueing network algorithm. If it appears that the average queue length at the various stations exceeds predefined limits, the batch is not released and the scheduling module considers another one.

The scheduling module is coded in OPS5, which is a rule-based, domain-independent, production language. This programming language offers the advantages of modularity and flexibility because the data, the knowledge, and the control are all separate entities. This encourages structured development through incremental addition and verification of new features and constraints. The second module, the queueing network analysis, requires programming techniques that are quite different. This module is coded in Fortran 77. A major consequence of the different programming languages is that two separate but consistent data structures have to be maintained.

18.4.3. A More General System for Flow Type Assembly

A more general system was developed at University College, Galway, Ireland, around 1987–1988. This scheduling system was developed for general flow type assembly. There are four main objectives: meeting the committed shipping dates, minimizing the overall assembly lead time, maximizing station utilization, and minimizing WIP.

The system consists of four modules:

1. The production profile database.
2. The rules base selector and manual selection.
3. The library of scheduling rules.
4. The interactive user interface.

From these four modules it is clear that, just like the Turin system, this system is a hybrid that uses rules as well as algorithms. However, unlike the Turin system, this system contains an entire library of algorithms, not just a single one.

The production profile segment contains data with respect to planned orders, resources, products, and materials. The rule based segment contains a number of elements that include a materials check, a capacity check, and an analysis of critical resources to detect potential bottlenecks.

The rules base also checks if a simple algorithm can be used to generate a feasible schedule [e.g., shortest processing time first (SPT), earliest due date first (EDD)]. If no easy solutions are applicable, the inference engine continues its search for a feasible schedule. In doing so it adopts an approach that is very similar to OPT (optimized production technology): It selects what appears to be critical stations—that is, stations subject to a heavy load—and then generates a schedule that, on the one hand, attempts to alleviate those loads and, on the other hand, attempts to keep those stations busy at all times.

The module that contains the library of scheduling rules includes the following options from which the user can make a selection:

1. The earliest due date heuristic.
2. The shortest processing time first heuristic.
3. The least slack time heuristic.
4. A priority-based scheduler.
5. A bottleneck scheduler.
6. A manual scheduler.

The priority-based scheduler allows the user to assign priorities to jobs in order to be able to expedite certain jobs. The bottleneck scheduler is based on the OPT philosophy and is used in conjunction with the rules base described earlier. The manual scheduler allows the user to construct the schedule entirely by hand.

The fourth and last module is the interactive scheduler. The input into this segment is the heuristically or manually generated schedule. The user now has the option to modify this first-pass schedule. All starting times are then computed and sent to the production profile database. The system has been implemented on an experimental basis in a plant of the Digital Equipment Corporation in Clonmel, Ireland. The results of this implementation have not yet been reported.

18.5. CONCLUDING REMARKS

During the last four decades a great amount of research has been done in the classical job shop scheduling domain. The problems studied usually involved only a finite number of jobs. It was not until the 1980s that FAS scheduling became a true research topic. A fundamental difference between job shop and FAS scheduling is that the FAS usually sees an unending stream of jobs, each one belonging to a finite number of job types. This is one of the main reasons why the objectives in FAS scheduling are somewhat different from the objectives in job shop scheduling. Nevertheless, the algorithmic concepts applicable to one type of scheduling problem can usually be adapted for use in the other. In spite of this, it appears that FAS scheduling is a somewhat more open field of research than job shop scheduling; a framework for these models has not yet been established the way it has for job shop scheduling.

The advent of the microcomputer on the factory floor in the early 1980s gave a strong impetus to the development of scheduling *systems* and many have been developed during the last decade. However, the number of systems actually implemented is significantly less than the number developed. There are various reasons for difficulties in the implementation, namely, (a) poor databases, (b) schedulers that are not cooperative, and (c) poor user interfaces.

In recent years the two approaches in system development—that is, the algorithmic approach and the rule-based approach—have started to converge. Most current system designers are well aware of both approaches. However, the history of system development and implementation has not yet reached the point where comprehensive system design recipes are established.

REFERENCES

1. Bruno, G., Elia, A., and Laface, P. (1986). A Rule-Based System to Schedule Production, *IEEE Computer*, Vol. 19, No. 7, pp. 32–40.
2. Burns, L., and Daganzo, C. F. (1987). Assembly Line Job Sequencing Principles, *International Journal of Production Research*, Vol. 25, pp. 71–90.
3. Copas, C. and Browne, J. (1990). A Rules-Based Scheduling System for Flow Type Assembly, *International Journal of Production Research*, Vol. 28, No. 5, pp. 981–1005.
4. McCormick, T., Pinedo, M., Shenker, S., and Wolf, B. (1989). Sequencing in an Assembly Line with Blocking to Minimize Cycle Time, *Operations Research*, Vol. 37, pp. 925–936.
5. Pinedo, M., Wolf, B., and McCormick, T. (1986). Sequencing in a Flexible Assembly Line with Blocking to Minimize Cycle Time. In: *Proceedings of the Second ORSA/TIMS Conference on Flexible Manufacturing Systems*, edited by K. Stecke and R. Suri, Elsevier, Amsterdam, pp. 499–508.
6. Rachamadugu, R., and Stecke, K. (1987). Classification and Review of FMS Scheduling Procedures, *Working Paper 481*, University of Michigan, Graduate School of Business Administration, Ann Arbor.
7. Wittrock, R. (1985). Scheduling Algorithms for Flexible Flow Lines, *IBM Journal of Research Development*, Vol. 29, No. 4, pp. 401–412.

CHAPTER NINETEEN

Navigation Problem in Manufacturing

JOHN CESARONE

Department of Mechanical Engineering, University of Illinois at Chicago

19.1. INTRODUCTION

Navigation is the process of combining *motion* and *intelligence* in such a way as to result in goal-oriented transportation. This implies that some physical material (machine, personnel, equipment, or information) occupies some initial location, and there exists a desire for it to occupy some different, perhaps imperfectly known or defined, location. There is also a set of constraints on how this disparity between current and desired system states may be resolved: limited transportation mechanisms, maximum velocities and local capacities, limited time and fuel, constrained pathways, scarce transportation resources, imperfect information concerning the current state of the system, and computational limits on the system's decision-making resources. Reconciling these constraints with the transportation goal is the basis of the navigation problem.

The navigation problem in manufacturing takes this general scenario and gears it toward a manufacturing setting. The material to be transported may be raw manufacturing material, work in process, completed product for inventory or shipping, packing material, tools, fixtures, or data storage media. The available transportation mechanisms may be conveyor systems, AGVs, manned vehicles, or gantry systems. They may be as simple as sending a human worker to the destination with the material in his/her hand, or as complex as a computer-generated route downloaded automatically to a radio-controlled AGV. The problem may be as well posed as moving one item from one fixed location to a fixed destination, or as open-ended as getting the nearest available spare part to the most critical down machine with the least disruption to an already overburdened trans-

Intelligent Design and Manufacturing, Edited by Andrew Kusiak.
ISBN 0-471-53473-0 © 1992 John Wiley & Sons, Inc.

portation network. The intelligent agent solving the navigation problem may be a shop supervisor telling the workers what to do, or a supervisory mainframe computer directing a large number of cell controllers, AGVs, and automatic storage/retrieval systems.

A variety of approaches to solving these types of problems have been documented. Many have come out of the manufacturing research community, while some come from pure mathematics and operations research, and others from the artificial intelligence and expert system arenas. These different approaches can roughly be classified along a continuum or spectrum that describes the degree of intelligence utilized. At one end of the spectrum are the strictly algorithmic techniques, which yield exact solutions and perfect optimization. Unfortunately, these techniques can only be used in situations with a relatively small number of variables and relationships. As the complexity of the problem increases, the computational burden of the solution increases much faster, often geometrically or exponentially. Even medium-sized problems often become unsolvable by these techniques.

At the other end of the spectrum are the purely heuristic approaches. These replace the strict mathematics and guaranteed optimality with less precise, but more intelligent, methods. The solutions are often suboptimal, but they are attainable in a reasonable amount of time, even for highly complex problems.

In-between these two extremes are the hybrid techniques, which use both algorithms and heuristics to fashion as ideal a solution as possible, or necessary, in a reasonable amount of time. In the following sections, approaches to the navigation problem from many points on this spectrum are described.

19.2. IMPORTANCE

Navigation is often considered a secondary aspect of the manufacturing endeavor. It does not directly add value to a product or work in process, nor does it have a direct impact on the bottom line. It is an element of overhead, such as rent or utilities. It is nevertheless a crucial part of the manufacturing operation, and becoming increasingly more so.

The reason for this is the trend toward simultaneous or concurrent engineering. One of the tenets of the concurrent engineering environment and the CIM revolution is that *information* is being processed as much as is raw material. This information includes facts about the current state of the manufacturing system, as well as plans to be implemented, and the anticipated future states of the system. The more powerful and far-reaching these plans, the more flexible and efficient the factory may become.

Navigation enables several aspects of this flexibility: It enhances the ability for materials, tools, and information to be where they can do the most good at the best time, and it allows for more complex, comprehensive, and useful plans to be implemented, thus increasing the flexibility of the entire operation. Automation and adaptability necessarily increase with navigational competence, bringing the factory closer to the goal of total integration.

19.3. PROBLEM DEFINITION AND ISSUES

The navigation problem is a complex one, impacting many areas of the factory and business operations. Consideration must be given to physical transportation devices, information processing capability, coordination of resources, and control and command schemes. One way to begin to understand the issues involved in the navigation problem is to break it down into subproblems. One rather arbitrary, yet useful, means of subdividing the issues is to consider navigation a combination of sensing, knowledge representation, search, and implementation.

Sensing can be defined as the combination of techniques that give the navigation system its understanding of the current system state. To make plans, the navigation system must have access to information concerning locations of material to be transported, locations of the devices to perform the transporting, availability of loading and unloading devices, and the destination of the material. It may also need to be constantly monitoring transportation pathways to avoid collisions, and monitoring system needs for possible changes in its mission due to unexpected system requirements or changes in priorities. The sensing problem is discussed in detail in the following section.

Knowledge representation can be defined as the method employed by the navigation system to store and organize its information about the factory once it has been acquired, either through sensing, original system definition, or updates from a supervisory information system. The information must be stored in a manner that facilitates planning and navigation and therefore must be organized in an intelligent manner. This is further explained in a following section.

Search is one method of finding a specific solution to a well-posed navigation problem. It is the final step, after sensing and knowledge representation, that results in the best possible navigation plan. The search method employed is intimately associated with the knowledge representation scheme in use, as is the type of plan generated. Various solution search techniques are presented later.

Finally, a navigation plan must be implemented. At first glance, this would seem a simple, nonintelligent process of translating the navigation plan into commands for the transportation actuators. However, the implementation method must be considered in the earlier phases of the problem. The knowledge representation scheme, the solution, and the plan for implementing the solution must all work together to ensure a plan that is feasible with the current, and available, hardware of the transportation system.

Filling in the details of this rather simple outline is, of course, a tricky process and dependent on a number of practical issues. One of the first issues to consider is the desired characteristics of the solution. Suppose, for instance, that the task under consideration is to load a pallet onto an AGV. This is a task that must be completed before other tasks (such as movement of the AGV) may begin, and therefore a solution is desired fairly rapidly. It is also a task that will be performed only once. This would indicate that there is no justification for careful calculations to determine the most time-optimal or energy-optimal method of performing it; the cost of the solution could easily outweigh the savings.

On the other hand, suppose the task is to manipulate components during an assembly operation; strictly, this is a navigation problem as well. However, it is a task that will be repeated many times. A small savings therefore, in time, energy, or control effort will have large benefits in the long run. In this situation, it is well worth expending the time and energy to develop a detailed and optimal navigation strategy.

Most navigation problems fall somewhere in the middle of this particular spectrum. For a savings in each repetition of the task, there is an additional cost in terms of working out a more detailed solution. In each case, it is important to consider how crucial optimality in the solution will be and to select an appropriate solution methodology in light of this engineering trade-off.

A similar issue, in terms of desired characteristics of the solution, is that of *short* solutions versus *safe* solutions. In some problems, the most important goal is to deliver the traveling material to its destination as quickly as possible, or by covering the least amount of distance. This is usually the case in time-critical or energy-intensive situations. Unfortunately, the shortest possible paths usually skim very close to walls or obstacles in the path, and this can constitute a danger. Sometimes it is far more important to find the safest possible path than the shortest, and to avoid walls and obstacles by the largest possible margin. These types of paths tend to stick to the centers of free spaces. A path of this type is most useful when information on obstacle location and size is poorly known or unreliable, when obstacles are in motion, or when payloads are particularly delicate.

This leads directly to the next issue to be considered: the completeness of the knowledge base for the problem. Certain facts will be known: dimensions and locations of objects and obstacles in the workzone; requirements, capabilities, and availabilities. These same types of data may also be unknown: varying sizes of workpieces, uncertainty in locations, unexpected obstacles, unpredictable or stochastic completion times for tasks, and so on. In selecting a methodology to solve a particular navigation problem, the completeness level of the knowledge must be considered. Some approaches are only valid for *deterministic*, or completely predictable, problem situations, while others can make allowance for probabilities of unknown factors. Still others take a hybrid approach, designing a deterministic solution to the most likely situation and modifying the solution on the fly as deviations are detected.

A further issue in a discussion of navigation problems is the level of detail desired in a solution. Most problems exist at many levels, all of which must ultimately be solved for a practical result to be achieved. In navigation literature, the modules for solving the various levels are often referred to as the *mission planner, navigator*, and *pilot*. Most of the solution methods discussed later focus on only one of these levels.

The mission planning level is the most intelligent level, and also the least likely to be automated. Solutions to this level of the navigation problem come almost exclusively out of the artificial intelligence research community, using expert system and automated reasoning techniques. Special computer programs, called *planning systems*, are used to plan tasks in terms of high-level subtasks and focus on such issues as sequencing, concurrency, and necessary and sufficient conditions. Outputs of planners are mostly logical in nature and consist of a sequence of simple commands. This is discussed toward the end of Section 19.6.

The navigation level is the most thoroughly researched of the three levels, especially by the engineering community. It focuses on planning the actual trajectory of the materials to be transported and is mostly geometric in nature. Outputs may be a series of *waypoints*, or intermediate locations on a trajectory, or may be a sequence of direction and velocity commands. These are discussed first in Section 19.6.1.

The pilot level of the navigation problem is not discussed in this chapter, as it is the least intelligent, and best understood, part of the overall problem. The pilot level implements the decisions of the navigator, just as the navigator implements the decision of the mission planner. The outputs of the pilot system are commands to the actuators of the navigation system: AGV drive systems, robot joint actuators, and other motion control servos.

As a final note to this introductory section, it should be recognized that most of the research in manufacturing navigation is performed at a simulation level only. While many detailed plans are designed, analyzed, and tested, much fewer are actually implemented in a physical laboratory or a working factory. Two basic factors contribute to this lack. One is the expense of physical implementation, in terms of time and equipment, as well as that of dedicating an expensive production facility to a research effort. The other factor is the fact that most navigation research has no *need* for physical implementation to demonstrate its value. Since navigation is primarily a matter of decision-making, the decisions tend to speak for themselves. In situations where the value of a method cannot be seen directly, discrete event simulation is generally sufficient to prove the efficacy of a navigation algorithm.

19.4. SENSING

As defined previously, sensing is the function that enables a navigation system to acquire information about the manufacturing system in which it operates. The navigation system requires this information to construct and control plans of action. It requires large amounts of past information to generate an overall plan, and it may require smaller amounts of constantly updated information so that the plan can dynamically evolve as the system state changes over time.

Many technologies are available for this sensing function. The simplest, which is sometimes sufficient, is to do no sensing at all. The navigation system can merely accept that the factory is configured as it should be, or as some supervisory computer claims it to be. A human operator/planner or the supervisory computer might command some component of the navigation system to pick up a pallet of parts in one location and take it to another location via a standard pathway. The navigation system might select the nearest AGV, route it to the specified work cell, and begin the loading operations, assuming the material is actually there. The rest of the transportation task is also completed in this same blind-faith manner. As long as nothing goes wrong, and the navigation system is not called on to perform impossible or unwise tasks, this open-loop control scheme can work perfectly well. However, it is not considered an ideal method in any but the simplest, and most predictable, scenarios.

To achieve a safer, more adaptable and efficient automated factory, it is better for the navigation system to be able to confirm its commands. This increases the robustness of the system, allowing for possible command errors to be detected and corrected, avoiding potential disaster. Sensing capability also allows for a more distributed command structure; a supervisory computer can command an AGV to a specific location with a general route, and the AGV can plan the details of its route as it travels. This frees up computing resources in the central computer, makes the AGV more adaptable, and allows for updates to the path when and if they are needed.

Sensing also alleviates the burden on the knowledge representation scheme, as discussed later. If all motion is to be conducted under open-loop control, then all obstacles must be incorporated into the motion plan. If the obstacles can be detected when they are approached, they need not be considered until collisions become imminent. In this way, reality itself bears part of the burden of storing the obstacle information.

Another tie between the sensing scheme and knowledge representation scheme is the form of the information to be sensed. The information acquired about the world must be in a form compatible with the computer's database of world information. For example, if the computer stores obstacle data in terms of edges and boundaries, the sensing system must detect the boundaries, possibly with a vision system. If the information is to be stored as range and direction data, ultrasonics may be more appropriate. If (x, y)

data on object location are most appropriate, an overhead sensing scheme may be the best approach.

One of the simplest sensing techniques is to sense *motion* rather than *location*. This is called *dead reckoning* and is often an appropriate approach when a mobile robot is the only moving object in the workspace. In this situation, the robot's own location is the most important information to determine.

Shaft encoders are one technology used for dead reckoning sensing. These are optical or electrical devices that count the number of revolutions, or fractions of revolutions, of an actuator shaft, such as an axle on a wheeled vehicle. This information and some simple mathematics yield the distance traveled by the robot or moving device. The advantage of encoders is their low cost, but the disadvantage is that they do not account for possible slippage in the drive mechanisms.

Other dead reckoning devices for measuring motion include accelerometers and gyroscopic inertial systems. These can provide more accurate information but tend to be too expensive for most applications.

Tactile, or proximity, sensing is the next most sophisticated method. A variety of technologies exist that allow a moving agent to detect an object when it has been reached. Electrical limit switches with probes are the simplest of these devices, although capacitance switches and electric eyes with light beams are also used. In each of these cases, the sensors are constantly asking the question, "Is there an object right here, right now?" and responding with a yes or no answer. There is no predictive capability or range finding. These types of sensors are acceptable for slowly moving vehicles in nondelicate environments only but have the advantage of being very inexpensive.

Still more sophisticated are rangefinding devices. The most popular of these is the ultrasonic sensor, which emits a short burst of high-frequency sound in a narrowly focused direction. The sound is reflected back by the nearest substantial object, and the reflected wave is detected. The time for the returning wave is directly related to the distance of the object, and the direction of the original emission gives the object's bearing. Accuracy of object location is on the order of a few inches, range of detection is great, and the information is acquired in a very short time. The ultrasonic systems are also quite inexpensive, which accounts for their popularity.

At the top of the sophistication scale is computer imaging technology, often called "computer vision." This is a combination of video technology and pattern recognition and therefore falls into the category of artificial intelligence. A video camera is used to produce a bit-map pattern of the scene to be sensed. This pattern will have some resolution, based on pixel density in the image, and some contrast level, based on the shades available in the video system. The vision system must then "extract" useful information from this bit-map, using various algorithms. A typical method is called edge detection, which scans the image for abrupt changes in shading and plots a line through the locations of these changes. These lines constitute the edges of some object in the field of view. The edges combine to form a shape, which the system compares to shapes of objects that it is expecting to detect. Research in pattern detection and vision algorithms is a continuing and exciting field, and a source of a vast body of literature (15, 19, 23, 33).

19.5. KNOWLEDGE REPRESENTATION

A navigation system is driven by knowledge. It must have knowledge about the manufacturing system that it serves, both long-term static knowledge such as locations of walls and machines, and short-term dynamic knowledge, such as current machine status, inventory levels, and resource utilization. It must have knowledge concerning its goals, both current immediate goals and long-range overall operational objectives. Furthermore, once the navigation system has formulated a plan, this plan constitutes a new piece of knowledge, which must be stored, both while it is in use and afterward for possible reuse.

While knowledge and data are both forms of information, there is an important distinction between them. To qualify as knowledge, information must be stored and structured in such a way as to make it useful by an intelligent algorithm or heuristic. It must be formed into a structure that is easily retrievable and directly applicable to the decision process which will be accessing it. *Knowledge representation* is the term for the manner in which the information is made useful.

The majority of the knowledge to be used by a navigation system will consist of the physical configuration of its domain. Other essential information will include goal locations, current locations, and the characteristics desired in the path to be generated to satisfy the commanded mission. The form of knowledge representation used will be directly dictated by the solution method employed, and thus this section is somewhat parallel to the following section on solution methods. However, it is instructive to consider the knowledge representation methods themselves first.

One of the most basic types of information that must be stored by the navigation knowledge base is the set of locations of positions to be avoided and the set of locations that can be traversed. This type of information is often called the *terrain map* of the navigation domain. In various scenarios, the navigation system may have a complete map of its domain, a partial map, or no map at all. In the latter case, part of its mission will be to map out the areas as motion occurs, utilizing its sensing capabilities. Given a map, the navigation system may or may not know the precise location of the moving object itself within the map, possibly requiring further sensing and analysis.

The map may be accepted as perfectly accurate, or it may include uncertainties. Strictly speaking, all physical measurements have some level of imprecision, so a better way of looking at this issue is in terms of accuracy versus resolution of the problem. If the navigation planner works with a resolution of 1 inch, map certainties of 1 inch or less can be considered perfectly accurate. Uncertainties greater than the desired resolution, in this case 1 inch, can similarly be thought of as imperfect knowledge.

So far, each terrain location has been considered either safe (meaning navigation through it is allowed) or unsafe (meaning that navigation through it is not allowed). However, this is often too simplistic a view to generate ideal navigation plans. Often there is a range of desirability to be considered, and any one location could rank at any level within this range. It is usually convenient to think of this level as a *penalty* level, with each location being assigned some penalty between zero and positive infinity.

One way of utilizing this penalty level method is to assign an infinite penalty to map areas that are absolutely forbidden. In this way, the solution method chosen to generate the path will avoid these areas at all costs. The very safest locations could be given a penalty of either zero or a very small positive number. The purpose of giving a small penalty to safe areas is to penalize distance, so that the cumulative penalty of a short path will be smaller than that of a longer, but equally safe, path. Areas that are less desirable to navigate through are given higher penalties, in proportion to the desire that they be avoided. In this way, these areas can be utilized if necessary but will be avoided if possible.

These penalties can reflect various aspects of each terrain location, depending on the desired solution characteristics. They could be proportional to uncertainty of the terrain knowledge, helping to steer the trajectories toward the well-known areas. They could reflect cost of traveling in each particular area, either in terms of fuel consumption over various types of surfaces or possibly in terms of disruption caused to other activities in the factory. Another use of these penalty levels is to set them to some function of the reciprocal of maximum speed in the area: In locations where rapid travel is possible, the penalty is low, while in locations where travel must be slower, penalties are higher. In this way, the overall path penalty reflects the time required to complete the path. In general, the penalty at any given location will often be some combination of these and other factors.

Once these regions are defined and penalty values set, they must be represented in a knowledge base in a form easily usable by the solution method. Many forms have been developed, each with their own sets of advantages and disadvantages. One of the simplest representation methods is the *regular grid*. In this rather obvious technique, the domain to be considered is divided into equally spaced segments in each dimension, as shown in Fig. 19.1. For each point in the grid, information about the terrain is stored. The navigation system plans paths through this grid, with transitions allowed only between adjacent grid points (6, 20).

The advantages of this regular grid method come largely from its simplicity and predictability. It is easily programmed, and its memory

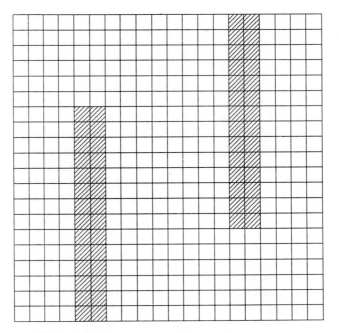

FIGURE 19.1. Terrain mapping with regular grid.

storage requirements are independent of the actual characteristics of the terrain. Paths through such a grid are easily solvable with operations research techniques, to be discussed in the next section. One disadvantage is that storage requirements and solution time can become quite large, especially in problems with more than two or three dimensions. The grid size must be chosen with care: Too coarse of a grid will not be able to represent all obstacles and pathways, yet too fine of a grid will require exorbitant computer resources. Whatever the grid size chosen, there will be a "digitization bias" in the system: Positions between grid points will not be allowed, and transitions at odd angles cannot be generated. For example, if the shortest path in reality would be a direct path at a 23° angle, the grid method may require a horizontal segment and a 45° diagonal segment, greatly increasing the path length. This is demonstrated in Fig. 19.2.

An enhancement of the regular grid is the irregular, or hierarchical, grid method. In a two-dimensional problem, this yields a "quad-tree" representation, or an "octree" in three dimensions. In the quad-tree method, the navigation domain is divided into four quadrants, as shown in Fig. 19.3, and each quadrant is examined. If a quadrant is composed entirely of free space, it is listed as such. If it is composed entirely of obstacles, this is also listed. If, however, it is composed of part free space and part obstacles, it is further divided into subquadrants. Each of these is in turn examined. This procedure is followed recursively until some final resolution is reached, with the

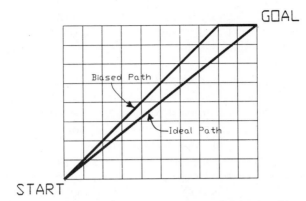

FIGURE 19.2. Path degradation from digitization bias.

smallest areas being listed as unsafe if they are not completely free of obstacles (16, 24).

The quad-tree method has the main advantage of reduced storage space for the same resolution as the corresponding regular grid. Alternatively, for the same storage requirements as the regular grid, vastly greater resolution can be achieved. The quad-tree suffers most of the same disadvantages of the regular grid, such as digitization bias, with the added problem of more

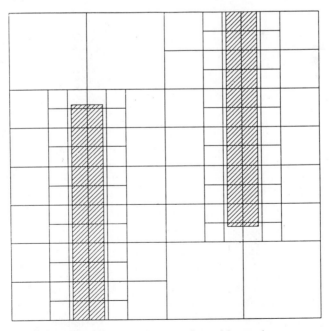

FIGURE 19.3. Terrain mapping with quad-trees.

complex solution strategies. It should also be noted that the storage requirements for a quad-tree are highly dependent on the number, size, and shape of the obstacles in the domain, and also on the orientation of the original quadrants.

The terrain need not be digitized to be represented, of course. Many methods store exact locations of obstacles and free spaces. These are often called *boundary representation* or *B-Rep* methods. A straightforward B-Rep technique is to store the geometric locations of either the edges or corners of each obstacle in the domain. Various algorithms are available to plan paths through B-Rep terrains. One constraint of this approach is that obstacles must be polygonal in shape. However, they are precisely defined in terms of size and location, and paths do not suffer from any digitization bias. Paths tend to be of the "taut-string" variety, being stretched from one obstacle edge or corner to the next.

A similar method is to store free spaces, rather than obstacles, as shown in Fig. 19.4. These can be described in various ways, such as *meadow maps* (1) and *generalized cylinders* (3, 18). Another option is the *Voronoi* diagram, the locus of points equidistant from the closest obstacles (4). Solving these types of maps is a matter of selecting which free zones will be traversed. Paths generated tend to be of the safest rather than shortest variety, sticking to the centers of the safe zones.

Another interesting method is called the *visibility graph*. This is stored as

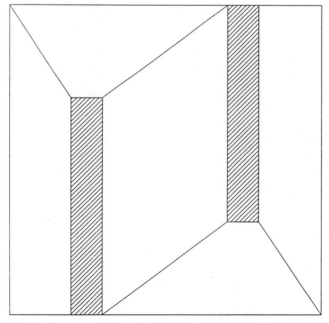

FIGURE 19.4. Terrain mapping with free-space map.

a linked list of obstacle vertices, once again assuming polygonal obstacles. For each vertex, a list is stored of vertices on other obstacles that can be seen (i.e., are not behind something). Paths are generated by traveling in straight lines from one vertex to the next. Shortest-path, taut-string solutions naturally result, just skimming the edges of the obstacles. This method is well suited to navigation coupled with knowledge acquisition, where a vision system is mapping the domain as it travels. Range and direction data for each visible vertex are added to the knowledge base as they are discovered.

19.6. SOLUTIONS

At this point, knowledge about the specific navigation problem has been acquired, organized, and stored in a format amenable to solution. To say that the problem has yet to be *solved* is somewhat misleading, as the previous steps are as responsible for the ultimate solution as is the final one. Some form of search, however, must be employed to select the best available navigation plan from those made possible by the previous steps of the process. This final "rabbit out of the hat" stage of the process is the subject of this section.

19.6.1. Exact Solutions

One class of solution methods is the strictly algorithmic or "exact" solutions. These use mathematical techniques to search all possible paths allowed by the knowledge representation and select the path that is optimal in terms of maximizing some objective function. The advantage of these techniques is that the solution is guaranteed to be the best one possible, within the context of the problem definition and the objective function. The disadvantage is that it cannot always be found, due to excessive computational burden. As such, these techniques are most appropriate for small-scale or low-dimensional problems, or in situations where there are sufficient time and need to justify a great search effort.

An interesting approach within the class of exact solutions includes those using direct geometry. As well as classical geometry, these techniques use algebra, algebraic geometry, computational topology, and combinatorics to generate solutions. Some techniques can be used in highly complex situations, where finding a feasible path is difficult, much less an optimal one. An example of this type of problem would be moving a large sofa through a small window, or a piano through a narrow doorway, where merely proving that it can be done is a difficult task. This class of problem is often called the "Piano Mover's Problem."

Geometric techniques begin by defining the degrees of freedom in the problem. For the simple case of a disk moving in a plane, this would be two:

translation in the x and y directions. Changing the disk to a polygon, such as a rectangle, adds a third degree of freedom to the problem: rotation of the polygon. A polyhedron moving in space would have six degrees of freedom: three translational and three rotational. A six-link robot arm bolted to the floor is another six degree of freedom problem: Each link has one degree of motion. If the robot is free to move in space, the problem generalizes to 12 degrees of freedom, and so on.

If the problem can be said to have k degrees of freedom, we can say that a solution exists in some k-dimensional free space F. Describing F is a difficult mathematical proposition, but in general it can be defined by some large number n of inequalities. It can be shown that a geometric solution can be generated with a computational burden that is polynomial in n or double exponential in k (26).

Obviously, this can get out of hand fairly easily. Geometric solutions are primarily of theoretical interest in proving averages and upper limits on computational burdens, and for designing submodules for more useful heuristic techniques. The procedures are fascinating but too lengthy to reproduce here. The interested reader is referred to Schwartz and Sharir (25) for a polygon moving in two dimensions, to Schwartz and Sharir (27) for three disks in two dimensions, to Sharir and Ariel-Sheffi (30) for multiarm linkages in two dimensions, and to Schwartz and Sharir (28) for motion of a "rod" (rectangle of finite length and insignificant width) moving in three dimensions.

19.6.2. Heuristic Solutions

The next class of solutions are those that use some sort of nearly exact mathematical search technique to generate the navigation plan. These are based more on the knowledge representation methods of the previous section than are the geometric techniques. The approximations involved in the knowledge representation method are what makes the solutions inexact and are also what make the solutions more feasible to calculate. These methods employ an operations research or heuristic search method to find the best of the available paths through a regular grid, quad-tree, visibility graph, or other knowledge base.

Finding an optimal or near-optimal path though a grid or graph of some type is mathematically described as the task of navigation through a *transition network*. A transition network is a series of *nodes* and *arcs*, as shown in Fig. 19.5. The nodes represent allowable system states; in this case, they are locations in space where the moving object may alter its path. Arcs are the allowable transitions between nodes; in a navigation system, they represent the path segments that may be used to make up the total path.

This concept is applicable to each of the knowledge representation method presented previously. In the regular grid case, each grid point is a

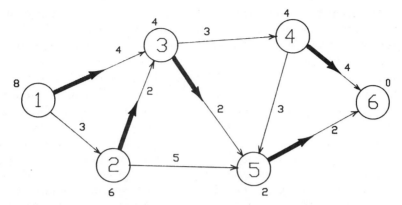

FIGURE 19.5. Transition network, with costs and solution.

node, and each node is connected by arcs to the adjacent grid points. If diagonal transitions are included, there will be eight arcs leaving each node in a two-dimensional grid; if diagonals are not allowed, there will be only four arcs per node, two in the x direction and two in the y direction. Paths generated will hop from node to node, traversing the area one grid point at a time.

In the vertex graph representation, each obstacle vertex represents a node. Each node is connected to every other node that can be viewed from it. In this representation, the number of arcs per node is highly variable. The length of each arc is also variable. In a Voronoi diagram, the arcs are the free-space line segments, and the nodes are their intersection points.

In all these representation methods, the resulting transition network can be thought of as a road map, with many roads connecting many cities. The navigation search procedure's task is to select the roads to take, and the cities to pass through, on the way from the current city to some distant goal city. Many approaches to this task exist in the operations research literature, such as dynamic programming and the "traveling salesman" problem (5, 6, 14, 22).

One of the most popular methods for finding a path through a transition network, and for solving many other tasks in artificial intelligence research, is called the A* search (13, 21). This is a heuristic technique and therefore *not* guaranteed to always find the optimal solution; however, it is a very good heuristic and is generally considered optimal enough in the vast majority of situations. The A* search algorithm recursively examines each node in the network to determine if that node belongs on the optimal path from the starting node to the goal node. This examination is accomplished by calculating an evaluation function, f^*, for the node under consideration. f^* is composed of two parts: an exact part, g^*, and a heuristic part, h^*. g^* is the exact value of the length of the best path from the start to the node currently under consideration, as previously determined in earlier iterations of the algorithm. h* is some sort of heuristic estimate of the length of the best path from the current node to the goal node. Examination of nodes proceeds from those adjacent to the start node and fans out into the network, until the goal is reached.

Under certain conditions, the A* algorithm is guaranteed to find the optimal path to the goal. These conditions can generally be met by judicious selection of the heuristic function, h*. In navigation work, with a cost based primarily on distance traveled, the usual function for h* is merely the remaining geometric distance from the current node to the goal node, ignoring the rest of the network and its constraints. Other heuristics are used in specific situations, where other costs outweigh the distance criterion, but these are based on experimentation for each specific instance (8, 17).

Whatever the heuristic or search method used, the result is a sequence of nodes in the transition network which comprise the optimal path to the goal. The navigation planner stores this sequence of nodes or locations and submits them to some lower level "pilot" program for translation into motion commands. It is desirable to be able to generate this type of plan in a short amount of time, because this allows it to be replanned should conditions change. For example, if an unexpected obstacle is encountered, the map of obstacles is updated, the current position becomes the new starting point, and a new plan is generated to account for the new system state.

19.6.3. Knowledge-Based Solutions

These heuristic approaches described previously are a definite improvement, in terms of computational efficiency, over the exact geometric methods, even at the cost of some optimality. However, they can still be quite intractable when dimensions get high and domain size is large in comparison to the precision required in the solution. Adding a dimension of artificial intelligence to the approach, using knowledge-based planning systems, brings the solution further up the scale toward complete system intelligence. Of course, optimality again suffers, but not beyond a reasonable amount.

General purpose planning systems have been a fertile research topic within the artificial intelligence community since the 1960s. They have focused on designing methods to plan logical sequences of operations in such a way as to meet a goal in an efficient manner. They deal with such issues as timing, sequencing, causality, constraints, and cooperation and interference between independent agents. They are not constrained to the largely geometric domain of the navigation problem but include the navigation scenario as a subset.

Planning systems, like the heuristics mentioned previously, deal with an initial system state, a goal state, and allowable transitions between states.

However, the transitions are now called *operators*. Each operator has a set of *preconditions*, which describe what the system state must be for the operator to be applied, and a set of *effects*, which describe how the operator can change the system state.

The task of the planning system is to determine a sequence of operators that will drive the system from the initial state to the goal state. Again, this is a matter of search. Since the domain is generally logical, rather than mathematical or numerical, heuristic logic techniques are used rather than operations research techniques.

A prototype planning system problem is the *blocks world*. In this situation, the domain consists of a set of blocks on a table top. Each block has a label, such as A, B, and so on. A block may be on the table or on another block. Operators include lifting a block from whatever it is on and placing it on the table, or on any block that currently has nothing on top of it. The initial state may be to have block A on top of block B, which is on top of block C; the goal state may be to have this order reversed.

Most planners begin by organizing a *search space*, or tree of sequences. If the goal is attainable at all, then at least one sequence in this tree will lead to the desired goal. Of course, this tree could be enormous, so that heuristic methods must be used to search it. The efficiency of these heuristic methods and their ability to minimize the search without missing the solution determine the power of the planning system.

Some planners use a state space approach to this organization and search through sequences of system states. More recent planners, however, have tended to organize the search space from the viewpoint of partial plans. That is, rather than search from one state to the next as in the A* search of the previous section, they search from one partial solution to the next, attempting to find a solution that reaches the prescribed goal.

Once organized, the search space must somehow be cut down to a manageable size, through some sort of *pruning*. Pruning can be based on eliminating low-priority goals or states, by considering how applicable partial plans are to the ultimate goal, and by eliminating plans that seem to be moving in the wrong direction. Much research is directed at intelligent search space pruning.

After reducing the size of the search space, some sort of intelligent search procedure is executed. Many are available from the artificial intelligence literature. One is depth-first backtracking, in which decisions are made sequentially, but the search state is saved whenever a decision is made. If a plan is found to fail, the state at the most recent decision is retrieved, and another choice is tried. Another is the beam-search method, which compares alternate solutions to specific subproblems, saving the most promising before going on to another subproblem.

Many of these techniques exist and are elaborated in the AI literature (2, 7, 9). Deciding which is best for a specific application is a highly domain-dependent issue. Since most decisions by a planning system are guesses, they must be trained to guess intelligently. The best guess is the one with the highest probability of being correct, and this will depend on the specific situation. Attempts to design "general purpose" planners, independent of the type of task to be planned, have met with limited success.

Applying these planning system approaches to the navigation problem in manufacturing consists of three issues: defining the possible factory conditions in terms of states; defining changes in factory conditions as operators, each with appropriate preconditions and effects; and determining appropriate heuristics for pruning and searching the search space, based on the probabilities of each operator being appropriate in each situation.

A planning system for factory navigation might be set up as follows: The locations of fixed machines and equipment must be known but do not affect the state of the system, since they do not change. They impact the system state only with reference to if they are in use, available, or in need of service. The AGVs, pallets, and other material handling devices impact the system state in terms of their utilization and location. Operators are commands that can be issued by the central computer controller and that initiate a task by one of the material handling devices. The preconditions for any one operator are that the device in question be available and functional and that the path it must take be clear. The effects are changes in location, loading and availability of the material handler, and changes in utilization of the fixed device it will be serving.

A famous example of manufacturing navigation by a planning system was the robot Shakey at Stanford Research Institute in the late 1960s. This system used the planning system called STRIPS (11). Further details of this and other planning systems can be found in Drummond and Tate (9), Fahlman (10), Fox et al. (12), Siklossy and Dreussi (21), and Tate (32).

As can be seen, the plans coming out of this planner are of a higher level than the physical navigation methods. These are more in the "mission planner" range. However, high-level motion commands can also be considered, such as the order in which to visit a number of cells calling for service. The planner can generate a cell sequence to minimize total distance traveled or total fuel consumed, or possibly a plan designed to minimize total due-date violations of scheduled completion times.

One of the most powerful facets of the planning system approach to the navigation problem is its ability to handle logical concepts. This allows the navigation problem to be combined with the other problems of factory automation, such as process planning, production scheduling, equipment selection, maintenance and downtime scheduling, and other tasks that must be performed in a logical sequence. Each of these tasks can be constructed as operators for the planning system, and total system performance can be planned at once. This leads to more complete computer integration of the manufacturing enterprise and more efficient overall performance.

19.7. THE FUTURE

The future is bright for navigation system in manufacturing. Much research is being conducted at all levels, from vision and sensing systems, and knowledge representation, to algorithmic and heuristic search procedures and AI planning systems. University and industry laboratories are doing more to bring experimentation out of the simulation arena and into actual hardware.

As individual technologies advance, navigation systems will become more powerful. As computers become smaller and faster, more complex plans can be computed and implemented in realistic amounts of time. As video hardware becomes cheaper and more effective, sensing will become more powerful. As material sciences advance, material handling will take less time and energy.

All of this is very fortunate, because as the state of the art in manufacturing navigation and its component technologies advance, so does the need for it. Greater navigational ability and quicker planning make a factory more flexible. This cuts leadtime, increases throughput, and enhances quality, making a factory more competitive in the world market. Just-in-time manufacturing control especially becomes more feasible as navigation ability increases, and small quantities of material can be where they are needed at the proper time.

Finally, manufacturing management must realized the importance of the navigation problem in manufacturing. While material handling and navigation hardware can properly be listed as overhead operating expense, it is an important piece of overhead and has great impact on the efficiency and flexibility of the factory and its ability to adapt over time. The navigation system should be overdesigned and over budgeted, as it will be one of the tightest bottlenecks to occur as a highly automated factory is pushed to the limits of its productivity.

REFERENCES

1. Arkin, R. (1989). Navigational Path Planning for a Vision-Based Mobile Robot, *Robotica*, Vol. 7, pp. 49–63.
2. Barr, A., and Feigenbaum, E. (1982). *The Handbook of Artificial Intelligence*, Vol. 3, William Kaufmann, Los Altos, CA.
3. Brooks, R. (1983). Solving the Find Path Problem by Good Representation of Free Space, *IEEE Transactions on Systems, Man, and Cybernetics*, Vol. 13, pp. 190–1197.
4. Canny, J. (1985). A Voronoi Method for the Piano-Movers Problem, *Proceedings of the IEEE International Conference on Robotics and Automation*, St. Louis, MO, pp. 530–535.

5. Cesarone, J., and Eman, K. F. (1988). Manipulator Collision Avoidance by Dynamic Programming, *Proceedings of the 16th NAMRC*, Champaign, IL, pp. 328–335.

6. Cesarone, J., and Eman, K. F. (1989). Mobile Robot Routing with Dynamic Programming, *Journal of Manufacturing Systems*, Vol. 8, No. 4, pp. 257–266.

7. Charniak, E., and McDermott, D. (1985). *Introduction to Artificial Intelligence*, Addison-Wesley, Reading, MA.

8. Chattergy, R. (1985). Some Heuristics for the Navigation of a Robot, *International Journal of Robotics Research*, Vol. 4, No. 1, pp. 59–66.

9. Drummond, M, and Tate, A. (1990). AI Planning. In: *Knowledge Engineering, Vol. I: Fundamentals*, McGraw-Hill, New York.

10. Fahlman, S. E. (1974). A Planning System for Robot Construction Tasks, *Artificial Intelligence*, Vol. 5, pp. 1–49.

11. Fikes, R. E., Hart, P. E., and Nilsson, N. J. (1972). Learning and Executing Generalized Robot Plans, *Artificial Intelligence*, Vol. 3.

12. Fox, M. S., Allen, B., and Strohm. G. (1981). Job Shop Scheduling: An Investigation in Constraint-Based Reasoning, *Proceedings of the International Joint Conference on Artificial Intelligence—1981*, Vancouver, British Columbia, Canada.

13. Hart, P. E., Nilsson, N. J., and Raphael, B. (1968). A Formal Basis for the Heuristic Determination of Minimum Cost Paths, *IEEE Transactions on Systems Science and Cybernetics*, Vol. 4, No. 2, pp. 100–107.

14. Hillier, F. S., and Lieberman, G. J. (1990). *Introduction to Mathematical Programming*, McGraw-Hill, New York.

15. Horn, B. K. P. (1986). *Robot Vision*, McGraw-Hill, New York.

16. Kambhampati, S., and Davis, L. S. (1986). Multiresolution Path Planning for Mobile Robots, *IEEE Journal of Robotics and Automation*, Vol. 2, No. 3, pp. 135–145.

17. Koch, E., Yeh, C., Hillel, G., Meystel, A., and Isik, C. (1985). Simulation of Path Planning for a System with Vision and Map Updating, *Proceedings of the IEEE International Conference on Robotics and Automation*, St. Louis, MO, pp. 146–160.

18. Kuan, D. T., Zamiska, J., and Brooks, R. (1985). Natural Decomposition of Free Space for Path Planning, *Proceedings of the IEE International Conference on Robotics and Automation*, St. Louis, MO, pp. 168–173.

19. Marr, D. C., and Hildreth, E. C. (180). Theory of Edge Detections, *Proceedings of the Royal Society of London*, Vol. 207, pp. 187–217.

20. Mitchell, J., and Keirsey, D. (1984). Planning Strategic Paths Through Variable Terrain Data, *Applications of Artificial Intelligence*, SPIE Vol. 485, pp. 172–179.

21. Nilsson, N. J. (1980). *Principles of Artificial Intelligence*, Tioga Press, Palo Alto, CA.

22. Parodi, A. M. (1985). Multi-Goal Real-Time Global Path Planning for an Autonomous Land Vehicle Using a High-Speed Graph Search Processor, *Proceedings of the IEEE International Conference on Robotics and Automation*, St. Louis, MO, pp. 161–167.

23. Rao, A. R., and Jain, R. (1988). Knowledge Representation and Control in Computer Vision Systems, *IEEE Expert*, Vol. 3, No. 1, pp. 64–79.

24. Sammet, H., and Webber, R. E. (1985). Storing a Collection of Polygons Using Quadtrees, *ACM Transactions on Graphics*, Vol. 4, No. 3, pp. 182–222.

25. Schwartz, J. T., and Sharir, M. (1983). On the Piano Movers Problem I: The Case of a Two Dimensional Rigid Polygonal Body Moving Amidst Polygonal Barriers, *Communications on Pure and Applied Mathematics*, Vol. 36, pp. 345–398.

26. Schwartz, J. T., and Sharir, M. (1983). On the Piano Movers Problem II: General Techniques for Computing Topological Properties of Real Algebraic Manifolds, *Advances in Applied Mathematics*, Vol. 4, pp. 298–351.

27. Schwartz, J. T., and Sharrir, M. (1983). On the Piano Movers Problem III: Coordinating the Motion of Several Independent Bodies, *Robotics Research*, Vol. 2, No. 3, pp. 46–75.

28. Schwartz, J. T., and Sharir, M. (1984). On the Piano Movers Problem V: The Case of a Rod Moving in Three-Dimensional Space Amidst Polyhedral Obstacles, *Communications on Pure and Applied Mathematics*, Vol. 37, pp. 815–848.

29. Schwartz, J. T., and Yap, C. (Eds.) (1987). *Algorithmic and Geometric Aspects of Robotics*, Erlbaum, Hillsdale, NJ.

30. Sharir, M., and Ariel-Sheffi, E. (1984). On the Piano Movers Problem IV: Various Decomposable Two-Dimensional Motion Planning Problems, *Communications on Pure and Applied Mathematics*, Vol. 37, pp. 479–493.

31. Siklossy, L., and Dreussi, J. (1975). An Efficient Robot Planner that Generates Its Own Procedures, *Proceedings of the International Joint Conference on Artificial Intelligence—1975*, Palo Alto, CA.

32. Tate, A. (1984). Planning and Condition Monitoring in a FMS. In: *International Conference on Flexible Automation Systems*, Institute of Electrical Engineers, London.

33. Torre, V., and Poggio, T. A. (1986). On Edge Detection, *IEEE Transactions PAMI*, Vol. 8, No. 2, pp. 147–163.

CHAPTER TWENTY

Intelligent Control of Manufacturing Systems

SANJAY B. JOSHI and JEFFREY S. SMITH

Department of Industrial and Management Systems Engineering,
Pennsylvania State University, University Park, Pennsylvania

20.1. INTRODUCTION

Computer integrated manufacturing (CIM) has existed, in concept, for several years. While it is still generally believed that CIM can have a major positive impact on productivity, it is far from commonplace in today's factories. The high costs of software development, maintenance, and integration are among the most prominent reasons for our slow evolution to CIM. This chapter discusses many of the software design and development issues associated with CIM systems. A CIM system integrates all the major functions within a manufacturing enterprise via the use of a computer system. These functions include design, planning and scheduling, manufacturing, and sales and marketing. The term *CIM architecture* is defined and an architecture for a shop floor control system (an integral part of a CIM system) is presented. Section 20.2 presents previous research on hierarchical and nonhierarchical control for manufacturing systems. Section 20.3 presents a theoretical basis for a shop floor control system (SFCS) along with the integration of such a system into an enterprise-wide CIM system.

CIM systems are being developed in response to the business trends toward shortened product life cycles and the need for reductions in the design-to-production transition period. These trends are forcing the manufacturing operations to become more flexible. Availability of timely information is one of the keys to achieving the necessary levels of flexibility. Designers need manufacturing information early in the design process to reduce the product lead time, manufacturing managers need sales and

Intelligent Design and Manufacturing, Edited by Andrew Kusiak.
ISBN 0-471-53473-0 © 1992 John Wiley & Sons, Inc.

marketing information in order to effectively schedule production, sales personnel need production and manufacturing information in order to accurately quote lead times, and so on. CIM systems are supposed to provide access to this information in a well-defined, automated fashion.

Integration is the key concept of a CIM system. Individual functions within an enterprise (e.g., sales, marketing, manufacturing, design) have been using computerized information systems for quite some time. However, these systems have existed independently of each other and therefore do not provide the cross-function information access or integration necessary to achieve the aforementioned flexibility objectives. Within the manufacturing function, this situation is commonly referred to as "islands of automation."

As one might expect, designing and developing a CIM system are extremely complicated tasks. It is highly unlikely that the tasks could be completed without significant planning and forethought. A *CIM architecture* is a plan outlining the design and functionality of a CIM system. Specifically, Andersen et al. (3) define a CIM architecture to be "a structural interrelationship of principles, functional models, and guidelines providing the framework needed for designing or modifying an integrated manufacturing system." Skevington and Hsu (33) provide the following definition of a manufacturing architecture:

> A manufacturing system architecture is the composite model of the virtual system configuration seen by each of the users (which may be internal or external). It is the specification of the functionality provided by the facility and information domains.

In general terms, Biemans and Blonk (6) state that "an architecture prescribes *what* a system is supposed to do, i.e., its observational behavior in terms of inputs and outputs and how these are related with respect to their time-ordering and contents."

These definitions illustrate that a CIM architecture is not the actual control software of a CIM system or even an implementation plan; rather, it is an overall framework that provides the basis for the development of an integrated system. The architecture provides the top–down view of the system, which is necessary during the bottom–up implementation of each subsystem to guarantee compatibility with the other subsystems.

An important distinction must be made between a CIM system and a shop floor control system (SFCS). Where a CIM system is an enterprise-wide system that provides information access, a SFCS is responsible for controlling the actual manufacturing operations. An effective SFCS is therefore a prerequisite for an overall CIM system. The majority of the work described in this chapter involves the development of a standard architecture for a SFCS. Shop floor control models are emphasized because, to date, a standardized SFCS model has not been developed and generally accepted.

20.2. BACKGROUND

The development of a standard CIM architecture has been a major research topic since the early 1980s (29). The majority of the developed CIM models follow some type of hierarchical model. These systems are based on the traditional hierarchical management systems in which the objectives at each level are broken down into more detailed objectives, each of which is the primary objective of one entity in the next lower level in the hierarchy. For example, a sales manager is responsible for all sales within an enterprise. Generally, the sales region is divided into districts, each of which is assigned to a district manager. Similarly, each district is further divided into sales territories, each with its own salesperson. In such hierarchical systems, each entity is controlled by a single higher level entity and, in turn, controls multiple entities in the next lower level. The hierarchical CIM models are discussed further in Section 20.2.1.

Recently, many researchers have investigated nonhierarchical CIM models. Such models are generally known as heterarchical. Typically, these systems are characterized by multiple, cooperating entities, which exhibit no control over any other entities. Each element in the system is a separate entity. The entities solicit and bid for services from the other entities in the system. These systems are similar to a competition-based, capitalistic economy, where entities freely compete for business and solicit services from the lowest capable bidder. Heterarchical CIM models are discussed further in Section 20.2.2.

20.2.1. Hierarchical CIM Models

In a CIM system that uses the hierarchical control model, the complexity of any one level can be controlled by the system designer. When an entity becomes so complex that it cannot be feasibly developed or maintained, the entity can be divided into a hierarchical set of entities. Figure 20.1 illustrates this principle using a machining workstation as an example. Figure 20.1a shows the workstation modeled as a single entity in the control system and

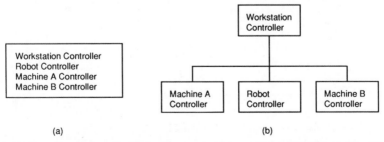

(a)   (b)

FIGURE 20.1. Example of splitting an entity into a hierarchical set of entities. (a) The workstation is modeled as a single entity. (b) The workstation is broken down into separate entities for each machine.

Fig. 20.1b shows the workstation modeled as a hierarchy where each machine in the workstation is modeled by an entity and all the machine entities are controlled by a single workstation entity. In Fig. 20.1a all the workstation control activities must be performed by the single controller. In a multiple machine workstation, the control and scheduling activities can get complicated. Therefore in Fig. 20.1b the activities have been broken down by machine. This frees the workstation controller from having to deal with the second-to-second control of each piece of equipment and allows it to concentrate on the workstation level activities such as scheduling and long-range planning.

The NIST Hierarchical Control Model. The most well-known hierarchical model for CIM was developed by the National Institute for Standards and Technology (NIST) at the Automated Manufacturing Research Facility (AMRF) (32). The AMRF is intended to be a fully functional flexible manufacturing system (FMS) designed to manufacture a complete family of parts identified through the use of group technology (12). Albus et al. (2) provide a detailed description of the underlying concepts of hierarchical control as applied to manufacturing systems.

The NIST hierarchy is made up of five levels or layers (see Fig. 20.2). Each controller in the hierarchy will have direct contact only with the next higher and lower levels in the hierarchy. Communication across multiple layers in the hierarchy will therefore be completed in multiple steps. Also, communication with other controllers on the same level must go through the first higher level controller common to both controllers. The term "controller" is used in this chapter to describe the software system used in a specific level of the hierarchical control structure. This should not be confused with the use of the term to describe the hardware system used to control a machine tool or robot. Each level will have a specific *planning horizon* going from several milliseconds at the lowest level to several months at the highest level. A planning horizon can be defined as the maximum

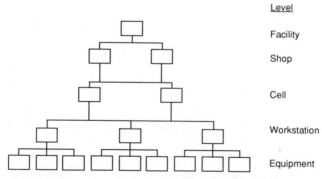

FIGURE 20.2. NIST five-level hierarchical control model.

amount of time for which the controller must plan and schedule the activities that fall under the domain of the controller. The individual levels in the NIST hierarchy are described in the following paragraphs.

The *facility level* is the highest level in the NIST hierarchy. It controls such long-range functions as cost estimation, inventory control, and labor rate determination. The facility level controller will also be responsible for long-range part selection decisions and associated machine selection decisions. The planning horizon for the facility level can be anywhere from several months to a year. Due to the nature of the task performed by the facility level controller, this level will have significant interaction with humans and other computer systems throughout the manufacturing organization.

The *shop level* is responsible for coordination activities between the manufacturing cells and allocating the required resources to the created cells. The NIST hierarchy uses *virtual cells* (17) or dynamically created cells to allow timesharing of workstations between cells. Therefore the virtual cell is not made up of a fixed group of workstations. Instead, workstations are allocated to the cell during shop level scheduling. The specific workstations assigned to a cell are determined by the parts being produced in that cell. Construction of the virtual cells is therefore one of the shop level controller's primary functions. Coordinating the materials handling requirements of the manufacturing cells will also be one of the primary functions of the shop level controller. The planning horizon for the shop level is anywhere between several weeks and several months.

The *cell level* is responsible for sequencing batch jobs through the workstations and supervising the activities of the workstations. This includes the determination of the necessary operations for each part and the allocation of the operations to the workstations making up the cell. Since the cells are virtual cells, the cell level controllers must be generic in nature in the sense that they can schedule and administer a variety of combinations of workstations. Since the workstations are dynamically allocated to the cells, the workstations in any given cell could be spread throughout the shop. Materials handling between the individual workstations will therefore be a significant responsibility of the cell controllers. The planning horizon for this level is anywhere between several hours and several weeks.

The *workstation level* controllers sequence and control the activities of the equipment controllers. A typical workstation in the NIST control hierarchy consists of a material handling robot, one or two logically connected machine tools, and a material storage buffer. The workstation controller determines the physical tasks required for each operation assigned by the cell controller, sequences these tasks on the machines in the workstation, and coordinates the material handling via the robot. The planning horizon for the workstation level is anywhere between several minutes and several hours.

The *equipment level* controllers are front-end computers for the machine

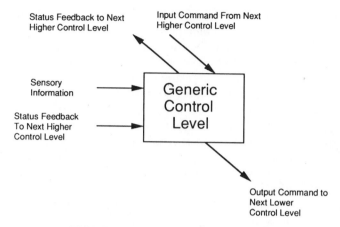

FIGURE 20.3. Generic control level.

tools and robots. They receive step-by-step commands from the workstation controller and convert them into the form required by the individual machine tools or robots. The equipment level controllers are also responsible for monitoring the machine tools they control and reporting their status back to the workstation level controller. The planning horizon for the equipment level controllers is between several milliseconds and several minutes (17).

Each level in the NIST control hierarchy operates with the same control logic. Figure 20.3 shows a generic control level with a schematic representation of the function of the controller. The commands entered from the next higher level in the hierarchy are broken down and interpreted. In conjunction with the sensory information and the status feedback from the lower levels, the controller determines the required actions and sends the required commands to the next lower level controllers (30). This type of operation has also been referred to as "sample inputs–process–generate outputs" operation (10). Utilizing this type of processing throughout the hierarchy creates a chain where the output of any controller becomes part of the input for the next controller in the hierarchy until, at the lowest level, the controller output goes directly into a machine tool or robot. For further descriptions of the AMRF and the NIST control hierarchy, see Barbera et al. (4), Fitzgerald and Barbera (9), Haynes et al. (14), McCain (24), and McLean et al. (25).

CIM–OSA. The European Community is also active in the development of a standard CIM architecture. CIM–OSA (computer integrated manufacturing–open system architecture) is a project of the ESPRIT program and is a consortium of 21 important European companies. Klevers (21) defines the objectives of CIM–OSA as follows:

The objective of ESPRIT project 688 "CIM–OSA" is to provide an open system architecture which covers all the information needs of all functions in a manufacturing enterprise. CIM–OSA provides a reference framework made up of two distinct parts: the CIM–OSA Reference Architecture and the CIM–OSA Particular Architecture.

The CIM–OSA Reference Architecture corresponds to the control architecture definition used in this chapter and the Particular Architecture includes the enterprise-specific information necessary for implementation of the architecture. Jackson and King (16) present a hierarchical reference architecture called ADEPT as part of the CIM–OSA project. The stated objectives of the ADEPT project include the development of a machine-independent model for fully integrated distributed plant control systems along with a fourth generation environment capable of supporting the implementation of such systems based on the model.

The control hierarchy is composed of three levels: factory, department, and cell. These levels roughly correspond to the top three levels in the NIST control hierarchy (facility, shop, cell). The model also calls for specific "observer" modules. These modules monitor the other system components, ensuring data and functional integrity. The observer modules provide the fundamental ideology behind the application fault tolerance. ADEPT also includes a system simulator to verify the design of the control system before actual implementation. The ADEPT system is currently under evaluation using a pilot system installed in a German printed circuit board (PCB) manufacturing plant.

20.2.2. Heterarchical Control Models

Hatvany (13) and Duffie and Piper (8) have proposed an alternative to the hierarchical control model called heterarchical control. A heterarchical control system is made up of multiple entities, none of which exhibits direct control of any others. Instead, the entities cooperate to meet the overall system objectives. The motivation for the heterarchical control system is the belief that supervisory decision-making should be located at the point of information gathering rather than in a central location (8).

Increased flexibility and improved fault tolerance are cited as the primary advantages of heterarchical control over hierarchical control. Duffie and Piper model each entity (including parts and machines) in the system with a running Pascal program. Part entities will request service and machine entities will seek clients. An agreement is made when a part requires processing and a capable machine is available. The important fact is that the agreement is made between the two entities (the part and the machine) only and does not involve a supervisory entity.

Since parts are modeled as running programs, all processing information for a part is stored within that part's program. Introducing a new part or machine to the cell can be accomplished by creating the entity's program. Modifications to the current system are not required. When the part is released, the program is started and will immediately start sending out request-for-processing messages. When a capable machine becomes available, it will respond by sending an agreement message to the part. Likewise, when a machine fails or is brought down for maintenance, there is no supervisory controller to modify. The entity will simply stop responding to request-for-processing messages.

20.3. THE PROPOSED SYSTEM ARCHITECTURE

A prototype hierarchical shop floor control system (SFCS) consisting of several levels is described in this section. The details of the cell, workstation, and equipment levels for a CIM system are also presented. System equipment includes NC machining centers, robot systems, AGVs, or any other computer controlled manufacturing equipment and related tooling resources. it is assumed that the cell controller periodically receives a list of jobs with associated due dates from a master production scheduler. For each job, a process plan will be retrieved either from a database or as a file transfer. The SFCS will determine the sequence of activities necessary to execute that process plan. This involves the selection of a specific process routing and activities, scheduling of all activities at both the cell and workstation levels, coordinating those activities across the equipment level, monitoring activities, job tracking, and some error recovery.

20.3.1. The Architecture

A "systems approach" to this problem is essential to achieve the level of integration that is required for CIM. A primary component of any system theory is the definition of the state of the system. For any complex system like manufacturing, the definition of the system state could involve thousands of variables, some discrete, some continuous, many stochastic. As the system evolves in time, those variables take on many different values. The goal is to define a set of decision rules that govern the way in which that evolution takes place. These rules can be generated "off-line" or "on-line" and are invoked to "control" how and when the system changes. The presented architecture is based on the work described in Joshi et al. (20), Joshi and Wysk (19), and Smith (34).

Since the complete problem is large and includes many complex interactions, a decomposition of the problem is necessary to create a series of well-defined solvable subproblems. The decomposition should result in the creation of components that perform tasks that are relatively independent within the context of the larger system. The need for decomposition creates the necessity of providing clear and well-defined "hooks" to integrate the various decomposed problems into an integrated whole. The ability to

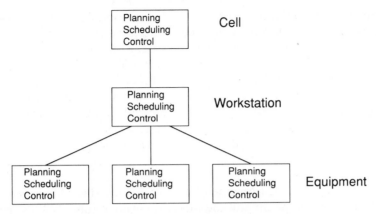

FIGURE 20.4. Proposed three-level hierarchy.

effectively decompose and define the necessary "functional hooks" provides scaleability to the architecture.

A three-level hierarchical decomposition of the SFCS is shown in Fig. 20.4. Each hierarchical controller module performs three main functions: planning, scheduling, and control. Typical architectural characteristics for each level in the hierarchy are shown in Table 20.1. Although the heterar-

TABLE 20.1. Typical Architectural Characteristics for Control Levels

Item	Equipment	Workstation	Cell
EXAMPLES Hardware	Lathe, Mill, T-10 Bridgeport Series I IBM 7545 Robot	Robot tended Machine Center, Cartrac Material Handling System	Variable Mission System, Several Integrated Workstations
Controller Hardware	Mark Century 2000, Accuramatic 9000, Custom-single-board system	Allen-Bradley PLC-4, IBM-PC/AT, etc.	VAX 11/750, SUN Workstation, etc.
Type Controller	Single-board processors, Machine tool controller, Servo-Controller, etc	PLC, PC Minicomputer	PC, Microcomputer Super-Minicomputer
Language Application	Assembler, Part programming, Robot programming, etc.	C, Lodder logic, Pascal and other sententual languages	C, LISP, FORTRAN and other high level languages
Memory/Size Requirements	8k - 128k RAM plus custom ROM, EPROM, etc.	256k - 1 Meg RAM 1 Meg - 80 Meg Hard drive	512k - 4 Meg RAM 10 Meg - 1 Gigabyte Hard drive
Response time	< 10exp(-3) sec	< 1 sec	< 20 sec
Machines/ Interconnects	1 - 1 connect	1 - many 1 - {1,8] Machine tools 1 - {1,50] Material handling	1 - many 1 - {1,15] Workstations

chical architecture described in Section 2 offers potential advantages over the traditional hierarchical control model in terms of flexibility and robustness, such systems lack the global view necessary for an integrated manufacturing system. For example, global optimization is difficult, if not possible, without exercising control over the job/machine/release-time decisions in a manufacturing system. Similarly, predictions of completion times, machine utilizations, and throughput are also very difficult in an environment where entities have no global knowledge. Jones and Saleh (18) also discuss the relative advantages and disadvantages of hierarchical and heterarchical control models.

The Equipment Level. Equipment controllers are the direct interfaces between the control system and the physical machine tools, robots, materials handling systems, and other shop floor equipment. Using dedicated equipment controllers allows the other levels of the control system to be developed independently of the specific brands and models of equipment being used. This independence is achieved by using generic control functions. The generic control functions used with a specific machine are determined by the type of machine. For example, consider the command dialogue between a workstation controller, a load/unload robot, and a machine tool required to load a part on the machine, process the part, and unload the part upon completion. Table 20.2 illustrates such a dialogue. The dialogue consists of the commands sent from the workstation controller to

TABLE 20.2. Command Dialogue Between a Workstation Controller, a Load/Unload Robot, and a Machine Tool

Robot Controller		Workstation Controller		Machine Tool Controller
		get status	==>	
			<==	idle, ready
	<==	move (part)		
move complete	==>			
		grasp(part)	==>	
			<==	fixture ready
	<==	release part		
gripper open	==>			
	<==	move (clear)		
move complete	==>			
		machine (part)	==>	
			<==	part complete
	<==	move (fixture)		
move complete	==>			
	<==	grasp(part)		
gripper closed	==>			
		release part	==>	
			<==	fixture ready
	<==	move(part)		
move complete	==>			

the equipment controllers for the robot and the machine tool and the respective responses to these commands. Note that the control dialogue contains exactly the same steps that a human operator would go through in a manually controlled workstation.

The control dialogue illustrated in Table 20.2 would be applicable to any workstation, regardless of the specific equipment, if the individual commands could be properly converted into the specific format required by the machines. This is the primary job of the equipment controllers. The equipment controllers are also responsible for interpreting the information returned by the machines and converting this information into the generic format used for return data. So, for example, although the *move part* command has the same meaning for all load/unload robots, the specific implementation of the command depends on the specific requirements of the robot being controlled. Table 20.3 gives examples of the generic control functions and return messages for use with machine tools, industrial load/unload robots, and automated storage and retrieval systems (AS/RS). Each equipment controller recognizes the set of generic commands and generates the corresponding return messages. The command recognizer software is therefore standard for all equipment controllers. The specific controllers call specialized subroutines to convert the generic commands into the format required by the specific equipment being controlled. These specialized subroutines are the equipment-dependent parts of the controllers. Therefore when a piece of equipment is replaced, only these subroutines will require modification.

Equipment modules are not limited to machine tools, load/unload robots, and AS/RS as listed in Table 20.3. An equipment module could be any piece of equipment that performs a specific set of tasks and can be used in conjunction with other equipment to form an autonomous workstation. For example, consider a vision-based inspection or coordinate measuring machine. The equipment would perform the function of inspecting parts and could be used with a robot to form an autonomous workstation. Note that the generic machine tool control functions listed in Table 20.3 would apply to an inspection station (with minor modifications to the return values to report the result of the inspection). The station must have some type of fixturing device that could be activated with the GRASP and RELEASE commands. The PROCESS command could provide all the operational information required by the individual device by using the calling parameter. Valid return values for the PROCESS command would include GOOD PART, REWORK, or SCRAP. Other examples of equipment level machines include production monitoring devices, packaging and palletizing equipment, pick and place equipment, and automatic cleaning devices.

The Workstation Level. The workstation level as defined in the hierarchy corresponds to several integrated pieces of equipment. At the simplest level, a workstation could be a robot tending a single machining center, along with

TABLE 20.3. Generic Control Functions for Equipment Level Controllers

Workstation -Equipment Commands	
Control Function	Response
Machine Tools	
GET_STATUS - Requests the operational status of the machine.	Block of data including information on the currently loaded part (if any), currently loaded NC file name, and currently loaded tools.
GRASP (part) - Instructs the machine to close the fixture or part chuck. The parameter part gives the part information required for flexible fixtures	FIXTURE CLOSED PART TYPE UNKNOWN ERROR
RELEASE (part) - Instructs the machine to open the fixture.	FIXTURE OPEN PART TYPE UNKNOWN ERROR
PROCESS (part) - Instructs the machine to start processing the part.	PROCESSING COMPLETE PART TYPE UNKNOWN MACHINE DOWN ERROR
STOP - Stops processing immediately	MACHINE STOPPED
Load/Unload Robots	
MOVE (part, loc1, loc2) - Instructs the robot to move part from location loc1 to location loc2. Part is defined for flexible grippers.	MOVE COMPLETED LOCATION INVALID PART TYPE UNKNOWN ERROR
MOVE (loc) - Instructs the robot to move to the location loc.	MOVE COMPLETED LOCATION INVALID ERROR
GRASP (part) - Instructs the robot to close the gripper.	GRIPPER CLOSED PART TYPE UNKNOWN ERROR
RELEASE (part) - Instructs the robot to open the gripper.	GRIPPER OPEN PART TYPE UNKNOWN ERROR
STOP - Stops the robot immediately	ROBOT STOPPED
Automated Storage and Retrieval System (AS/RS)	
RETRIEVE (part) - Instructs the system to locate part and bring it to the load/unload station. Part could represent finished goods or raw materials.	PART RETRIEVED PART NOT FOUND PART TYPE UNKNOWN ERROR
STORE (part) - Instructs the system to find part's storage location and either bring the location to the load/unload station or take the part to the storage location from the load/unload station.	PART STORED SPACE NOT AVAILABLE PART TYPE UNKNOWN ERROR
LOCATE (part) - Requests the current storage location for part.	PART NOT FOUND PART TYPE UNKNOWN ERROR
ALLOCATE (part) - Instructs the system to allocate space for part.	SPACE ALLOCATED SPACE NOT AVAILABLE PART TYPE UNKNOWN

the requisite fixtures, buffers, and sensors. A more complex workstation could be a single robot tending several machines. As an example, a rotational machining workstation could consist of a robot and several machines (NC turning centers, NC grinders, etc.) performing various rotational operations. The definition of what constitutes a *unit operation* determines the configuration, size, and individual characteristics of a workstation. The minimum requirement of a workstation is that a part be coupled and uncoupled with at least one piece of processing equipment.

Unit operations are defined in terms of the processing capabilities of an entity in the control hierarchy. For example, the unit operations of a NC turning center include turning operations. Similarly, the unit operations of a materials handling robot include moving parts within the work volume of the robot. The unit operations of a workstation include all the unit operations of the equipment within the workstation. Therefore the unit operations of a workstation containing a NC turning center, a vertical mill, and a load/unload robot include turning operations, milling operations, and part movement within the workstation. However, when the workstation is considered as an entity within the control hierarchy, the unit operations can be described in terms of the parts being processed. In other words, from the cell level, the processing operations of a workstation can be considered to be a single operation on the parts. Therefore the unit operations are hierarchical in nature and form a one-to-one mapping into the physical hierarchy of the control system.

As with the equipment controllers, the workstation controllers are controlled using a generic command set. Table 20.4 shows the workstation level command set developed for this control system. The main purpose of the SEQUENCE( ) command is to provide information to the cell controller to

TABLE 20.4. Generic Control Functions for Workstation Level Controllers

Cell - Workstation Commands	
Command	Returned Values
GET STATUS - Requests the current status of the workstation.	Block of information detailing the status of the workstation. Includes information on the current batch, expected completion time, and inoperative machines.
SEQUENCE (batch) - Requests the workstation to sequence the parts in batch. This function will be used by the cell controller when determining a viable batch configuration.	Runtime of the batch INVALID PART TYPE INVALID BATCH ERROR
MANUFACTURE (batch) - Requests the workstation to start processing batch.	Expected completion time INVALID PART TYPE INVALID BATCH ERROR

aid in the formation of part batches for the workstations. With this structure, the scheduling methodology of the cell level is independent of the scheduling modules used by the constituent workstations. When the cell controller is developing the batch sequence for each workstation, the controller will "ask" the workstation how long it will take to process a given batch. Based on the response provided by the workstation, the cell controller will either schedule the batches or develop another batch. The independence of the cell and workstation scheduling models is critical since cells with be constructed dynamically (virtual cells). This concept is analogous to the cell controller considering each workstation to be a "black box" with an associated processing time for each part. The sequence of operations within the workstation is therefore irrelevant to the cell controller. Instead, the cell controller is concerned with the total time the part is in the workstation.

The Cell Level. A cell is viewed as several integrated workstations, coupled by material transport workstations. A cell can be considered a virtual entity and exists to couple and uncouple the unit operations of the workstation level. Cells are constructed in real time based on the requirements of the parts being produced at the specific time of interest. The primary activity of the cell is to integrate workstation activities into a broader system.

Since cells are dynamically created to respond to the current part mix, workstations are not statically assigned to any one cell. This is a different view from the traditional manufacturing cell concept, where machines are *physically* grouped together. Although the basic idea of providing the equipment necessary to provide the processing for a part family is the same with the proposed architecture as with the traditional view, the equipment is grouped in a *logical* sense only under the proposed architecture. Significant materials handling capabilities will therefore be required and the control of those material handling services will be a major responsibility of the cell level controller.

Planning, Scheduling, and Control Functions. A key element of this architecture is that each level controller module performs the same functions: planning, scheduling, and control. For the proposed SFCS, *planning* is the activity responsible for selecting/updating the process plan or strategy to be used in executing assigned jobs and generating revised completion times for those jobs. Cell level process plans contain routing summaries for each job. Each workstation plan contains the operations to be done and the information necessary to generate part programs or their equivalent for each processing step. At each level, *scheduling* is responsible for evaluating candidate process plans and generating/updating a sequence of expected start and finish times (the schedule) for the activities in the selected plan. *Control* is responsible for interfacing with all subordinates and other applications and executing scheduler commands. It initiates start-up and shutdown, issues commands to perform assigned activities, uses feedback to monitor the execution of those activities, and oversees error recovery. Note that these functions are executed at different frequencies within the same level and across separate levels. Table 20.5 shows a functional breakdown of the activities at each level in the hierarchy.

Execution of commands by each controller occurs from a top–down execution. In case of failure and error recovery, the information and requests flow in a "bottom–up" manner, both within each controller and between the different levels as shown in Fig. 20.5.

The key to developing this type of distributed system lies in the generation of "robust schedules." Robust means that the schedule remains close to optimal (relative to some performance measures) over a realistic range of uncertainties. This means that the schedule is rigorous enough to absorb minor perturbations. A robust schedule will limit the number of reschedulings required when something goes wrong. Another important factor in the development of distributed systems is error recovery. In systems of this size errors will occur and must be handled without crashing the entire system.

TABLE 20.5. Typical Functions of Each Control Level

Functions \ Level	Equipment	Workstation	Cell
Planning	Tool selection, parameter specification, tool path refinement, G,M,T code, tool assignment to slots, job setup planning	-Resource allocation jobs -Batch splitting and equipment load balancing	Batching, Workload balancing between workstations, Requirements Planning -Task allocation to workstations
Planning Horizon	Millisecs - Minutes	Minutes - Hours/Days	Hours - Days/Weeks
Scheduling	-Operation sequencing at individual equipment	-Sequence equipment level subsystems -Deadlock detection and avoidance -Gantt chart or E. S. based scheduling -Buffer management	-Assignment of due dates to individual workstations -Look ahead ES/ simulation based scheduling -Optimization based tech -Batch sequencing
Control	-Interface to workstation controller -Physical control (motion control at NC and robot pick and place level) -Execution of control programs (APT, AML, etc.)	-Monitor equipment states and execute part and information flow actions based on states -Synchronize actions between equipment (e.g. robot and machine while loading/unloading parts) -Ladder logic execution	Organizational control of workstations, interface with MPS, generation of reports, etc.

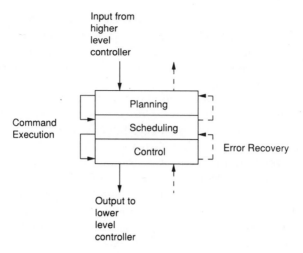

FIGURE 20.5. Execution schema.

Error recovery in this context means dealing with the impact of delays. This is a two-step process: determining the impact on the current schedule and updating the schedule. The first step requires a robust data structure for the schedule, which allows one to determine the "ripple effect" of a delay on the current start and finish times. Based on this analysis, the recovery can be a quick fix (in which a few times get revised), a major rescheduling, and/or a major replanning.

Due to the complexity of the scheduling problem in FMS (the job shop scheduling problem is, in its general form, a NP-hard problem), expert systems are commonly used to effectively schedule production. Expert systems can incorporate all relevant knowledge and constraints in the construction of schedules. This knowledge can be specified in terms of rules and these rules can be added, deleted, and edited at any time. Besides the multipass expert scheduling system presented in Section 20.3.4, descriptions of expert system-based scheduling systems can be found in Kusiak and Villa (22), Steffen (35), and Sauve and Collinot (31).

20.3.2. Process Plan Representation

Process plans play an important role in the definition of the control structure as well as in defining alternative sequences of operations that must be known by the planning and scheduling modules. The need for alternative processing options manifests itself in two important ways:

1. They provide dynamic schedules with various alternatives available at any instant.

2. The process of creating control software to implement the scheduling actions requires that alternatives be known a priori so that control software can provide for the necessary links in the control architecture.

A process plan representation that is capable of representing all multilevel interactions and possible precedence that occur among the planning and processing decisions is required. An AND/OR precedence graph-based representation is proposed as a compact representation of the process plan (27). The hierarchical nature of CIM control makes the graph representation even more attractive. Figure 20.6 shows the use of hierarchical AND/ OR graphs to represent the different levels of process planning required. The cell level process plan represents the facts that the part must visit four workstations and that two alternative sequences exist ({W1-W2-W3-W4} and {W1-W3-W2-W4}) based on the precedence requirements. Within each workstation there may also be alternate routes through machines. This is represented by the expanded graph of the cell level node. In the example in Fig. 20.6, if the part is processed at workstation W1, the two alternative machine sequences are {M1-M2-M3} and {M1-M2-M4}. The individual machine nodes at the workstation level can be further expanded to represent the equipment level alternatives (e.g., tools and tool sequences). For example, the part illustrated in Fig. 20.6 can be processed at machine M1 using either of the two tool alternatives {T1-T2} or {T3}.

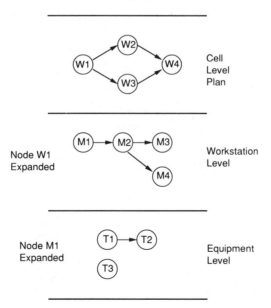

FIGURE 20.6. AND/OR graph process plan representation.

The process plan decomposition utilized in the architecture parallels that of the control architecture and thus ensures that the routing alternatives are provided for decision-making at an appropriate level of detail suitable for each level in the architecture.

20.3.3. Detailed Architecture of a Controller Module

Efficient control of the shop floor control system is accomplished using a multipass expert control system philosophy. This philosophy is illustrated in Fig. 20.7. The philosophy is based on the premise that a good set of manufacturing knowledge can be collected and stored in an expert system that can be used for a manufacturing control module. The major problem associated with planning, scheduling, and control is the intractable size of most problems; however, with the architecture presented, the problem can be decomposed into smaller solvable subproblems.

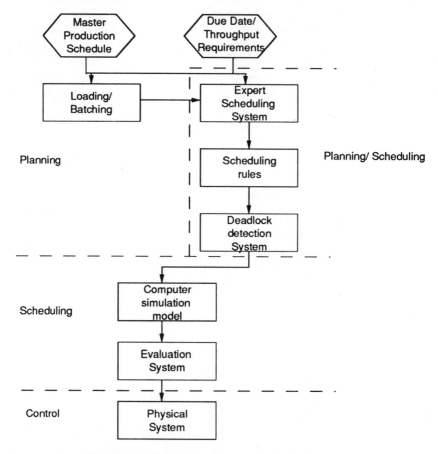

FIGURE 20.7. Multipass expert control philosophy.

The workstation controller module is illustrated in Fig. 20.8. As can be seen in the figure, it receives commands from the cell controller and plans the activities of the various equipment to successfully ensure completion of activities in the desired time frame. It should be noted that the expert scheduling system and a deadlock detection system operate at both the cell and workstation levels. The expert scheduling system for the workstation control module maintains the necessary knowledge base for the level's activity. It uses a simulation model coupled with expert system logic to determine future courses of action. Thus the effects of the decision can be previewed before execution. The simulation is based on actual shop status maintained in the database, incorporates flexible process plans, and utilizes deadlock detection/avoidance (see Section 20.3.4.3) modules to determine "best" part movement strategy. This enables dynamic routing/rerouting of parts in cases of machine failure or other unforeseen circumstances.

The error recovery module houses requisite error recovery algorithms/ procedures and is invoked when commands cannot be completed as requested. The error recovery module determines the type of error and the viable actions for recovery, interfaces with the scheduler to determine efficiency/performance of the recovery strategy, and initiates actions to effect the requisite part and information flow required to complete the recovery process.

The dynamic deadlock detection, avoidance and recovery provides algorithms and procedures required to maintain the system in a deadlock-free state where possible, and to initiate recovery from an unavoidable situation. It should be noted that many of the same functions and activities will be performed at both the workstation and cell levels.

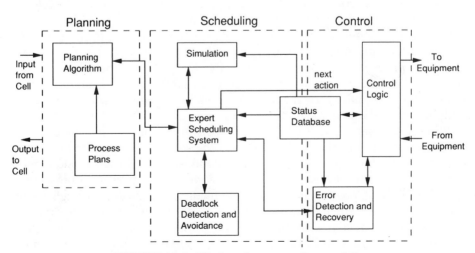

FIGURE 20.8. Workstation controller module.

20.3.4. Software Elements for Intelligent Control

The software required in a shop floor control system consists of several components:

1. Control software to effect part and information flow.
2. Scheduling software to effect part flow and timing.
3. Software for contingency planning (deadlock detection and recovery, error recovery, etc.).
4. Data management and data collection software.

In the following section, issues and algorithms relating to the development of software for intelligent control as pertaining to (a) automatic generation of control software, (b) on-line procedures to detect and avoid system deadlocks, and (c) expert scheduling are presented.

Automatic Generation of Control Software. The development of software for control systems is characterized by the need to carefully model machines on the factory floor, the interaction of parts on those machines as specified by a process plan, and the specific data updates that occur within computers constituting the control system. To focus on automatic software generation, equivalence of formal models of the manufacturing system and of software must be achieved. Loosely speaking, analogous techniques have been used in compiler construction (1), but in order to apply those techniques to manufacturing control systems, strict equivalence of the properties of manufacturing systems and the corresponding messages and formal models used to control the system must be created.

Automatic software generation significantly increases the likelihood of producing correct control software with greatly reduced software development time. Since the control elements of the software will be generated from a formal model of the system and the process plans that execute on the system, automatic upgrades will be possible when changes to the system layout or product mix occur. Additionally, the use of a formal model to generate run-time software can be used to generate the preconditions and postconditions required for theoretical proofs of correctness associated with each action in the manufacturing system.

Several authors (5, 6, 28) have developed formal descriptive models of the factory floor with the intent of supporting software generation but have not shown methods for systematically extending their formal models into running software. Other authors have developed rapid software prototyping systems (23) that provide only for the structural specification of cell level control software, still demanding enormous development of hand-crafted code to realize effective systems. In order to overcome the current impediments to the adoption of FMS technologies by small- to medium-sized firms, we believe that reliable, affordable factory floor level control software generators deserve consideration.

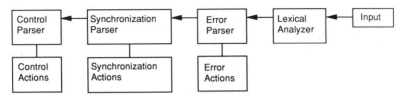

FIGURE 20.9. Layered parser model of a controller.

The idea of controlling a FMS as a by-product of grammar recognition is developed in Refs. 20a and 26a. Each controller (cell, workstation, and equipment) is modeled as a layered parser (Fig. 20.9) capable of satisfying the varying levels of command and processing required. Within each control module the highest level of control is executed by a control parser that is an automatically constructed push down automata. The parser recognizes a control grammar, which is generated by searching a system state space graph. The control parser is responsible for recognizing control elements of the input and activating the appropriate control actions, which are typically associated with the movement of parts in the system, consultation of schedulers, or preparation of reports. The next layer of actions includes the synchronization actions, which are activated upon recognition of grammatical constructs reserved for synchronization of machine interaction (e.g., when a robot holding a part in a machine fixture requests that the machine grasp the part). Synchronization is defined in a separate parser layer since the states in the underlying model are not affected by synchronization actions. Error actions are executed upon recognition of error conditions and implemented via the use of the error parser. The parsers and associated context-free control grammars are generated automatically from the formal models used to describe the system.

System Deadlock: Detection and Avoidance. In order to achieve true flexibility in a FMS, the parts must be capable of being routed through the cell in a manner that does not inhibit part flow. System deadlock is a situation where machines in a manufacturing system have been allocated parts in such a way that part flow is impossible. Deadlocking can arise even in the simplest of automated manufacturing systems. Consider a two-machine cell consisting of machines A and B, processing two parts P1 and P2, and serviced by a robot (Fig. 20.10a). The routing for part P1 is A → B and for part P2 is B → A. Suppose part P1 arrives and is loaded onto machine A, and then part P2 arrives and is loaded onto machine B. Any further part flow is now inhibited, resulting in a system deadlock. A deadlock may involve more than one machine, as shown in Fig. 20.10b. Although a system deadlock can eventually be resolved if there is a storage to buffer at least one individual part, a deadlock can still occur if the part storage is not properly used, as illustrated in Fig. 20.10c. Once a deadlock occurs, automated control for the manufacturing system may have to evoke

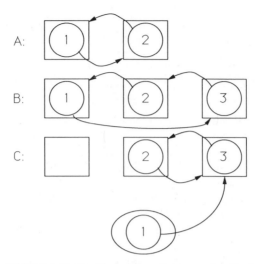

FIGURE 20.10. Examples of system deadlocks.

manual recovery procedures, resulting in the degraded performance of automated manufacturing systems. Hence the detection and avoidance of system deadlocks should be an important integral part in any automated manufacturing system.

Two factors that affect system deadlock are complexity of part routing and availability of part storage. A flow line FMS will not have a deadlock problem. However, if each part has a different routing through the FMS, the possibility of frequent deadlocks can be expected. The complexity of the problem can be illustrated by the following example. A manufacturing system with six machines and parts capable of moving from any one machine to another can have a total of 57 deadlock possibilities ($C^6_2 + C^6_3 + C^6_4 + C^6_5 + C^6_6$), where C^6_2 corresponds to the number of 2 machine deadlocks possible in a 6 machine system, C^6_3 corresponds the number of 3 machine deadlocks, and so on. Several deadlocks could exist simultaneously. Availability of storage space reserved for deadlock situations can make recovery possible if properly used. However, some manufacturing systems may not be equipped with any storage at all. In such a case, deadlock must be avoided. In this section two approaches are presented: (a) a deadlock detection and recovery approach when special storage is available and (b) an avoidance approach for a system with no storage.

System deadlocks can have an adverse effect on performance of automated systems and have been ignored in most scheduling studies, which typically assume infinite capacities and few limits on physical constraints imposed by material handling equipment. Conventional implementations of FMSs allocate large amounts of storage space to avoid deadlocking and rely on creating batches of parts using linear flow characteristics (7), which results in inefficient use of resources and major loss in flexibility of systems.

Deadlocking can be prevented in three ways: (a) providing large in-process queues and storage area (*planning*); (b) restricting flow of parts, such that they only have a positive undirectional flow direction (*planning and scheduling*); and (c) controlling parts flow to inhibit deadlocking (*scheduling*). Co (7) showed that if the maximum number of parts in a system is n and there are $n - 1$ queuing stations, then deadlocking cannot occur. In practice, several queueing stations are installed in a FMS to plan for worst case operation.

Deadlocking problems have been studied by researchers in computer science in the context of multiple coprocesser operating systems (11, 15, 36) and distributed systems (10a, 26). These studies cannot be applied directly to manufacturing systems but provide insight into deadlock resolution in manufacturing systems. The deadlock detection and avoidance process interfaces with the scheduling system to assist in sequencing part flow in a manner that does not lead to a deadlock situation. The approach presented here implements the detection and avoidance in three steps.

Step 1: Development of Detection Procedure

This step is required to isolate combinations of parts and machines that could lead to a potential deadlock in the context of a running system. A graph theoretic approach is proposed for the identification of system dead-locks, where a graph $G = (V, E)$ is created such that $V = \{$set of nodes corresponding to the machines in the system$\}$, and $E = \{$set of directed arcs corresponding to the part routing$\}$. Based on the definition of a system deadlock, a circuit is a necessary condition for a system deadlock. Hence the next step is to create a procedure for detecting circuits in the graph.

A string multiplication technique developed by Boffey (6a) is used to identify all circuits in the graph. However, it is well known that identification of all circuits is a NP-hard problem, but for our application the following observation will make the problem size manageable by providing an upper bound on the search procedure for closed circuits. A system deadlock in a manufacturing system composed of M machines and N parts cannot possibly involve more than M machines or N parts. Hence the string multiplication technique for identifying circuits will terminate after x iterations where $x = \min\{M, N\}$.

Not all circuits correspond to system deadlocks. Attributes and characteristics of system deadlocks are correlated with properties of the graph to further narrow down the search for circuits that have potential for dead-locks. Further details on deadlock detection are presented in Wysk et al. (38).

Step 2: Avoidance of Deadlocks

The avoidance problem deals with problems of detecting and preventing a system deadlock from materializing in a running system. The complete route of each newly arriving part as well as the remaining routes of all parts already in the system are converted into a graph, and the graph is examined for all circuits that could potentially lead to deadlocks. These circuits are stored in a database, which provides the scheduling system the capability of directing part flow without causing a system deadlock, and are updated dynamically as parts flow through the system.

This process can be further complicated by the interaction of two or more circuits to create a deadlock situation. Consider Fig. 20.11, which shows four machines A, B, C, D and three part routes marked 1, 2, 3. Part 1 is at machine A, part 2 at machine B, and part 3 is at machine D. Each of the circuits by itself cannot cause a system deadlock, but as soon as either part 2 or part 3 moves, one of the circuits will result in a deadlock. Hence the avoidance system must also be capable of preventing such situations from arising. Such circuit interactions are identified by identifying higher level closed circuits created by collapsing the circuits into a single higher level node and retaining the arcs emanating from/to the collapsed circuit as arcs in the higher level graph (39). The higher level circuits identification procedure is the same as the one discussed in step 1.

Step 3: Recovery from Deadlock

This approach is applied in a manufacturing system that reserves special storage for a deadlocking situation. Here deadlocks are allowed to occur and the recovery procedure is activated to resolve it. In this case, the circuits provide both the necessary and sufficient conditions for system deadlocks. Furthermore, circuits do not interact with each other, and all circuits interact independently.

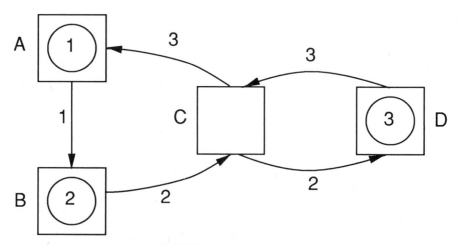

FIGURE 20.11. Deadlock.

The first step in the correction procedure is to move one of the deadlocked parts to the specially reserved storage. Interaction with the scheduling module is used to determine which part should be moved to storage. The remaining parts are processed on requested machines in an order to prevent subsequent deadlocks (if only one storage location is provided). The last step is to reinsert the part from storage back to the machines for processing. It can be seen that if the recovery approach is not carefully planned, it can result in unproductive time spent in moving parts between storage and machines.

Multipass Expert Scheduling System. All decisions on part movement will be made by the scheduler, and the control system merely serves to execute the commands from the scheduler. The multipass expert scheduling system integrates with the structure as shown in Fig. 20.7.

Multipass scheduling is an approach to resolving factory scheduling problems and has shown promise recently. Wu and Wysk (37) showed that significant throughput improvements can be made using a simulation model to determine the future course for a manufacturing system. Essentially, the procedure works as follows. A simulation model of the system is resident on the system control computer. At each decision point, a deterministic simulation is run to see what control policy (from a series of rule-based policies) impacts the current system most favorably. This control policy is then chosen and the appropriate control response is signaled for execution on the system.

One unfortunate drawback of this procedure is that FMSs are not static (the types of parts, demand, tooling, etc. change over time). Each time the specifics change, the simulation model must be rewritten or updated, which has to be done manually. Although the benefits can be significant, the implementation costs can be quite high.

The basic idea behind the multipass expert scheduling system is to use deterministic simulation as a short-term predictive tool for alternate control strategies in a manufacturing cell. There are many controllable (endogenous variables that must be affected). There are also many uncontrollable (exogenous) factors that will impact a system. A deterministic simulation can be run as a Gantt chart like the scheduler to analyze short-term effects. Decisions based on interference, machine utilization, and other system performance measures can be assessed for short-term sequencing and operational decisions. This allows for use of a reasonably simple set of rules that are evaluated to determine performance.

The simulation does not examine exogenous factors. The system will respond to exogenous factors by creating a new simulation experiment/model when unforeseeable events occur. However, the scheduler need only respond to these events rather than try to minimize conflict form them. The system interacts with cell databases, on-line deadlock detection, avoidance and recovery module, and error recovery modules.

The multipass expert scheduling system consists of three major components:

1. An intelligent scheduling module.
2. A simulator.
3. Interfaces to other modules, databases, and cell controllers.

Intelligent Scheduling Module (ISM). Scheduling is the principal vehicle for utilizing the resources efficiently, responding to managerial objectives rapidly, and satisfying the system constraints effectively. The scheduling module enables generation of various schedules based on input from the factory control system along with the actual status at the shop floor. ISM utilizes expert methodologies to assist in shop scheduling.

The ISM consists of the following components:

1. The acquisition module for transferring the knowledge and expertise of scheduling into facts and rules. This module is constructed so that ISM can continuously be expanded since scheduling knowledge is domain specific and will vary based on scheduling environments of different machine/part configurations with different levels of complexity. In addition, the acquisition module employs machine learning techniques to learn from feedback from the shop floor.

2. The knowledge base consists of primary and meta knowledge. The primary knowledge comprises current cell status, deadlock circuits information, dynamic dispatching rules, and scheduling heuristics. The meta knowledge consists of knowledge to apply scheduling rules and heuristics to previously unexperienced circumstances and of general criteria for selecting scheduling rules and learned heuristics.

3. An inference engine manipulates the rules and facts in the database to generate an alternate space consisting of several "good" alternative scheduling rules that are simulated to evaluate their performance potential.

Simulation Model. The major function of the simulation model is to evaluate alternative control policies. The simulation model queries the current databases and determines the future system status by making a pass of deterministic simulation according to a rule(s) defined in the rule module. The system performance predicted by each pass of simulation is a measure of closeness to an objective function from the higher level factory control system. Thus, at the end of all passes of the simulation, the best schedule is then applied to the physical manufacturing system. The simulation model will operate on parts only currently available to the system (exogenous arrivals will not be permitted). This creates a deterministic system that can easily be analyzed.

An important issue that must be considered is the length of the simulation window, which determines how long a simulation should look ahead.

Intuitively, this value is critical to overall performance of the multipass scheduling mechanism. Further research will determine the most desirable simulation window based on its effect on performance of the scheduling mechanism, computational burden on the system, and other factors that will arise.

20.4. SUMMARY

In this chapter, a scaleable architecture for shop floor control has been presented. The key elements of this architecture are that each controller performs the same functions (planning, scheduling, and control), and the degree of complexity of the controllers can be scaled without affecting the control elements. Essential shop floor control software elements (process plan representation, control software generation, deadlock detection/avoidance algorithms, and scheduling systems) have been described. These software elements have been implemented and tested at the Pennsylvania State University's Computer Integrated Manufacturing Laboratory. The implementation experience looks very promising and this SFCS should provide the basis for the implementation of an enterprise-wide CIM system.

REFERENCES

1. Aho, A. V., Sethi, R., and Ullman, J. D. (1985). *Compilers—Principles, Techniques and Tools*, Addison-Wesley, Reading, MA.
2. Albus, J. S., Barbera, A. J., and Nagel. R. N. (1981). Theory and Practice of Hierarchical Control, *Proceedings of the IEEE 1981 COMPCON Fall*, Washington, DC.
3. Andersen, A., Jenne, T., and Mikkilineni, K. (1990). CIM Architecture: One Perspective. In: *Proceedings of CIMCON '90*, edited by Albert Jones, NIST Special Publication 785, pp. 506–524.
4. Barbera, A. J., Fitzgerald, M. L., and Albus, J. S. (1982). Concepts for a Real-Time Sensory-Interactive Control System Architecture, *Proceedings of the 14th Annual Southeastern Symposium on System Theory*, pp. 121–126.
5. Baxter, R. D. (1987). A Brief Discussion of Formally Based Approaches to Rapid Prototyping, *British Aerospace Technical Report BAE-WAA-RP-RES-SWE-440* (unclassified), March 5.
6. Biemans, F., and Blonk, P. (1986). On the Formal Specification and Verification of CIM Architectures Using LOTOS, *Computers in Industry*, Vol. 7, pp. 491–504.
6a. Boffey, T. B. (1982). *Graph Theory in Operations Research*, MacMillian Computer Science Series, MacMillian, NY.
7. Co, H. C. (1984). *Design and Implementation of FMS—Some Analysis Concepts*, Ph.D. Thesis, Virginia Polytechnic Institute and State University, Blacksburg.

8. Duffie, N. A., and Piper, R. S. (1987). Non-Hierarchical Control of a Flexible Manufacturing Cell, *Robotics and Computer Integrated Manufacturing*, Vol. 3, No. 2, pp. 175–179.

9. Fitzgerald, M. L., and Barbera, A. J. (1985). A Low-Level Control Interface for Robot Manipulators, *NBS–Navy NAV/SIM Workshop on Robots Standards*.

10. Fitzgerald, M. L., Barbera, A. J., and Nagel, R. N. (1985). Real-Time Control Systems for Robots, *1985 SPI National Plastics Exposition Conference*, Chicago.

10a. Gray, J. N. (1978). Nots on Data Base Operating Systems, *Operating Systems – An Advanced Course*, Vol. 6, Bager, Trahm, and Segmuller, eds. Springer–Verlag, N.Y..

11. Habermann, A. N. (1971). Prevention of System Deadlocks, *Communications of the ACM*, Vol. 14, p. 373.

12. Ham, I., and Gongaware, T. (1981). *Application of Group Technology Concept for Higher Productivity of NBS Shop Operations*, Report for NBS, Pennsylvania State University, University Park.

13. Hatvany, J. (1985). Intelligence and Cooperation in Heterarchic Manufacturing Systems, *Robotics and Computer Integrated Manufacturing*, Vol. 2, No. 2, pp. 101–104.

14. Haynes, L. S., Barbera, A. J., and Albus, J. S. (1984). An Application Example of the NBS Robot Control System, *Robotics and Computer Integrated Manufaturing*, Vol. 1, No. 1, pp. 81–95.

15. Holt, R. C. (1971). Comments on Prevention of System Deadlocks, *Communications of the ACM*, Vol. 14, p. 36.

16. Jackson, P., and King, M. (1989). ADEPT—A Way Forward, *Proceedings of the 5th CIM Europe Conference*, May 17–19, pp. 121–131.

17. Jones, A. T., and McLean, C. R. (1985). A Proposed Hierarchical Control Model for Automated Manufacturing Systems, *Journal of Manufacturing Systems*, Vol. 5, No. 1, pp. 15–25.

18. Jones, A. T., and Saleh, A. (1989). A Decentralized Control Architecture for Computer Integrated Manufacturing Systems, *IEEE International Symposium on Intelligent Control*, pp. 44–49.

19. Joshi, S. B., and Wysk, R. A. (1990). Intelligent Control of Flexible Manufacturing Systems, *Modern Production Concepts, Theory and Applications*, Fandel, G. and Zapfel, G. eds., Springer–Verlag, Berlin, pp. 416–437.

20. Joshi, S. B., Wysk, R. A., and Jones, A. (1990). A Scaleable Architecture for CIM Shop Floor Control. In: *Proceedings of CIMCON '90*, edited by Albert Jones, NIST Special Publication 785, pp. 21–33.

20a. Joshi, S., Mettala E., and Wysk R. CIMGEN: A Computer Aided Software Engineering Tool for Development of CIM Software, *IIE Transactions* (To Appear).

21. Klevers, T. (1989). The European Approach to an Open System Architecture for CIM, *Proceedings of the 5th CIM Europe Conference*, May 17–19, pp. 109–120.

22. Kusiak, A., and Villa, A. (1987). Architectures of Expert Systems for Scheduling Flexible Manufacturing Systems, *IEEE 1987 International Conference on Robotics and Automation*, pp. 113–117.

23. Maimon, O. Z., and Fisher, E. L. (1988). An Object-Based Representation Method for a Manufacturing Cell Controller, *Artificial Intelligence in Engineering*, Vol. 3, No. 1.

24. McCain, H. G. (1985). A Hierarchically Controlled, Sensory Interactive Robot in the Automated Manufacturing Research Facility, *IEEE International Conference on Robotics and Automation*, pp. 931–939.

25. McLean, C., Mitchell, M., and Barkmeyer, E. (1983). A Computer Architecture for Small Batch Manufacturing, *IEEE Spectrum*, May, pp. 59–64.

26. Menasce, D. A., and Muntz, R. R. (1979). Locking and Deadlocked Detection in Distributed Data Bases, *IEEE Transactions on Software Engineering*, Vol. SE-5, No. 3, May.

26a. Mettala, E. G., (1989). *Automatic Generation of Control Software in Computer Integrated Manufacturing*, Ph.D. Thesis, Pennsylvania State University, University Park.

27. Mettala, E., and Joshi, S. (1990). A Compact Representation of Process Plans for FMS Control Activities, *Pennsylvania State University Industrial Engineering Working Paper Series*.

28. Naylor, A. W., and Maletz, M. C. (1986). The Manufacturing Game: A Formal Approach to Manufacturing Software, *IEEE Transactions on Systems, Man, and Cybernetics*, Vol. SMC-16, May/June, pp. 321–344.

29. O' Grady, P. J., and Menon, U. (1986). A Concise Review of Flexible Manufacturing Systems and FMS Literature, *Computers in Industry*, Vol. 7, pp. 155–167.

30. Rippey, W., and Scott, H. (1983). Real Time Control of a Machining Work Station, *20th Numerical Control Society Conference*, Cincinnati, OH.

31. Sauve, B., and Collinot, A. (1987). An Expert System for Scheduling in a Flexible Manufacturing System, *Robotics and Computer Integrated Manufacturing*, Vol. 3, No. 2, pp. 229–223.

32. Simpson, J. A., Hocken, R. J., and Albus, J. S. (1982). The Automated Manufacturing Research Facility of the National Bureau of Standards, *Journal of Manufacturing Systems*, Vol. 1, No. 1, pp. 17–31.

33. Skevington, C., and Hsu, C. (1988). Manufacturing Architecture for Integrated Systems, *Robotics and Computer Integrated Manufacturing*, Vol. 4, No. 3/4, pp. 619–623.

34. Smith, J. S. (1990). *Development of a Hierarchical Control Model for a Flexible Manufacturing System*, Master's Thesis, Pennsylvania State University, University Park.

35. Steffen, M. S. (1986). A Survey of Artificial Intelligence-Based Scheduling Systems, *Proceedings of the 1986 Fall Industrial Engineering Conference*, Boston, MA, pp. 395–405.

36. Tsutsui, S., and Fujimoto, Y. (1987). Deadlock Prevention in Process Computer Systems, *The Computer Journal*, Vol. 30, No. 1.

37. Wu. S. D., and Wysk, R. A. (1990). An Applicaiton of Discrete-Event Simulation to On-Line Control and Scheduling in Flexible Manufacturing Systems, *International Journal of Production Engineering*.

38. Wysk, R. A. Yang, N. S., and Joshi, S. B. (1990). Detection of System

Deadlocks in Flexible Manufacturing Cells, Pennsylvania State University, Industrial Engineering Working Paper Series.

39. Yang, N. S. (1989). *Resolution of System Deadlock in Real Time Control of Flexible Manufacturing Systems*, Ph.D. Thesis, Pennsylvania State University, University Park.

PART IV

QUALITY ENGINEERING AND MANAGEMENT

CHAPTER TWENTY ONE

Machine Diagnostics

IOANNIS O. PANDELIDIS*

Manufacturing Systems Research, The Gillette Company, Boston Massachusetts

21.1. INTRODUCTION

Unmanned machining is an elusive goal that cannot be met without signifi-cant advances in machine diagnostics. The function of a complete diagnostic system in unmanned machining is to monitor the machining system, evaluate the relevant system states, and provide remedial action in response to any deterioration in the system.

Diagnostic methods presented in this chapter address subsets of the general diagnostic problem in machining, where the status of the following aspects of the machining system can be monitored:

- Workpiece
- Machine tool
- Cutting tool
- Cutting process

Dimensional sensors to check workpiece quality requirements such as dimensional tolerances and surface roughness are discussed by Tlusty and Andrews (31) and are not treated here. Similarly, the functioning of the machine tool components, such as the control system, feed drives, spindle bearing, and guideways, is not discussed. This chapter focuses on the determination of the status of the cutting tool, and the detection of chatter in the cutting process.

The cutting tool undergoes deterioration in the process of cutting, which may be classified as various types of tool wear or tool breakage. Tool wear

---

*This work was completed while the author was at the University of Maryland.

Intelligent Design and Manufacturing, Edited by Andrew Kusiak.
ISBN 0-471-53473-0 © 1992 John Wiley & Sons, Inc.

and tool breakage have been dealt with in the past by changing the cutting tool at regular intervals. However, the variability in tool life is large, resulting in tools being replaced either too early or too late, when damage to the workpiece has already occurred. A typical drill, for instance, has an average drill life of 1800 holes but might fail under normal operations anywhere between 500 and 5000 holes (30).

Tool breakage is a catastrophic failure condition that requires immediate retraction of the cutting tool to prevent further damage to the workpiece. In this event, the diagnostic method employed must act in a short period of time. Tool wear is a more gradual phenomenon and can be determined over a relatively longer period of time, which may nevertheless be too short for some computationally intensive methodologies. Computational speed is therefore an important consideration in any on-line diagnostic scheme.

Chatter is the condition of self-excited vibrations between the cutting tool and the workpiece. This can often cause damage to the workpiece, the tool, or the machine. In the past, chatter has been avoided by running conservative cutting conditions, which limit the productivity of the machine tool. When the cutting conditions are known, well established prediction methods can be used (14, 20–22, 32). However, since the cutting conditions are rarely known, on-line detection methods are required.

A general diagnostic framework is presented in Section 21.2. Sensor feedback for diagnostics is discussed in Section 21.3. Sections 21.4–10 present various diagnostic methodologies accompanied by applications, and future developments are discussed in Section 21.11.

21.2. GENERAL DIAGNOSTIC FRAMEWORK

A diagnostic module can be conceived as part of a general diagnostic framework, as shown in Fig. 21.1. The overall system is an asynchronous, event-driven feedback control system. The diagnosis module receives feedback information from the machining system, evaluates the signal, and produces a process status as output. The supervisory controller receives the status of the machining system as input and produces supervisory action as output. The action can be a STOP, CONTINUE, or some other remedial

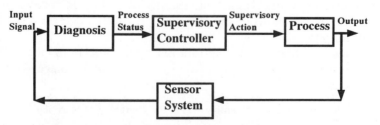

FIGURE 21.1. General diagnostic system.

command to the machining system. In most present machining systems, the supervisory controller function is carried out by the operator, but in unmanned machining of the future, it will be automatic.

Any diagnostic module may be distinguished by its function, the sensor input, and the diagnostic method employed. The function of the module defines the set of conditions recognized in the machining system and therefore defines the set of possible process status output signals. For instance, a diagnostic module for detecting chatter might have either a binary output denoting chatter or no chatter condition, or some continuous measure of chatter as output. The sensor input defines the available on-line information to the module.

21.3. SENSOR FEEDBACK

In machining, the most important signals for diagnostics are acoustic emission (AE), cutting and thrust force, tool interface temperature, and vibration feedback. A variety of sensor inputs have been employed for diagnostic tasks, either alone or in combination with other signals. An excellent review of available sensors for machining is given by Tlusty and Andrews (31). From practicality considerations, vibration and tool interface temperature feedback is not often used, whereas force and acoustic emission feedback is very useful.

21.3.1. Acoustic Emission (AE) Feedback

The cutting process produces sound waves that can be used as feedback to monitor the state of the machining system. The sound waves are primarily a result of the deformation process occurring within the metal during cutting. It is well known that machinists use acoustic feedback for that purpose. However, the main attraction of AE is that most significant information lies above the audible range, namely, between 50 kHz and 1 MHz. Once the AE signal is passed through a high-pass filter, it is largely uncontaminated by machine vibration and other low-frequency noise.

For single-point cutting, as in turning, it is fairly easy to mount a piezo-electric transducer on the cutting tool side for continuous monitoring of the process. It has been much more difficult to do so in multipoint cutting such as milling, where the AE signal is generated by more than one tool point at a time. A significant advancement has been achieved by Inasaki et al. (13), who developed a "magnetic sensor coupling device" that can be mounted on the cutting tool side, with good disturbance rejection. Based on this sensor, many AE signal processing methods developed for turning can now be applied to milling as well (2). Other significant developments in AE sensor technology for turning and milling are reported by Ramalingam and Frohrib (26) and Ramalingam et al. (27).

An excellent account of various applications of acoustic emission signals and other integrated signals is given by Dornfeld (9). These methodologies are based on the event count analysis, the RMS (root mean square) of the signal, frequency analysis, and parametric monitoring. In this chapter, the latter two approaches and their combination are examined.

21.3.2. Force Feedback

Changes in tool wear and tool fracture state have been shown to be highly correlated with changes in thrust and cutting forces. Thus force feedback constitutes an important diagnostic signal that can be measured in a variety of ways. In lathes, dynamometers can be mounted on the tool holder, while in milling and drilling machines, dynamometers are mounted on the table or are built into the spindle bearings. An alternative is to estimate cutting force by measuring spindle motor current, voltage, and speed.

21.3.3. Sensor Fusion

The various attempts at creating reliable tool condition monitoring based on a single sensor such as AE or cutting force have not been sufficiently successful to be used reliably for a wide range of materials and cutting conditions. This is in large part due to the limited information that any one sensor can reveal, both from a physical point of view as well as from signal-to-noise considerations.

Cutting force alone, for instance, may be inadequate to determine tool wear, because cutting force also depends on the cutting velocity (33). In addition, even though flank wear tends to increase the cutting force, crater ware tends to decrease it.

On the other hand, the combination of different sensors into an intelligent sensor system (4) promises a new era of robust diagnostics. Sensors can serve complementary functions. For instance, cutting force is related primarily to macroscopic phenomena (vibrations), whereas acoustic emission is related to microscopic phenomena (deformation process). Not only is the information available richer, but some redundancy increases the diagnostic system robustness. Applications of sensor fusion are shown in this chapter, where AE, current, and force measurements are combined for LDF and neural network diagnostics.

21.4. DIAGNOSTIC METHODOLOGIES

In on-line diagnostics the identification of patterns from input signals can be classified as a pattern recognition problem. Pattern recognition is the set of operations used to associate input patterns with some classification. In machine diagnostics, these patterns are typically related to tool wear or tool breakage conditions. Pattern recognition may also be thought of as data reduction, since it evolves mapping of a large number of input data into a lower dimensional output space.

Pattern recognition schemes using supervised learning have two stages: the training stage and the classification stage. The purpose of the training stage is to establish the classification parameters that will be used during the subsequent classification stages. The classification scheme is established by providing the system with typical input signals such as cutting force or acoustic emission (AE). Each signal is associated with a particular known classification such as worn tool or sharp tool. Based on these associations, the classification parameters are established.

The classification stage follows and uses the training parameters to perform a series of transformations on the signal with the output being the estimated status of the tool.

In the following sections, a number of diagnostic methodologies along with applications in machining are presented.

21.5. FUZZY PATTERN RECOGNITION

In fuzzy logic, the state of the process is indicated by the "grades of member ship." For example, assuming the state of wear tool condition is classified as "initial," "small," "normal," and "severe," the grade of membership given by

$$\underline{u} = [u_1, u_2, u_3, u_4] = (.01, .02, .07, .9) \tag{1}$$

will indicate that the grade of membership for "severe" tool wear condition is .9, whereas the grade of membership for "initial" tool wear condition is .01. Based on the identified grade of membership vector, the tool would need to be replaced (Fig. 21.2). For any state, the grades of membership vector has the constraint that

$$\sum_{i=1}^{n} u_i = 1, \tag{2}$$

similar to a probability relation.

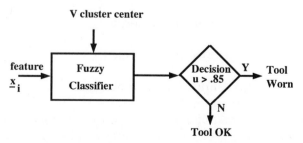

FIGURE 21.2. Fuzzy classifier.

Fuzzy pattern recognition has two phases—the classification and the training phases as discussed later for the case of drill wear monitoring (17).

21.5.1. Fuzzy Classification Phase of Drill Wear States

During the classification phase, input signals of thrust force and torque are mapped onto a grade of membership vector indicating the drill wear state. If the fuzzy classification indicates a significant grade of membership for the "severe" condition, action is taken to replace the tool.

Given the current observation vector

$$x_c = [\text{thrustforce, torque}], \tag{3}$$

the fuzzy classification process depends on previously determined "cluster centers" and is given by

$$u_{cj} = \frac{1}{\displaystyle\sum_{k=1}^{4} \left(\frac{\|x_c - v_j\|}{\|x_c - v_k\|} \right)^2}, \qquad j = 1, 2, 3, 4, \tag{4}$$

where v_i is the cluster center for the ith wear state classification.

The tool would be considered worn if u_{c4}, the grade of membership for "severe," is above .85 or some other prespecified number.

21.5.2. Training Phase of Drill Wear States

The cluster center of all fuzzy grades,

$$V = [v_1, v_2, v_3, v_4], \tag{5}$$

where

$$v_1 = (\text{thrust force, torque})_{\text{initial}}, \tag{6}$$

$$v_2 = (\text{thrust force, torque})_{\text{small}}, \tag{7}$$

$$v_3 = (\text{thrust force, torque})_{\text{normal}}, \tag{8}$$

$$v_4 = (\text{thrust force, torque})_{\text{severe}}, \tag{9}$$

is determined during the training phase. Given a set of observations $x_1, x_2, \ldots, x_n$, the cluster center is calculated by an iterative algorithm that minimizes the loss function

$$J(U, V) = \sum_{i=1}^{n} \sum_{j=1}^{4} (u_{ij}^2)\|x_i - v_j\|^2 \tag{10}$$

subject to the constraint that

$$\sum_{j=1}^{4} u_{ij} = 1, \qquad i = 1, \ldots, n, \tag{11}$$

where $U = [u_{ij}]$ is the membership function, and u_{ij} is the fuzzy grade assigned to the jth wear state of the ith observation.

Note that during optimization both U and V must be found. A search procedure iterates between estimates of U and estimates of V, until the estimates change less than a predetermined level.

21.6. LINEAR DISCRIMINANT FUNCTION (LDF) ANALYSIS

In what follows, the LDF approach for pattern recognition is discussed first using acoustic emission (AE) as input signal, and then using sensor fusion to combine force and AE signals.

LDF pattern recognition consists of the classification phase and the training phase. During the classification phase, the sampled signals go through a series of predetermined operations, as shown in Fig. 21.3. The input signal is first transformed from the pattern space to a feature space. The feature signal is now considered as a point in a multidimensional feature space. Through feature extraction, only a few of the relevant features are selected. The lower dimensional feature space signal is transformed again by a linear transformation to a decision space whose coordinate axes are the possible tool condition states. In effect, this constitutes a projection of the higher dimensional feature vector upon the decision space. The effect of the transformations is to cluster similar input patterns close to each other in the decision space. The tool condition state that has the largest projection of the feature vector is classified as the identified tool condition. The last two stages in this procedure are accomplished by a single LDF linear transformation, and classification involves choosing the maximum of these functions.

During the training state, controlled experiments for each tool condition state are conducted and the resulting feature signals are recorded. The

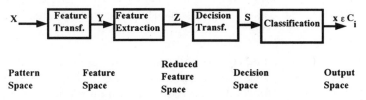

FIGURE 21.3. LDF pattern recognition.

relevant features that are to be used as input to the LDF transformation are identified by the some feature selection criterion. The parameters of the LDF linear transformation are then determined by using the resulting reduced feature vectors as input and the controlled tool condition as target output signal.

21.6.1. AE Monitoring Using LDF Analysis

In this LDF diagnostic application (10a), the sensor input is the AE signal, which as was pointed out earlier is an important diagnostic signal. During the classification phase, the AE signal is sampled for a short period of time, and the time domain data are transformed to the feature space by fast fourier transform (FFT), which identifies the power of the time domain signal at different frequencies. During feature extraction, only the highly relevant spectral components are chosen, while the remaining frequency components are ignored. Thus from a sampled signal in the time domain,

$$X = [x_1, x_2, \ldots, x_N], \tag{12}$$

a FFT transform of complex spectral components is given by

$$Y = [y_1, y_2, \ldots, y_{N/2}]. \tag{13}$$

By feature selection, spectral components are selected from this vector to form the reduced feature space vector

$$Z = [z_1, z_2, \ldots, z_K]. \tag{14}$$

Assuming n_C distinct classes in the decision space, $\{C_1, C_2, \ldots, C_{n_C}\}$, a set of scalar linear discriminant functions $\{g_1(Z), g_2(Z), \ldots, g_{n_C}(Z)\}$ are computed. The incoming feature vector Z belongs to the class k associated with the maximum g_k of these functions. Note that the LDF combines transformation of the feature vector to the decision space with an implied calculation of distance of the vector from each decision class.

A two-dimensional decision space is shown in Fig. 21.4. Signals associated with worn tool condition are indicated as squares, and signals associated with sharp tool (tool OK) condition are shown as circles. Points are clustered around ideal classification points, shown as filled points. The arrow points to a misclassified signal, which will be classified with the wrong cluster based on LDF analysis.

During the training phase, controlled experiments are conducted for each of the identifiable classifications so as to have minimal interference from other conditions. The important features are chosen on the basis of a feature selection criterion that minimizes the scatter within a class and maximizes

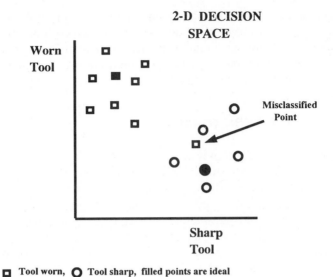

FIGURE 21.4. Clusters in two-dimensional decision space.

the variability between classes. The parameters of the linear discriminant functions are then determined from the reduced feature space data and the known classifications.

21.6.2. Principal Component Analysis

One of the problems in LDF analysis using FFTs is that many assumptions about the statistical properties of the feature components, such as independence, are violated in practice, resulting in suboptimum pattern recognition. In the research conducted by Houshmand and Kannatey-Asibu (12), these problems are overcome by using principal component analysis. The goal of principal component analysis is to transform the high-dimensional feature vector into a lower-dimensional space through a series of operations. The new axes in the lower-dimensional space have the property of being orthogonal to each other. In addition, these principal components account for most of the variance exhibited in the original data. In this research, for example, six principal components accounted for 91% of the total variance of the 46 feature components. Since principal components are orthogonal (and therefore uncorrelated) to each other, each component may also represent a particular characteristic of the tool wear condition.

21.6.3. AE and Force Sensor Fusion

Sensor fusion is the incorporation of two or more feedback signals into an integrated approach. The combination of two independent sources of information increases both the reliability and the performance capability of the diagnostic system. In the method developed by Balakrishnan et al. (1), AE and force signals were simultaneously transformed into the feature space, where after feature extraction, the LDF analysis was applied for pattern recognition of tool wear condition.

21.7. PARAMETRIC MONITORING

A static model predicting tool wear as a function of sensed signals can be the basis for diagnostics. Alternatively, a dynamic parametric process model that predicts future output based on past values of the output and/or the input signals could be estimated. When the underlying process changes due to tool wear, the model parameters also change. Based on the parameter changes, a decision is made as to the status of the machining system. Another approach is to base the tool status decision on the predicted value of the tool wear estimate.

A typical static model is an algebraic relation of inputs to outputs. Some dynamic models are time series models, as in the case shown for chatter detection and chipping and breaking in milling, or state space models, as in the case of monitoring tool wear.

21.7.1. Static Modeling

A static model for tool wear prediction has the form

$$y = f(x_1, x_2, \ldots, x_n),\tag{15}$$

where $x_1, x_2, \ldots, x_n$ are observed variables, and y is the resulting tool wear.

Chryssolouris and Domroese (5, 6) compared linear and nonlinear static models, along with neural networks, for their effectiveness in predicting simulated and also experimentally obtained tool wear values. Their conclusion was that neural networks, which are discussed in later sections, performed better in this task, particularly when the relation between input and output was nonlinear.

21.7.2. Modeling Based on Time Series

Many applications in machining of the time series analysis are based on the dynamic data series (DDS) analysis (35). A time series $\{x_i\}$ is a sequence of observations ordered by an index related to time or space. A simple model for a time series is an autoregressive (AR) model that relates observation x_i in terms of its past values and a white noise term:

$$x_k = a_1 x_{k-1} + a_2 x_{k-2} + \cdots + a_n x_{k-n} + n_k,\tag{16}$$

where a_i is the ith autoregressive coefficient and n_k is discrete white noise.

The set of coefficients $\{a_1, a_2, \ldots, a_n\}$ constitute the model parameters and characterize the process. For a given set of observations $\{x_1, x_2, \ldots, x_N\}$, the AR coefficients can be determined based on least-squares estimation. The estimation can be done either by accessing a whole block of data or by updating an estimate of the coefficients after every observation.

21.7.3. Chipping and Breaking in Milling

A method to detect chipping and breaking of cutting teeth in face milling by time series AR modeling was given by Matsushima et al. (19). An AR model with 28 coefficients was fitted to predicted spindle motor current in terms of its past measured values.

The residual error e_k may be calculated by subtracting the expected value $\hat{x}_k$ from the observed value x_k as

$$e_k = x_k - \hat{x}_k , \tag{17}$$

where

$$\hat{x}_k = a_1 x_{k-1} + a_2 x_{k-2} + \cdots + a_n x_{k-n} . \tag{18}$$

The residual error or the AR(28) model was monitored to indicate when the expected current obtained from the model varies significantly from the actual measured current. A large residual was shown to occur even for a small breakage of a tooth.

21.7.4. Tool Wear Monitoring in Turning

In the following approach (18), the parameters of the AR model were continuously updated after each observation, and selected coefficients were the basis of decision for the flank wear condition of the tool. An AR model with six parameters was estimated, but only the first and fourth parameters were monitored. In the two-dimensional plane formed by the parameters a_1 and a_4, points above $[a_1, a_4] = [.046, .007]$ were associated with worn tool condition.

21.7.5. Chatter Detection Based on ARMA Modeling

A more general time series model is an autoregressive-moving average ARMA(n, m) model, where

$$
\begin{aligned}
x_k = a_1 x_{k-1} + a_2 x_{k-2} + \cdots + a_n x_{k-n} + n_k + b_1 n_{k-1} \\
+ b_2 n_{k-2} + \cdots + b_m n_{k-m} .
\end{aligned} \tag{19}
$$

Based on the parameters of this model, it is possible to obtain model parameters of the system, such as natural frequencies and damping ratios.

Measures of process stability can be associated with the underdamped modes that have a large contribution to the total variance of the signal. Since chatter can be associated with the onset of instability in the process, these measures can be used as indicators of the onset of chatter (10).

21.7.6. Tool Wear Models Based on State Space Observers

This approach by Danai and Ulsoi (7) develops a state space dynamical model relating the input feed to the output cutting force. The input–output observations are the basis of estimating the "internal" tool ware process states. A state space observer estimates these tool ware states as a function of time.

A state space model is a representation of a dynamical system using a set of first-order differential equations. For a linear dynamical system with a single input u and a single output y, the state space representation is of the form

$$\dot{\underline{x}} = A\underline{x} + Bu , \tag{20}$$

$$y = C\underline{x} + Du , \tag{21}$$

where A, B, C, and D are appropriately dimensioned matrices defining the parameters of the dynamical system, and $\underline{x}$ are the states of the system.

Since only the input u and the output y are accessible for observation, the states of the system must be estimated based on the input–output observations. The set of equations that produces an estimate of the states as a function of time is called a state estimator.

In this case, the observed output of the system is the cutting force, and the input is the feed. Here we assume that depth of cut and screw rpm are held constant, as is the usual case in cutting. The states of the system are x_{f1}, the flank wear due to abrasion, x_{f2}, the flank wear due to diffusion, and x_c, the crater wear.

Figure 21.5 shows an observer coupled with a parameter estimator that

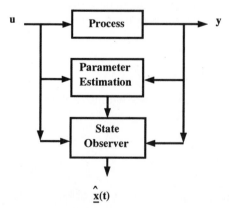

FIGURE 21.5. State observer for tool wear monitoring.

uses the input–output observations of the cutting process to estimate the tool wear states. The parameter estimator uses the same observations also to update estimates of the coefficient matrices that characterize the state space model.

21.8. EXPERT SYSTEMS

Expert systems are programs based on artificial intelligence (AI) that attempt to function like a human expert in a specialized field. Off-line diagnostic problems are well suited for AI applications, where symbolic manipulation is required in order to arrive at an appropriate conclusion. Several commercial applications have appeared, ranging from medical diagnosis to oil exploration and many others. A summary of commercially available expert system applications can be found in Schutzer (29a).

Expert systems for diagnosis have also been developed in other manufacturing processes such as injection molding. A rule-based system based on the AI language Prolog, for instance, was developed to diagnose the causes of defects in injection molded parts as an off-line diagnostic tool (24). With appropriate sensor input development, the system is suitable for on-line diagnostics as well. However, the reaction times required in injection molding are considerably slower than in machining. Applications in other manufacturing problems can be found in Wright and Bourne (34).

The quick reaction time required for on-line diagnostics of machining makes the application of expert systems in this area difficult. This is particularly true if complex relations and large databases are involved in the decision process. A more promising area presently seems to be neural network diagnosis.

21.9. NEURAL NETWORKS

Neural networks are computer models that are based on a crude model of the human nervous system. A special type of neural network, called the feed-forward multilayered neural network, is shown in Fig. 21.6. It consists of a multilayered system of input–output units called neurons. Each neuron in the input layer receives signals from outside and produces an output to the next layer. A neuron in the output layer receives signals from the previous layer of neurons and produces outputs to the external world. Neurons in the inner (hidden) layers have no direct connection to the external world. In a single hidden layer system, an inner neuron receives signals from the input neurons and presents outputs to the output neurons.

A simulated neuron is an input/output unit, as shown in Fig. 21.7, which consists of the following elements:

> *Inputs.* The input signals $x_1, x_2, \ldots, x_n$ are received either directly from the external world or from outputs of other neurons.

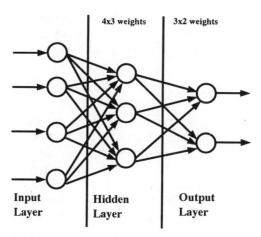

FIGURE 21.6. Three-layer neural network.

Summation Function. The input signals are summed according to their connection weights, which could be either positive (excitatory) or negative (inhibitory). The output of the summation function is

$$s = \sum_{i=1}^{n} w_i x_i \,. \tag{22}$$

Threshold Function. This function produces an output signal y based on the activation input s, so it has the form

$$y = f(s; t) \,. \tag{23}$$

The simplest threshold function has an output of 0 or 1 depending on a threshold value t. Otherwise, it is an S-shaped sigmoid function, that saturates at high input values of s. One such function, for instance, is given by

$$y = f(s; t) = \frac{1}{1 + \exp(-s - t)} \,. \tag{24}$$

Output. The output y is accessed directly by the external world or becomes input to the next layer of neurons.

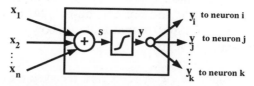

FIGURE 21.7. Model of simulated neuron.

A good review of the different neural algorithms, applications (in robotics), and architectures can be found in Kung and Hwang (15). Neural networks have been used for tool condition monitoring by Ranwala and Dornfeld (28) and Dornfeld (8), while the learning capabilities of neural networks have been studied by Chryssolouris and Domroese (5). The latter paper conducted simulation studies that compared a number of estimation methods with neural networks. The conclusion of the research was that neural networks performed better than other approaches, particularly in nonlinear relations.

Neural networks can be used for pattern recognition by inputing a set of observed features to the network and obtaining an output corresponding to a recognized classification. The knowledge about the relation of input feature pattern to output classification pattern is embodied in the weights of the network, which are obtained through a prior training stage. Both input feature patterns and known classification patterns are presented, and the network adjusts its own weights after every input–output pair based on some algorithm. A breakthrough in training neural networks occurred when a scheme was proposed by Rumelhart et al. (29) and similar schemes by Le Cun (16) and Parker (25). The training scheme is referred to as the generalized delta rule or as back propagation (BP).

In back propagation, the present weights are used to propagate forward the training input pattern from the input nodes to the output nodes. The output signal is then compared with the desired output signal, which is supplied during the training stage. The error between the desired output and the actual output is then propagated back through the network layers while the weights are adjusted at each layer so as to minimize the observed error. The weight adjustment is done at each node based on a localized hill climbing algorithm, which attempts to minimize the error locally by moving in the direction of steepest descent. When the adjustment reaches the input nodes, the training proceeds with a new pair of input–output pattern, until all training pairs are exhausted.

One of the advantages of neural networks over other methods is robustness. Unlike serial systems, embodied knowledge is distributed over the whole network, so that a partial failure of some neurons might still produce correct output response. Another advantage is that neither an explicit mathematical model nor explicit rules are required as in expert systems.

21.10. COMBINED PARAMETRIC AND NONPARAMETRIC METHODS FOR TOOL CONDITION MONITORING

This section discusses a neural network that combines sensor input from AE, current, and force signals (8). The feature space consists not only of spectral components derived from the FFT transform of the input signals, but includes the coefficients of the multivariate autoregressive models for the input signals as well. Thus a rich amount of information is available as input to the neural network. The system is shown in Fig. 21.8 from Dornfeld (8). Feature selection limits the number of input features to the number of input units of the neural network. The input layer of the neural network has six input nodes, while the single hidden layer has four hidden nodes. The output of each neuron is a zero or one, depending on the weighted input and the associated threshold. A single output gives a binary signal signaling whether the tool is worn or not.

During the classification phase, three data blocks of sampled AE, current, and force signals are input to the FFT transformation and the AR estimation modules. The AR and FFT features of high relevance determined in the training phase are then input to the neural network, which produces a binary output signal.

The training phase consists of first deciding which M features out of the possible N features will be used and then determining appropriate neural weights. Choosing a large number of features as input to the neural network increases implementation and computational cost, while it does not necessarily improve performance. The goal is to find the M features that show minimum variability within the particular class and maximum variability between classes. Relevant features are thus selected according to the interclass criterion

$$J = \frac{\text{trace}(S_b)}{\text{trace}(S_w)} , \tag{25}$$

where S_w is the within class scatter matrix, and S_b is the between class scatter matrix. Note that the class mean scatter criterion J, which is a measure of feature relevance, increases when the between class variability increases and increases also when the within class variability decreases.

The neural connectivity weights are determined next by the back propagation scheme discussed earlier.

21.11. FUTURE DEVELOPMENTS

Advances in the diagnostics field for machining are needed in three areas, namely, in sensor development, theoretical off-line modeling, and empirical on-line diagnostics.

Sensors that are sensitive to the process parameters, reliable, and easy to mount are greatly needed. Some promising recent developments along those lines were discussed in Section 21.3. Another limitation in the present systems is that they must be trained under both normal and abnormal conditions using actual cutting experiments. This is a significant limitation, since it requires both time and effort as well as wasted parts.

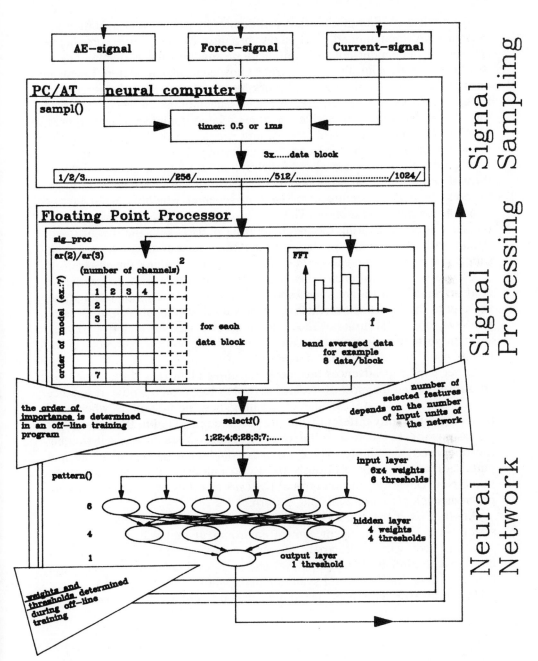

FIGURE 21.8. Neural network for tool condition monitoring. (From ref. 8, with permission.)

Theoretical off-line models that will accurately predict tool condition are presently in their infancy but show promise in the future. Predictive accuracy is dependent not only on adequate fundamental understanding of the failure modes but also on accurate knowledge of parameters based on knowledge of material, tool, and workpiece and structure characteristics. Large databases will be required for such calculations. However, much of this knowledge will be available for other related manufacturing purposes. Thus in an integrated system, tool failure prediction could become part of the simultaneous engineering and manufacturing analysis and design.

Theoretical models can also be helpful for on-line diagnostic systems, through establishing appropriate model structures for on-line parametric systems. In addition, pattern recognition schemes can be trained on the basis of computer simulation. This would eliminate costly experiments, at the same time integrating the manufacturing process with the design stage. Along with the cutting process parameters, the manufacturer could be supplied with simulation data that can be used for training the system.

Of course, this approach relies heavily on the accuracy of the predictive models. Significant progress has occurred over the last few years that points the way to the future. Nevertheless, models used for diagnostic purposes do not have to be as accurate as those that might be used for adaptive control, so even with some uncertainty about the cutting process parameters, the results could be adequate for diagnosis.

Other expected advances are along the lines of further development of empirical on-line diagnostics. Neural networks based on adaptive resonance theory (ART) (3) are capable of creating new neuron units when new patterns significantly different from the past are presented to the network. It is thus possible to create neural network diagnostic systems with self-training (unsupervised) capability. These systems could then distinguish abnormal patterns that have not previously been encountered during normal conditions. New categories of non normal patterns can thus be identified without previous explicit training on that pattern.

Another issue that must be addressed is the integration of the diagnostic system in a larger intelligent control environment that will integrate diagnosis methods with multivariate adaptive control (11). A related area of interest is the integration of low-level and high-level constructs so that both low-level data and high-level symbolic manipulation are performed.

The future of unmanned machining depends greatly on the success of such research efforts.

ACKNOWLEDGMENTS

I would like to thank Drs. K. Eman, D. Dornfeld, E. Kannatey-Asibu, S. Kapoor, and I. Minis for their valuable input in the writing of this chapter.

REFERENCES

1. Balakrishnan, P., Trabelsi, H., Kannatey-Asibu, E. Jr., and Emel, E. (1989). A Sensor Fusion Approach to Cutting Tool Monitoring, *Advances in Manufacturing Systems Integration and Processes, Proceedings of the 15th Conference on Production Research and Technology*, SME, University of California, Berkeley, pp. 101–108.

2. Blum, T. (1990). Tool Failure Detection in Milling Operations Using an Intelligent Tool Monitoring System, Laboratory for Manufacturing Automation, Research Reports 89/90, University of California, Berkeley, pp. 35–47.

3. Carpenter, G. A., and Grossberg, S. (1987). ART2: Self-Organization of Stable Category Recognition Codes for Analog Input Patterns, *Proceedings of the IEEE ICNN 87*, San Diego, CA, pp. II 727–II 736.

4. Chiu, S. L., Morley, D. J., and Martin, J. F. (1987). Sensor Data Fusion on a Parallel Processor, *Proceedings of the 1987 IEEE International Conference on Robotics and Automation*, IEEE, Raleigh, NC, pp. 1629–1633.

5. Chryssolouris, G., and Domroese, M. (1988). Sensor Integration for Tool Wear Estimation in Machining, *Proceedings of the Winter Annual Meeting of the ASME, Symposium on Sensors and Controls for Manufacturing*, pp. 115–123.

6. Chryssolouris, G., and Domroese, M. (1989). An Experimental Study of Strategies for Integrating Sensor Information in Machining, *Annals of the CIRP*, Vol. 38, No. 1, pp. 425–428.

7. Danai, K., and Ulsoi, A. G. (1987). A Dynamic State Model for On-line Tool Wear Estimation in Turning, *ASME Journal of Engineering and Industry*, Vol. 109, pp. 396–399.

8. Dornfeld, D. A. (1990). Neural Network Sensor Fusion for Tool Condition Monitoring, *Annals of CIRP*, Vol. 39, p. 1.

9. Dornfeld, D. A. (1990) Unconventional Sensors and Signal Conditioning for Automatic Supervision, *III International Conference on Automatic Supervision, Monitoring and Adaptive Control in Manufacturing*, Poland, pp. 197–233.

10. Eman, K., and Wu, S. M. (1980). A Feasibility Study of On-line Identification of Chatter in Turning Operations, *ASME Journal of Engineering and Industry*, Vol. 102, pp. 315–321.

10a. Emel, E., and Kannatey-Asibu, E. (1988). Tool Failure Monitoring in Turning by Pattern Recognition Analysis of AE Signals, ASME Journal of Engineering for Industry, Vol. 110 No. 2, pp. 137–145.

11. Hallamasek, K. (1990). Architecture for Multiple Sensor Process Monitoring, Laboratory for Manufacturing Automation, Research Reports 89/90, University of California, Berkeley, pp. 56–59.

12. Houshmand, A. A., and Kannatey-Asibu, J. E. (1989). Statistical Process Control of Acoustic Emission for Cutting Tool Monitoring, *Mechanical Systems and Signal Processing*, Vol. 3, No. 4, pp. 405–424.

13. Inasaki, I., Blum, T. Suzuki, I., Itagaki, H., and Sato, M. (1988). A Practical Mounting Device for an Acoustic Emission Sensor for the Failure Detection of Multipoint Cutting Tools, *Proceedings of the U.S.A.–Japan Symposium on Flexible Automation, Vol. 2, pp. 1017–1024*.

14. Koenigsberger, I., and Tlusty, J. (1971). *Structures of Machine Tools*, Pergammon Press, New York.

15. Kung, S., and Hwang, J. (1989). Neural Network Architectures for Robotic Applications, *Transactions on Robotics and Automation*, Vol. 5, No. 5, pp. 641–657.

16. Le Cun, Y. (1985). A Learning Procedure for Asymmetric Threshold Networks, *Proceedings of Cognitiva*, Paris.

17. Li, P. G., and Wu, S. M. (1988). Monitoring Drilling Wear States by a Fuzzy Pattern Recognition Technique, *ASME Journal of Engineering and Industry*, Vol. 110, pp. 297–300.

18. Liang, M. S., and Dornfeld, D. A. (1989). Tool Wear Analysis Using Time Series Analysis of Acoustic Emission, *ASME Journal of Engineering and Industry*, Vol. 111, No. 3, pp. 199–205.

19. Matsushima, K., Bertok, P., and Sata, T. (1982). In-process Detection of Tool Breakage by Monitoring the Spindle Current of a Machine Tool, *Proceedings of ASME Winter Annual Meeting*, Phoenix, AZ, pp. 145–154.

20. Merritt, H. E. (1965). Theory of Self-Excited Machine Tool Chatter, *Transactions of the ASME*, Series B, pp. 87–94.

21. Minis, I. E., Magrab, E. B., and Pandelidis, I. O. (1990). Improved Methods for the Prediction of Chatter in Turning, Part I: Determination of Structural Response Parameters, *ASME Journal of Engineering and Industry*, Vol. 112, pp. 12–20.

22. Minis, I. E., Magrab, E. B., and Pandelidis, I. O. (1990). Improved Methods for the Prediction of Chatter in Turning, Part II: Determination of Cutting Process to Parameters, *ASME Journal of Engineering and Industry*, Vol. 112, pp. 21–27.

23. Minis, I. E., Magrab, E. B., and Pandelidis, I. O. (1990). Improved Methods for the Prediction of Chatter in Turning, Part III: A Generalized Linear Theory, *ASME Journal of Engineering and Industry*, Vol. 112, pp. 28–35.

24. Pandelidis, I. O., and Kao, J. (1990). Detector: An Expert System for Injection Molding Diagnostics, *Journal of Intelligent Manufacturing*, Vol. 1, No. 1, pp. 49–58.

25. Parker, D. B. (1985). Learning-Logic, TR-47, Center for Computational Research in Economics and Management Science, MIT, Cambridge.

26. Ramalingam, S., and Frohrib, D. A. (1987). Real Time Tool Condition Sensing with a New Class of Sensor-Transducer System. In: *Interdisciplinary Issues in Materials Processing and Manufacturing*, ASME, New York, pp. 277–284.

27. Ramalingam, S., Shi, T., Frohrib, D. A., and Moser, T. (1988). Acoustic Emission Sensing with an Intelligent Insert and Tool Fracture Detection in Multitooth Milling, *Proceedings of the 16th North American Manufacturing Research, Conference, SME*, University of Illinois, Champaign, pp. 245–255.

28. Ranwala, S., and Dornfeld, D. (1987). Integration of Sensors Via Neural Networks for Detection of Tool Wear States, *Proceedings of the Winter Annual Meeting of the ASME*, PED 25, pp. 109–120.

29. Rumelhart, D. E., Hinton, G. E., and Williams, R. J. (1986). Learning Internal Representations by Error Propagation. In: *Parallel Distributed Processing*, MIT Press, Cambridge, MA.

29a. Schutzer, D. (1987). Artifical Intelligence and End User Computing, *Proceedings of Conference on Expert Systems in Business*, pp. 209–217.

30. Subramanian, K., and Cook, N. H. (1977). Sensing of Drill Wear and Prediction of Drill Life, *ASME Journal of Engineering and Industry*, Vol. 99, pp. 295–300.

31. Tlusty, J., and Andrews, G. C. (1983). A Critical Review of Sensors for Unmanned Machining, *Annals of the CIRP*, Vol. 32, pp. 563–572.

32. Tobias, S. S. (1965). *Machine Tool Vibrations*, Wiley, New York.

33. Wright, P. K. (1983). Physical Models of Tool Wear for Adaptive Control in Flexible Machining Cells. In: *Computer Integrated Manufacturing*, edited by Martinez, M. R. and Lev, M. C. PED-Vol. 8, ASME, New York, pp. 19–31.

34. Wright, P. K., and Bourne, D. A. (1988). *Manufacturing Intelligence*, Addison-Wesley, Reading, MA.

35. Wu, S. M., and Pandit, S. M. (1983). *Time Series and Systems Analysis with Applications*, Wiley, New York.

CHAPTER TWENTY TWO

Quality Control

ALEX MARKOVSKY and SUBHASH C. SINGHAL

AT&T Bell Laboratories, Holmdel, New Jersey

22.1. INTRODUCTION

Quality is one of the most important aspects in manufacturing, service, or any other kind of industry. Therefore *quality control* is getting more and more attention. In this chapter, some of the traditional as well as some more recent approaches of quality control are discussed. The recent approaches are discussed in more detail. A number of references provide the reader with additional information on related topics.

22.2. PROCESS CONTROL

22.2.1. Basic Concepts of Control Charts

Statistical process control (SPC) is the use of statistical techniques for measuring and analyzing process status and taking necessary actions toward maintaining and improving process capability. During the 1920s, Walter A. Shewhart of Bell Telephone Laboratories developed the theory of statistical quality control. Based on his theory, the output of each process will exhibit some kind of fluctuation over time. This fluctuation is called the *variability* of the process. Some processes will fluctuate more than others.

The variations in a process can be classified into two main categories: natural and unnatural variations. Each process has some inherent or natural variability. For example, some of the factors influencing inherent variability can be different operators, machines, or time of day. The causes contributing to this type of variability are called random or chance causes. If the variation in a process is only due to random or chance causes, the process is

Intelligent Design and Manufacturing, Edited by Andrew Kusiak.
ISBN 0-471-53473-0 © 1992 John Wiley & Sons, Inc.

considered to be in the state of statistical control. Unnatural variation in a process is attributed to some special or assignable causes. For example, some of the factors causing unnatural variation can be defective raw materials or operator errors. If variation in a process is due to special or assignable causes, the process is considered to be out of control. These assignable causes introduce additional variability in the performance of the process and should be discovered and eliminated. The variation in the process must be kept to the level of only natural variation.

The purpose of process control is to identify and eliminate the assignable causes as they occur and limit the variability to random or chance causes only. Control charts are commonly used to detect if the variation in the process is due to assignable causes. If so, corrective action to eliminate this variation can be taken before a large number of nonconforming units are produced.

Based on some basic statistical principles, it is possible to calculate the limits that define the natural variation of the process. These limits are called the *upper control limit* and the *lower control limit*. If a process operates within these limits, it is considered to be in the state of statistical control and only natural variation is present. On the other hand, if process performance falls outside these limits, the process is considered to be out of control and assignable causes of variation are present. At this time, an investigation must be carried out to discover the cause of the variation and corrective action must be taken to remove it.

There are two types of control charts: control charts for variables data and control charts for attributes data. The variables type of data is measured on a continuous scale, such as thickness, width, or strength. The attributes type of data is some kind of count of discrete measurements such as good or bad, defective or nondefective, conforming or nonconforming.

A sample control chart is shown in Fig. 22.1. Normally, the x axis of the chart represents time or order of production. Since the purpose of the control chart is to detect significant shifts in the process over time, samples

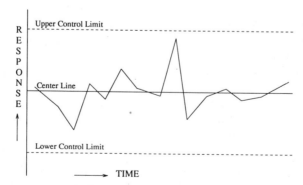

FIGURE 22.1. Sample control chart.

reflecting the status of the process are pulled over different time periods. The y axis of the chart represents the quality characteristic being measured. For the variables type of data, it may be the measured value of the response. However, for the attributes type of data, it may be fraction defective or number of defects. The center line on the chart represents the mean value of the quality characteristic. The two dashed horizontal lines represent the *upper control limit* (*UCL*) and *lower control limit* (*LCL*). Normally, control limits are three standard deviations from the center line. In the absence of variation due to assignable causes, almost all the sample values will fall within three standard deviations of the mean value of the quality characteristic. Depending on the chart, some statistics are calculated and a point is plotted for each sample. If the point is outside the control limits, the process is considered to be out of control.

22.2.2. Control Charts for Variables

Control charts for variables are based on the assumption that the distribution of the measured quality characteristic is normal. If the distribution is not normal, a control chart based on the averages of a set of samples can be used to plot points on the chart. The central limit theorem states that irrespective of the distribution of the universe of individual observations, the distribution of the averages of a set of samples will tend toward a normal distribution. The larger the sample size, the closer is the approximation to the normal distribution. Therefore these charts are applicable irrespective of the distribution of the individual observations. The three main control charts that can be used for variables type of data are:

1. Control chart for average and range ($\bar{X}$ and R).
2. Control chart for average and standard deviation ($\bar{X}$ and S).
3. Control chart for individual measurements and the moving ranges (X and MR).

Control Chart for Average and Range ($\bar{X}$ and R Charts). Each subgroup (sample) will contain more than one (typically four or five) observations. A process capability study (see Section 22.3.2) can be performed to compute the control limits for the $\bar{X}$ and R charts. Based on the process capability study, calculate the average of the sample means, $\bar{\bar{X}}$, and average of the sample ranges, $\bar{R}$. The control limits can then be computed as follows:

$$\text{Center line } (\bar{X}) = \bar{\bar{X}} \, ,$$

$$\text{Upper control limit } (UCL_{\bar{x}}) = \bar{\bar{X}} + A_2\bar{R} \, ,$$

$$\text{Lower control limit } (LCL_{\bar{x}}) = \bar{\bar{X}} - A_2\bar{R} \, .$$

$$\text{Center line } (R) = \bar{R} \text{ ,}$$

$$\text{Upper control limit } (UCL_R) = D_4\bar{R} \text{ ,}$$

$$\text{Lower control limit } (LCL_R) = D_3\bar{R} \text{ .}$$

The values of A_2, D_3, and D_4 depend on the sample size of the subgroup and are available in most statistical quality control books (17, 34).

Plot both $\bar{X}$ and R charts on the same sheet. Samples should be randomly taken at different time periods to monitor the performance of the process. For each sample, calculate the sample mean and sample range. Plot these points on the chart. If any of the plotted points is outside the control limits, it means that the process is out of control. If a point on the R chart is out of control, it indicates that the variability of the process has increased and there may be some assignable causes that are affecting the uniformity of the process. On the other hand, an out of control point on the $\bar{X}$ chart means that there may be a shift in the process level. In process monitoring, one should also look for unnatural patterns. The *AT&T Statistical Quality Control Handbook* (34) provides a comprehensive discussion on these patterns and tests that can be used to identify these patterns.

Control Chart for Average and Standard Deviation ($\bar{X}$ and S Charts). Sometimes, it may be preferable to estimate the variability of the process from the standard deviation rather than from the range of the sample. In these situations, the $\bar{X}$ and S control charts can be used. In applications where the sample size is greater than 10, it is desirable to use $\bar{X}$ and S charts. The $\bar{X}$ and S charts can be used in a manner similar to the $\bar{X}$ and R charts, but the main difference is the calculation of control limits. Based on the process capability study, calculate the average of the samples, $\bar{\bar{X}}$, and the average of the sample standard deviations, $\bar{S}$. Control limits can be computed as follows:

$$\text{Center line } (\bar{X}) = \bar{\bar{X}} \text{ ,}$$

$$\text{Upper control limit } (UCL_{\bar{x}}) = \bar{\bar{X}} + A_1\bar{S} \text{ ,}$$

$$\text{Lower control limit } (LCL_{\bar{x}}) = \bar{\bar{X}} - A_1\bar{S} \text{ .}$$

$$\text{Center line } (S) = \bar{S} \text{ ,}$$

$$\text{Upper control limit } (UCL_S) = B_4\bar{S} \text{ ,}$$

$$\text{Lower control limit } (LCL_S) = B_3\bar{S} \text{ .}$$

The values of A_1, B_3, and B_4 depend on the sample sizes of the subgroup and are available in most statistical quality control books (17, 34).

Control Chart for Individual Measurements and Moving Range (X and MR Charts). These X and MR charts plot the individual observations and can be used when only one observation per sample is available and when the underlying distribution is close to normal. These charts plot individual measurements and the range between consecutive individual measurements. For the ith sample, the moving range MR_i will be defined as $|x_i - x_{i-1}|$, where x_{i-1} and x_i are the values for $(i-1)$th and ith samples, respectively. Based on the process capability study, calculate the mean of the individual observations, $\bar{X}$, and the mean of the moving ranges of successive observations, $\overline{MR}$. Control limits can be calculated as follows:

$$\text{Center line } (X) = \bar{X},$$

$$\text{Upper control limit } (UCL_X) = \bar{X} + 2.66\overline{MR},$$

$$\text{Lower control limit } (LCL_X) = \bar{X} - 2.66\overline{MR}.$$

$$\text{Center line } (MR) = \overline{MR},$$

$$\text{Upper control limit } (UCL_{MR}) = 3.267\overline{MR},$$

$$\text{Lower control limit } (LCL_{MR}) = 0.$$

Example 1 illustrates the use of control charts ($\bar{X}$ and R) for variables to monitor a process producing amplifiers.

Example 1. Amplifiers were produced and five observations were taken in each sample. The $\bar{X}$ and R charts were used to monitor the performance of the process. A process capability study was conducted to determine the control limits of the process (see Example 2). Based on this process capability study, the control limits for the $\bar{X}$ chart were 36.940 and 36.132, and control limits for the R chart were 1.479 and 0. Samples were pulled at every hour. Table 22.1 shows the data for 20 samples.

Figure 22.2 shows the $\bar{X}$ and R charts for the data in Table 22.1.

The last point on the chart is out of control. The last point on the R chart is within control limits, but the last point on the $\bar{X}$ chart is outside control limits. This means that there has been a shift in the process level. At this point, an investigation should be made to discover the cause of this shift and a corrective action should be taken to remove it.

22.2.3. Control Chart for Attributes

Many quality characteristics cannot be measured in numerical terms but can simply be classified into one of two categories: either conforming or nonconforming to the specifications. A defect is a failure to meet an individual requirement. A unit containing one or more defects is called a defective unit. Depending on the situation, it may be better to deal with the

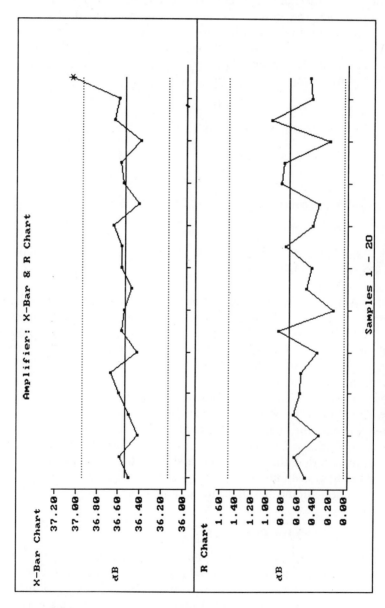

FIGURE 22.2. $\bar{X}$ and R chart: amplification data.

TABLE 22.1. Amplification Data

Sample	Observations					$\bar{X}$	R
1	36.50	36.25	36.75	36.55	36.45	36.50	0.50
2	36.33	36.50	36.73	36.96	36.40	36.58	0.63
3	36.33	36.55	36.54	36.23	36.44	36.42	0.32
4	36.11	36.75	36.64	36.42	36.61	36.51	0.64
5	36.37	36.92	36.60	36.72	36.36	36.59	0.56
6	36.40	35.45	36.83	36.95	36.74	36.67	0.55
7	36.18	36.52	36.46	36.53	36.43	36.42	0.35
8	36.21	36.44	36.64	36.53	37.05	36.57	0.84
9	36.52	36.48	36.57	36.62	36.55	36.55	0.14
10	36.25	36.73	36.49	36.36	36.55	36.48	0.48
11	36.32	36.45	36.74	36.74	36.59	36.57	0.42
12	36.23	36.57	36.64	36.45	36.98	36.57	0.75
13	36.53	36.74	36.65	36.85	36.45	36.64	0.40
14	36.55	36.33	36.23	36.54	36.39	36.41	0.32
15	36.52	36.93	36.12	36.57	36.63	36.55	0.81
16	36.64	36.98	36.21	36.49	36.56	36.58	0.77
17	36.43	36.49	36.39	36.31	36.37	36.40	0.18
18	36.15	36.55	36.73	36.67	37.08	36.64	0.93
19	36.77	36.64	36.35	36.59	36.63	36.60	0.42
20	37.22	37.10	36.95	36.78	37.14	37.04	0.44

number of defects or defectives. Control chart for defectives are based on the binomial distribution, and control charts for defects are based on the Poisson distribution. Four main control charts can be used for the attributes type of data:

1. Control chart for fraction defective (p).
2. Control chart for number defective (np).
3. Control chart for number of defects (c).
4. Control chart for defects per unit (u).

Control Chart for Fraction Defective (p). The *fraction defective*, p, is defined as the ratio of the number of defective units to the total number of units inspected. Based on the process capability study, calculate the average fraction defective, $\bar{p}$. The control limits (34) for the p chart are:

$$\text{Center line } (p) = \bar{p} \ ,$$

$$\text{Upper control limit } (UCL_p) = \bar{p} + 3\sqrt{\bar{p}(1-\bar{p})/n} \ ,$$

$$\text{Lower control limit } (LCL_p) = \bar{p} - 3\sqrt{\bar{p}(1-\bar{p})/n} \ .$$

For each sample or subgroup, compute p and plot a point on the chart. If a point is outside the control limits, it means that the process is out of control.

Control Chart for Number of Defectives (np). The *number of defectives*, np, is the number of defective units in a sample. If the sample size is constant, it is easier to plot the number of defectives found in each sample rather than calculating the fraction defective. Based on the process capability study, calculate the average number of defectives, $n\bar{p}$. The control limits (34) for this chart are:

$$\text{Center line } (np) = n\bar{p} \,,$$

$$\text{Upper control limit } (UCL_{np}) = n\bar{p} + 3\sqrt{n\bar{p}(1 - \bar{p})} \,,$$

$$\text{Lower control limit } (LCL_{np}) = n\bar{p} - 3\sqrt{n\bar{p}(1 - \bar{p})} \,.$$

For each sample of subgroup, compute np and plot a point on the chart. If a point is outside the control limits, it means that the process is out of control.

Control Chart for Number of Defects (c). This control chart is used for total number of defects, c, in a sample. The sample size could be one or more units but it must be constant for each sample. Based on the process capability study, calculate the overall average number of defects, $\bar{c}$. The control limits (34) for this chart are:

$$\text{Center line } (c) = \bar{c} = \frac{\text{total\#defects}}{\text{total\#samples}} \,,$$

$$\text{Upper control limit } (UCL_c) = \bar{c} + 3\sqrt{\bar{c}} \,,$$

$$\text{Lower control limit } (LCL_c) = \bar{c} - 3\sqrt{\bar{c}} \,.$$

For each sample or subgroup, compute c and plot a point on the chart. If a point is outside the control limits, it means that the process is out of control.

Control Chart for Defects per Unit (u). This control chart is used for average number of defects per unit, u, in a sample. This chart does not require the constant sample size, n, for each sample. Based on the process capability study, calculate the overall average number of defects per unit, $\bar{u}$. The control limits (34) for this chart are:

$$\text{Center line } (u) = \bar{u} \,,$$

$$\text{Upper control limit } (UCL_u) = \bar{u} + 3\sqrt{\bar{u}/n} \,,$$

$$\text{Lower control limit } (LCL_u) = \bar{u} - 3\sqrt{\bar{u}/n} \,.$$

For each sample or subgroup, compute u and plot a point on the chart. If a point is outside the control limits, it means that the process is out of control.

For p and u charts if the sample size is not constant, the chart may have variable control limits, which means that the control limits will vary inversely with the square root of the sample size n (24).

22.2.4. Special Control Charts

The control charts discussed in Section 22.2.2 and 22.2.3 are called Shewhart control charts and may not be very effective in detecting the small shifts in the process. There are other control chart that are more appropriate in specific applications.

Cumulative Sum Control Chart. The cumulative sum control chart will detect a small but sustained shift in the process. This chart was first developed by Page (25). The plotted point depends on the observations in the latest sample as well as the observations from the prior samples. By nature, the plotted point is a cumulative sum of deviations of the mean value of i observations from the target value. The use of cumulative charts can be justified through the procedure of *hypothesis testing* (39). Based on *sequential analysis* (39), we define three hypothesis in the sense of partition of the process:

H_0 = process is in control (*acceptable region*) ,

H_1 = process is out of control (*critical region*) ,

process is in *intermediate region*, meaning that the number of measurements is insufficient.

Beginning from $n = 1$ (i.e., from one set of measurements), one should either make a decision (H_0 or H_1) or assert that the number of the measurements is insufficient. If any hypothesis is taken, the test is finished; otherwise, the test continues. Defining a *penalty function* (typically associated with the cost of the experiment) (39) and minimizing it (typically leads to a minimization of the *mean* number of measurements corresponding to each hypothesis), one obtains a set of rules that geometrically is interpreted as a *V*-mask illustrated in Fig. 22.3.

Suppose that, after a certain period of operation in control at the level μ_0, the mean shifts to a new level μ_1 above μ_0. Each new point being added to the cumulative sum of $\bar{x}_i - \mu_0$ ($\bar{x}_i$ is an average in ith sample) will force the sum to increase and will result in a general upward trend on the chart. Similarly, if the average shifts downward, the cumulative sum will show a downward trend. Therefore the *V*-mask consists of two "arms" defining the lower and upper control limits and a "vertex" defining the location of the mask. The set of parameters needed to build a *V*-mask are the following:

α Risk of a false statement that the process shifts from the target level [error of the first kind (39)]

β Risk that any change in process shift is not detected [error of the second kind (39)]

μ_0 Acceptable (target) process level

μ_1 Rejectable process level

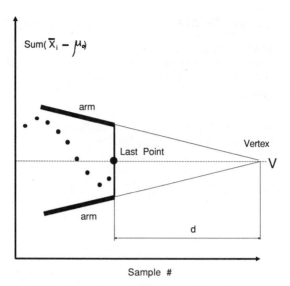

FIGURE 22.3. *V*-mask for cumulative sum chart.

D $|\mu_0 - \mu_1|$ the smallest shift in process level, whose detection is desired; that is, $D = \delta\sigma_{\bar{x}}$, where δ is a scaling number indicating the width of the error distribution that should be detected; it is important to mention that typically $\sigma_{\bar{x}}$ is not known and one should replace it with an estimate $\sigma_{\bar{x}} = \sigma_{\bar{x}_i}$, where $\sigma_{\bar{x}_i}$ is standard error of the given process with i measurements

The V-mask is constructed by plotting the sum of all data collected up to a point versus time. The vertical location (coordinate) of the vertex coincides with the vertical location of the last point. The two remaining parameters of the *V*-mask—horizontal location of the vertex and the angle between the "arms" (control limits)—are found by minimizing a penalty function. These are

$$d = 2 \ln\left[(1 - \beta)/\alpha\right]/\delta^2 = -2 \ln \alpha/\delta^2 \quad \text{if } \beta \text{ is small},$$

which is a distance between the last measurement and the vertex (i.e., horizontal coordinate of the vertex with respect to the last point), and

$$\tan \theta = \delta/2a,$$

where a is a scaling factor in multiples of standard error [$a = 2$ is recommended (25)].

If all measurements S_i defined as

$$S_i = \sum_{j=1}^{i} (\bar{x}_j - \mu_0)$$

are within the arms, then the process is in control; otherwise, the process is out of control. Several examples of cumulative sum charts applications can be found, for instance, in Montgomery (24).

Control Charts Based on Exponentially Weighted Moving Average.
This special type of chart was introduced by Wortham and Ringer in 1971; see also Montgomery (24) for applications in the chemical industry as well as for financial institutions, where single observations are more practical than subgrouping. The performance of exponentially weighted moving average (EWMA) control charts is approximately equivalent to the cumulative sum control charts but it is easier to set up and operate.

The exponentially weighted (moving) average is defined as (24)

$$\bar{x}_t = (1 - \lambda)\bar{x}_{t-1} + \lambda x_t ,$$

where $\bar{x}_t$ is exponentially weighted moving average at the present time t and $\bar{x}_0 = x_0$; $\bar{x}_{t-1}$ is the exponentially weighted moving average at the previous time step; x_t is the present observation; and $0 < \lambda < 1$ is the weighting factor for the present observation.

Using the definition of weighted average recursively, we obtain

$$\bar{x}_t = \lambda \sum_{j=0}^{t-1} (1 - \lambda)^j x_{t-j} + (1 - \lambda)^t \bar{x}_0 ,$$

where the "total weight" $\lambda(1 - \lambda)^j$ of all previous samples decreases geometrically with the age of a sample. Since the average is taken through all past and current observations, the total weight makes it very insensitive to the normality assumption. It is therefore an ideal control chart to use with individual measurements.

If all averages are independent with variance equal to σ^2/n, then the variance of $\bar{x}_t^2$ is (24)

$$\sigma_{\bar{x}_t}^2 = \frac{\sigma^2}{n} \frac{\lambda}{2 - \lambda} [1 - (1 - \lambda)^{2t}] .$$

For large t this expression converges to

$$\sigma_{\bar{x}}^2 = \frac{\sigma^2}{n} \frac{\lambda}{2 - \lambda} ,$$

where $\bar{x}$ is a limiting value of $\bar{x}_t$: that is, $\bar{x}$ is an overall mean. Consequently, the upper and lower control limits are (24)

$$UCL = \bar{x} + 3\sigma\sqrt{\frac{\lambda}{(2 - \lambda)n}} ,$$

$$LCL = \bar{x} - 3\sigma\sqrt{\frac{\lambda}{(2 - \lambda)n}} ,$$

respectively, if the sample number, n, is moderately large. See also Montgomery (24) for examples and further development.

22.2.5. Precontrol

Many processes, once set up, will perform well enough so that they need be subject to only occasional inspection. For these processes, control charts would be an unnecessary waste of time and resources. Precontrol is a natural approach in such processes. It is based on a worst-case scenario of a normally distributed process centered in the middle between specification limits and just capable of meeting specifications (i.e., precontrol is only effective in an environment in which observations fall far from the specifications limits).

Assuming that the specification spread (i.e., a distance between low and high *specification* limits) is equal to 6σ and choosing 3σ as a distance between the low and high *control* limits (Fig. 22.4), which are sometimes called precontrol limits, one can determine the probability of a case when two successive points (observations) can fall outside the precontrol limits.

The recommended set of precontrol rules can be found in Bhote (1) and Ott and Schilling (25). These are:

1. Set precontrol lines (upper and lower control limits—UCL and LCL in Fig. 22.4); the area between these lines (zone A) is called the green zone; the two areas between each control limit and each specification limit (zone B) are called the yellow zones; and the two areas (zone C) outside the specification limits are called the red zones.

2. Begin the process and choose an initial sample (usually five consecutive units from the process); if all are in zone A, the process is in control; if even one unit is in zone B, the process is not in control; conduct an investigation.

3. During the production cycle take two consecutive units periodically:
   (a) If both are in zone A, continue.
   (b) If one unit is in zone A and another one is in zone B, still continue.

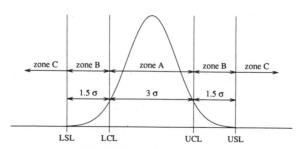

FIGURE 22.4. Precontrol.

(c) If both are in zone B, stop production and investigate the cause of variation.

(d) If even one unit is in zone C, stop production and investigate the cause of variation.

4. When the cause of variation is eliminated after each stop, the initial sample rule (e.g., five units) should be reapplied before the production resumes.

We also suggest using regression lines (trends) to predict intersection with control lines and to correct the frequency of sampling.

22.3. PROCESS CAPABILITY

What is *process capability*? The *AT&T Statistical Quality Control Handbook* (34) defines process capability as "the natural behavior of the process after unnatural disturbances are eliminated." Process capability refers to the normal behavior of the process when operating in a state of statistical control. Normally, process capability (process spread) is defined as six times the standard deviation of the process when the process is under a state of statistical control. If the process has a normal distribution, then 99.73% of the process output will fall between three standard deviations to either side of the process mean.

22.3.1. Process Capability Indices

The performance of a process is normally measured by its capability to meet the preset specification limits. These specification limits are usually referred to as the *upper specification limit (USL)* and the *lower specification limit (LSL)*, Kane (19) developed the process capability indices that can be used to quantify the capability of a process. C_p is a measure of process variability relative to the specification limits and is defined as

$$C_p = \frac{\text{specification spread}}{\text{process spread}} = \frac{USL - LSL}{6\sigma},$$

where σ is the standard deviation of the process. If the process spread is equal to the specification spread, then $C_p = 1.0$. A C_p value of 1.0 indicates that a process is capable of performing within the specification limits. As the process variability decreases, the corresponding value of C_p increases. Assuming that the process is centered on the target, the percent yield goes up as the value of C_p goes up. On the other hand, PPM (parts per million) defective goes down as the value of C_p goes up. Table 22.2 gives PPM defective and percent yield for different values of C_p.

TABLE 22.2. PPM Defective and percentage
Yield for Different Values of C_p

C_p	PPM Defective	% Yield
0.08	803,000	19.7
0.25	453,000	54.7
0.50	134,000	86.6
0.67	45,500	95.45
0.90	6,900	99.3
1.00	2,700	99.73
1.33	66	99.993
1.67	0.57	99.999943
2.00	0.002	99.9999998

Although C_p takes into account the magnitude of process variation, it ignores the effect of the deviation of the process mean from the target value. C_{pk} takes into account the magnitude of the process variation as well as the departure of the process mean from the target value. C_{pk} is defined as follows:

$$CPU = \frac{\text{allowable upper spread}}{\text{actual upper spread}} = \frac{USL - \mu}{3\sigma},$$

$$CPL = \frac{\text{allowable lower spread}}{\text{actual lower spread}} = \frac{\mu - LSL}{3\sigma},$$

and

$$C_{pk} = \text{minimum}(CPU, CPL),$$

where μ is the process mean.

Processes with a C_{pk} value below 1.00 are normally considered to be poor performing processes. When the process is centered, the C_p, CPU, CPL, and C_{pk} are equal. For an off-center process, the value of C_{pk} will always be less than the value of C_p. If the process is drifting toward the upper specification limit, the value of CPU decreases and the value of CPL increases, and hence the value of C_{pk} decreases. If the process is drifting toward the lower specification limit, the value of CPL decreases and the value of CPU increases, and hence the value of C_{pk} decreases.

Figure 22.5 illustrates different cases of a process performance and corresponding process capability indices. In each case, the variability of the process is the same. In case (a), the process is centered and the values of C_p and C_{pk} are the same: 1.33. In cases (b) and (c), the value of C_{pk} is less than the value of C_p. As the process moves further from the center, the value of C_{pk} decreases.

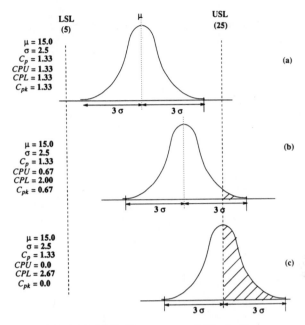

$\mu = 15.0$
$\sigma = 2.5$
$C_p = 1.33$
$CPU = 1.33$
$CPL = 1.33$
$C_{pk} = 1.33$

$\mu = 15.0$
$\sigma = 2.5$
$C_p = 1.33$
$CPU = 0.67$
$CPL = 2.00$
$C_{pk} = 0.67$

$\mu = 15.0$
$\sigma = 2.5$
$C_p = 1.33$
$CPU = 0.0$
$CPL = 2.67$
$C_{pk} = 0.0$

FIGURE 22.5. Process capability indices.

22.3.2. Process Capability Study

A process capability study is a scientific and systematic procedure for estimating process capability. The estimate of the process capability may be mean, standard deviation, control limits, or even the form of the distribution. The main purpose of the process capability study is to detect and eliminate the unnatural variation in the process. After performing the process capability study, you should know how the process should perform. You can estimate the central tendency (mean), variability (standard deviation), the statistical control limits, and capability indices of the process. These control limits can be used to monitor the performance of the process. A process capability study should be conducted under the normal operating conditions and should contain a sufficient number of observations so that the sampled data are representative of the process.

A number of techniques can be used to conduct a process capability study. Three main techniques are:

- Control charts
- Histograms/probability plots
- Designed experiments

Using control charts is the most common technique for performing a process capability study. Control charts can be used to detect the unnatural

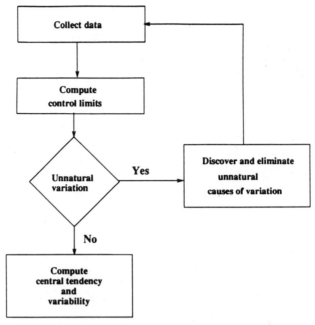

FIGURE 22.6. Process capability study.

variation, that is, to determine if the process is in statistical control or not. For variables type of data, it is preferable to use the $\bar{X}$ and R charts. For attributes type of data, p (or np) or c (or u) charts can be used.

Figure 22.6 shows the steps for performing a process capability study using control charts. In a process capability study, you collect data and compute the statistical control limits. The next step is to plot the control limits and data on a control chart to detect unnatural variation in the process, that is, if any plotted point is outside the control limits. If you find there is any unnatural variation, you should discover and eliminate the cause of it. After eliminating the causes of the unnatural variation, collect another set of data and repeat the same process. If there is no indication of unnatural variation, then the process is in a state of statistical control and the remaining variation is due to random causes. At this point, you can compute the mean and standard deviation of the process. Once the mean and standard deviation of the processes are known, process capability (control limits and process capability indices) can easily be calculated. Donnell and Dellinger (8) discuss performing process capability studies using control charts.

Example 2 illustrates the use of $\bar{X}$ and R control charts for conducting the process capability study to determine the capability of a process.

Example 2. Voice amplifiers were produced to the following specifications: upper specification limit = 37.5 dB and lower specification limit = 35.5 dB. A

TABLE 22.3. Sample Amplification Data

Sample			Observations			$\bar{X}$	R
1	36.71	36.41	36.50	36.81	37.45	36.776	1.04
2	36.65	35.64	35.75	36.74	36.75	36.306	1.11
3	36.65	36.25	36.42	36.60	36.35	36.454	0.40
4	36.35	36.24	36.15	36.24	36.58	36.312	0.43
5	36.45	36.42	36.25	36.87	36.80	36.558	0.62
6	36.25	36.39	36.25	36.23	36.35	36.294	0.16
7	36.53	36.28	36.22	36.42	36.85	36.460	0.63
8	37.30	36.30	36.42	36.45	36.47	36.588	1.00
9	36.52	36.12	36.14	37.22	36.81	36.562	1.10
10	35.90	36.22	36.73	36.54	36.44	36.366	0.83
11	36.11	36.43	37.22	36.53	36.80	36.618	1.11
12	36.23	36.54	36.76	36.89	36.55	36.594	0.66
13	36.45	36.45	36.76	36.76	36.84	36.652	0.39
14	36.44	36.53	36.12	36.32	36.81	36.444	0.69
15	37.22	36.54	36.78	36.54	36.76	36.768	0.68
16	37.41	36.52	36.53	36.45	36.66	36.714	0.96
17	36.34	36.45	36.65	36.52	36.77	36.546	0.43
18	36.92	36.56	36.87	36.43	36.58	36.672	0.49
19	36.43	36.54	36.57	36.87	36.12	36.506	0.75
20	36.54	36.46	36.86	36.44	36.35	36.530	0.51
Totals						730.72	13.99
Averages						36.536	0.70

process capability study was conducted to determine the capability of the process. Table 22.3 shows the amplification data for 20 samples with 5 observations in each sample. The calculations for the control limits of $\bar{X}$ and R charts are as follows:

R Chart:

$$\text{Center line} = \bar{R} = 0.700 \,,$$

$$\text{Upper control limit} = UCL_R = D_4\bar{R} = 2.113 \times 0.700 = 1.479 \,,$$

$$\text{Lower control limit} = LCL_R = D_3\bar{R} = 0 \times 0.700 = 0 \,.$$

$\bar{X}$ Chart:

$$\text{Center line} = \bar{\bar{X}} = 36.536 \,,$$

$$\text{Upper control limit} = UCL_{\bar{X}} = \bar{\bar{X}} + A_2\bar{R} = 36.536 + 0.577 \times 0.700 = 36.940 \,,$$

$$\text{Lower control limit} = LCL_{\bar{X}} = \bar{\bar{X}} - A_2\bar{R} = 36.536 - 0.577 \times 0.700 = 36.132 \,.$$

The values of A_2, D_3, and D_4 can be obtained from the *AT&T Statistical Quality Control Handbook* (34).

Figure 22.7 shows $\bar{X}$ and R charts for data in Table 22.3. All points on the chart fall within the control limits, and hence both charts are in the state of statistical control. The estimated process parameters are:

$$\text{Process mean } (\bar{X}) = 36.536 \,,$$

$$\text{Process standard deviation } (s) = \frac{\bar{R}}{d_2} = \frac{0.700}{2.326} = 0.312 \,.$$

The process capability indices are:

$$\hat{C}_p = \frac{USL - LSL}{6s} = \frac{37.5 - 35.5}{6 \times 0.312} = 1.068 \,,$$

$$C\hat{P}U = \frac{USL - \bar{X}}{3s} = \frac{37.5 - 36.536}{3 \times 0.312} = 1.0299 \,,$$

$$C\hat{P}L = \frac{\bar{X} - LSL}{3s} = \frac{36.536 - 35.5}{3 \times 0.312} = 1.1068 \,,$$

$$\hat{C}_{pk} = \text{minimum}(C\hat{P}U, C\hat{P}L) = \text{minimum}(1.0299, 1.1068) = 1.0299 \,.$$

Since the values $\hat{C}_p$ and $\hat{C}_{pk}$ are greater than 1.0, the process capability seems to be marginal.

As discussed earlier, if some points on the control chart are out of control, then an investigation must be conducted to discover the unnatural cause of variation. These assignable causes should be eliminated and a new set of data should be collected to determine the capability of the process.

Process capability studies can also be conducted using histograms, probability paper, or designed experiments. Histograms can be helpful in estimating the central tendency and variability of the process. Probability plots can be used in determining the shape, central tendency, and dispersion of the distribution. Designed experiments can be used for determining the optimum values of the process variables that minimize the variation in the process. The application of these techniques for performing a process capability study is discussed in more detail in Montgomery (24).

Process capability studies have a large number of applications. Some of these applications are:

1. Establishing statistical control limits of a process.
2. Selection of workers/operators.
3. Assisting developers in the design of a product.
4. Specifying requirements of new or reconditioned manufacturing equipment.
5. Scheduling work on different machines.

A more detailed discussion of these applications is provided in Feigenbaum (10).

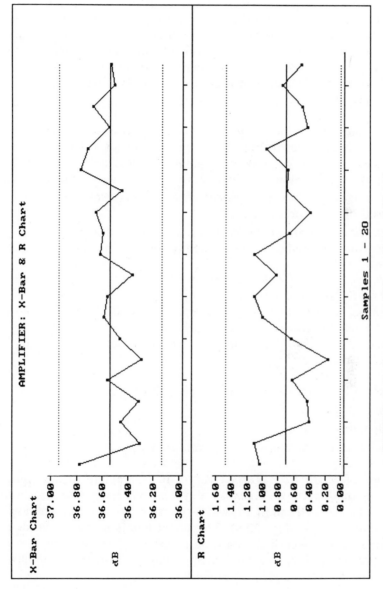

FIGURE 22.7. Process capability study: amplifier data.

22.3.3. Multiprocess Analysis Chart[1]

Control charts are widely used for monitoring individual manufacturing processes on a routine basis and are essential to the control and improvement of these processes. However, in a multiprocess environment, it can be a difficult and time-consuming task for a shop supervisor or engineer to analyze the performance or capability of all charts on a manufacturing line to evaluate the overall status of a shop. One shop in a factory may use hundreds of different control charts. Singhal (33) developed a graphical tool to illustrate the performance of a group of processes. This chart performs the following functions:

1. *Illustrates the overall status* of a group of processes in a multiprocess environment. Departure of the process mean from target value, process variability, process capability indices, and expected fallout outside specification for all processes are shown on the same chart.
2. *Prioritizes quality improvement efforts* in complex operations composed of many processes.
3. *Quantifies improvements* resulting from reduction in process mean departure and reduction in variability for each process.

The *multiprocess analysis chart* shown in Fig. 22.8 can be used to illustrate and analyze the status of a process or a group of processes. The x axis of the chart represents $C\hat{P}U$, and the y axis represents $C\hat{P}L$. The $C\hat{P}U$ and $C\hat{P}L$ values of each in-process parameter are plotted on this chart. For each parameter, the *multiprocess analysis chart* provides the following four performance measures:

1. Departure of mean from target value (T).
2. Variability.
3. Capability indices.
4. Expected fallout outside specifications.

Departure of Mean from Target Value. If a parameter mean equals its target value (*and specification limits are symmetrical around the target value*), then the values of $C\hat{P}U$ and $C\hat{P}L$ will be equal and the plotted point will lie on the target line (see Fig. 22.8). The target line is a line drawn at a 45° angle with respect to both the abscissa (x axis) and the ordinate (y axis). Accordingly, any $(C\hat{P}U, C\hat{P}L)$ data point lying on a target line will have the property that its mean equals its target value.

When the parameter mean is greater than its target value, the $C\hat{P}U$ value is less than the $C\hat{P}L$ value. This area is illustrated in Fig. 22.8 as the area in

---

[1] U.S. Patent Pending.

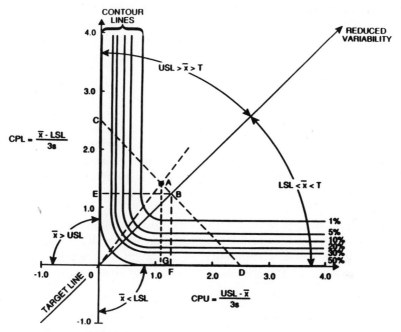

FIGURE 22.8. Multiprocess analysis chart.

the first quadrant that is above the target line. Accordingly, any parameter with a $(C\hat{P}U, C\hat{P}L)$ data point above the target line will have the property that its mean is greater than its target value. Note that when the parameter mean is equal to USL, then the $C\hat{P}U$ is equal to zero and the corresponding data point for the parameter will lie on the y axis. If the parameter mean is greater than USL, then the $C\hat{P}U$ will be negative and the data point for the parameter will lie on the second quadrant of the chart.

When the parameter mean is less than its target value, the $C\hat{P}U$ is greater than $C\hat{P}L$. This area is illustrated in Fig. 22.8 as the area in the first quadrant that is below the target line. Accordingly, any parameter with a $(C\hat{P}U, C\hat{P}L)$ data point below the target line will have the property that its mean is less than its target value. Note that when the mean is equal to LSL, then the $C\hat{P}L$ is equal to zero and the corresponding data point for the parameter will lie on the x axis. If the parameter mean is less than LSL, then the $C\hat{P}L$ will be negative and the data point for the parameter will lie on the fourth quadrant of the chart.

The relative departure of a parameter is defined as the ratio between the absolute departure of the parameter mean from the target value and the difference between the *upper or lower specification limit and the target value.*

This measure may readily be generated from the chart in Fig. 22.8. Consider the $(C\hat{P}U, C\hat{P}L)$ data point labeled A in Fig. 22.8. To determine the relative departure, construct a line through A in such a geometric manner that this line is perpendicular to and intersects the target line at B. This line also intersects the y axis at C and the x axis at D. Relative departure is equal to the ratio of the length of line segment AB to the length of line segment CB, or mathematically

$$\frac{AB}{CB},$$

where AB is the length of line segment AB and CB is the length of line segment CB. Note, through simple trigonometric relations, that the relative departure is also equal to the tangent of the angle, formed by the intersection of line OA and the target line, where OA is a line drawn though point A and through the origin O of the chart.

Variability. It is also possible to estimate the variability of a parameter relative to the spread between the upper specification limit and the lower specification limit. Relative variability will be taken to mean the ratio of the standard deviation s to the spread, USL–LSL, or mathematically

$$\frac{s}{USL - LSL}.$$

In Fig. 22.8, for every point on line CD, the ratio

$$\frac{USL - LSL}{3s}$$

is a constant and is equal to the $\hat{CPL}$ value at point C of the y axis and is equal to the $\hat{CPU}$ value at point D of the x axis. The meaning of the constant ratio is that all data points on line CD have equal variability relative to the spread between upper and lower specification limits.

These facts taken together lead one to conclude that as line CD "moves away" from the origin, that is, as the intersecting point B moves along the target line and away from the origin of the chart, intersecting points C and D move in a positive direction on the y axis and x axis, respectively. That means that the constant ratio is increasing, which for a fixed value of spread means that the standard deviation s is decreasing and the relative variability of the parameters is decreasing. This conclusion is expressed in Fig. 22.8 by the arrow on the target line pointing in the direction of the words "Reduced Variability."

The closer to the origin that intersecting points like point B occur on the target line, the higher the relative variability of the corresponding parameter. Conversely, the further away from the origin on the target line that the intersection occurs, the lower the relative variability of the corresponding parameter. It is clear that the origin is the point of maximum relative variability.

Capability Indices $\hat{C}_p$ and $\hat{C}_{pk}$. This chart provides the following capability indices:

$$\hat{C}_p = \frac{(USL - LSL)}{6s}$$

and

$$\hat{C}_{pk} = \text{minimum } (C\hat{P}U, C\hat{P}L).$$

To determine the value of $\hat{C}_p$, draw the line CD (see Fig. 22.8) as discussed earlier. The value of $\hat{C}_p$ is equal to one-half the value of $C\hat{P}U$ when evaluated at point D and is equal to one-half the value of $C\hat{P}L$ when evaluated at point C. To determine $\hat{C}_p$, construct lines BE and BF from point B so that the constructed lines BE and BF are perpendicular to the y axis and x axis, respectively. The value of $\hat{C}_p$ for parameter A can easily be read from the chart as the value of $C\hat{P}U$ when evaluated at point F or as the value of $C\hat{P}L$ when evaluated at point E, both values being equal to each other.

The measure $\hat{C}_{pk}$ is based on both departure and variability and is also available from the chart. $\hat{C}_{pk}$ is defined as the minimum of both $C\hat{P}U$ and $C\hat{P}L$.

A parameter that has a data point about the target line will have a value of $C\hat{P}U$ less than its value of $C\hat{P}L$. Therefore the minimum of $C\hat{P}U$ or $C\hat{P}L$ for that data point will always be $C\hat{P}U$. On the other hand, any parameter that has a data point below the target line will have a value of $C\hat{P}U$ greater than its value of $C\hat{P}L$. Therefore the minimum of $C\hat{P}U$ or $C\hat{P}L$ for the data point will always be $C\hat{P}L$.

The value of $\hat{C}_{pk}$ therefore is the value of the $C\hat{P}U$ for data points above the target line and is the value of $C\hat{P}L$ for data points below the target line. Any data point on the target line will have identical values of $C\hat{P}U$ and $C\hat{P}L$, and its $\hat{C}_{pk}$ will be either $C\hat{P}U$ or $C\hat{P}L$. Since point A is above the target line, its $\hat{C}_{pk}$ value is the same as its $C\hat{P}U$ value (point G).

Expected Fallout Outside Specifications. The contour lines in Fig. 22.8 represent a specific expected percentage of fallout for different pairs of $C\hat{P}U$ and $C\hat{P}L$ values. *These contour lines are based on the assumption that parameter values follow a normal distribution and the specification limits are symmetrical around the target value.* The contour lines in Fig. 22.8 are drawn for expected fallout ranging from 1.0 to 50.0%. The methodology for drawing these contour lines is discussed in Singhal (33). The expected fallout for a parameter may be estimated based on the relative location of the contour line with respect to the parameter data point on the chart.

By plotting points for many processes on one chart, this tool provides a useful graphical summary of a group of processes. It has many applications and uses for a shop supervisor and engineer:

1. Evaluating individual processes.
2. Setting priorities.
3. Quantifying improvements.
4. Communicating.
5. Identifying inconsistencies in a process.
6. Identifying special trends/conditions.

These applications are discussed in more detail in Singhal (33). This chart can also be used to illustrate a concise graphical summary of the inspection/ testing results for multiple numbers of parameters of a single lot. These could be lots of incoming material or a final product.

Singhal (33) presents an example to demonstrate the use of this chart for analyzing the performance of a group of process in a multiprocess environment.

22.4. DESIGN OF EXPERIMENT IN QUALITY IMPROVEMENT

22.4.1. Relation Between Quality and Design of Experiment

The objective of engineering design is to produce all necessary information to manufacture products satisfying customer requirements. The word *product* is immediately associated with the word *quality*. However, the last word means different things for different people (5, 10, 17, 26). According to Phadke (26), *quality* is a measure of a product performance with respect to customer requirements. As a reference point, the concept of an *ideal quality* is introduced. The ideal quality corresponds to a product that delivers the target performance each time the product is used throughout its intended life under any prescribed operating condition.

In real life, product performance deviates from one unit to another, performs differently in different operating conditions, deteriorates in time, and so on; in other words, product *loses* its quality. Mathematically, this change can be modeled by a *penalty* function of *quality loss* function that can be defined by means of deviation from a certain "ideal" level.

Quality is directly associated with the cost of the product, so that typically high quality is conflicting with low product cost. There are three main categories of cost (26) to deliver a product.

1. *Manufacturing Cost.* This part includes equipment, machinery, raw materials, labor, and repair. A typical estimate in the category is a *unit manufacturing cost* that can be minimized by using low cost material, nonexpensive equipment, and low skilled labor, but producing high quality product. The only way to achieve that is to minimize the process sensitivity to the manufacturing environment and distur-

bances, that is, to design a product and manufacturing process robustly.

2. *Operating Cost.* This category includes the cost of energy, environmental control, maintenance, and inventory of units. Product can be sensitive to the special storage conditions requiring some additional heating and/or refrigerating equipment. Again, minimization of the product cost, that is sensitivity to all the above conditions, would also require robust design.

3. *R&D Cost.* This category includes time to develop/improve a new product and the amount of engineering and laboratory resources. The major goal is to keep the two previous categories as low as possible with respect to cost. Robust design applied here (and defined below) is one of the most important tools to achieve this goal.

Manufacturing and R&D costs mainly define a purchase price of the product. Operating cost mainly defines the quality of the product. Higher quality means lower operating cost and is an important indicator for a customer to continue buying the same product from the same manufacturer.

Designing a manufacturing process to produce a product is a complicated activity. First, the designer should choose an architecture of the process and technology, that is, *concept design* (sometimes called system engineering) (3, 4, 13, 26, 30). Second, the designer should consider a *parameter design* to determine the best settings for the control factors that, at the same time, do not affect manufacturing costs. In other words, one should minimize the quality loss. This leads to the concept of the *design of experiments* that can be summarized as follows (14, 36).

1. Experiment
   (a) Statement of problem
   (b) Choice of response or dependent variables (outputs, or output variables)
   (c) Selection of factors to be varied (input, or input variables)
   (d) Choice of levels of these factors
      (i) Quantitative or qualitative
      (ii) Fixed or random (are levels of the factors to be set at certain fixed values or are such levels to be chosen at random from among all possible levels)
   (e) How factor levels are to be combined (e.g., see orthogonality conditions)
2. Design
   (a) Number of observations to be taken
   (b) Order of experimentation (e.g., the sequence of experiments can be chosen in accordance with a table of random numbers)

(c) Mathematical model to describe the experiment

(d) Hypothesis to be tested

3. Analysis

(a) Data collection and processing

(b) Computation of test statistics

(c) Interpretation of results

Robust design and its associated methodology deals with parameter design. In general, the goal of the parameter design is to force the quality loss to stay within the limits of specifications and to minimize sensitivity to noise. We typically assume that the manufacturing cost is fixed at a certain low level. However, it is not known a priori whether the actual minimum of manufacturing cost is achieved as the result of the design. Therefore it is necessary to perform the *tolerance design*. The idea of the tolerance design is to find a trade-off between reduction in the quality loss due to performance variation and the increase in manufacturing cost due to the usage of better materials (26).

For complicated products the following hierarchy of engineering design is considered.

Overall system design

Subsystem design

Component design

Each of these categories requires applications of concept design, parameter design, and tolerance design.

22.4.2. Principles of Quality Engineering

The cost of a product can be divided into two parts: manufacturing costs (before sale) and usage costs (after sale) associated with the quality loss. *Quality engineering* (6, 18, 26, 36) should reduce both of these costs and therefore should deal with engineering design, manufacturing operations, R&D, and economics.

First, one should introduce a mathematical model of the quality loss. There are many different models depending on how a product is defined to meet customer requirements. One of the simplest models is a *step function* that is based on the *fraction defective* concept. This function is set as zero if percent defective is within certain (assigned) limits and is set as a repair (or total loss) cost otherwise. However, in this particular case, simplicity of this measure is incomplete and misleading.

A more convenient model is a *quadratic loss function* centered over the number of "ideal (target) quality" units denoted x_0. Defining the *functional limit* δ as a (symmetric) deviation from the center (center is typically a number of units failed in a certain percentage of applications) and the cost of repair C, one can write this function explicitly:

$$L(x) = \frac{C}{\delta^2} (x - x_0)^2 . \tag{1}$$

Analogously, other types of quadratic quality loss functions can be defined (15, 26). They include "smaller-the-better" function (usually to minimize percent defective), "larger-the-better" (usually to maximize yield), "asymmetric" (usually to differ the directions), and others. In order to associate quality of a product with a quality loss function, one introduces a *quality characteristic* $Q(L)$ as a product response. This approach allows one to define system identification formalization, stating that the response of the product is caused by the certain input factors that are called *input variables*, *control parameters*, and so on. Now one can use a well developed and powerful technique describing the system Input–Black Box ("Machine")–Output. In this scheme, input -variables are classified into three categories:

1. *Signal*. The set of parameters (provided by the user or operator) necessary to achieve the proper response of a product. These parameters are selected by the design engineers during the product development.
2. *Noise*. The set of parameters that cannot be controlled (or are very difficult to control) by the designer and/or the manufacturer. These factors typically vary from unit to unit, time to time, and from one environment to another. Noise causes a deviation from the target response and leads to quality loss.
3. *Control Parameters* (*Input variables*). The set of parameters that can be specified (and easily changed) during product manufacturing. Values of these parameters are called *levels* and directly influence the quality of product (usually through percent defective and/or yield) and therefore the quality loss function.

Product quality characteristics vary depending on uncontrollable noise variables. If noise variables are not correlated, the variance of quality characteristics can be written in the form

$$\sigma^2 = a_1 \sigma_1^2 + a_2 \sigma_2^2 + \cdots + a_n \sigma_n^2 ,$$

where a_i is a *sensitivity coefficient* (equal to the square of the partial derivative of the system transfer function with respect to a given noise variable), and where σ_i^2 is a variance of a given noise variable.

The *robust process* is a search of the minimization procedure for the sensitivity coefficients. In that case, noise variance is minimized and the manufacturing process is stable in the sense that small deviations of the noise parameters will lead to a small deviation of quality characteristics. The robust design method described in the next section is the way of finding a robust process.

The idea is based on following. Due to unit-to-unit variation, one should interpret x_i as a realization of a random variable in each (from n) experiment and μ as an average number of units through all the experiments [i.e., $\mu = \text{mean}(x) = (x_1 + x_2 + \cdots x_n)/n$] to obtain the "average" quality characteristic:

$$Q(L) = (L(x_1) + L(x_2) + \cdots + L(x_n)) = \text{constant}\left((\mu - x_0)^2 + \frac{n}{n-1}\sigma^2\right),$$

(2)

where σ^2 is a variance of unit-to-unit (lot-to-lot) variations from its mean value μ. Because of the noise, the quality loss function deviates from its "target value" due to deviation of the quality product from its mean value of *signal* (μ) by the *variance* (σ^2). It is natural to define $\xi^2 = \mu^2/\sigma^2$ as the signal-to-noise ratio. Assuming that the number of measurements n is large enough, one can rewrite the relation for $Q(L)$ in the form

$$Q(L) = \text{constant}[(\mu - x_0)^2 + \mu^2/\xi^2].$$

(3)

Mathematically, the problem of optimization can be formalized as maximization of the signal-to-noise ratio. It is important to understand that any factor changing the mean also changes the noise variance. Therefore it is not always possible to minimize mean and standard deviation simultaneously. Roughly speaking, maximization of the signal-to-noise ratio sometimes allows one to avoid computational difficulties of an optimization procedure.

22.4.3. Robust Design of Experiment

Process variability is always part of the manufacturing process. One of the most powerful tools to understand and explain variability is to run a series of controlled experiments to analyze and create a hierarchy of subprocesses in order of their effect on the output. In reality, however, there are too many variables that could be candidates for consideration, and the answer must be obtained as soon as possible. Another difficulty that designers encounter is how one should choose the number of levels and the values of input variables on these levels. There are different design methodologies that address this question (2, 14, 20). Here we briefly describe factorial design with *orthogonal arrays* or *orthogonal contrasts* (26, 29, 32, 37), represented in the form of a matrix experiment.

The following example illustrates this approach. In the manufacturing of circuit packs on a soldering machine there is a typical defect—"solder cross"—that depends on two parameters: temperature of a preheater and height of a solder wave. The output variable is the number of solder crosses

TABLE 22.4. Control Factors on Levels "−" and "+" and Their Values

Variables	Level "−"	Level "+"
Preheating temperature (T)	475	500
Solder wave height (H)	0.24	0.26

on each board. The objective therefore is to find the optimal set of (two) variables to minimize the number of solder crosses.

Table 22.4 lists the control factors and two levels each for the setup experiment. The determination of the control factors and the test levels are based on engineering judgment.

Table 22.5 shows the design matrix. It consists of four individual experiments corresponding to the four rows. The first column is the experiment number, or run number. Columns 2 and 3 correspond to the levels of the two control factors of preheat temperature and solder wave height ("−" corresponds to the low level and "+" corresponds to the high level). The last column represents the results (number of solder crosses) of the individual experiments for each particular setting.

We illustrate how one can estimate the effect of each input variable from the observed values of crosses. First, the *overall mean* value of crosses is equal to

$$m(C) = \tfrac{1}{4}\,(C_{11} + C_{12} + C_{21} + C_{22}) = 3.25$$

The effect of the temperature (T) variable at each level is estimated on the base of common sense as

$$m(T_-) = \tfrac{1}{2}\,(C_{11} + C_{12}) = 4 \quad \text{and} \quad m(T_+) = \tfrac{1}{2}\,(C_{21} + C_{22}) = 2.5\,.$$

Similarly, the effect of the solder wave (H) variable at each level is estimated as

$$m(H_-) = \tfrac{1}{2}\,(C_{11} + C_{21}) = 2 \quad \text{and} \quad m(H_+) = \tfrac{1}{2}\,(C_{12} + C_{22}) = 4.5\,.$$

TABLE 22.5. Design of Experiment Matrix Consisting of Four Individual Experiments

Run	Parameter (Variable) Pheat T	Height H	Output Cross
1	−	−	$C_{11} = 3$
2	−	+	$C_{12} = 5$
3	+	−	$C_{21} = 1$
4	+	+	$C_{22} = 4$

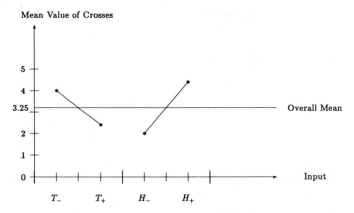

FIGURE 22.9. Effect of input variables.

We shall show later that the commonsense approach can be justified through the use of a linear model where the overall effect of all variables is estimated as $m(T) - m(C)$ and $m(H) - m(C)$ on each level. All these averages (on both levels identified with positive and negative subscripts) are shown graphically in Fig. 22.9.

The effect of each factor is typically called the *main effect*. Our primary goal is to find a setting that minimizes the number of solder crosses. From Fig. 22.9, it follows that minimal number of crosses corresponds to the settings T_+ and H_-. The result seems to be trivial and could be found directly from Table 22.5. The explanation of this triviality is that our design of experiment counted *all possible combinations* of the variables. This type of design is usually called a *full factorial design*. In general, however, a full factorial design is either not possible or not economical. In the general case, one can use only partial combination of factor levels (a *fractioned factorial design*), and therefore the predicted best setting need not correspond to any of the rows in the matrix experiment. Fortunately, in the most practical situations, the additive model provides a good approximation.

We describe the procedure of the linear modeling assuming that response of the system (output) is the linear transfer function (operator) and the error is normally distributed. In this case, response of the system (output) is approximated as

$$C_{ij} = \mu + t_i + h_j + \varepsilon_{ij}, \qquad (4)$$

where t_i and h_j are the deviations from the (overall) mean value of the output caused by the settings of T and H, respectively, and where ε_{ij} is an error (caused by an assumption on the mathematical model and errors of measurements).

For the normally distributed processes (31), the sum of all deviations from the mean should be equal to zero:

$$t_1 + t_2 = 0, \qquad h_1 + h_2 = 0. \tag{5}$$

Now our intuitive estimate of effects of each variable can be justified (26) through the model, Eq. (4). For instance, the mean value of the temperature on the level "−" is

$$m(T_-) = \tfrac{1}{2}\,(C_{11} + C_{12}) = \tfrac{1}{2}\,[(\mu + t_1 + h_1 + \varepsilon_{11}) + (\mu + t_1 + h_2 + \varepsilon_{12})].$$

Due to conditions (5), $h_1 + h_2 = 0$, we obtain that the average temperature on level "−" precisely corresponds to the effect of setting the temperature at level "−". This means that $m(T_-)$ is an estimate of $(\mu + t_1)$ within a certain error:

$$m(T_-) = \mu + t_1 + \tfrac{1}{2}\,(\varepsilon_{11} + \varepsilon_{12}) = \mu + t_1 + \text{error}.$$

The result shows the importance of condition in Eq. (5). Formally, any set of numbers a_i, $i = 1, 2, \ldots, n$, satisfying the condition

$$a_1 + a_2 + \cdots + a_n = 0, \tag{6}$$

is called *contrast*. Now the term *orthogonal array* can be associated with the formal approach in linear algebra, where orthogonality is defined as a zero internal product of two vectors. We illustrate orthogonality using Table 22.5. Considering the first and second columns as vectors—that is, $T = (t_1, t_1, t_2, t_2)$ and $H = (h_1, h_2, h_1, h_2)$—and calculating their internal product, one obtains

$$t_1 h_1 + t_1 h_2 + t_2 h_1 + t_2 h_2 = (t_1 + t_2)(h_1 + h_2) = 0,$$

by virtue of Eq. (5). Orthogonality is sometimes called the *balancing property*, meaning that all combinations of levels occur an equal number of times.

Choosing the optimal level (in our case, this is the level that minimizes the number of solder crosses), one can predict (to be within the additive linear model) the value of crosses under the optimal setting. The result is

$$C_{\min} = m + (m(T_+) - m) + (m(H_-) - m) = 1.25.$$

On average, the optimal setting will produce 1.25 solder crosses.

Further analysis (such as *analysis of variances*) on the estimate of variances of the "signal" and noise can be found, for instance, in Gabel and Roberts (11), Phadke (26), and Scheffe (28).

After determining the optimal conditions and predicting the response, one should produce an additional *verification (confirmation) experiment(s)* and compare the observed result with the predicted result. If the predicted and observed values are close to each other in the sense of given precision, then the problem is solved. If, however, observations are strongly different from the prediction, then one should conclude that the linear model is not adequate to analyze the process. Avoiding the concept of *variables interactions* (14, 26), we introduce another method, useful when one needs to come back to the experiment matrix and redefine the number of levels and/or values of the variables on each level. The idea is based on the concept of "interpolation between the levels" and the use of the initial experiment matrix to determine a "transfer function" in the scheme of *input–operator–output*, and determination of the input through the knowledge of a "target" output. This procedure is called *inverse design* (22, 23).

22.4.4. Inverse Design

Robust design, described previously, is an important tool for improving product quality and for increasing process stability. Since 1980, when Taguchi introduced this approach, the robust design problem and Taguchi's method of solving it have received serious attention in theoretical development as well as in practical applications including software development (e.g., "Robust Design" software has been developed by AT&T Bell Laboratories).

In certain classes of problems, the output is often known since it represents a *target* variable or, at least, a *target* subset of this variable. The values of the input variables that should produce a *desirable* output are normally not known and should be determined or estimated a priori. In other words, *inverse* problem (Fig. 22.10) should be solved.

Dealing with the inverse design approach, we shall try to use robust design as an initial step. The main concept is based on the attempt to determine the optimal setting of inputs, if the target values or the target intervals of the output variable are known a priori. The scheme consists of two parts.

First, in the chain *multiple input–operator–single output*, the approximation of operators is determined based on one or another modification of a regression method defined over the experiment matrix. Second, by interpolating output at the level or target interval, the "optimal" set of values of the input is determined. The inverse method is not sensitive to nonlinearities or interactions between the variables and naturally introduces the concept of interpolation. However, such formulation leads to an ill-posed problem (38).

FIGURE 22.10. Concept of inverse design.

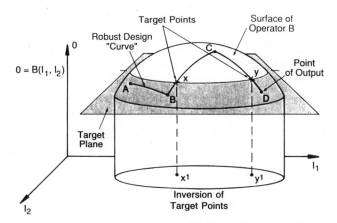

FIGURE 22.11. Geometric interpretation of inverse design.

It is easier to look at the geometrical interpretation of this problem (see Fig. 22.11). Any variable can be treated as a finite discrete subset of the interval $[X_j^{min}, X_j^{max}]$, where j is the index of a variable, in our example $X_1 = T$, $X_2 = H$. An output $[C(T, H)$ in our example] in this model is a single value function defined in an N-dimensional space. Geometrically, an output represents a set of points belonging to a certain $(N + 1)$-dimensional surface. When the output is given as a desirable value, this means that a plane parallel to the N-dimensional subspace of input variables, defined in $(N + 1)$-dimensional space, can be drawn at the level of the target value so that the intersection of this plane with the surface is a solution of our problem.

In other words, the experiment can be interpreted geometrically as an intersection between an imaginary surface of the operator with the generalized cylindrical surface defined by the incomplete (fractioned) design. The result of intersection is a certain (discrete) curve belonging to the surface of the operator. As we shall see, success in solving the inverse problem is related to the ability to build an analytic representation of the operator and parameterization of the curve(s).

In our example, each row in the experiment matrix can be interpreted as a point in a $(2 + 1)$-dimensional space, that is point

$$A = (475, 0.24, 3), \qquad B = (475, 0.26, 5), \tag{6a}$$

$$C = (500, 0.24, 1), \qquad D = (500, 0.26, 4). \tag{6b}$$

Here, we substituted values of variables from Table 22.4 into Table 22.5. Sequence of points $ABCD$ is a certain (generally speaking, randomized) curve in the $[T, H, C]$ space.

Using linear regression $(7, 35)$ with data from the experiment matrix, the following equation is obtained:

$$C(T, H) = 1.25 - 0.06T + 125H .\qquad(7)$$

This equation represents a plane as an image of the (matrix) operator, where variable T is defined over the interval $475 \le T \le 500$, and where variable H is defined over the interval $0.24 \le H \le 0.26$. Although variables T and H are defined now over the intervals, we can verify the precision of Eq. (7) only at four points from Table 22.5. Comparing C_{ij} obtained from the equation with corresponding C_{ij} from Table 22.5 and choosing the maximal difference, we obtain that the error of this equation is equal to $\delta = 0.25$.

First, we partition the curve $ABCD$ into three intervals:

$$AB = B - A = (0, 0.02, 2) , \qquad BC = C - B = (25, -0.02, -4) ,$$
$$CD = D - C = (0, 0.02, 3) .$$

Second, using the equation of the straight line interesting two given points (x_1, y_1) and (x_2, y_2) in the form

$$\frac{x - x_1}{x_2 - x_1} = \frac{y - y_1}{y_2 - y_1} = s ,$$

we obtain three different parametric representations of the curves AB, BC, and CD, respectively:

$$\frac{T - 475}{0} = \frac{H - 0.24}{0.02} = s \in [0, 1]$$
$$\text{so that } T = 475 , \quad H = 0.02s + 0.24 ;\qquad(8a)$$

$$\frac{T - 475}{25} = \frac{H - 0.26}{-0.02} = s \in [0, 1]$$
$$\text{so that } T = 25s + 475 , \quad H = -0.02s + 0.26 ;\qquad(8b)$$

$$\frac{T - 500}{0} = \frac{H - 0.24}{0.02} = s \in [0, 1]$$
$$\text{so that } T = 500 , \quad H = 0.02s + 0.24 .\qquad(8c)$$

Now we should map the output C into the interval $s \in [0, 1]$. Geometrically, this equation defines curve $A - B - C - D$ (when the parameter s changes continuously from 0 to 1) on the plane defined by Eq. (7). It is important to point out that *any change of sequence of points* changes the parameterization of the curve. Now we illustrate how to solve the inverse problem.

Let us assume that the target value of C is equal to 2. Looking at the interval $A - B$ we can see that C changes from 3 to 5 and there is no solution there (since $C = 2$ is outside this interval).

At the interval $B - C$ output C changes from 1 to 5 and we can find a solution. We substitute parametric expressions (8b) on BC into Eq. (7):

$$C = 1.25 - 0.06(25s + 475) + 125(0.02s + 0.26) \quad \text{or} \quad C = 5.25 - 4s \,.$$
$$(9a)$$

Similarly, at the interval $C - D$ output C changes from 1 to 4 and we can also find a solution. Again, we substitute parameteric expressions (8c) on CD into Eq. (7):

$$C = 1.25 - 0.06(500) + 125(0.02s + 0.24) \quad \text{or} \quad C = 1.25 + 2.5s \,.$$
$$(9b)$$

Taking $C = 2$ and using Eq. (9a), we obtain $s = 0.8125$ on BC so that

$$T = 25s + 475 = 495 \quad \text{and} \quad H = -0.02s + 0.26 = 0.244 \,.$$

Again, taking $C = 2$ and using Eq. (9b), we obtain $s = 0.3$ on CD so that

$$T = 500 \quad \text{and} \quad H = 0.02s + 0.24 = 0.246$$

As the result, one can see that the process of wave soldering is highly unstable in the sense that a small deviation of wave height (from 0.24 to 0.244 or to 0.246) leads to a large and nonlinear deviation of the number of solder crosses (from 1 to 2 on BC or from 3 to 2 on CD).

22.4.5. Generalization of Inverse Design

There are two possible directions in which inverse design can be generalized. First, instead of inverting the operator along the chosen (typically completely randomized) curve, one can prefer to define *all possible curves* by means of the $n!$ transpositions of points defined by each row. In this case, the target value of the output will define a set of points from the intersection of the target plane with the surface of the operator. All these solutions can be classified with respect to (a) proximity to each other from the defined precision, (b) signal-to-noise maximization, and (c) easiness of engineering realization.

Second, the precision of the analytic/numeric representation of an operator can be improved by means of nonlinear regression and/or splines (16, 23, 27).

It is likely that the inverse design technique may be a useful adjunct to designed experiment methods, particularly when the response surface is partially known or of particular interest, or when there are multiple design objectives.

22.5. SUMMARY

This chapter has introduced the basic concepts of control charts. For variables type of data, the $\bar{X}$ and R charts, $\bar{X}$ and S charts, and X and MR charts can be used. For attributes type of data, the p chart, np chart, c chart, and u chart can be used. Other types of charts that may be preferred for certain specific applications are also discussed. Two of these specific charts are the cumulative sum chart and the exponentially weighted moving average chart. A brief discussion of the precontrol approach is also provided.

The concept of process capability, which refers to the normal behavior of the process when operating in a state of statistical control, is discussed. We then present process capability indices and their use to quantify the capability of a process. Finally, a process capability study is conducted using control charts. The newly developed *multiprocess analysis chart* is discussed in detail. This chart can be used to analyze the performance of group processes.

The second major topic discussed is the design of experiments methodology, which is used for improving a process through the optimization procedure of selecting the best set of control parameters. In addition to the classical robust design, the newly developed concept of the inverse design is introduced. Applicability of the inverse design and internal relationship between the two designs are illustrated with an example.

REFERENCES

1. Bhote, K. R. (1988). *World Class Quality*, American Management Association, New York.

2. Burr, I. W. (1976). *Statistical Quality Control Methods*, Marcel Dekker, New York.

3. Clausing, D. P. (1988). Taguchi Method Integrated into the Improved Total Development, *Proceeding of the IEEE International Conference on Communications*, pp. 826–832.

4. Cochran, W. G., and Cox, G. M. (1957). *Experimental Design*, Wiley, New York.

5. Crosby, P. (1979). *Quality Is Free*, McGraw-Hill, New York.

6. Davies, O. L. (1954) *Design and Analysis of Industrial Experiments*, Hafner Publishing, New York.

7. Dixon, W. J., and Massey, F. J. (1957). *An Introduction to Statistical Analysis*, McGraw-Hill, New York.

8. Donnell, A. J., and Dellinger, M. E. (1990). *Analyzing Business Process Data: The Looking Glass*, Customer Information Center, AT&T, Holmdel, NJ.

9. Faddeev, D. K., and Faddeeva, V. N. (1963). *Computational Methods of Linear Algebra*, Physics–Mathematics Publishing, Moscow.

10. Feigenbaum, A. V. (1983). *Total Quality Control*, McGraw-Hill, New York.

11. Gabel, R. A., and Roberts, R. A. (1980). *Signals and Linear Systems*, Wiley, New York.

12. Grant, E. L., and Leavenworth, R. S. (1988). *Statistical Quality Control*, McGraw-Hill, New York.

13. Hauser, J. R., and Clausing, D. (1988). The House of Quality, *Harvard Business Review*, Vol. 66, No. 3, pp. 63–73.

14. Hicks, C. R. (1982). *Fundamental Concepts in the Design of Experiments*, Holt, Rinehart & Winston, New York.

15. Jessup P. (1985). The Value of Continuing Improvement, *Proceedings of the IEEE International Conference on Communications*, ICC-85, Chicago, IL.

16. John, P. W. M. (1971). *Statistical Design and Analysis of Experiments*, Macmillan, New York.

17. Juran, J. M. (1979). *Quality Control Handbook*, McGraw-Hill, New York.

18. Kackar, R. N. (1986). Taguchi's Quality Philosophy: Analysis and Commentary, *Quality Progress* Vol. 12, pp. 21–29.

19. Kane, V. E. (1986). Process Capability Indices, *Journal of Quality Technology*, Vol. 18, No. 1, January, pp. 41–52.

20. Kempthorne, O. (1952). *The Design and Analysis of Experiments*, Wiley, New York.

21. Markovsky, A., Soules, T. F., Chen, V., and Vukcevich, M. R. (1987). Mathematical and Computational Aspects of a General Viscoelastic Theory, *Journal of Rheology*, Vol. 31, No. 8, pp. 785–813.

22. Markovsky, A. (1988). Development and Applications of Ill-Posed Problems in the USSR, *Applied Mechanics Reviews*, Vol. 41, No. 6.

23. Markovsky, A. (1991). Inverse Design, Concept and Algorithm, to be published

24. Montgomery, D. C. (1991). *Introduction to Statistical Quality Control*, Wiley, New York.

25. Ott, E. R., and Schilling, E. G. (1990). *Process Quality Control*, McGraw-Hill, New York.

26. Phadke, M. S. (1989). *Quality Engineering Using Robust Design*, Prentice-Hall, Englewood Cliffs, NJ.

27. Saks, S., and Zygmund, A. (1971). *Analytic Functions*, Elsevier Publishing, London.

28. Scheffe, H. (1959). *Analysis of Variance*, Wiley, New York.

29. Seiden, E. (1954). On the Problem of Construction of Orthogonal Arrays, *Annals of Mathematical Statistics*, Vol. 25, pp. 151–156.

30. Sellivan, L. P. (1986). Quality Function Deployment, *Quality Progress*, June, pp. 39–50.

31. Shapiro, S. S. (1980). How to Test Normality and Other Distributional Assumptions. In: *The ASQC Basic References in Quality Control*, Vol. 3, ASQC.

32. Shoemaker, A. C., and Kacker, R. N. (1988). A Methodology for Planning Experiments in Robust Product and Process Design, *Quality and Reliability Engineering International*, Vol. 4, pp. 95–103.

33. Singhal, S. C. (1990). A New Chart for Analyzing Multiprocess Performance, *Quality Engineering*, Vol. 2, No. 4, pp. 379–390.

34. Small, B. B. (1956). *AT&T Statistical Quality Control Handbook*, AT&T, New York.

35. Snedecor, G. W., and Cochran W. C. (1967). *Statistical Methods*, Iowa State University Press, Ames.

36. Taguchi, G. (1978). Off-Line and On-Line Quality Control System, *International Conference on Quality Control*, Tokyo, Japan.

37. Taguchi, G., and Konishi, S. (1987). *Orthogonal Arrays and Linear Graphs*, ASI Press, Dearborn, MI.

38. Tichonov, A. N., and Arsenin, V. L. (1974). *Methods of Solving Ill-Posed Problems*, Nauka, Moscow.

39. Wald, A. (1950). *Statistical Decision Functions*, Wiley, New York.

CHAPTER TWENTY THREE

Inspection Systems

KWEI TANG

Department of Quantitative Business Analysis, Louisiana State University

23.1. INTRODUCTION

The basic functions of inspection in a manufacturing system are to eliminate or reduce nonconforming (defective) items and to provide information for manufacturing fault diagnosis and process improvement. The two basic forms of inspection operations are sampling and screening (100% inspection). In the past, screening was not considered an efficient inspection method because it was expensive, time consuming, and error prone, especially when the inspection was performed by humans. This has led, in the last several decades, to extensive research and implementation of sampling methods in both incoming material inspection and process control.

However, because of recent advances in automated testing equipment (ATE) using laser, pattern recognition, and other sophisticated machine vision techniques, screening is becoming an attractive practice. In electronics industries, for example, many sophisticated ATEs have been used in assembly lines to test integrated circuit (IC) chips, subassemblies, and finished items. This equipment can efficiently process a large number of items and produce consistent and accurate results. In fact, producers have been trying aggressively to incorporate inspection as an inherent part of manufacturing, so that all the measurements of an item in different manufacturing stages are maintained throughout the course of manufacturing, and so that corrective actions can be taken immediately after processing (25, 31). For example, in many manufacturing systems in the Toyota Group, products are measured immediately after each important manufacturing operation. Before the next item is processed, the operation is adjusted if a deviation from the target (ideal) value is found (35).

To incorporate the inspection function into a manufacturing system,

Intelligent Design and Manufacturing, Edited by Andrew Kusiak.
ISBN 0-471-53473-0 © 1992 John Wiley & Sons, Inc.

manufacturing personnel face a dilemma. They need timely and correct diagnoses of system faults to avoid producing excess scraps or nonconforming items; however, at the same time, they do not want manufacturing to be slowed down or interrupted by inspection operations. This problem can be resolved, in some situations, by using an efficient inspection instrument. For example, in a pickup assembly line at General Motors, only 2 seconds were available for inspecting 90 holes in a truck's front end for correct punching (44). If a system fault occurred without being detected, about 1200 defective front ends could be produced before the next inspection period. Using a machine-vision system, AUTOVISION II, all the front ends were checked and the process was stopped if three consecutive nonconforming front ends were found. Similar applications have been found in various other industries.

In addition to using more efficient inspection instruments, inspection efficiency can also be improved by using statistical methods. For example, a surrogate variable can be used when inspection on the quality characteristic of interest is costly, time consuming, or destructive. Owen et al. (20) gave an example, where the quality characteristic of interest was the strength of welding by which an automobile seat was attached to the frame. Measuring the welding strength required destructive testing. However, the x-ray penetration was known to be negatively correlated with the welding strength and thus could be used as a surrogate variable for inspection.

To address the issue associated with a single inspection operation is relatively less complicated than that of designing an inspection system for a manufacturing system, which requires a systematic approach to address the question concerning how to deploy inspection functions in the whole system. The purpose of this chapter is to discuss the basic issues associated with inspection operations, as well as those in complex manufacturing environments.

In the next section, basic issues pertaining to the design of an inspection operation are discussed. These issues include the formulation methods for inspection decisions, inspection error, and the methods of improving inspection efficiency. In Section 23.3, more complicated situations—including designs of multicharacteristic and multistage inspection systems—are addressed. In Section 23.4, statistical process control (SPC) methods and an expert systems approach to manufacturing fault diagnosis are discussed.

23.2. BASIC INSPECTION ISSUES

In this section, several fundamental issues concerning the design of an inspection plan are discussed. These issues include two formulation concepts associated with inspection decisions, inspection error and its effects on inspection plans, and several statistical methods for improving inspection efficiency.

23.2.1. Formulation of Inspection Decisions

There are two basic approaches to formulating a model for an inspection decision. The first approach is called the statistically based method, which considers the stochastic factors associated with the inspection outcomes and selects an inspection strategy to meet given statistical goals. The second approach is called the economically based method, which considers both the economic and stochastic factors associated with the decision and finds an inspection strategy to optimize the expected economic consequence. These formulation concepts are separately illustrated as follows.

The selection of a lot-by-lot attribute acceptance sampling plan is used to explain the statistically based method. Suppose there is a production lot consisting of N items to be inspected. A sample of size n is randomly drawn from the lot and inspected. If the number of nonconforming items in the sample is larger than the predetermined acceptance number c, the lot is rejected. Otherwise, the lot is accepted. Acceptable quality level (AQL), defined as the maximal lot nonconforming rate that is acceptable to the purchaser, is often used as the criterion to select a sampling plan. However, a lot with a nonconforming rate smaller than or equal to AQL may be rejected because of sampling variation. The chance of rejecting an acceptable lot is called the producer's risk, which should be controlled at a low level to protect the producer.

When the lot size is relatively much larger than the sample size, the binomial distribution is used to compute the likelihood of each possible sampling outcome. Specifically, the probability of observing x nonconforming items in the sample for a given nonconforming rate p is obtained by the following equation:

$$f(x) = \binom{n}{x} p^x (1-p)^{n-x}, \qquad x = 0, 1, 2, \ldots, n. \tag{1}$$

For a given sampling plan (n, c), the probability of rejecting a lot is

$$p(\text{rej}) = 1 - \sum_{x=0}^{c} \binom{n}{x} p^x (1-p)^{n-x}. \tag{2}$$

As a result, the producer's risk is obtained by using $p = $ AQL in the last expression.

Example 1. Consider a sampling plan with $n = 20$ and $c = 1$. If AQL is 1%, then the producer's risk is given by

$$p(\text{rej}) = 1 - \sum_{x=0}^{1} \binom{20}{x} p^x (1-p)^{n-x} \approx 1.69\%. \tag{3}$$

This suggests that, when 100 acceptable lots with a defective rate of 1% are submitted for inspection, an average of 1.7 lots will be rejected. Based on this computation procedure, an appropriate combination of n and c can be selected to control the producer's risk at a given level for a selected AQL.

Many statistically based acceptance sampling and process control methods have been well established and are widely used today. Discussion of these methods can be found in most standard quality control textbooks, such as Duncan (6), Feigenbaum (7), and Montgomery (17).

The second approach to designing an inspection plan is based on economic criteria. It is known that most inspection decisions in a manufacturing system are binary; that is, to decide whether corrective actions should be taken. For example, the decision in process control is to determine whether a manufacturing process should be stopped for possible adjustments or repairs. Three costs are therefore considered in this decision. The first cost, which is often referred to as the inspection cost, is the cost of collecting information for this decision. This cost includes the expense of testing materials, labor, equipment, and so on. The second cost is incurred because of corrective actions such as scrapping, reworking, or selling rejected items at reduced prices. The third cost is that incurred because of the imperfect quality when no corrective action is taken. The exact cost structure is dependent on product, manufacturing, and sales environments. For example, if the nonconforming items are moved to the next manufacturing stage for further processing, the third cost is the investments on these items in the downstream stages if the items are not repairable and cannot be used for other purposes.

The economic model for single-item inspection (screening) is used to illustrate this approach. Consider an inspection decision to determine whether an item should be accepted and moved to the next manufacturing stage, or rejected and subjected to corrective actions.

To determine the tolerance limits for a continuous quality characteristic for this go or no-go decision, a mathematical model can be formulated. Let τ be the target value of the quality characteristic of interest, and let Y denote the deviation of the quality characteristic from the target value. Let L and U be the tolerance limits and the item having a Y value outside the interval $[L, U]$ is rejected. The probability of rejecting an item is determined by

$$p(\text{rej}) = 1 - \int_{L}^{U} f(y)\,dy , \qquad (4)$$

where $f(y)$ is the probability density function of Y. Let C_r denote the per-item cost associated with the disposition of the rejected item; then the expected cost of rejection (ECR) is

$$\text{ECR} = C_r p(\text{rej}) . \qquad (5)$$

Let $C_a(y)$ be the cost incurred because of imperfect quality of accepted items, which is a function of y. The per-item expected cost of acceptance (ECA) is

$$\text{ECA} = \int_{L}^{U} C_a(y)f(y)dy \, . \tag{6}$$

Let s denote the cost of inspection; then the per-item expected total cost (ETC) associated with the decision is

$$\text{ETC} = \text{ECR} + \text{ECA} + s \, . \tag{7}$$

It was shown (37) that the optimal tolerance limits L^* and U^*, which minimize ETC, are the values where

$$C_a(y) = C_r \, . \tag{8}$$

To explain this solution, the following quadratic loss function advocated by Taguchi (34) is used:

$$C_a(y) = ky^2 \, , \tag{9}$$

where k is a positive constant. Then

$$[L^*, U^*] = [-\sqrt{C_r/k}, \sqrt{C_r/k}] \, . \tag{10}$$

As illustrated in Fig. 23.1, the decision is simply to reject an item when the acceptance cost is larger than the rejection cost, and to accept the item otherwise. Note that this decision is independent of the distribution $f(y)$.

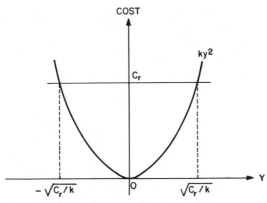

FIGURE 23.1. The optimal tolerance limits for single-item inspection.

Economic models have been developed to minimize the expected cost or to maximize the expected profit in various acceptance sampling, process control, and screening plans (16, 14, 46).

23.2.2. Inspection Precision

Just like manufacturing operations, inspection instruments and systems have inherent variability (error). For attribute inspection, there are two types of error. A type I error occurs when a conforming item is classified as nonconforming, and a type II error occurs when a nonconforming item is classified as conforming. Define

p_1 The probability of a type I error

p_2 The probability of a type II error

For a given nonconforming rate p, the probability that an item is classified as nonconforming is given by

$$\tilde{p} = (1 - p)p_1 + p(1 - p_2) . \tag{11}$$

Consider the acceptance sampling situation. Using $\tilde{p}$ in Eq. (3), the producer's risk associated with p is obtained. It is obvious that, using the same sampling plan, the producer's risk changes when inspection errors are present. For example, if p_1 and p_2 are .01 and .05, respectively, then $\tilde{p}$ corresponding to AQL = 1% is 1.94%. The producer's risk in Example 1 becomes 5.68%, which is much larger than the original value, 1.69%. Consequently, when the effects of inspection errors cannot be ignored, statistically based procedures have to be modified to give producers the same protection. A survey on this issue was conducted by Dorris and Foote (5).

For variable inspection, inspection error is characterized in terms of bias and imprecision, where bias is the difference between the true value of the quality characteristic of an item and the average of a large number of repeated measurements of the same item, and imprecision is the dispersion among the measurements of the same item (15).

Let X denote the observed value of the quality characteristic Y. A common approach to model the inspection error of variable inspection is to define the relationship between X and Y by the conditional density function $l(x|y)$. For example, assuming $l(x|y)$ is a normal density function with mean y and variance σ_ϵ^2 implies that the inspection is unbiased and that σ_ϵ is the imprecision level. As shown in Fig. 23.2, when an item with $Y = y$ is within the tolerance limits $[L, U]$, the observed value has a chance of being outside the tolerance limits. If Y follows a normal density function with mean 0 and σ_Y, then the observed value X follows a normal distribution with mean 0 and variance $\sigma_Y^2 + \sigma_\epsilon^2$. Note that the variance of the observed value is larger than that of the original quality characteristic. Consequently, using the tolerance limits developed without inspection error consideration results in a larger probability of rejection (as shown in Fig. 23.2).

To determine the tolerance limits for screening, the model formulation (7) needs some modifications. Since the probability of rejection is a function of the pdf of X, $p(\text{rej})$ is given by

$$p(\text{rej}) = 1 - \int_L^U g(x)dx , \tag{12}$$

where $g(x)$ is the probability density function of X. Because the cost of acceptance is a function of Y, but the decision rule is based on X, the per-item cost of acceptance is given by

$$\text{ECA} = \int_L^U \int_{-\infty}^{\infty} C_a(y, \tau)l(x|y)f(y)dy \, dx . \tag{13}$$

It was derived that, when the quadratic loss function (9) is used, the optimal tolerance interval is

$$[L_\epsilon^*, U_\epsilon^*] = [-\sqrt{(C_r/k - \sigma_\epsilon^2)(\sigma_\epsilon^2/\sigma_Y^2 + 1)}, \sqrt{(C_r/k - \delta_\epsilon^2)(\sigma_\epsilon^2/\sigma_Y^2 + 1)}] . \tag{14}$$

The tolerance limits obtained from Eqs. (14) and (10) may be significantly

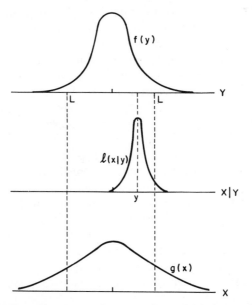

FIGURE 23.2. Inspection error in variable inspection.

different when σ_ϵ^2 is large. Note that if σ_ϵ^2 is so large that $(C_r/k - \sigma_\epsilon^2)$ becomes negative, then inspection is not economical. Therefore inspection error should be considered in designing an inspection plan when its effect cannot be ignored.

23.2.3. Improving Inspection Efficiency

When inspection is incorporated into a manufacturing system, an inspection operation is treated as a manufacturing operation. In this situation, issues such as line balancing and scheduling are also applied to the inspection operations. As a result, inspection efficiency becomes an important factor in manufacturing decision-making. Inspection efficiency can be improved by using better inspection instruments, more efficient software in generating machine vision, and so on. However, inspection efficiency can also be improved by using statistical methods. In this section, several existing statistical methods of improving inspection efficiency are introduced. These methods are sometimes product-specific or process-specific and thus may not be applicable in every application.

Selection of Inspection Precision Level. In many situations where inspection error is present, the inspection precision level can actually be considered as a decision variable. For example, the inspection precision can be improved by using the result of multiple tests on the same item. This practice has been used to test IC chips in the computer industry. Another example is that in many machine vision methods, vision precision level is dependent on the software and running time. The determination of inspection precision level is often based on the trade-off of inspection cost and the costs incurred because of inspection errors.

Tang and Schneider (40) discussed a variable inspection problem where the average of repeated measurements $\bar{y}$ is used to make an inspection decision. If n independent measurements are taken on the same item, the conditional variance of $\bar{y}$ becomes σ_ϵ^2/n, implying a higher precision level associated with a larger number of repeated measurements. However, the inspection cost is a linear function of n. The costs of acceptance and rejection are similar to those defined in Eqs. (13) and (5). The optimal number of repeated measurements and its corresponding tolerance limits are simultaneously determined by minimizing the total expected cost.

A similar approach can be used in attribute inspection, where the decision rule is based on the number of times that an item is classified as conforming in multiple inspections. As an example, suppose three independent inspections are taken on an item. Then the probabilities of observing x results showing that the item is nonconforming, given that the item is actually conforming and nonconforming, are given, respectively, as follows:

x	Conforming Item	Nonconforming Item
0	$(1-p_1)^3$	p_2^3
1	$3(1-p_1)^2 p_1$	$3p_2^2(1-p_2)$
2	$3(1-p_1)p_1^2$	$3p_2(1-p_2)^2$
3	p_1^3	$(1-p_2)^3$

Suppose the decision rule is to classify an item as conforming if $x \le 1$ and as nonconforming if $x = 2$ or 3. Then the type I error of this decision rule is equal to $3(1-p_1)p_1^2 + p_1^3$, and the type II error is equal to $p_2^3 + 3p_2^2(1-p_2)$. For numerical illustration, let $p_1 = 1\%$ and $p_2 = 1.5\%$; then the probabilities of type I and type II errors are 2.98×10^{-4} and 6.68×10^{-4}, respectively. These probabilities are much smaller than those associated with single inspection. To economically determine the decision rule, one should consider the economic consequences associated with the two types of errors and inspection cost.

Inspection Using Correlation Structure. There are two situations when using a known correlation structure can improve inspection efficiency. The first situation is using a surrogate variable that is correlated with the quality characteristic of interest as the inspection variable. The second situation is when a product has multiple quality characteristics that are correlated with one another. These situations are discussed separately as follows.

The practice of using a surrogate variable has widely been used in various industries and is attractive when inspection on the quality characteristic is not feasible, costly, or time consuming. As a simple example, the weight of a can is the quality characteristic considered in a bottling process. However, because of high manufacturing speed, the weight is not measurable. Instead, the liquid height is used as a surrogate variable, which is measurable by using an x-ray scanner. The height and the weight do not have a perfect relationship because of the vibration of the conveyor. Suppose the relationship between the height and the weight is represented by the ellipse in Fig. 23.3. A carefully determined inspection limit can be designed so that the nonconforming rate is controlled. For example, if the weight has a lower limit of 9 ounces, then all the items below this limit are considered as nonconforming. Let δ be the lower limit for the height, and all the items having a height smaller than δ are excluded from shipment; then the nonconforming rate is

$$P_d = p(\text{weight} < 9 | \text{height} \ge \delta), \tag{15}$$

which is the ratio of area III to the sum of areas II and III. Note that area I represents the decision error that conforming items are excluded, and area III represents the decision error that nonconforming items are accepted. To

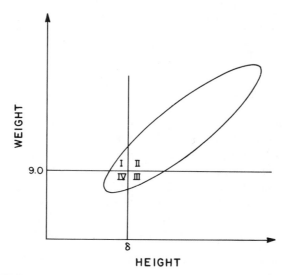

FIGURE 23.3. Inspection using a surrogate variable.

design such an inspection plan to control P_d, the joint distribution of the quality characteristic and the surrogate variable has to be known or estimated by training sample. Values of δ for different combinations of population parameters and desired nonconforming rates were tabulated by Owen et al. (21).

Another approach to selecting δ is based on a consideration of the cost associated with the disposition of rejected items and the cost of accepting nonconforming items (36). Let C_a and C_r denote the costs of accepting a nonconforming item and rejecting an item, respectively. If both C_a and C_r are constants, then the per-item expected total cost is

$$\text{ETC} = C_a p(\text{weight} < 9 | \text{height} \geq \delta) + C_r p(\text{height} \leq \delta) + s . \qquad (16)$$

The optimal limit δ^* is obtained by minimizing ETC. Note that C_a may be a function of the weight. The solution procedures for obtaining δ^* when C_a is a quadratic, linear, or step function were given in Tang (36). A reference list for related studies was given by Tang and Tang (41). When inspection on the quality characteristic is not destructive, a two-stage procedure can be developed. In Tang (37), a two-stage procedure is proposed, where in the first stage, an item is inspected on the surrogate variable. If a decision cannot be made, the item is inspected on the quality characteristic. It is shown that the two-stage procedure is attractive when the surrogate variable and the quality characteristic do not have a high correlation.

In the second situation (i.e., inspection efficiency can be improved by using a correlation structure), a product has several correlated quality characteristics to be inspected. For example, tensile strength and com-

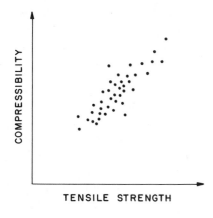

FIGURE 23.4. Some results of testing an alloy on tensile strength and compressibility.

pressibility are two important characteristics of an alloy. Testing the tensile strength is much more expensive than testing the compressibility. Suppose some test results for an alloy are plotted in Fig. 23.4. It seems that a positive relationship exists between these two characteristics. Consequently, it is possible to use the test result of compressibility to predict the result of tensile strength. Although the prediction is not perfect, a procedure can be designed to control the conforming rates of both characteristics. This approach is attractive when the characteristics are highly correlated and a large variation exists among the inspection costs of the characteristics.

Group Testing. Group testing is used when all the nonconforming items have to be removed from a production lot and it is possible that a (group) test on a pool of items can be used to detect whether the items in the pool are free of defect. If the group test indicates that the items are not free of defect, the items are retested individually, and nonconforming items are identified and removed from the lot. This inspection procedure is called the two-stage group testing procedure. The decision of designing a group testing procedure is to select an appropriate group size that minimizes total inspection efforts. The apparent trade-off of selecting a group size is that using a large brown size reduces the frequency of group tests but increases the chance of retests. It is evident that the lot defective rate is an important factor in making this decision.

A well known example for group testing is the blood testing problem given by Dorfman (4): A large number of blood samples are to be tested for contamination. Portions of blood samples can be tested together. These samples are tested individually only when the group test is positive. Hwang (10) identified many other industrial applications, including pollution test, leakage test, and flow test.

In these studies, it is assumed that there is a lot consisting of N items and that the lot defective rate, denoted by p, is a known constant. Let n denote the group size; then the expected total number of group tests and retests is equal to $N/n + N(1 - p)^n$. Samuels (27) showed that the optimal group size n^* that minimizes the expected total number of tests is

$$n^* = 1 + p^{-1/2} \,. \tag{17}$$

A cost model for this two-stage procedure was developed by Schneider and Tang (28). There are many other forms of group testing. For example, the group that was found to contain nonconforming items may be divided into subgroups, and group tests can be applied to subgroups before testing the items individually. Furthermore, group testing procedures may be modified according to the manufacturing environment. When it is applicable, group testing techniques can significantly reduce the inspection effort needed to eliminate nonconforming items.

23.3. DESIGN OF COMPLEX INSPECTION SYSTEMS

In a multistage manufacturing environment, the quality of the finished product is affected by the variation of its components as well as variations in the manufacturing operations. Consequently, an effective approach to ensure that a quality product is produced is to develop a system of inspection (monitoring) plans to control these variations. However, quality may not be the sole criterion in addressing these issues. Related manufacturing costs also have to be taken into consideration.

23.3.1. Tolerance Design

Suppose a finished product consists of n components. Let Y be the quality characteristic of the finished product, which is determined by $X_1, X_2, \ldots, X_m$, the values of m components. Tolerance design essentially involves studying the effect of the component variation on the variation of the finished product, and assigning tolerance limits to each component. Let $Y = \Psi(X_1, X_2, \ldots, X_m)$ define the relationship between Y and the components' values, where the form of Ψ may be unknown. There are several different approaches to estimate this relationship, including (a) Taylor-series methods, (b) simulation methods, (c) numerical integration methods, and (d) experimental design methods. These methods were discussed in D'Errico and Zanino (3).

When the function Ψ is available, individual tolerances can be assigned to the components so that the tolerance requirement of Y is met. This process can be illustrated by using the classical "stack tolerancing" problem, where $Y = X_1 + X_2 + \cdots + X_m$. If $X_1, X_2, \ldots, X_m$ are statistically independent and these components are assembled randomly, then the variance of Y is the sum of the individual components' variances. To numerically illustrate the procedure, suppose there are three components and the tolerance limits of

Y are 9 ± 0.09. It also requires that the length of the tolerance interval is equal to 5.15 standard deviations of Y. In other words, there is, at most, a 1.0% chance that Y is outside the tolerance interval. Therefore σ_Y^2, the variance of Y, must be smaller than or equal to 0.03^2. Furthermore, it is also known that the means of the three components μ_1, μ_2, and μ_3 are equal to 2, 3, and 4, respectively. Because the components are obtained from different suppliers and the components are assembled randomly, the mean of Y is equal to $2 + 3 + 4 = 9$, and the variance of Y is the sum of the components' variances:

$$\sigma_Y^2 = \sigma_1^2 + \sigma_2^2 + \sigma_3^2 = 0.03^2 . \tag{18}$$

A common approach is to evenly allocate the tolerance to the components:

$$\sigma_i = \sigma_Y/\sqrt{3} = 0.01732 , \quad \text{for } i = 1, 2, \text{ and } 3 . \tag{19}$$

The individual tolerance limits then are determined by

$$
\begin{aligned}
X_1 : \quad & \mu_1 \pm 2.575\sigma_1 = 2 \pm 0.044560 ; \\
X_2 : \quad & \mu_2 \pm 2.575\sigma_2 + 3 \pm 0.044560 ; \\
X_3 : \quad & \mu_3 \pm 2.575\sigma_3 = 4 \pm 0.044560 .
\end{aligned}
\tag{20}
$$

A similar approach has been applied to the situation where Ψ is a nonlinear function (see refs. 17 and 35 for further discussion).

If the natural variances of the components are identical to σ_i^2, then the nonconforming rate of the finished items is 1.0%. If all the components are inspected before assembly and the components outside the tolerance limits are excluded, then the nonconforming rate becomes smaller. A simulation study shows that the final nonconforming rate is about 0.6%.

In fact, the allocation is not unique. For example, the following allocation gives approximately the identical results:

$$
\begin{aligned}
X_1 : \quad & \mu_1 = 2.375\sigma_1 = 2 \pm 0.04114 ; \\
X_2 : \quad & \mu_2 + 2.575\sigma_2 = 3 \pm 0.04460 ; \\
X_3 : \quad & \mu_3 \pm 2.775\sigma_3 = 4 \pm 0.04806 .
\end{aligned}
\tag{21}
$$

In this allocation, the first two components have the largest rejection rate and the third component has the least rejection rate. This allocation is attractive when producing the third component is more costly.

One important assumption used to obtain the tolerance limits in Eqs. (19) and (20) is that the components' variances are equal to those allocated. However, this assumption is not valid in most applications. Consequently, it is possible to modify this method according to the natural variances of the components. Furthermore, the tolerance interval of a component may be set so wide that essentially the component is not inspected because of the high cost of inspection.

Consequently, the assignment of tolerance limits should take into account the components' variances and inspection costs. In a multistage manufacturing system, the sequence of assembling the components is also an important factor in tolerance assignments because scrapping or repairing an item in a later stage may cause a much larger loss than in an earlier stage. In addition, in an automated manufacturing environment, if the measurements of the components are recorded, then selective assembly becomes feasible. These two situations are discussed separately.

Manufacturing Cost Consideration. Suppose, in the previous example, that the three components are assembled in three consecutive operations and that subassemblies may be scrapped at any stage of operation. In the first stage, the first component is inspected according to predetermined tolerance limits. The conforming items are used in the second stage where the first and the second components are assembled. The subassembly then is inspected and only the accepted items are moved to the third stage. After the third stage, the finished items are inspected according to the tolerance limits of the finished product. In this situation, scrapping an item at a later stage is more costly because additional component(s) and operation costs have been added to the item.

Consider the following two sets of tolerance limits for the three-stage inspection:

	Set 1	Set 2
Stage 1 X_1	[1.9554, 2.0446]	[1.9571, 2.0429]
Stage 2 $X_1 + X_2$	[4.9369, 5.0631]	[4.9369, 5.0631]
Stage 3 $X_1 + X_2 + X_3$	[8.9100, 9.0900]	[8.9100, 9.0900]

Note that the second set has tighter tolerance limits in the first stage. A simulation study of 100,000 components of each type was used to compare these two sets of tolerance limits. The numbers of rejected items at the three stages were found as follows:

Stage	Set 1	Set 2	Difference
1	960	1296	-336
2	786	728	58
3	640	580	60

The results show that the second set of tolerance limits has a larger rejection rate in the first stage and smaller rejection rates in the remaining two stages. It is easy to find cases where the total manufacturing cost of using the second set is smaller. For example, when the scrapping costs at the three stages are $5, $15, and $20, respectively, the total expected cost of using the second set is $390 less than that of using the first set. Developing a set of optimal tolerance limits requires model formulation and optimization efforts. An example of this approach was given by Tang (38).

Selective Assembly. As opposed to random assembly where components are assembled arbitrarily, selective assembly matches components according to certain rules. For example, the units of each component may first be divided into several groups and then the units in a group of a component are assembled only with the units of a selected group of another component. Using this approach, nonconforming rate and product variation are reduced.

To illustrate this approach, let us consider the stack tolerancing problem with two components; that is, $Y = X_1 + X_2$. For simplicity, suppose that both X_1 and X_2 follow a normal distribution with a mean of 2 and a variance of 0.2^2, and that the tolerance limits of Y are 4.0 ± 0.424. Using random assembly, the distribution of Y is normal with a mean of 4 and a standard deviation of 0.283, resulting in a nonconforming rate of 13.36%.

As an alternative to random assembly, the units of each component are first divided into two groups. Group 1 consists of the units with values larger than the mean, and group 2 consists of the remaining units. Then the units in group 1 of the first component are randomly assembled with those in group 2 of the second component. Likewise, the units in group 2 of the first component are assembled with those in group 1 of the second component. The exact form of the distribution of Y resulting from this approach is complicated to derive. A simulation of 100,000 units showed that the nonconforming rate was only 1.8% and that the standard deviation of Y was 0.22.

This result shows that selective assembly is promising in reducing nonconforming rate and product variation. There are other possible rules that can produce even better results. Of course, when using selective assembly in a manufacturing environment, one needs to consider constraints on storage space (e.g., limited number of buffers) and scheduling (limited time for selection). At present, this method has received very little attention from researchers. However, it should become an interesting issue because of the increased use of automated inspection and manufacturing systems.

23.3.2. Inspection Effort Allocation

In a multistage manufacturing system, "where to inspect" is an important decision for both controlling manufacturing costs and providing fast diagnostics on process conditions. Several general rules concerning inspection location include (18):

1. Inspect after operations that are likely to produce nonconforming items.
2. Inspect before costly operations.

3. Inspect before operations in which nonconforming items may damage or jam machines.
4. Inspect before operations that cover up defects.
5. Inspect before assembly operations where rework is very costly.

This issue has been investigated by using mathematical programming methods. The general formulation concept can be explained as follows. Let the nonconforming rate generated by the first manufacturing stage be p_1. If there is an inspection station after this stage, all the nonconforming items produced by this stage are detected and repaired. Therefore the items entering the second stage are free of defect ($p = 0$). If no inspection takes place after the first stage, then the nonconforming rate of the items entering the second stage is $p = p_1$. Let p_2 be the nonconforming rate of the second stage. The nonconforming rate after stage 2 (before inspection) is either p_2 or $p_1 + (1 - p_1)p_2$, depending on whether the items are inspected after the first stage.

Since there are two possible choices associated with each stage, inspect or not inspect, there are totally 2^m possible actions (see Fig. 23.5), where m is

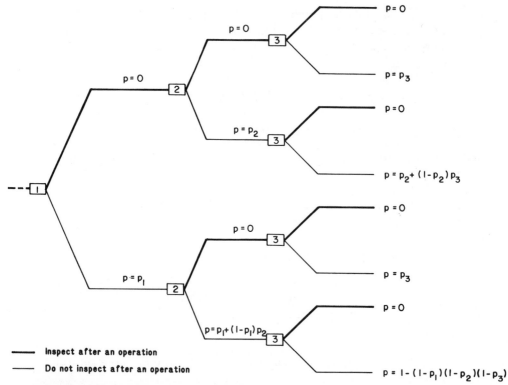

FIGURE 23.5. Decision tree for allocating inspection efforts in a three-stage inspection.

the number of stages. When inspection and repair costs are considered, the problem can be formulated as a dynamic programming problem. The optimal solution can easily be obtained by using a standard dynamic programming or integer programming algorithm. A review of the studies in this issue was given by Raz (26).

23.4. MANUFACTURING FAULT DIAGNOSIS

Manufacturing fault diagnosis includes the methods of detecting possible faults in the system and finding the causes of the faults. In this section, two useful methods are discussed. The first one is the traditional statistical process control (SPC) method, which has been used frequently to provide producers with statistical evidence about process conditions. The second method is the expert system approach, which uses recent developments in decision analysis and computer information systems to develop a causality structure between symptoms and the causes of process faults so that a user can employ the observed symptoms to find the causes of process faults.

23.4.1. Statistical Process Control Method

Items produced by a manufacturing system are not exactly equal to the predetermined target (ideal) value because of the natural variability of the system caused by the variation in manufacturing conditions, labor, materials, and other factors. Therefore observing some off-target or nonconforming items should not always lead to a conclusion that the system has a fault(s). However, it is necessary to develop methods to determine whether the deviations or nonconforming items are caused by a fault(s) in the system or by natural variability.

The SPC method is a collection of graphical and statistical tools, including process-control charts, histograms, Pareto charts, flowcharts, fishbone (cause-and-effect) diagrams, check sheets, and scatter diagrams. *Quality Progress*, a monthly publication of the American Society for Quality Control, features a series of detailed discussions about these methods, starting in June 1990. These methods are also discussed in Montgomery (17), Ott (19), and many other books. Among these tools, the process-control chart is the one most frequently used in monitoring process stability. A brief discussion of this tool follows.

Process-control charts are commonly used in industries to detect a shift in process mean and/or a shift in process variability. Process-control charts are classified, according to data type, into variable charts and attribute charts. Variable charts are used when the exact measurement of the quality characteristic can be taken. The most frequently used variable charts are the $\bar{x}$ chart, range chart, and standard deviation chart. Attribute charts are used for controlling the nonconforming rate of the process, including mainly the p chart and np chart.

The concept of process-control charts is based on a simple statistical distribution theory, which is used to model the inspection outcome of a sample randomly drawn from a process. The basic concept of the $\bar{x}$ chart is explained as follows. Let the quality characteristic of a manufacturing process be adequately described by a normal distribution with the probability density function

$$f(y) = 1/\sqrt{2\pi}\, \exp\{[(y - \mu)/\sigma^2/2\}, \qquad -\infty \le y \le \infty, \qquad (22)$$

where μ is the process mean and σ^2 is the process variance. $f(y)$ is called the process density function, which is estimated empirically by using past data. Usually, the process mean is set equal to the target value except in situations such as a tool-wearing process, where the process mean decreases over the number of items processed (29) or a canning situation where the material cost is a function of the process mean (39).

It is known that the sample mean $\bar{x}$ of n items randomly drawn from the process follows a normal distribution with mean μ and variance σ^2/n. Note that the variance of the simple distribution of $\bar{x}$ is smaller when n is larger. As shown in Fig. 23.6, if the process is "in control," the value of the sample mean has a 68.26% chance of falling between $\mu - \sigma/\sqrt{n}$ and $\mu + \sigma/\sqrt{n}$, a 95.46% chance of falling between $\mu - 2\sigma/\sqrt{n}$ and $\mu + 2\sigma/\sqrt{n}$, and a

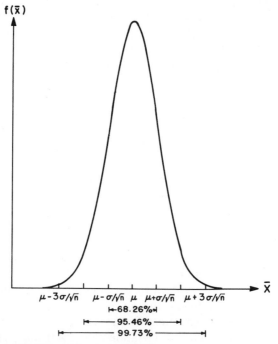

FIGURE 23.6. The sampling distribution of $\bar{x}$.

99.73% chance of falling between $\mu - 3\sigma/\sqrt{n}$ and $\mu + 3\sigma/\sqrt{n}$. In other words, the chance that the sample mean will fall outside the interval $[\mu - 3\sigma/\sqrt{n},\ \mu + 3\sigma/\sqrt{n}]$ is only 0.27% (about three times out of a thousand), which represents an unlikely result when the process is in control and suggests that an unusual situation may have happened in the process. This constitutes the popular 3-sigma rule for determining the upper control limit (UCL) and the lower limit (LCL) of the control chart:

$$UCL = \mu + 3\sigma/\sqrt{n} \tag{23}$$

and

$$LCL = \mu - 3\sigma/\sqrt{n}. \tag{24}$$

Using the control chart in Fig. 23.7, n items are drawn from the process and inspected. If the sample mean is outside the control limits, an alarm is sent out to manufacturing personnel. However, it is clear that there is a 0.27% chance of sending out a false alarm, which may cause not only a cost of finding possible faults but also a loss in manufacturing time. On the other hand, great numbers of nonconforming items may be produced before the next inspection if the process is not in control and the sample mean is still within the control limits. To illustrate this error, suppose the process mean shifts to μ_1 and the process variance stays the same. The shaded area in Fig. 23.7 represents the decision error. This may result in a large amount of scrap or may require costly rework. To avoid decision errors, larger samples and/or higher sampling frequency are required. However, increasing inspection efforts means higher inspection costs. Furthermore, sampling frequency is also affected by the frequency of out-of-control states.

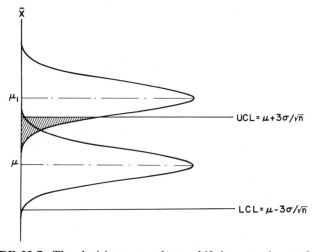

FIGURE 23.7. The decision error that a shift in mean is not detected.

Consequently, there are three parameters in designing a process control procedure: sample size, control limits, and sampling frequency (or interval between two samples). Many factors are involved in determining these parameters. The basic trade-off is between inspection cost and the economic consequence of wrong decisions.

Other variable charts for process mean include the median chart, mid-range chart, and CUSUM chart. Variable charts for process variation are the range chart, standard deviation chart, and CUSUM chart. It has been suggested that a control chart for process mean should be used, accompanied by a chart for the process variation.

Attribute charts include the np chart, p chart, c chart, and CUSUM chart. Let p_0 be the nonconforming rate when the manufacturing process is in control. If n items are randomly drawn from the process, the number of nonconforming items X is described by a binomial distribution, with a mean of np_0 and a variance of $np_0(1 - p_0)$. If the 3-sigma rule is used, the control limits for the np chart are

$$np_0 \pm 3\sqrt{np_0(1 - p_0)}. \tag{25}$$

More complicated situations occur when the process mean is not constant but has a known pattern of change over time. Special charts such as the autocorrelation (43) and tool-wearing control charts (8) are needed. A detailed discussion of both statistically based and economically based control chart designs is included in Montgomery (16, 17). The SPC tools can easily be computerized. In fact, many mainframe and PC versions of software are available on the market. A list of commercial software for SPC and other quality control methods was given in the March 1991 issue of *Quality Progress*.

23.4.2. Expert System Approach

If there is evidence indicating that the process may be in an out-of-control state, it is necessary to check the system for problems. Usually, a diagnostic expert or a group of manufacturing personnel are called on to examine the situation. Although the human diagnostic process may not be rigorously formulated, it basically involves a trace from the observed symptoms to the possible causes through a causality structure learned from experience. This causality structure may be very loose and not well organized. For a complex manufacturing system, the diagnostic process requires extensive knowledge and experience. It may take human experts a long time to identify the problems. In these situations, expert systems, which are essentially the computer models of experts' knowledge and experience, may be an effective solution.

The most common expert systems are rule-based systems, which consist of three major components: the knowledge base, a blackboard, and an inference engine. The knowledge base consists of a set of "if A then B" statements used to describe the causality relationship among various observed performance measures, operating conditions, and manufacturing faults. The blackboard is a list of possible faults in the manufacturing system, and the inference engine is a set of procedures for searching the knowledge base and the blackboard to find the possible faults for given symptoms.

The rules in the knowledge base are obtained by relating observed symptoms and the process faults. Establishing the knowledge base is the most costly and important task in developing an expert system. In general, two approaches have been suggested. The first is the inductive method, which essentially analyzes and summarizes some training examples or empirical data into decision trees by using statistical or decision analysis methods. This approach is relatively simple and requires less intensive study of the physical systems. However, it is less flexible to modify if there is a change in the manufacturing process, and it is also difficult to incorporate new faults in the knowledge base. DELTA is an example, which was developed by General Electric to assist locomotive personnel in identifying the causes of malfunctions and the procedure to correct the problems (13). Other applications were reviewed by Pau (23). Tools are available to construct inductive diagnostic rules, including RuleMaster and TIMM for large-scale systems and Expert-Ease and 1st-Class for small systems (9).

The second approach is the qualitative reasoning approach. De Kleer and Brown (2) suggested that human experts usually use qualitative measures rather than exact quantitative values to reason the behavior of a physical system. For example, an expert may know that a low or a high operating temperature of a reactor causes a low yield in a chemical process, instead of knowing the exact functional relationship. Consequently, it is appropriate to use adequate qualitative measures to describe the knowledge about a manufacturing system. In most existing studies, three qualitative values $\{+, 0, -\}$ have been used to describe the input, output, and operating conditions of a manufacturing operation. If $[a, b]$ is the acceptable region for a continuous variable V, then the qualitative measure is equal to $[0]$ if V is in $[a, b]$, is equal to $[+]$ if V is larger than b, and is equal to $[-]$ if V is smaller than a.

Using qualitative values, one can establish a qualitative and deterministic relationship between operation conditions and observable performance measurements in the systems. The relationship is called the schematic, which consists of behavior rules describing, for each manufacturing operation, the output condition responding to input condition and operation conditions. For example, consider the reactor example mentioned in the last paragraph. Suppose $[1009°, 1011°]$ is the range of the normal operating temperature inside the reactor; then T, the corresponding qualitative values, are

$$T = [0] \quad if \ 1009° \leq T \leq 1011°$$
$$= [-] \qquad\quad T < 1009°$$
$$= [+] \qquad\quad T > 1011°.$$

If the normal yield level is in the range $[25\%, 26\%]$, then the corresponding qualitative values are

$$Y = [0] \quad \text{if } 25\% \leq Y \leq 26\%$$
$$= [-] \qquad Y < 25\%$$
$$= [+] \qquad Y > 26\% .$$

Then the behavior rules associated with the reactor are:

If $T = [0]$, then $Y = [0]$.
If $T = [-]$, then $Y = [-]$.
If $T = [+]$, then $Y = [-]$.

These behavior rules have to be developed for all the important operations in the system. The diagnosis process is a trace from abnormal qualitative values $[-]$ or $[+]$ of observable performance measurements to the abnormal operating variables. A complete enumeration of all possible combinations may be impossible for a large system; therefore some assumptions may be needed to simplify the diagnosis process. For example, Shaw and Menon (30) used a single-fault assumption, which assumes that only one fault occurs at a given time. In addition, efficient procedures for searching the abnormal operating variable are also needed. An application of this approach in a torque-converter manufacturing process was given by Shaw and Menon (30).

Successful development and implementation of an effective expert system require careful planning, effective knowledge acquisition, reliable software and skillful computer programming, and extensive testing and validation. A general procedure for developing a diagnostic expert system is suggested as follows. The first step is to analyze the manufacturing system and develop the system structure. In a large manufacturing system, a better strategy is to use a distributed system that partitions the diagnostic problems into several smaller subsets of problems according to product, manufacturing line, manufacturing operation, quality characteristic, defect, and inspection station. The main reasons are (a) it is expensive to develop and maintain a large-scale expert system, and (b) successful applications of expert systems are often found in narrowly defined areas. This step is critical to the success of the whole effort and requires extensive interaction with users, manufacturing engineers, and quality engineers. In some cases, statistical analysis of the system using statistical tools, such as experimental design (1), can help system designers to study and develop a system structure.

Based on the partition defined in the system structure, prototypes are developed for selected subsystems and tested with real or simulated cases. In the process of developing prototypes, the feasibility of the whole project, the software, hardware, personnel, and methods of collecting, inputting, and maintaining inspection results are reviewed. If the prototypes are successfully developed, they are expanded to other subsystems. The complete systems are then validated and tested. The detailed procedures and tools of developing expert systems have been discussed in many books on artificial intelligence and expert systems, for example, Harmon et al. (9) and Tauban (33).

23.5. CONCLUSION

In this chapter, the issues of designing inspection plans and diagnostic systems in manufacturing systems were discussed. Although many successful applications of these methods have been found in practice, many methods are still under development. In particular, the expert systems approach to manufacturing faults diagnoses still has limited success. The difficulty may be attributed to the restrictive methods used in knowledge representation. Recently, Pandelidis and Kao (22) used fuzzy set theory to represent the relations between causes and defects in an injection-modeling process. Using the fuzzy set theory, inexactness in knowledge, such as ambiguity and vagueness, can be represented in the knowledge base. This may be a promising method. Another important issue is the integration of inspection and diagnostic functions into the manufacturing management system. The issues of integrating inspection systems with inventory systems and scheduling systems have been addressed theoretically (14, 24). However, the interactions between inspection functions and other manufacturing functions in automated manufacturing environment have not been addressed.

ACKNOWLEDGMENT

The author was supported, in part, by National Science Foundation Grant No. DDM-8857557.

REFERENCES

1. Box, G. E. P., Hunter, W. G., and Hunter, J. S. (1978). *Statistics for Experimenters*, Wiley, New York.

2. De Kleer, J., and Brown, J. S. (1984). A Qualitative Physics Based on Confluences, *Artificial Intelligence*, Vol. 24, No. 1–3, pp. 7–83.

3. D'Errico, J. R., and Zanino, N. A. (1988). Statistical Tolerancing Using a Modification of Taguchi's Method, *Technometrics*, Vol. 30, No. 4, pp. 397–405.

4. Dorfman, R. (1964). The Blood Testing Problem, *Applied Statistics*, Vol. 13, pp. 43–50.

5. Dorris, A. L., and Foote, B. L. (1978). Inspection Errors and Statistical Quality Control: A Survey, *AIIE Transactions*, Vol. 10, No. 2, pp. 183–192.

6. Duncan, A. J. (1986). *Quality Control and Industrial Statistics*, Irwin, Homewood, IL.

7. Feigenbaum, A. V. (1983). *Total Quality Control*, McGraw-Hill, New York.

8. Gibra, I. N. (1967). Optimal Control of Processes Subject to Linear Trends, *Journal of Industrial Engineering*, Vol. 18, No. 9, pp. 35–41.

9. Harmon, P., Maus, R., and Morrissey, W. (1988). *Expert Systems: Tools and Applications*, Wiley, New York.

10. Hwang, F. K. (1984). Robust Group Testing, *Journal of Quality Technology*, Vol. 16, No. 4, pp. 189–195.

11. Kackar, R. N. (1985). Off-Line Quality Control, Parameter Design, and the Taguchi Method, *Journal of Quality Technology*, Vol. 17, No. 4, pp. 176–188.

12. Kivenko, K., and Oswald, P. D. (1974). Some Quality Assurance Aspects of Automatic Testing, *Journal of Quality Technology*, Vol. 6, No. 3, pp. 138–145.

13. Kumara, S. R. T., Joshi, S., Kashyap, R. L., Moodie, C. L., and Chang, T. C. (1986). Expert Systems in Industrial Engineering, *International Journal of Production Research*, Vol. 24, No. 5, pp. 1107–1125.

14. Lee, H., and Rosenblatt, M. (1987). Simultaneous Determination of Production Cycle and Inspection Schedules in a Production System, *Managment Science*, Vol. 33, pp. 1125–1136.

15. Mei, W. H., Case, K. E., and Schmidt, J. W. (1975). Bias and Imprecision in Variable Acceptance Sampling: Effect and Compensation, *International Journal of Production Research*, Vol. 13, No. 4, pp. 327–340.

16. Montgomery, D. C. (1980). The Economic Design of Control Charts: A Review and Literature Survey, *Journal of Quality Technology*, Vol. 12, No. 2, pp. 75–87.

17. Montgomery, D. C. (1991). *Introduction to Statistical Quality Control*, Wiley, New York.

18. Moore, F. G. (1973). *Production Management*, Irwin, Homewood, IL.

19. Ott, E. R. (1975). *Process Quality Control*, McGraw-Hill, New York.

20. Owen, D. B., Li, L., and Chou, Y. M. (1981). Prediction Intervals for Screening Using a Measured Correlated Variate, *Technometrics*, Vol. 23, pp. 165–170.

21. Owen, D. B., McIntire, D., and Seymour, E. (1975). Tables Using One or Two Screening Variables to Increase Acceptance Product Under One-Sided Specifications, *Journal of Quality Technology*, Vol. 7, No. 3, pp. 127–138.

22. Pandelidis, I. O., and Kao, J. F. (1990). DETECTOR: A Knowledge-Based System for Injection Molding Diagnostics, *Journal of Intelligent Manufacturing*, Vol. 1, No. 1, pp. 49–58.

23. Pau, L. F. (1986). Survey of Expert Systems for Fault Detection, Test Generation and Maintenance, *Expert Systems*, Vol. 3, No. 2, pp. 100–111.

24. Peters, M., Schneider, H., and Tang, K. (1988). Joint Determination of Optimal

Inventory and Quality Control Policy, *Management Science*, Vol. 34, No. 8, pp. 991–1004.

25. Plossl, K. (1988). Production in the Factory of the Future, *International Journal of Production Research*, Vol. 26, No. 3, pp. 501–506.

26. Raz, T. (1986). A Survey of Models for Allocating Inspection Effort in Multistage Production Systems, *Journal of Quality Technology*, Vol. 18, No. 4, pp. 239–247.

27. Samuels, S. M. (1978). The Exact Solution to the Two-Stage Group Testing, *Technometrics*, Vol. 20, pp. 497–500.

28. Schneider, H., and Tang, K. (1990). Adaptive Procedures for the Two-Stage Group Testing Problem Based on Prior Distribution and Costs, *Technometrics*, Vol. 32, pp. 379–405.

29. Schneider, H., Tang, K., and O'Conneide, C. (1990). Optimal Control of a Production Process Subject to Random Deterioration, *Operations Research*, Vol. 38, pp. 1116–1122.

30. Shaw, M., and Menon, U. (1990). Knowledge-Based Manufacturing Quality Management, *Decision Support Systems*, Vol. 6, pp. 59–81.

31. Stile, E. M. (1987). Engineering the 1990s Inspection Function, *Quality Progress*, Vol. 10, No. 11, pp. 70–71.

32. Suresh, N. C., and Meredith, J. (1985). Quality Assurance Information Systems for Factory Automation, *International Journal of Production Research*, Vol. 23, No. 3, pp. 479–488.

33. Tauban, E. (1988). *Decision Support and Expert Systems*, Macmillan, New York.

34. Taguchi, G. (1984). *Quality Evaluation for Quality Assurance*, American Supplier Institute, Romulus, MI.

35. Taguchi, G., Elsayed, E. A., and Hsiang, T. (1989). *Quality Engineering in Production Systems*, McGraw-Hill, New York.

36. Tang, K. (1987). Economic Design of a One-Sided Screening Procedure Using a Correlated Variable, *Technometrics*, Vol. 29, No. 4, pp. 477–485.

37. Tang, K. (1988). Design of a Two-Stage Screening Procedure: A Loss Function Approach, *Naval Research Logistics*, Vol. 35, No. 5, pp. 513–533.

38. Tang, K. (1990). Design of Multi-Stage Screening Procedures for a Serial Production System, *European Journal of Operational Research*, Vol. 52, pp. 280–290.

39. Tang, K., and Lo, J. (1991). Determination of the Process Mean When Inspection Is Based on a Correlated Variable, *IIE Transactions*, to be published.

40. Tang, K., and Schneider, H. (1988). Selection of the Optimal Inspection Precision Level for a Complete Inspection Plan, *Journal of Quality Technology*, Vol. 20, No. 3, pp. 153–156.

41. Tang, K., and Tang, J. (1989). Design of Product Specifications for Multi-Characteristic Inspection, *Management Science*, Vol. 35, No. 6, pp. 743–756.

42. Trippi, R. (1974). An On-Line Computational Model for Inspection Resource Allocation, *Journal of Quality Technology*, Vol. 6, No. 4, pp. 167–174.

43. Vasilopoulos, A. V., and Stamboulis, A. P. (1978). Modification of Control Chart Limits in the Presence of Data Correlation, *Journal of Quality Technology*, Vol. 10, No. 1, pp. 20–35.

44. Villers, P. (1984). Intelligent Robots: Moving Toward Megassembly. In: *The AI Business*, edited by P. H. Winston and K. A. Prendergast, MIT Press, Cambridge, MA, pp. 205–222.

45. Wadsworth, H. M., Stephens, K. S., and Godfrey, A. B. (1986). *Modern Methods for Quality Control and Inprovement*, Wiley, New York.

46. Wetherill, G. B., and Chiu, W. K. (1975). A Review of Acceptance Sampling Schemes with Emphasis on the Economic Aspect, *International Statistics Review*, Vol. 43, pp. 191–210.

PART V

STANDARDIZATION IN DESIGN AND MANUFACTURING

CHAPTER TWENTY FOUR

Standardization in Design and Manufacturing

JUAN R. PIMENTEL*

Universidad Politecnica de Madrid, DISAM, Madrid, Spain

MACIEJ ZGORZELSKI

Department of Mechanical Engineering, GMI Engineering & Management Institute, Flint, Michigan

24.1. INTRODUCTION

Traditionally, a *standard* has been defined as that which is established by authority as a rule or convention for the measurement of quality, weight, extent, value, or quantity. In this chapter, however, we are concerned with such rules as the ones involved in the interaction, interface, performance, processing, control, and communication of design and manufacturing elements. There are two major types of standards: de jure and de facto. Whereas the former are established by standard making organizations such as ANSI in the United States and ISO internationally, the latter are conventions established as a result of common use of an industry sector. In the decade of the 1990s not only the value of standards is increasing, but standard developments are also proceeding at a faster rate. There are currently many standards that have reached maturity and are being used in modern, computer-based industrial automation while others are in the beginning stages. Standards have tremendous implications in today's highly competitive environment. The objects of this chapter are to discuss the status, applications, and future of standards in the areas of advanced design and manufacturing systems. It is impossible to provide a comprehensive description of all or even the majority of standards in design and manufac-

---

*This work was completed while the author was at GMI Engineering & Management Institute, Flint, Michigan

Intelligent Design and Manufacturing, Edited by Andrew Kusiak.
ISBN 0-471-53473-0 © 1992 John Wiley & Sons, Inc.

turing. The approach taken in this chapter is to discuss a few fairly well advanced standards directly related to the present trends in manufacturing system development and briefly mention standards in other areas.

Standards for advanced design and manufacturing are not fully developed yet. There are many areas where the work is just beginning (e.g., manufacturing application languages); in others (e.g., MAP/TOP, IGES/PDES), significant progress has been made. Because of their potential future impact, the standards chosen for discussion here are IGES/PDES, MAP/TOP, the ISO Reference Model for Shop Floor Control, and the CIM–OSA architecture for CIM. IGES is a standard for representing product data such as those typically developed by CAD systems. Whereas IGES has been in use for some time, its successor PDES is in the final stages of standardization. The ISO reference model provides the nomenclature, function, interfaces, and relationships for *shop floor production* in the context of other CIM functional elements. CIM–OSA defines an architecture for integrating the various CIM components. Both the ISO reference model and the CIM–OSA architecture were finalized around mid-1990. MAP/TOP is a set of specifications for computer communications in the manufacturing, design, and office environments and they are based on ISO–OSI communication standard protocols. The current MAP/TOP specification was completed in 1987. The chapter also makes references to applications where these standards are being used. Finally, a discussion of future issues in standards for modern design and manufacturing is included.

24.2. WHY STANDARDS?

The use of standards brings the following benefits for users and vendors of manufacturing equipment:

- Facilitates and enables integration. This is perhaps the single most important benefit for users and manufacturing equipment. Standards allow the integration of equipment from different vendors even when the underlying technologies used are vastly different. Standards will enable manufacturing processes that were previously fairly isolated from one to another to work in unison.
- Facilitates the development of powerful engineering tools. The success of CIM depends on the existence of powerful engineering tools for analysis, design, operation, and management. Effective engineering tools built to standards will be widely accepted.
- Performance improvement. The improvement of performance is measured through several indicators, for example, a reduction in lead time for production, a reduction of development time for the product, and shortened system debug time.

- Ease product procurement. Users can choose products that conform to a certain standard from different vendors as long as their products meet the standards specifications.
- Ease maintainability of products. Whether they are de jure or de facto, standards are well documented and well known to professionals working in the area. Typically, a variety of tools exist for testing, debugging, diagnosing, monitoring, and managing functions that are standardized. Thus it is relatively easy to maintain products that are built to standards.
- Increase marketability of equipment built to standards. This advantage applies mostly to equipment vendors. Standards are something desirable; otherwise why bother with them. By making products that the industry wants, vendors can increase their marketability.
- Increase cost effectiveness. Developing a standard is a formidable effort typically lasting several years. For the case of de jure standards, most of the work is coordinated and managed by the standards making organizations with the bulk of the work being done by many volunteers from companies interested in the standards. Thus the cost of fully developing a standard is shared. Better yet, some companies just wait until the dust is settled and simply buy the already developed standard for a minimum charge; thus the cost of developing the standard is saved.

24.3. WHO'S WHO IN STANDARDS

We begin by listing the major organizations developing or coordinating the development of standards for design and manufacturing.

The most significant worldwide and U.S. national standards organizations are:

International Organization for Standardization (ISO). This worldwide voluntary standards association, based in Geneva, is composed of 77 countries' standards-setting bodies. It has over 150 technical centers, writing standards in many areas. One task, for example, is to create OSI (Open Systems Interconnection) standards through Subcommittee 21 of its Technical Committee 97 (ISO TC97/SC21) for Information and Data Processing.

International Electrotechnical Commission (IEC). Founded in 1906. This international standards organization has a membership comprising 43 countries that produce over 95% of the world's electric energy. It develops standards covering the entire field of electrotechnology, including electric power apparatus, electronics, appliances, radio communications, and transportation equipment.

International Consultative Committee for Telephone and Telegraph (CCITT). The national telecommunication companies around the world gathered in this organization, chartered by the United Nations, to set telecommunication standards. CCITT has a great deal of input into ISO standards, and final standards issued by the two groups in areas of common interest are frequently identical. The U.S. State Department heads the U.S. delegation to CCITT.

National Electric Manufacturers Association (NEMA). This old and influential trade association of electrical equipment manufacturers develops standards for products produced by its members.

American National Standards Institute (ANSI). This group, based in New York City, is the U.S. member of the International Organization for Standardization (ISO). It coordinates the writing of many U.S. domestic standards and approves them.

National Institute of Standards and Technology (NIST) (formerly *National Bureau of Standards, NBS*). NIST is the U.S. government organization responsible for basic physical and measurement standards in the United States, and for research and development of scientific measuring methods. It also develops computer and information systems standards used by the U.S. government, called Federal Information Processing Standards.

Institute of Electrical and Electronics Engineers (IEEE). In its role as the technical and professional association for electrical engineers, the IEEE recommends standards in many areas of electrotechnology. It acts internationally through the U.S. National Committee for the IEC, and through ANSI (the U.S. member-body) for ISO.

Corporation for Open System (COS). This consortium of vendors and users was formed in 1985 to accelerate the development and commercial availability of interoperable computer and communications equipment and services that conform to ISDN, the Open System Interconnection (OSI) model, and related international standards.

Electronics Industries Association (EIA). This trade association represents manufacturers of electronic equipment and components, telecommunications equipment, radios, and televisions.

European Committee for Standardization–European Committee for Electrotechnical Standardization (CEN—CENELEC). Major equipment manufacturers join European standard-setting bodies and user groups in this organization for Common Market members. It has the main standardization role for computers in Europe.

Manufacturing Automation Protocol & Technical and Office Protocols (MAP/TOP) User Group. This group includes professionals and corporations involved with communications integration in the areas of engineering, office, and manufacturing automation. It became a technical group for the Society of Manufacturing Engineers (SME) in 1985. It works to develop existing and emerging international communications standards and accepted operating practices.

24.3.1. Steering Committee on Industrial Automation (SCIA)

As expected, ISO and IEC have a close cooperation through their *Joint Technical Planning Committee* (JTPC), which includes the Steering Committee on Industrial Automation (SCIA). Basically, the SCIA coordinates the standards developments within ISO TC184, IEC TC65, and IEC TC44. We briefly describe those technical committees in turn.

24.3.2. ISO/TC 184: Industrial Automation Systems and Integration

Of particular importance to design and manufacturing is the ISO TC 184, which is concerned with standardization in the field of industrial automation systems and integration concerning discrete parts manufacturing and encompassing the application of multiple technologies (e.g., information systems, machines and equipment, and communications). ISO/TC 184 is currently composed of the following five subcommittees:

1. *TC 184/SC 1: Physical Device Control.* This includes but is not limited to codes, formats, axis and motion nomenclatures, data structures, command languages and related system aspects, programming methods, and requirements for information exchange for the control of physical devices.
2. *TC 184/SC 2: Robots for Manufacturing Environment.* The areas within the scope of this subcommittee include definition, characterization, terminology, performances and performance testing methods, safety, mechanical interfaces, programming methods, and requirements for information exchange.
3. *TC 184/SC 3.* (Not currently active).
4. *TC 184/SC 4: Manufacturing Data and Languages.* Concerned with standardization in the field of data and languages including neutral representation, methodology, testing, and implementation for manufacturing applications.
5. *TC 184/SC 5: Architecture and Communications.* In addition to architecture and communication, this subcommittee is concerned with an industrial automation glossary and the requirements for global programming languages environment.

As already noted, the following IEC technical committees are also of interest because of their role in industrial automation.

24.3.3. IEC/TC 44: Electrical Equipment of Industrial Machines

Standards for electrical equipment (e.g., numerical controls) of industrial machines (e.g., sewing machines) fall under IEC/TC 44.

24.3.4. IEC/TC 65: Industrial Process and Control

Standards developed by IEC/TC 65 are in the area of data communications for measurement and control systems, message data formats, programmable measuring apparatus, electromagnetic compatibility, system elements (e.g., temperature sensors), and system considerations (e.g., system considerations and safe software).

24.4. SIGNIFICANT STANDARDS

As already noted, we cannot discuss all the significant standards in just one chapter. We limit our discussion to standards for modern, computer-based product and design data (IGES/PDES), computer communications in the manufacturing environment (MAP/TOP), and computer integrated manufacturing (CIM) architectures and control. In the standards-making world, there are currently three major activities related to CIM architectures and control (13):

1. ISO/TC184/SC5/WG1 Reference model, parts 1 and 2.
2. ISO/TC184/SC5/WG1 CIM framework for integration (under development).
3. CEN/CENELEC CIM–Open System Architecture (CIM–OSA).

Whereas the ISO Reference model identifies standard areas that have been or need to be standardized, CIM–OSA identifies suitable structures for CIM integration. As far as CIM architectures and control are concerned, we limit our discussion to the ISO Reference model (parts 1 and 2) and CIM–OSA.

24.4.1. IGES/PDES: The Product Data Exchange Standards

In all advanced manufacturing systems, product definition data originate from computer aided design (CAD). The design activity leads to the generation of a computer file (or files) in a proprietary (frequently even undisclosed) format, native to the CAD system used. As the number of different CAD systems and their versions, being used worldwide, amounts to several hundreds, there exists an extremely complex issue of compatibility between these systems and their respective file formats. This is usually referred to as the product data exchange problem. It is particularly complicated in the case of large companies and governmental organizations, which usually have to work with multiple suppliers and customers, exchanging product and product-related (e.g., tool and fixture) data in both directions with these external sources. The U.S. military services were among the first to realize the importance and complexity of the problem and started the first efforts to standardize, as early as the mid-1970s. Large automotive and aerospace companies, both in the United States and in Europe, soon followed. From these early efforts, two major standardization efforts, of particular relevance to the topic of this book and thus to be described briefly, emerged: IGES and PDES.

Initial Graphics Exchange Specification (IGES)

Brief History. Coordinated effort to produce a standard to exchange graphical data between dissimilar CAD systems started in 1979 as a Department of Defense initiative, with support from the Air Force ICAM program, the NASA IPAD project, and the National Bureau of Standards. The first IGES specification was published in 1980, and in 1981 it was approved by ANSI as an American National Standard Y14.26M. Ever since, these efforts have been going on; the current version of the standard is 5.0 (NBSIR 1988).

IGES Capabilities. IGES currently enables the exchange of product data models in the form of 2D and 3D wireframe representations, fully surfaced representations, and solid model representations. Applications supported by IGES include the traditional engineering drawing, 3D mechanical part models for analysis or for numerical control machining, and electronic part models for printed circuit boards and cabling and wiring systems. For these purposes, IGES provide what is usually referred to as a *neutral file format*, that is, one that various CAD systems are expected to be able to translate to and from. This function is provided by the CAD system vendors themselves, in the form of IGES preprocessor and postprocessor software (usually an option a customer may decide to add to the system if the data exchange problem seems to be of importance).

   An IGES file consists of an ordered set of five sections: the *start* section, the *global* section, the *directory entry* section, the *parameter data* section, and a *terminate* section. The key parts are the directory entry section specifying *entities* entered into the drawing (i.e., lines, arcs, surfaces, etc.) an the *parameter data* section, which contains geometric data (coordinates, etc.) unique to each instance of the entity in the drawing.

Translation Problems. Given the number of CAD systems currently in the marketplace, with their own, frequently unique entity constructs and their different internal representations, it is virtually impossible for one neutral file format to support all systems. Fortunately, there is a large subset of entities that can easily be mapped between dissimilar systems using IGES.

However, in spite of over a decade of development, one still cannot rely in 100% on IGES translation, unless special precautions are made (e.g., on the sending system side, entities that are known to cause problems in translation are never used by the designers) or manual clean-up of the resulting file is undertaken. Nevertheless, IGES is currently used in everyday practice of hundreds of manufacturing companies in the United States which face the everyday need of electronic exchange of drawings.

Alternatives to IGES. Because of the problems with IGES translation, alternate approaches to the whole problem have been developed. Among the most popular ones, two are worth mentioning. One is the DXF (Data Exchange File) format, originally developed by the Autodesk Company, for their main CAD product, AutoCAD. Because this package has been the most popular PC-based CAD software (over 60% of the market) for several years, most other vendors developed DXF translators. It thus became a de facto industry standard, although again it is not free from the problems related to IGES, as these are of a more general nature, inherent to any neutral file approach. The second alternative is the use of *direct translators*, developed and tuned to work correctly for pairs of systems. Several companies develop and market these software products. They usually guarantee a very high quality of translation, but in the frequent case of multiple, dissimilar systems a large number of such translators may have to be used and this is obviously both a costly and a troublesome alternative.

International Aspects. While IGES has enjoyed a relatively wide acceptance in the United States, it has never become an ISO standard and the general issue of international CAD graphics exchange standardization so far has not found a satisfactory solution. Several European countries developed national standards; there are also in existence national de facto standards.

Relationship to Computer Graphics Standardization Efforts. Almost totally independently of the CAD graphics exchange standardization efforts, the "general" computer graphics community has been working, both locally in the United States and internationally, on a broad range of general computer graphics standards (such as Graphical Kernel System—GKS, Computer Graphics Metafile—CGM, and Programmers Hierarchical Interactive Graphics System—PHIGS). These efforts had little impact on practical resolution of the very specific CAD graphics exchange problems (15). Nevertheless, they certainly do have some influence on graphical display technology (some CAD systems employ, for instance, PHIGS as their standard approach to image processing functions).

Product Data Exchange Specification: PDES

Background. For quite some time it has already been obvious that IGES, although constituting some step forward, really will never be able to address some of the issues of the presently emerging modern, highly automated and integrated manufacturing enterprises. In computer integrated manufacturing, heterogeneous "paperless' CAD/CAM environments, and intelligent manufacturing systems, the simple exchange of drawings no longer suffices. We see the need to allow shared access to product databases and automated processing of problems, which presently require large amounts of manual, highly skilled, analytic work. The IGES technology assumes that there are people at both ends of the pipeline, to encode meaning into the drawings at the sending end and decode this meaning at the receiving end (even when the IGES translation is 100% correct). This is due to the fact that an engineering drawing in the form we use today, while readily understandable by an engineer, is nearly totally incomprehensible to the computer. For instance, an engineer is needed to look at a drawing to determine how many holes are in a flange, because the concept of a "hole" does not exist within the framework of most present-day CAD systems and is not reflected in IGES technology. An IGES representation of a hole may consist of, say, two circles and several lines, or a cylindrical surface, with each of these being indistinguishable from a solid bolt or shaft.

If downstream manufacturing applications are eventually to be automatically inferred from product design data, much more sophisticated computer-based representation techniques have to be developed; exchange of these representations (part and product models) rather than present-day exchange of drawings thus becomes the objective. These are the premises of the Product Data Exchange Specification (PDES) project (14).

PDES Project Organization. The PDES project evolved naturally from the IGEs project and is currently carried out under the auspices of the IGES/PDES Organization led by the National Institute of Standards and Technology (previous NBS). Most of the standardization work within the IGES/PDES Organization is performed by voluntary participation of members from various interested companies, research institutions, and governmental agencies. Although the IGES effort is currently and will continued to be in parallel with PDES initial efforts, probably the last IGES version ever published will be Version 5.0; following that, IGES functionality will be incorporated into PDES. Specifications developed by the IGES/PDES group are submitted for ANSI approval through the Y14.26 Subcommittee of ASME and for international standard approval through ISO (PDES, different from IGES, is being developed in close contact with ISO, as discussed later). In parallel with these activities, since 1988 a consortium of several major companies (both users and software developers) operates under the name of PDES, Inc., with the mission of accelerating research and implementation of PDES.

PDES Project Approach and Contents. PDES is aimed at transferring between CAD/CAM systems (and other applications) part and product models, rather than purely geometric information. In addition to geometry,

PDES concepts support a wide range of data of nongeometric nature, such as design features, material properties, manufacturing specifications (such as tolerancing, surface finish, and heat treatment). PDES model is thus capable in principle of supporting functions such as computer aided process planning, NC part program generation, and inspection and robotic assembly program generation.

PDES modeling methodology is based on a three-layer architecture: an applications layer, a logical layer, and a physical layer. Applications layer models have been developed mostly using two graphical information modeling techniques: IDEF1x and Nijssen Information Analysis Methodology (NIAM). The resulting models are analyzed for consistency at the logical layer and translated into a conceptual schema expressed in EXPRESS—a data specification language developed specifically for the purpose of PDES (within this project). At the physical layer the conceptual schemes is either used to create a database structure or formed into an EXPRESS based ASCII file format for exchange. The First Working Draft of PDES (Version 1.0) was published in 1988 (NTIS 1988).

STEP: The International Effort. The PDES project, which should be seen more as cooperative research rather than a simple "straightforward" standardization undertaking, constitutes the base of an international effort known as STEP—Standard for the Exchange of Product Model Data. It is channeled through ISO TC184/SC4, with 23 countries participating in this Subcommittee's work. Although the international standard acceptance procedure is rather lengthy, it is important to note that a general agreement exists that a common international product data exchange standard is indeed needed, and STEP/PDES is currently the only viable candidate for that role.

Practical Applications of Product Data Exchange Technology. Although problems with the present status of the product data exchange technology certainly do exist (with IGES being not 100% reliable and not internationally accepted and PDES still in its very early phase of development), serious practical applications in modern manufacturing do exist. For instance, in a recent survey of about 100 major automotive supplier companies in the United States, undertaken by the Automotive Industries Action Group—AIAG (AIAG 1990), over 50% indicated that they are using IGES (as the format *required to do business*) for engineering data exchange with various manufacturing partners (the remaining companies indicating various proprietary data formats). As far as PDES is concerned, certainly of primary importance is the U.S. Department of Defence CALS (Computer-aided Acquisition and Logistic Support) initiative, a huge effort to go "paperless" in product data exchange for military equipment acquisitions and servicing. CALS depends very strongly on timely and successful development of PDES and its practical implementations. The U.S. Navy RAMP (Rapid Acquisition of Manufactured Parts) project, which is one of the CALS initiative constituents, may serve as an example of a PDES-based manufacturing system already in place (3).

24.4.2. MAP/TOP

This section is intended to provide an overview of the different architectures that have been developed for MAP and TOP. We only describe the latest architectures since earlier ones have been mostly used as stepping stones to produce more stable architectures. Furthermore, the architectures described in this chapter have been used for actual implementations. The architectures described are MAP 2.2, MAP 3.0, MAP–EPA, mini-MAP, and TOP 3.0. Earlier architectures include MAP 2.0, MAP 2.1, and TOP 1.0.

General Motors (GM) and Boeing were the early developers and promoters of MAP and TOP, respectively. Further developments involved greater cooperation of GM and Boeing particularly in those standards that are common to both MAP and TOP (e.g., FTAM). Currently, the MAP and TOP specifications are being administered by the MAP/TOP users group.

MAP is a specification for a set of communication protocol standards which enables communication among computer and other intelligent devices interconnected by a local area network in manufacturing environments. TOP is a similar specification primarily intended for technical (e.g., design) and office automation environments. MAP and TOP are based on the famous OSI (open system interconnection) reference model developed by ISO in the early 1980s. Basically, the OSI reference model partitions the overall communication functions into seven groups. The communication groups are organized hierarchically, and each group of functions constitutes a *layer* of communication functions, as illustrated in Fig. 24.1. A brief explanation of the functions of each layer follows. For additional details the reader is referred to Pimentel (12).

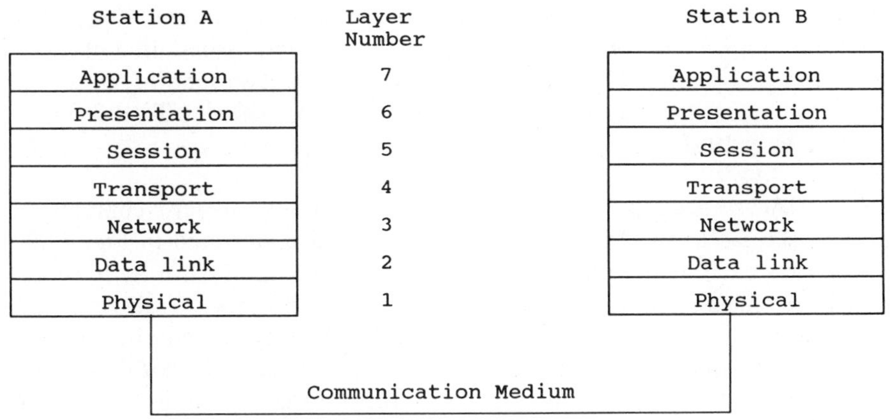

Station A	Layer Number	Station B
Application	7	Application
Presentation	6	Presentation
Session	5	Session
Transport	4	Transport
Network	3	Network
Data link	2	Data link
Physical	1	Physical

Communication Medium

FIGURE 24.1. OSI reference model.

Application Layer. Provides all services directly comprehensible to application programs. For example, MAP/TOP uses a file transfer protocol known as FTAM and a protocol specifically designed for manufacturing applications referred to as MMS (manufacturing message specification).

Presentation Layer. Restructures data to/from standardized format used within the network.

Session Layer. Synchronizes and manages dialogues between application programs.

Transport Layer. Provides transparent, reliable data transfer from end node to end node. MAP/TOP uses a well known transport protocol standard developed by ISO known as TP4 (Transport Protocol class 4).

Network Layer. Performs packet routing and congestion control for data transfer between nodes.

Data Link Layer. Provides access to shared medium and improves the error rate for frames moved between physically adjacent nodes. MAP uses a well known time-token protocol known as the *token bus* protocol and TOP uses a version of the famous *Ethernet* protocol.

Physical Layer. Encodes and physically transfers bits between adjacent nodes.

The term *architecture*, in the context of this section, is used to indicate station architecture. A station architecture specifies the number, name, configuration, functions, protocols, and protocol options of all layers that, taken together, allow network stations to communicate with one another. The layer architecture should also specify how the functions of each layer relate to one another, and also the relationship to the OSI reference model. Other terms that are sometimes used to denote *station architecture* are *node architecture*, *layer profile*, or *protocol stack*.

Historically, the initial MAP document was issued in October 1982, with major additions incorporated in 1984. MAP 2.0 was issued in February 1985, MAP 2.1 in March 1985, MAP 2.2 in August 1986, and MAP 3.0 in April 1987. TOP 1.0 was released in October 1985, whereas TOP 3.0 was issued in May 1987.

MAP 2.2 Architecture. The MAP 2.2 architecture is upward compatible and very similar to the MAP 2.1 specification and is depicted in Fig. 24.2. Two options exist at the physical layer: an AM/PSK modulator at 10 Mbps or an FSK phase coherent modulator at 5 Mbps. The MAC sublayers uses the token bus protocol as described in the IEEE 802.4 standard, and the LLC sublayers specifies the use of the IEEE 802.2 with class 1 or 3 options. The network layer corresponds to that of the ISO connectionless network standard. The connection-oriented, ISO transport class 4 standard is used at

OSI LAYER

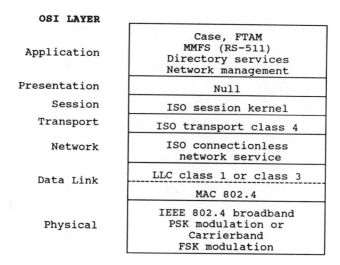

OSI LAYER	
Application	Case, FTAM MMFS (RS-511) Directory services Network management
Presentation	Null
Session	ISO session kernel
Transport	ISO transport class 4
Network	ISO connectionless network service
Data Link	LLC class 1 or class 3
	MAC 802.4
Physical	IEEE 802.4 broadband PSK modulation or Carrierband FSK modulation

(a)

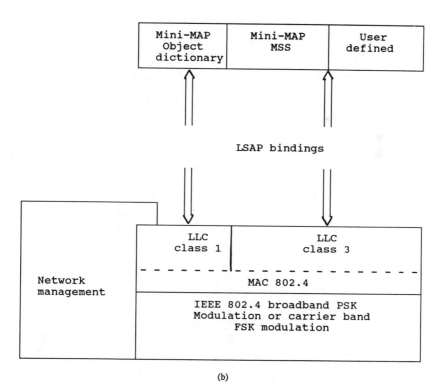

(b)

FIGURE 24.2. (a) MAP 2.2 architecture. (b) Mini-MAP architecture. (c) MAP/EPA architecture.

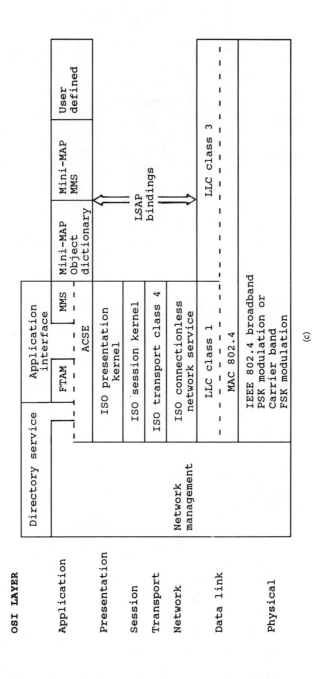

FIGURE 24.2. (*Continued*)

(c)

the transport layer. The session layer used the kernel functionality of the ISO session layer standard. Whereas the presentation layer is null (i.e., empty), the application layer specifies CASE, FTAM, MMFS, Directory Services, and Network Management.

MAP 3.0 Architecture. Whereas MAP 2.2 and MAP 2.1 were very similar, MAP 3.0 provides significant enhancements and it is not upward compatible with MAP 2.2. In addition to enhancements in virtually all layers, perhaps the major enhancements over MAP 2.2 are the inclusion of the Manufacturing Message Specification (MMS), the use of the Association Control Service Element (ACSE), the incorporation of the ISO presentation layer, and the specification of an application layer interface with application processes. The MAP 3.0 architecture is depicted in Fig. 24.3.

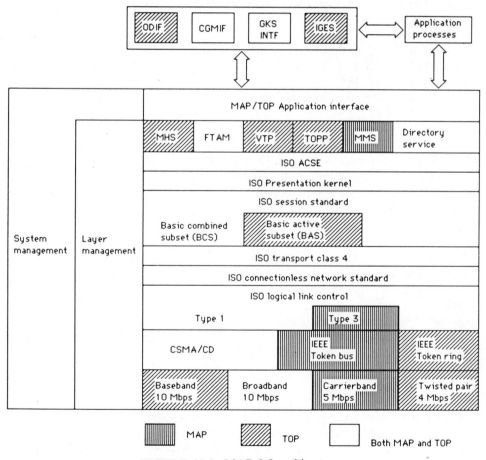

FIGURE 24.3. MAP 3.0 architecture.

In addition to the network management, directory service, and application layer interface specifications, the MAP 3.0 specification also contains a detail specification for broadband transmission systems including topology, coaxial cables, system design, performance, installation, test, and maintenance procedures. This specification is also applicable for TOP 3.0. Currently, there are many commercial implementations that support MAP 3.0. Perhaps the most significant application of this technology is in the Saturn Corporation of General Motors at Spring Hill, Tennessee. The plant-wide broadband network at Saturn encompasses over 30 miles of coaxial cable.

MAP/EPA Architecture. One approach to achieve a simpler architecture (for performance reasons) and still maintain ISO compatibility is to use a dual architecture. As depicted in Fig. 24.2, the MAP/EPA architecture follows this approach. In addition to the dual architecture, perhaps the most significant feature of MAP/EPA is the use of an LLC class 3 sublayer. Class 3 operation allows a responder to issue an acknowledgment immediately, without the need to wait for the token (immediate acknowledgment). Crucial to the operation of a MAP/EPA station is a version of MMS called mini-MAP MMS and its direct interface to the data link layer, which is known as LSAP bindings. Optionally, a user may employ a mini-MAP object dictionary or a user-defined application layer protocol (instead of mini-MAP MMS).

Normally, a MAP/EPA station communicates with similar stations using the three-layer path (i.e., application, data link, and physical). When the station communicates with an ISO node, the seven-layer path is used. As noted, the dual architecture allows the MAP/EPA station to behave as an ISO station, but only when needed.

The carrierband modulator accepts a data rate of 5 Mbps and delivers a signal into a 75-ohm coaxial cable.

Mini-MAP Architecture. If it is not important to be ISO compatible but still have a single architecture, then a mini-MAP architecture could be used. As depicted in Fig. 24.2, the mini-MAP architecture is basically one of the portions of the dual architecture of MAP/EPA; thus the observations made on the MAP/EPA architecture also apply here. One difference, however, is that network management in a mini-MAP environment (at least for the time being) works in an environment consisting of the physical and data link layers only.

Network Architecture. In addition to the station architecture, one should also specify a network architecture. By network architecture we mean the topological configuration, transmission system devices, transmission media, and station configuration that, taken together, allow end user applications to interwork even when the applications are on different physical segments.

Possible topologies for MAP includes a bus, rooted tree, and a star.

Transmission system devices include head ends, amplifiers, taps, bridges, routers, gateways. The transmission medium is broadband and baseband (for carrierband) coaxial cable. Stations can be configured using the MAP 2.2, MAP 3.0, MAP/EPA or mini-MAP architectures.

TOP 3.0. Since the Technical and Office Protocol (TOP) specification basically addresses applications (i.e., design and office automation) different from those of MAP, their application layer protocols differ. Likewise, because of real-time considerations, the lower two layers also differ. Accordingly, as shown in Fig. 24.3, the TOP 3.0 specification basically shares the network, transport, session, and presentation layers with the MAP 3.0 specification. In addition, MAP 3.0 and TOP 3.0 also share the LLC class 1 sublayer, ACSE, FTAM, application layer interface, directory services, and network management.

Although not defined in the OSI architecture, MAP and TOP architectures can also share a broadband coaxial cable as the medium since both networks can coexist on the same cable.

As depicted in Fig. 24.3 the TOP option for medium access control is the CSMA/CD-based IEEE 802.3 specification, which can run on top of baseband or broadband coaxial cable, delivering a data rate of 10 Mbps. Other options of the TOP architecture at the MAC sublayer and physical layer are the use of the IEEE 802.5 at 4 Mbps into a twisted pair of wires, the IEEE 802.4 at 10 Mbps on a broadband coaxial cable, CSMA/CD at 10 Mbps on a broadband coaxial cable (10 BROAD 36), and a TOP packet switching interface based on the X.21 standard.

At the application layer the TOP architecture specifies the message handling system (MHS), file transfer access and management (FTAM), virtual terminal (VT), directory services, and network management. Application-specific standards defined on top of the TOP 3.0 architecture include the product definition interchange format (PDIF), office document interchange format (ODIF), computer graphics metafile interchange format (CGMIF), computer graphics application interface (GKS INTF), and remote file transfer application interface.

In addition, TOP 3.0 groups subsets of standards into the following building blocks:

- CSMA/CD subnetwork access
- Token passing ring subnetwork access
- X.25 packet switching subnetwork access
- MAP token passing bus subnetwork access
- Remote terminal access
- Remote file access
- Electronic mail
- Network directory

- Network management
- Computer Graphics Metafile Interchange format
- Product definition interchange format
- Office document interchange format
- Computer graphics application interface
- Remote file transfer application interface

24.4.3. ISO Reference Model for Shop Floor Production

Part 1. In 1984, ISO TC 184 SC5 addressed the need to develop relevant standards in areas that were under development at the IEC (programmable controllers), CAM-I (discrete parts manufacturing), AFNOR (multi-user requirements), IBM (tools and interfaces for CIM), and NIST (CIM control architectures). As a result, Working Group 1 (WG1) was formed and charged with the specific task of *developing a basic reference model, specifically to create a multidimension open-ended reference model as a basis for long-term planning.* During mid-1990, WG1 issued a technical report called *ISO Reference Model for Shop Floor Production Standards.* The work was issued as a report rather than a standard because it was not possible to reach consensus on a standard that would be complete and precise and that would not be too restrictive in this rapidly changing field. The report has two parts with the first one describing the reference model and the second one discussing a methodology to identify potential areas for shop floor standards.

The ISO reference model for industrial automation is concerned with concepts of a generic model for shop floor production using the following *base technologies*: information, materials/products, product/design engineering, instrumentation/control, computer and communications, and human interface. Although not formally part of the model, WG1 concentrated on a limited production scope (LPS) and adopted a hierarchical model known a the factory automation model (FAM). The reference model views a CIM system as an integrated system of functions in three major categories: enterprise, facility, and shop floor production. The function categories are listed following:

- *Enterprise.* Corporate management, finance, marketing and sales, and research and development.
- *Facility.* Product design and production engineering, production management, procurement, shipping, waste material treatment, resources management, and maintenance management.
- *Shop Floor Production.* Shop floor production. Examples of shop floor production functions include material store, material transport, material processing (transform), incoming inspection, in-process gauging and testing, in-process audit, and product audit.

Four types of flows can be identified in a manufacturing facility: control data, data, materials, and resources, which are generically termed *subjects*. *Control information* includes command, status, request, and response information.

Each subject (e.g., material) can undergo some operations referred to as *actions*. Four actions are identified: transform (Tf), transport (Tp), verify (VE), and store (ST), *Transform* is defined as the act of changing control information, data, material, or resources from one form or state to another form or state. Transform includes encoding or parsing information, decomposing commands, and cutting, forming, assembling, or adjusting materials or resources. The act of moving control information, data, material, or resources from one point of the enterprise to another is referred to as *transport*. *Verify* involves certifying the compliance of all transformed control information, data, material, and resources to determine its conformance to a specification. Finally, *store* is defined as the act of retaining control information, data, material, or resources at a specified location within the *shop floor production* or facility until it is required to be transported.

The reference model basically consists of three items: a hierarchical model, a set of *activities*, and a *subject–action* interaction depicted in a *generic action model* (GAM). Several hierarchical control models have appeared in the literature (2, 4). The ISO model recognizes the following automation levels:

- Section/area
- Cell
- Station
- Equipment

with the following activities:

- Execute
- Command
- Coordinate
- Supervise

Only certain types of activities can be performed at a given automation level as listed in Table 24.1. At each automation level, all subject flows and possible actions (e.g., transform, transport, verify, and store) are represented as a generic activity model (GAM) as depicted in Fig. 24.4.

Part 2. ISO has identified a set of more detailed potential standards to be developed in the future. The reference model recognizes the following *viewpoints* that will guide the development of standards: safety, environment, compatibility, performance, operability, maintainability, reliability, qualifications, and description.

TABLE 24.1. Shop Floor Production Model (SFPM)

	Level	Subactivity	Responsibility
4	Section/area	Supervise shop floor production process	Supervising and coordinating the production and supporting the jobs and obtaining and allocating resources to the jobs
3	Cell	Coordinate shop floor production process	Sequencing and supervising the jobs at the shop floor production process
2	Station	Command shop floor production process	Directing and coordinating the shop floor production process
1	Equipment	Execute shop floor production process	Executing the job of shop floor production according to commands

Based on the relationship among subjects, actions, functions, and interfaces within a GAM, five interactions (also referred to as *Procedures*) have been identified as described next. In the following Subject = {control, data, materials, resources} and Action = {Transform, Transport, Verify, Store}.

Procedure A1: Subject–Action Interrelationships. Standards relating an action with a subject at any hierarchical level. Examples of standards in this area involve standards for the processing of materials, inspection of materials, signal conversion, data buffering, data transmission, and data verification.

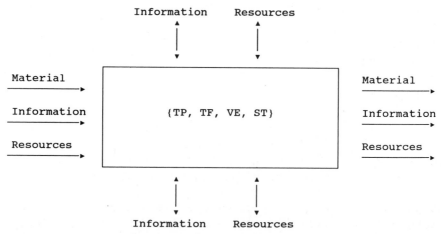

FIGURE 24.4. Generic activity models (GAM) of the ISO Reference model for SFP.

Procedure A2: Subject–Subject Interrelationships. Standards relating a subject with another subject at any hierarchical level. Examples of standards in this area involve standards for bar code representation, data and control presentation and specification, and material and resource definition.

Procedure A3: Action–Action Interrelationships. Standards relating an action with another action at any hierarchical level. Examples of standards in this area involve data transport standards at levels 2, 3, and 4 of Table 24.1 corresponding to gateways or bridges of communication networks.

Procedure B1: Horizontal (Contextual) Interrelationships. Standards relating SFP with other manufacturing functions (e.g., production management) in terms of a subject (e.g., data). Examples of standards in this area involve standards for *production description data* as it goes from *product design and production engineering* to SFP.

Procedure B2: Vertical Interrelationships. Standards relating to a subject (except materials) as it is transported across levels. Examples of standards in this area involve standards for the exchange of commands and status, sensor and operational data, and software updates between the different hierarchical levels of Table 24.1. (ISO TC/184/SC5/ WG1 N80, 1990).

Table 24.2 lists some important standards under the coordination of the ISO/IEC Steering Committee on Industrial Automation in terms of the *actions, subjects, procedures, hierarchical level, viewpoints,* and *base technologies* as defined by the ISO reference model.

24.4.4. CIM–OSA

The need for an *Open System Architecture* for CIM was recognized by the European Community within the ESPRIT program. ESPRIT, the European Strategic Programme for Research and Development in Information Technology, was launched in 1984 as a 10-year program (8). An open system architecture provides the following advantages:

- A general definition of the scope and nature of CIM
- Guidelines for CIM implementation
- A description of constituent systems and subsystems
- A modular framework complying with international standards

To develop the architecture, an ESPRIT consortium made of 20 European companies and known as AMICE was established in 1986 to work on ESPRIT project 688. The project's objectives are to develop a CIM *Open System Architecture* referred to as CIM–OSA to encompass not only the

TABLE 24.2. Significant Industrial Automation Standards

Organization	Committee	SC	WG	Committee Document	Scope of Activity or Standards Document	Example Standards Areas	(Procedure) Level Cell	Viewpoint	Base Technology
IEC	TC 44		WG5	Rep/Act	Interface between Numerical Controls and Industrial Machines	Resource/resource	(B2) 1/2 R	Description	Computer and communications
IEC	TC 65	SC A	WG6	Rep/Act	Language for programmable controller—Application Languages (including Part 3)	Transform control information	(A1) 2,3,4 TfC	Compatibility	Information
IEC	TC 65	SC A	WG6	Rep/Act	PLCs—Part 3: Programming Languages	Transform control information	(A1) 2,3,4 TfC	Compatibility	Control
IEC	TC 65	SC A	WG6	Rep/Act	PLCs—Part: Messaging Service Specification	Transport control information/data	(A1) 2,3,4 TpC&D; (B2) 3/2, 4/3 C&D	Compatibility	Communication
IEC	TC 65	SC B		Rep/Act	IEC 546 Controller performance	Verify resource	(A2) 2,3,4 VR	Performance, reliability, compatibility	Control
IEC	TC 65	SC C	WG1	Rep/Act	Development of Real-Time Architecture (from ISA DS 72.03)	Transport control information/data	(A1) 2,3,4 TpC&D; (B2) 2/1, 3/2 C&D	Compatibility	Communication
IEC	TC 65	SC C	WG1	Rep/Act	PMS–MMS Companion Standard for Process Control	Transport control information/data	(A1) 2,3,4 TpC&D; (B2) 2/1, 3/2, 4/3 C&D	Compatibility	Communication
IEC	TC 65	SC C	WG3	Rep/Act	Investment of a Common Application Interface for 625-2 or not MMS	Transform control information	(A1) 1 TfC&D	Compatibility	Communication
IEC	TC 65	SC C	WG6	Rep/Act	Fieldbus Standard	Transport control information/data	(A1) 1,2 TpC&D; (B2) 2/1 C&D	Compatibility	Communication
ISO	TC 184	SC 1	WG1	Committee	Extended Formats and Data Structure	Transform control information	(A1) 2 TfC	Compatibility	Information
ISO	TC 184	SC 1	WG3	Committee	Companion Standard to MMS (ISO 9506)	Transport control information	(B2) 2/3; 3/4 C	Compatibility	Information
ISO	TC 184	SC 2	WG1	Committee	Terminology and Graphical Representation	Data to data	(A2) All DD	Description	Human interface
ISO	TC 184	SC 2	WG4	Committee	Robot Programming Languages—Application Languages	Transform control information	(A1) 1,2 TfC	Compatibility	Information
ISO	TC 184	SC 2	WG6	Committee	Robot Companion Standard to MMS (ISO 9506)	Transport control information and data	(B2) 2/3, 3/4 C&D	Compatibility	Communication
ISO	TC 184	SC 2		Committee	Performance Testing of Robots	Resource verification	(A1) 1,2 VR	Performance	Production
ISO	TC 184	SC 2		Committee	Robots for Manufacturing Environment	Transport, transform, verify and store material and resource	(A1) 1 (Tp, Tf, V, S)* {M,R}	Safety	Production

ISO	TC 184	SC 3	WG1	Committee	Programming Languages for Automatically Controlled Equipment	Transform control information and data	(A1) 1,2 TfC	Compatibility	Information
ISO	TC 184	SC 3	WG1	Rep/Act	Revision of ISO 4342 CLDATA. 4343 ATP – Application Languages	Transform control information and data	(A1) 2,3 TfC	Compatibility	Production
ISO	TC 184	SC 3	WG2	Committee	Requirements for a Global Programming Language Environment	Beyond SFRM			
ISO	TC 184	SC 3		Committee	Manufacturing Application Languages	SC3 Title only			
ISO	TC 184	SC 4		Committee	External Representation of Product Definition Data	SC4 Title only			
ISO	TC 184	SC 5	WG1	Committee	Reference Models	Beyond SFRM			
ISO	TC 184	SC 5	SG2	Committee	Communications and Interconnections	Transform control information and data	(B2) 2/3, 3/4 C&D	Compatibility	Communication
ISO	TC 184	SC 5		Committee	System Integration and Communication	SC5 title only			
ISO	TC 184	SC 1	WG1	Rep/Act	Extended Format and Data Structure (Rev: TR 6132–1981)	Transform control information	(A1) 2 TfC	Compatibility	Information
ISO	TC 184	SC 1	WG3	Rep/Act	NC Companion Standard to MMS (9506/4)	Transport control information	(B2) 2/3, 3/4 C	Compatibility	Information
ISO	TC 184	SC 1		Rep/Act	Numerical Control of Machines—Symbols	Resource to resource	(A2) 1 RR	Description	Human interface
ISO	TC 184	SC 1		Rep/Act	Part 2: Coding and Maintenance of Preparatory Functions G and Misc Functions M	Transform control information	(A1, 1, 2 TfC	Compatibility	Production
ISO	TC 184	SC 1		Rep/Act	Part 3: Programming of Miscellaneous Functions M (Classes 1 to 9)	Transform control information	(A1) 1,2 TfC	Compatibility	Production
ISO	TC 184	SC 2	WG4	Rep/Act	Robot Programming Languages	Transform control information	(A1) 1 TfC	Compatibility	Production
ISO	TC 184	SC 4	WG1	Rep/Act	Data Exchange and Transfer Standard Specification (STEP)	Dta with product design	(B1) 3, 4 F04D	Compatibility	Information
ISO	TC 184	SC 5	WG1	Rep/Act	Basic Reference Model for Automated and CAD Systems	Beyond SFRM			
ISO	TC 184	SC 5	WG2	Rep/Act	Functional Standards on I-LANs	Transport control information and data	(B2) 1/2, 2/3, 3/4 C&D	Compatibility	Communication
ISO	TC 184	SC 5	WG2	Rep/Act	Manufacturing Message Service—Protocol Specification	Transport control information and data	(B2) 1/2, 2/3, 3/4 C&D	Compatibility	Communication
ISO	TC 184	SC 5	WG2	Rep/Act	Manufacturing Message Service—Service Definition	Transport control information and data	(B2) 1/2, 2/3, 3/4 C&D	Compatibility	Communication
ISO	TC 184	SC 5		Rep/Act	Production management standard for ISO 9506	Data and control information with production management	(B1) All F06 C&D	Compatibility	Communication
ISO	TC 184	SC 5		Rep/Act	Real-Time LANs (proposed new work item)	Transport control information and data	(B2) 1/2, 2/3, 3/4 C&D	Compatibility	Communication

requirements definition and building blocks but also a method for migrating to the new architecture. In addition, standardization activities are part of the objectives of the consortium.

CIM–OSA is concerned with building a total CIM modeling framework ranging from the business requirements to a physical CIM system. Concepts developed by CIM–OSA aim at supporting not only the development but also the running of CIM systems.

In computer integrated manufacturing, it is widely acknowledged that mastering and effectively supporting the *action, resource,* and *information* flow will improve the quality of products (1). As time evolves, CIM–OSA aims at achieving integration at the following levels: physical system, application, and business. Thus the final objective of the CIM–OSA project is to ensure integration of the enterprise at the level of business functions. For a more detailed technical account on CIM–OSA readers are referred to Jorysz and Vernadat (5, 6) and Klittich (7).

To achieve this objective, CIM–OSA uses three approaches: top–down, bottom–up, and some means to support the building of CIM systems. In the top–down approach, the business requirements of the enterprise are captured in the *enterprise model* (i.e., a *modeling framework*) and starting from the business requirements, the *implementation model* (which is the physical CIM system) is derived. The bottom–up approach focuses on the development of a set of services common to most CIM systems, referred to as the *integrating infrastructure*, which realizes application integration and physical system integration. Finally, to deal with the dynamic behavior of CIM systems, CIM–OSA supports the building of CIM systems by defining three architectures: *generic, partial,* and *particular*, as well as an appropriate set of tools. The three approaches are described next.

Modeling Framework. The modeling framework is the top–down approach used by CIM–OSA to ensure integration of the enterprise. The life cycle of a CIM project typically goes through the following stages: requirements, design, and implementation. Accordingly, CIM–OSA defines three modeling levels, one for each of the previous stages. The *requirements modeling level* describes what needs to be done within the enterprise in the context and terminology of the business. Design steps such as structuring and optimizing the business and system constraints are captured in the *design modeling level*. Finally, the *implementation modeling level* specifies an integrated set of components for effective realization.

Each of the modeling levels can be described in terms of the following four different viewpoints referred to as *functional views*: function, information, resource, and organization. The *function view* focuses on the functional structure of the enterprise. The *information view* deals with the structure an the content of the information. Whereas the *resource view* describes and organizes the enterprise resources, the *organization view* fixes the organizational structures of the enterprise. The modeling levels are depicted in Fig. 24.5 and further described next.

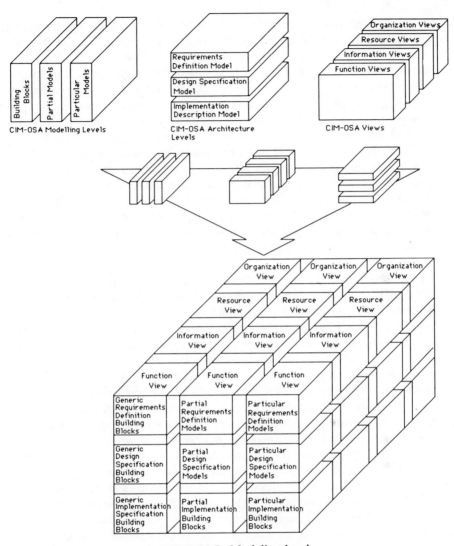

FIGURE 24.5. Modeling levels.

Requirements Modeling Level. The requirements modeling level builds a description of the business requirements of the enterprise in terms of *business processes, enterprise activities,* and *procedural rule sets.* A *business process* constitutes the user's view of what tasks are required to achieve the enterprise objectives. The *enterprise activity* defines the functionality of the enterprise, whereas the *procedural rule set* defines the flow of control between business processes. The requirements modeling level is finalized by expanding each enterprise activity into its constituent parts: *inputs, outputs,* and *transfer function.* The *transfer function* describes the actions required to produce the outputs from the inputs. Inputs and outputs are categorized into the following three types: objects, constraints, and resources.

Design Modeling Level. From the *information view*, the design modeling level utilizes a three schema approach: external, conceptual, and internal schema. From the *resource view*, the concept of a *logical cell* is introduced, and from the *organization view*, the responsibilities for resources, information, and business processes are organized. The *conceptual schema* defines the meaning of information in an unambiguous way and is developed according to the *entity relationship* approach in terms of entities, relationships, and attributes for both entities and relationships. Whereas the *external schema* identifies how the information described by the conceptual schema is perceived by its user, the *internal schema* specifies how the information described in conceptual schema is structured and physically stored inside the computer system. The primary purpose of the *logical cell* is to identify collections of equipment and resources that are candidates for having a high degree of integration because they support groups of functions that require close or frequent interaction.

Implementation Modeling Level. Implementation modeling involves the description of the physical CIM including operational processes and physical components. A physical CIM is described by specifying how the functionality is to be provided, how and where the information is to be stored, and how organizational aspects are to be implemented. As already noted, the business requirements of the enterprise are described in terms of *business processes* and *enterprise activities*. From this description, which is implementation independent, the function view at the implementation modeling level is created. Business processes and enterprise activities are translated into a hierarchical structure of *implemented business process*, a nonhierarchical pool of *implemented enterprise activities*, and a transfer function of each implemented enterprise activity. This complex structure has to be mapped to a physical environment consisting of machines, computers, and people, linked by multiple communication networks (i.e., a distributed environment). To describe the functionality of the distributed environment in an abstract fashion, a generic element referred to as a *functional entity* is introduced. A *functional entity* is an abstract functional object able to send, receive, process, and optionally store information. Functional entities communicate with one another via transactions with a *physical environment* described by a collection of communicating functional entities and referred to as the *implemented functional entities*. The following types of *implemented functional entities* can be distinguished: machine, human, application, data storage, and communication.

Certain functions such as execution control, resource management, distributed information handling, and communication handling are common to most CIM systems. The generic functions constitute the generic functionality offered by the *integrating infrastructure* as described next.

Integrating Infrastructure. As noted, the *integrating infrastructure* is the bottom–up approach used by CIM–OSA to ensure integration of the enterprise and constitutes the *implemented functional entity* common to all nodes of the distributed CIM environment. A set of implemented functional entities is referred to as an *integrated data processing environment*. The functionality of the integrating infrastructure is offered as a set of services with a *client agent* issuing the service requests and a *service agent* providing the service. Clients agents and service agents interact with one another through an *access protocol*. It is possible that a service agent that receives a request service from a client agent cannot directly provide the requested service. In this case, the service agent interacts with another service agent through an *agent protocol* in order to satisfy the service request.

Services in the integrating infrastructure belong to the following groups: front end, communication, business process, and information, as depicted in Fig. 24.6. *Front end services* are the ones that directly deal with the CIM application (real world) and include such components as machine front end, human front end, and application front end. The communication services

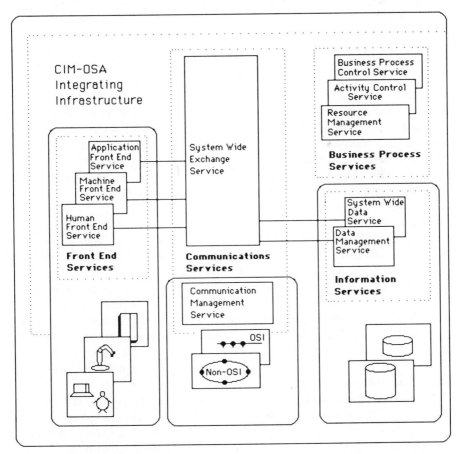

FIGURE 24.6. Integrated enterprise engineering tools.

enable the implementation of *access* and *agent protocols* as well as access to OSI and non-OSI networks. Service components of communication services include *system-wide exchange* and *communication management*. Services in the *business process* group are made up of the following: business process control, activity control, and resource management. Finally, *information services* provides data access protocols and ensures consistency, data integrity, and integration across the system. Service components of information services include *system-wide data* and *data management*.

Supporting the Building of CIM Systems.

Supporting the building of CIM systems is the third and final approach used by CIM–OSA to ensure integration of the enterprise. CIM systems do not remain static; rather, they need to be changed or updated whenever a change occurs in the business requirement or in the physical environment. As noted earlier, CIM–OSA introduces generic building blocks such as *enterprise activity* and *business process* at the requirements level, *logical cells* and *enterprise objects* at the design level, and *information infrastructure* and *implemented functional entities* at the implementation level. Likewise, one can envision building blocks that are more particular (i.e., specific). Accordingly, to cope with changing and updating a CIM system in a dynamic environment, CIM–OSA defines the following architectures: generic, partial, and particular. Basically, a generic architecture is made of generic building blocks.

According to the European Prestandard (ENV 40 003, 1990), a level of genericity is a level of abstraction representing the genericity of the architectural entities described at that level. As the name implies, a *generic architecture* contains elements for a wide range of applications, whereas a *partial architecture* is concerned with a specific type of industrial activity (e.g., an industry sector such as aerospace, automotive, and electronics). On the other hand, a *particular architecture* is concerned entirely with one specific enterprise.

In order to help with the construction of a detailed and specific architecture, CIM–OSA provides three processes: *instantiation*, *derivation*, and *generation* (ENV 40 003, 1990). The process of an orderly architectural development from a generic level via a partial level to the particular level is referred to as *stepwise instantiation*. *Stepwise derivation* is the process of deriving models of the implemented components for the *implementation model level* from the *requirements model level* via the *design model level*. *Stepwise generation* is the process of generating the contents of the enterprise views, by identifying successively requirements, designs, and implementation needs for the enterprise views, in an appropriate order and iterating as necessary to achieve optimal solutions.

In addition to the three architectural levels (i.e., generic, partial, and particular), CIM–OSA proposes a second aid to deal with the dynamic behavior of a CIM system: the *Integrated Enterprise Engineering tools* (1). The tools are made up of the *Enterprise Engineering Environment* (build time support), *Enterprise Operation Environment* (run time support), *Enterprise Engineering Software*, and *Enterprise Operation Software* as depicted in Fig. 24.7. These tools provide two advantages. First, the tools guide the system designer through the entire design process, decreasing the design time and improving the quality of the product. Second, the tools ensure consistency between the initial business requirements defined in the *requirements definition modeling framework* and the resulting physical CIM system described in the *particular implementation model*.

In summary, CIM–OSA defines a CIM architecture for the building of a computer supported integrated manufacturing enterprise which covers the requirements, design, and implementation phases. The architecture has three major components: a *modeling framework*, an *integrated infrastructure*, and *support for system building*. The integrating infrastructure supports the execution, monitoring, and controlling of the manufacturing system. The support of system building includes the enterprise engineering environment

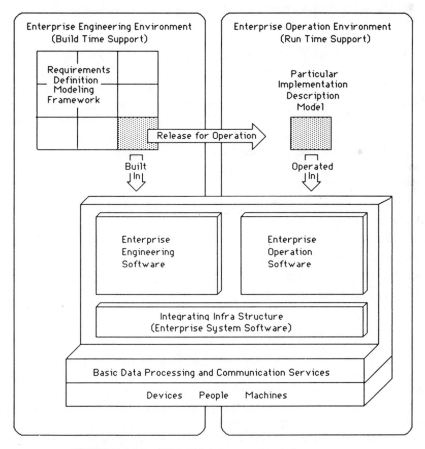

FIGURE 24.7. CIM–OSA integrating infrastructure.

and enterprise operation environment. The modeling framework defines three architectures (generic, partial, and particular), three modeling levels (requirements, design, and implementation), and four functional views (function, information, resource, and organization). The resource functional view is modeled as the virtual manufacturing device (VMD) and MMS (10) in that it is considered as a server in a distributed environment. The integrating infrastructure is concerned about partial models at the implementation description of the modeling level. It is concerned about the following generic constructs: functional entities, services, agents, protocols. The following categories of services are considered: system–wide data, data management (DBMS), human front end, machine front end, application front end, business process control, activity control, resource management, system-wide exchange, and communications management. Of particular importance is the machine front end, which is intended to provide an MMS-based interface to communicate with plant floor devices (e.g., robots, NC machines). For a thorough introduction to OSI, MAP/TOP networks, and MMS, in particular, readers should consult Pimentel (12).

24.5. OTHER SIGNIFICANT STANDARDS

CEN/CENELEC has identified the standardization requirements for information technology to support advanced manufacturing technology (AMT) applications in Europe (ISO/TC 184 N 223, 1990). The requirements will be met by utilizing international standards whenever possible. CEN/ CENELEC provides the following standards classification:

M1. Internetworking
M2. Data
M3. Processing
M4. Control Equipment
M5. Human Aspects
M6. Mechanical Aspects
M7. General Aspects

24.5.1. Internetworking

Standards in this class cover the general framework for CIM from the viewpoint of architecture and communication. Whereas the architecture deals with the overall requirements for achieving interoperation between different computer environments, the problems of intercommunication are addressed by using the Open System Interconnection (OSI) seven-layer reference model and the associated OSI standards to provide the necessary communications environment.

24.5.2. Data

Data standards cover the definition of information and related application data, which is stored, exchanged, or used by applications within an industrial enterprise. Three main classes of data are identified:

1. Administrative data, including prices, accounting, inventory, and ordering.
2. Technical and product definition data.
3. Manufacturing process data.

24.5.3. Processing

The main objective of processing standards is to allow industrial users to choose freely both the hardware and software data processing elements best suited to their problem and to assemble and interconnect these elements in order to obtain the desired system without modifying individual components or adding special interface devices.

24.5.4. Control Equipment

The scope of this activity is the standardization of the control equipment interfaces for manufacturing systems, industrial machines, transportation systems, testing devices, process controls, and storage systems.

24.5.5. Human Aspects

Standards in this class cover all aspects of the integration of human operators into the advanced manufacturing technology environment. Two main areas of work can be distinguished:

1. The dialogue between humans and the system, using a multimedia interface involving symbols, text, graphics, or voice.
2. The ergonomics of the manufacturing system or its elements and its use by human operators.

24.5.6. Mechanical Aspects

Mechanical aspects involve primarily issues of the physical nature of machines, production equipment, and accessories such as tooling and handling equipment. Typical issues that are appropriate for standardization include performance, capability and testing, main dimensions, mechanical interfaces between machines and tools, modularization, operating characteristics, and processing characteristics.

24.5.7. General Aspects

The general aspects concern activities or functions that are applicable throughout the manufacturing system.

24.6. FUTURE DEVELOPMENTS

As noted, work on standards for advanced design and manufacturing is just beginning and there is considerable planning being done by CEN/ CENELEC, the steering committee on industrial automation of ISO and IEC, and other organizations to develop standards. For example, ISO is planning work on a *framework for CIM systems integration* to provide a means of describing enterprises in an executable (i.e., computer process- able) description and to manage change in the highly dynamic manufactur- ing environment. To this end, they will rely on the CIM–OSA architecture described in this chapter.

In addition to the standards activities mentioned thus far, specific efforts currently being pursued or to be pursued in the future include CALS, a machine tool-independent BCL (Binary Cutter Location/data), SERCOS (a standard for the interface of machine tools and their drives), EIA RS-511 (Static Performance measurement of Robots), SLIM (a Standard Language for Intelligent Manufacturing developed in Japan), NASREM, and NGC (Next Generation Controller). NASREM is rapidly becoming a de facto standard for robot control systems.

The Next Generation Controller project is part of a national initiative to revitalize the American machine tool industry. Experts from the U.S. manufacturing industry have formulated economic and technical goals for the program over a series of workshops and studies during the last $2\frac{1}{2}$ years.

To meet the identified goals and objectives of NGC, the following tasks are being performed:

- Determine needs
- Establish requirements
- Identify gaps in enabling technologies
- Develop necessary enabling technology
- Develop specifications for an Open System Architecture Standard (SOSAS)
- Develop validation methods and tests for SOSAS compliance
- Achieve market consensus
- Establish organization to evaluate product conformance to SOSAS
- Develop design guidelines for NGC systems and modules
- Establish international consensus and standards based on SOSAS
- Foster a wide range of SOSAS-compliant products

The most important component of NGC will be the Specification for an Open System Architecture Standard. It will permit independent designers to create interchangeable and interoperable NGC components. The standard will govern the interaction and functioning of NGC-compliant products and will provide a basis for developing and marketing many compatible controls-related products.

In international standardization efforts, additional emphasis will likely be placed on characterizing end user needs, performance metrics for systems and subsystems (e.g., vision systems), and an increased U.S. participation in international standards developments.

To summarize the contents of this chapter, the emerging standards seem to cover basically three areas, directly affecting the future developments in advanced design and manufacturing systems:

1. *Product data standards*, with PDES/STEP being the main line of effort; this should lead in the future to a complete replacement of engineering documentation as we know it today by digital data. These computer-based representations will contain not only product geometry and manufacturing process specifications but also elements as sophisticated as rationale for design or product performance characteristics.

2. *Factory communication standards*, with MAP/TOP being the main specifications; there is a need for communication integration involving internetwork communication among MAP, TOP, mini-MAP, and other fieldbus networks (12) including network management. Other areas of future standardization effort include the use of optical fiber media and other protocol stacks to meet performance and time-critical requirements.

3. *Advanced manufacturing system architectures*, with CIM–OSA being currently the most substantial effort, leading probably in the future to a wide range of standardization efforts on "building blocks" of the architectures of these systems, such as access protocols, agent protocols, and details services of the integrating infrastructure. The true effectiveness of the CIM–OSA effort will be dictated by the effectiveness of the "build time support tools" and "run time support tools" built based on CIM–OSA concepts.

ACKNOWLEDGMENTS

The authors wish to thank the following individuals for helpful discussions and for the information provided: W. Kozikowsky (NEMA), L. Haynes (Intelligent Automation), K. Goodwin (NIST), R. Panse (IBM–Germany), B. H. Squier (CAM-I), and K. Kosanke (AMICE).

REFERENCES

1. Beekman, D. (1989). CIM–OSA: Computer Integrated Manufacturing Open System Architecture, *International Journal of Computer Integrated Manufacturing*, Vol. 2, No. 2, pp. 94–105.

2. Biemans, F. P. M. (1990). *A Reference Model for Manufacturing Planning and Control*, Manufacturing Research and Technology 10, Elsevier Science Publishers, New York.

3. Holcomb, C. (1990). An Application Study of PDES in Manufacturing, *Proceedings of the 2nd Annual User/Industry Conference on CALS/Concurrent Engineering*, CAD/CIM Alert, Boston, MA.

4. Jones, A. T., and McLean, C. R. (1986). A Proposed Hierarchical Control Model for Automated Manufacturing Systems, *Journal of Manufacturing Systems*, Vol. 5, pp. 15–25.

5. Jorysz, H. R., and Vernadat, F. B. (1990). CIM–OSA Part 1: Total Enterprise Modeling and Function View, *International Journal of Computer Integrated Manufacturing*, Vol. 3, No. 3, pp. 144–156.

6. Jorysz, H. R., and Vernadat, F. B. (1990). CIM–OSA Part 2: Information View, *International Journal of Computer Integrated Manufacturing*, Vol. 3, No. 3, pp. 157–167.

7. Klittich, M. (1990). CIM–OSA Part 3: CIM–OSA Integrating Infrastructure— The Operational Basis for Integrated Manufacturing Systems, *International Journal of Computer Integrated Manufacturing*, Vol. 3, No. 3, pp. 168–180.

8. MacConaill, P. (1990). Introduction to the ESPRIT Programme, *International Journal of Computer Integrated Manufacturing*, Vol. 3, No. 3, pp. 140–143.

9. Merchant, M. E. (1988). The Past, Present, and Future of Manufacturing— Intelligent or Otherwise. In *Intelligent Manufacturing System II*, Elsevier Science Publishing, New York, pp. 205–220.

10. Panse, R. (1990). *A Vendor Independent CIM Architecture*, CIMCON'90, NIST special publication # 785, Gaithersburg, MD, May, pp. 177–196.

11. Pimentel, J. R. (1989). Communication Architectures for Fieldbus Networks, *Control Engineering*, Vol. 36, No. 11, pp. 74–78.

12. Pimentel, J. R. (1989). *Communication Networks for Manufacturing*, Prentice-Hall, Englewood Cliffs, NJ.

13. Shorter, D. N. (1990). *Progress Towards Standards for CIM Architectural Frameworks*, CIMCON'90, NIST special publication # 785, Gaithersburg, MD, May, pp. 216–231.

14. Smith, B. M. (1989). *Product Data Exchange: The PDES Project Status and Objectives*, National Institute of Standards and Technology, Publ. No. NISTIR 89-4165, Gaithersburg, MD.

15. Zgorzelski, M. (1987). On the Need of New Standardization Efforts in CAD, *SAE Transactions*, Paper No. 870876.

STANDARD AND OTHER DOCUMENTS

AIAG (1990). Product Data Exchange Survey Results, Automotive Industry Action Group Publ. No. DESurv 7/25/90, Southfield, MI.

CEN/CENELEC, ENV 40 003 (1990). Computer Integrated Manufacturing—Systems Architecture—Framework for Enterprise Modelling.

ISO TC/184/SC5/WG1 N80 (1990). Reference Model for Shop Floor Production Standards, Part 1, The Reference Model, Technical Report No. 10314.

ISO TC/184/SC5/WG1 N160 (1990). Reference Model for Shop Floor Production Standards, Part 2, Application of the Reference Model, Technical Report No. 10314.

ISO TC/184 N 223 (1990). Directory of European Standardization Requirements for Advanced Manufacturing Technology and Programme for the Development of Standards.

ISO/DP 9506 (1987). Manufacturing Message Specification (MMS), Part 1: Service Definition.

ISO/DP 9506 (1987). Manufacturing Message Specification (MMS), Part 2: Protocol Specification.

MAP/TOP Users Group (1988). Map Specification, Ann Arbor, MI.

NBS (1988). Initial Graphics Exchange Specification (IGES) Version 4.0, National Bureau of Standards Publ. No. NBSIR 88-3818, Gaithersburg, MD.

NTIS (1988). Product Data Exchange Specification, First Working Draft, Publ. No. PB-89-144794, Springfield, VA.

PART VI

ARTIFICIAL INTELLIGENCE TOOLS AND TECHNIQUES

CHAPTER TWENTY FIVE

Search Techniques

SURANJAN DE

Department of Decision and Information Sciences, Santa Clara University, Santa Clara, California

25.1. INTRODUCTION

The use of intelligent systems for design and manufacturing decisions often requires the solution of discrete optimization problems; typical problems include the design of manufacturing cells or VLSI circuits and the generation of production schedules or a location pattern of distribution centers over a geographic area. Such problems often are combinatorially complex and require the examination of a large number of alternatives for reasonably large problems. This chapter examines the role of search techniques in addressing design and manufacturing problems in the context of intelligent systems development.

This chapter is based on the premise that a good quality solution is not generated by an intelligent system by the application of search techniques alone; the generation of such solutions also requires the acquisition of domain knowledge and expert knowledge and the use of that knowledge to generate better quality solutions. Hence this chapter focuses on an integrated handling of search techniques and other mechanisms to address the issues referred to above.

25.2. PROBLEM REPRESENTATION AND SEARCH

Search strategies are part of a larger intelligent problem solving system. In artificial intelligence, such problem solving systems are viewed as consisting of at least three components: (a) a set of *states*, where each state describes a problem configuration at a given stage of the solution process; (b) a set of

Intelligent Design and Manufacturing, Edited by Andrew Kusiak.
ISBN 0-471-53473-0 © 1992 John Wiley & Sons, Inc.

operators, where each operator transforms one problem state to another; and (c) a *control strategy* (or search strategy) that determines the sequence in which operators should be applied to transform an initial state to a goal state. Problem representation consists of the specification of the set of states and operators that are relevant to a problem or a class of problems. A problem is specified by stating the initial state and (often unspecified) goal state; the problem is viewed as solved when the goal state is reached. The focus of this chapter is on the control strategy. However, the appropriateness of search strategies often depends on the choice of problem representation. The same problem can be represented differently and different search strategies can be applied to solve the problem; a clever choice of problem representation may often simplify the solution process. A variety of problem representations have been proposed in the artificial intelligence literature. Two of the representations that have been widely used are state space representation or a problem reduction representation. Both representations can be stated as graphs, in which case the determination of the control strategy is transformed into a graph search problem; in this chapter problems are represented as graphs. However, other representations are possible. For example, we have represented FMS scheduling problems using (a) a problem reduction representation (AND/OR graphs) captured as expressions of first-order predicate calculus (2) and (b) a state space representation captured using a frame-based language (3, 4).

The state space representation consists of two important entities, states and operators. The set of all attainable states is called the state space. The complete specification of a problem using a state space representation requires the specification of one or more initial states, one or more goal states, and a set of operators. Hence a search problem using a state space representation requires the determination of the sequence of operators that will transform an initial state to a goal state; the solution strategy uses a forward reasoning approach. As described by De and Lee (3, 4), the initial state of a FMS scheduling problem can describe the status of the workstations at the start of the scheduling process, the fact that each of jobs to be scheduled for production are available to be loaded onto the FMS, and so on. The goal state for the FMS scheduling problem describes, for example, the set of manufacturing (and/or assembly) operations that have been completed and when. The scheduling problem is transformed into a state space search problem and solved using beam search (3, 4); beam search is discussed in Section 25.3.

The problem reduction representation requires the specification of (a) an initial problem description, (b) a set of operators, each of which transforms a single problem into a set of subproblems, and (c) a set of primitive problems that can be solved immediately by available techniques. The initial problem description is specified as a set of goals that must be achieved. Each goal is transformed into several subproblems by the application of an operator; to achieve the goal each subproblem must be solved and the solutions to the subproblems must be combined. The initial goal can be achieved by the successive reduction of goals into subgoals until primitive problem descriptions are reached; hence a backward reasoning approach is required. The search strategy involves the determination of appropriate operators that can be applied at every stage of the solution process. The problem reduction approach has been used effectively to schedule jobs in a FMS (2).

25.3. BASIC SEARCH STRATEGIES AND THEIR APPLICATION

A search strategy is a prescription for determining the order in which nodes should be generated. In this chapter, we examine a variety of search strategies that can be applied to state space search problems. Some of these strategies can be modified so as to be applicable to problem reduction problems.

25.3.1. Hill Climbing

Hill climbing is a simple search strategy based on local optimizations. The strategy amounts to repeatedly expanding a node, examining the generated successors, and choosing and expanding the best (based on some evaluation function) among them, while removing all references to the parent and child nodes. It seeks iterative improvement of a structure by searching in the neighborhood around the structure; at each stage of the search process, the move that provides maximum improvement in the quality of the solution generated is applied until no more uphill moves are available. The scheme is simple and very easy to implement. The scheme requires the representation of a structure (such as a set of states and a set of applicable operators). The goal states are those with minimum costs (assuming cost minimization problems). The hill climbing attempts to achieve a minimum cost state as quickly as possible. The problem with hill climbing is that if the search process leads to a local optima, it cannot get out of that state and proceed toward global optima. Some of the newer heuristic search techniques (see Section 3.6) specifically attempt to circumvent local optima (e.g., tabu search, simulated annealing). From the point of view of the effectiveness of a search strategy, the interest in hill climbing results from viewing it as a minimum standard against which any general strategy should be compared.

25.3.2. Uninformed Systematic Search Strategies

The search strategies considered in this subsection are applied to a state space denoted by a directed graph in which each node is a state and each arc represents the application of an operator transforming a state into a successor state. A solution is a path from a start state to a goal state. The search strategies described in this section do not use domain-specific information to determine where the solution is likely to lie and explores potential solution paths using an arbitrary order.

In each of the methods described in this section, the state space graph is a tree. The implication is that there is only one start state and that the path from the start node to any other node is unique. There is a procedure for finding all the successors of a given node. Such a procedure is said to *expand* the given node. Whenever a node is expanded, creating a node for each of its successors, the successor nodes contain pointers back to the parent node. When a goal node is finally generated, this feature makes it possible to trace the solution path.

Depth-First Search. Depth-first search can be characterized by the expansion of the most recently expanded node first. After each node expansion, one of the newly generated successor nodes is selected for expansion. This process is repeated until it reaches a state with no successors, in which case the search process resumes from the deepest of all nodes left behind. Unfortunately, this strategy is unacceptable if the state space tree has a depth too large to be explored within a reasonable time. Hence a depth bound is usually specified and any node at that depth is treated as if it had no successors. The depth-first search strategy is stated as follows:

1. Put the start node on a list, called OPEN, of unexpanded nodes. If the start node is a goal node, a solution has been found.
2. If OPEN is empty, no solution exists.
3. Remove the first node, n, from OPEN and place it in a list, called CLOSED, of expanded nodes.
4. If the depth of node n is equal to the depth bound, go to step 2.
5. Expand node n, generating all its successors. Place all successors at the beginning of OPEN. If node n has no successors, to to step 2.
6. If any of the successors of node n is a goal node, a solution has been found. Otherwise go to step 2.

As an illustration of blind search strategies, consider the solution of a traveling salesman problem. A traveling salesman problem (TSP) involves finding a shortest tour (i.e., a Hamiltonian circuit) in a given graph $G = (V, E)$ in which every edge $(v_i, v_j) \in E$ has length d_{ij}; V is the set of vertices in G and E is the set of edges in G. If G is directed, $d_{ij} \neq d_{ji}$ is generally assumed and the problem is called the asymmetric traveling salesman problem. if G is undirected, $d_{ij} = d_{ji}$ and the problem is called the symmetric traveling salesman problem. Figure 25.1 describes the graph of an example symmetric traveling salesman problem (in which $d_{12} = d_{21} = 10$, etc.).

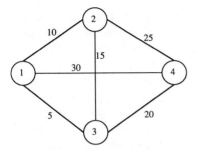

FIGURE 25.1. An example traveling salesman problem.

The solution to the TSP of Fig. 25.1 using the depth-first search is shown in Fig. 25.2. Node 1 is the start node and corresponds to the starting city 1. The successors of each node are generated by applying operators GTi, which corresponds to "Go to city i from present city." The two entries adjacent to each successor node correspond to the length of the path from the start node (i.e., city 1) to the present node and the path from the start node to the present node. Hence node 2 is 2 from 1 is 10, and the path to city 2 is $1-2$. The shaded nodes are terminal nodes. The solution obtained by applying the operator GT2 (i.e., Go to city 2 from present city) is not

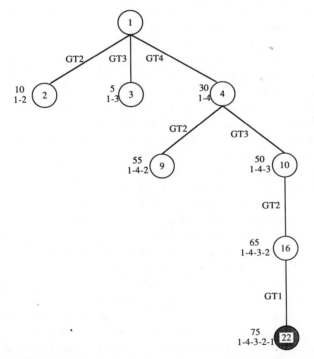

FIGURE 25.2. A depth-first search tree for TSP of Fig. 32.1.

optimal. Only the nodes that do not lead to subtour formation are shown; a subtour is formed when the traveling salesman travels from a node, say node 1, and returns to node 1 without traveling to all nodes in the graph.

Breadth-First Search. Compared to depth-first search, breadth-first search is a more conservative search strategy in which the search begins at the start node and progresses level by level; it systematically searches all paths of length one, followed by all paths of length two, and so on, until it reaches the goal node. Although the search may be an extremely long one, it is guaranteed eventually to find the shortest possible solution sequence if any solution exists. The breadth-first search strategy is stated as follows:

1. Put the start node on a list, called OPEN, of unexpanded nodes. If the start node is a goal node, a solution has been found.
2. If OPEN is empty, no solution exists.
3. Remove the first node, n, from OPEN and place it in a list, called CLOSED, of expanded nodes.
4. Expand node n generating all its successors. If it has no successors, go to step 2.
5. Place all successors of node n at the end of OPEN.
6. If any of the successors of node n is a goal node, a solution has been found. Otherwise, go to step 2.

Breadth-first search is viewed as effective with the number of alternatives at the choice points is not too large (16).

The solution to the TSP of Fig. 25.1 using breadth-first search is shown in Fig. 25.3. Only the nodes that do not lead to subtour formation are shown. The same conventions as in Fig. 25.2 are used here. There are two optimal tours: 1-2-4-3-1 and 1-3-4-2-1. Given that this is a symmetric TSP, it is not surprising that one optimal tour is simply in a reverse order of the other.

In practice, a breadth-first search procedure is more space bound than time bound (i.e., it is more likely to use up available memory before it uses up substantial amount of computation time); a depth-first search procedure, on the other hand, is more time bound than space bound. Since a breadth-first search procedure expands all nodes at a given depth before expanding any nodes at a greater depth, it will always find the path of the shortest length. The first solution generated by depth-first search, in contrast, may not be an optimal one (as illustrated by the example tour shown in Fig. 25.2 in Section 3.2.1).

A generalization of breadth-first search is an approach called uniform cost search. The idea is to solve the problem of finding the cheapest path from the start node to a goal node. A nonnegative cost $c(i, j)$ is associated with every arc joining two nodes i and j. The cost of the solution path is then the sum of the arc costs along the path. The cost of a path from the start

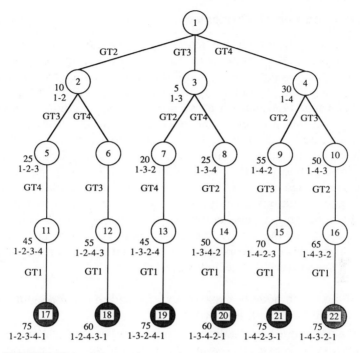

FIGURE 25.3. A breadth-first search tree for the TSP of Fig. 32.1.

node to any node i is denoted by $g(i)$. An uniform cost search procedure is stated as follows:

1. Put the start node, s, on a list, called OPEN, of unexpanded nodes. If the start node is a goal node, a solution has been found. Otherwise, set $g(s) = 0$.
2. If OPEN is empty, no solution exists.
3. Select from OPEN a node i such that $g(i)$ is minimum. If several nodes qualify, choose node i to be a goal node is there is one; otherwise, choose among them arbitrarily. Move node i from OPEN to a list, CLOSED, of expanded nodes.
4. If node i is a goal node, a solution has been found.
5. Expand node i. If it has no successors, go to step 2.
6. For each successor node j of node i, compute $g(j) = g(i) + c(i, j)$ and place all the successor nodes j in OPEN.
7. Go to step 2.

Both breadth-first search and uniform cost search can be applied to AND/OR graphs as well.

25.3.3. Informed Search Strategies

Both depth-first and breadth-first search strategies are "blind" in the sense that they use exhaustive approaches that do not attempt to evaluate the quality of the paths being explored until they arrive at the goal. Hence these approaches spend substantial computational effort in exploring paths that are not promising. However, in many problems, some guiding information about the particular problem domain (heuristic information) is available and, in those cases, such information can be used effectively to reduce the search. In design and manufacturing, such information can be obtained, for example, from a theory of the problem domain. In our earlier research, we have used knowledge of scheduling theory to determine paths that are likely to lead to good solutions quickly (2–4). Such knowledge can be formally captured by specifying an evaluation function on every node of the search graph that estimates the relative benefit or relative cost of continuing the search from that node. The points at which heuristic information can be applied in a search include the following (1):

1. Deciding which node to expand next, instead of expanding in a strictly breadth-first or depth-first order.
2. In the course of expanding a node, deciding which successor or successors to generate—instead of blindly generating all possible successors at one time.
3. Deciding that certain nodes should be discarded from the search tree.

We refer to search strategies that use domain information to improve the quality of the solution generated and/or improve the performance of the search process as informed search strategies.

Evaluation Function. A key feature of informed search strategies is to evaluate the promise of a node before expansion; the measure by which the promise of a node is estimated is called an *evaluation function f^**; $f^*(n)$ is the estimate of the actual cost of the solution path from the start node to the goal node through node n. The evaluation function f^* is defined such that the more promising a node is, the smaller is the value of f^*.

A variety of evaluation functions have been used in practice. One particular form of the evaluation that we have found very useful in our work on scheduling (3, 4) is described below. Since f^* evaluates nodes in the light of the need to find a minimum cost solution, it considers the value of each node n as having two components: the cost of reaching n from the start node and the cost of reaching a goal from node n. Accordingly, f^* is defined by

$$f^*(n) = g^*(n) + h^*(n) ,$$

where g^* estimates the minimum cost of a path from the start node to node n, and h^* estimates the minimum cost from node n to a goal. The value

$f^*(n)$ thus estimates the minimal cost of a solution path passing through node n. The actual costs, which f^*, g^*, and h^* only estimate, are denoted by f, g, and h, respectively.

The function g^*, applied to a node n being considered for expansion, is calculated as the actual cost from the start node to node n along the cheapest path found so far by the algorithm. Since the state space is a tree in our case, g^* gives a perfect estimate, since only one path from the start node to node n exists. The function h^* is the carrier of the heuristic information. The manner in which we define these functions depends on the nature of the problem. The only requirement is that f^* should be an underestimate of f (1). When the scheduling objective function changes, the only change in using the algorithm is in the evaluation function used. As discussed by De and Lee (3), only six measures need be considered: $C_{\max}$ (makespan), $\Sigma\, C_j$ (sum of completion times), $\Sigma\, w_j C_j$ (weighted sum of completion times), $L_{\max}$ (maximum lateness), $\Sigma\, T_j$ (sum of tardinesses), and $\Sigma\, w_j T_j$ (weighted sum of tardinesses). For each of the measures, we use a distinct evaluation function:

Objective Function	Evaluation Function
$C_{\max}$	$\displaystyle\max_{j\in\{\text{lob}\}} [g_j(n) + h_j^*(n)]$
$\displaystyle\sum_j C_j$	$\displaystyle\sum_{j\in\{\text{lob}\}} [g_j(n) + h_j^*(n)]$
$\displaystyle\sum_j w_j C_j$	$\displaystyle\sum_{j\in\{\text{lob}\}} w_j[g_j(n) + h_j^*(n)]$
$L_{\max}$	$\displaystyle\max_{j\in\{\text{lob}\}} [g_j(n) + h_j^*(n) - d_j]$
$\displaystyle\sum_j T_j$	$\displaystyle\sum_{j\in\{\text{lob}\}} \max(0, [g_j(n) + h_j^*(n) - d_j]$
$\displaystyle\sum_j w_j T_j$	$\displaystyle\sum_{j\in\{\text{lob}\}} w_j \max(0, [g_j(n) + h_j^*(n) - d_j]$

Here d_j is the due date of job j.

If we are interested in minimizing the makespan, the evaluation function is defined as

$$f^*(n) = \max_{j\in\{\text{lob}\}} [g_j(n) + h_j^*(n)],$$

where at node n

{job} is the set of jobs considered,

$g_j(n)$ is the completion time of the current operation scheduled for job j, and

$h_j^*(n)$ is the average total processing time for the remaining operations of job j.

It should be noted that in a FMS alternative routes are possible for each job with different total processing time for each route. $h_j^*(n)$ estimates the average of the total processing times of all the routes. In estimating the evaluation function at each node, we are in effect computing the completion time for each job assuming that the remaining operations can be performed without any interaction; that is, all the jobs can be performed concurrently. This also guarantees that f^* is an underestimate of the true makespan of the set of jobs.

In dealing with due date oriented schedules, one can attempt to minimize lateness related measures or tardiness related measures. An example of an evaluation function in such a case is

$$f^*(n) = \max_{j \in \{job\}} [g_j(n) + h_j^*(n) - d_j],$$

where at node n

{job} is the set of jobs considered,

$g_j(n)$ is the completion time of the current operation scheduled for job j,

$h_j^*(n)$ is the average total processing time for the remaining operations of job j, and

d_j is the due date of job j.

$g_j(n) + h_j^*(n)$ gives an estimate of the completion time of job j. Hence $g_j(n) + h_j^*(n) - d_j$ is an estimate of the lateness of job j. It is an underestimation of the true lateness of a job because the completion time of a job is underestimated as conflicts between jobs and transfer times between machines for any job are ignored.

Beam Search. *Beam search* is like breadth-first search in the sense that the search process searches paths of length k before searching paths of length $k + 1$. Unlike breadth-first search, however, beam search only moves downward from the best w nodes at each level (w is called the *beam width*). The other nodes are ignored. The choice of the best nodes is based on an evaluation function f^*. The basic beam search method is summarized as follows:

1. Put the start node s on a list, called OPEN, of unexpanded nodes. Calculate $f^*(s)$ and associate its value with the node s.
2. If OPEN is empty, exit with failure; no solution exists.
3. Select from OPEN the w best nodes for expansion (i.e., the nodes with the w least values of f^*; resolve ties arbitrarily); w is the beam width. Move these nodes from OPEN to a list, CLOSED, of expanded nodes.

4. If any of the w nodes is a goal node, exit with success; a solution has been found.

5. Expand the w nodes, creating nodes for all its successors. For every successor node j of i:

   (a) Calculate $f^*(j)$.

   (b) If j is neither in list OPEN nor in list CLOSED, then add it to OPEN, with its f^* value. Attach a pointer from j back to its predecessor i.

   (c) If j was already on either OPEN or CLOSED, compare the f^* value just computed for j with the value previously associated with the node. If the new value is lower, then

   (i)   Substitute it for the old value.

   (ii) Point j back to i instead of its previously found predecessor.

   (iii) If node j was on the CLOSED list, move it back to OPEN.

6. Go to step 2.

A variation of the beam search strategy is filtered beam search in which only k of the successors (instead of all successors) are generated for each node, further reducing the computational effort. k is referred to as the *filter width*. Hence, in step 5 above, instead of generating all successors of a node only k of them are generated. It may be noted that the size of the search tree increases linearly with the beam width and the filter width. If w is the beam width and k is the filter width, the number of nodes at any level is $(w \times k)$.

Filtered beam search is a heuristic search method aimed at carrying out a backtrack-free search quickly. Since a node on the optimal path may not necessarily be selected for the beam, an optimal solution may not be obtained by using this method. Moreover, since no backtracking is allowed, a node once excluded from the beam is not considered again during the search process. Since the quality of the solution obtained depends on the accuracy (or inaccuracy) of the evaluation function, the effectiveness of this method is enhanced by considering a number of promising paths concurrently. Filtered beam search has been viewed as good when there is a natural measure of goal distance and a good path is likely to be among the good-looking partial paths at all levels (16). In scheduling using filtered beam search, promising paths were chosen by De and Lee (3, 4) based on rules stored in the knowledge base.

As an illustration of the operation of filtered beam search, consider a FMS with one loading dock, one unloading dock, and two machines: M1 and M2. Each machine is connected with another machine as well as the loading dock and the unloading dock by a dedicated path (e.g., by using an AGV). We assume that, at the manufacturing level, the system has three operations: OP1, and OP2, and OP3. Besides, the loading and unloading operations are labeled as LOAD and UNLOAD, respectively. The operation times in minutes for various operations on different machines are given as follows:

	OP1	OP2	OP3
M1	20	23	19
M2	22	18	20

The time to load or unload a part is assumed to be 1 minute. The time to transfer a job between a machine and the loading dock (or unloading dock) or another machine is assumed to be 2 minutes. Two jobs, J1 and J2, need to be scheduled such that the makespan is minimized. Each job has a strict sequence of operations to be performed. The order of operations for each job is as follows:

J1:	OP2	OP3	
J2:	OP1	OP2	OP3

The solution of the example problem is illustrated by using a beam width $w = 2$ and a filter width $k = 2$. In other words, at each level of the search tree only two nodes are considered for expansion (since $w = 2$), and for each node expanded only two successors are generated (since $k = 2$). Given that the objective function is to minimize makespan, the evaluation function at a node is the estimate of the maximum completion time of all jobs at that node; this estimate is revised as we go down the search tree. In estimating the completion time, we use the average processing time for each operation (i.e., the processing time of an operation averaged over all the machines on which the operation can be performed). The average processing time for the operations are

$$\text{OP1: } 21 \; ; \quad \text{OP2: } 20.5 \; ; \quad \text{OP3: } 19.5 \; .$$

Moreover, the loading and unloading times are taken into account in computing the estimated completion time but the transfer times are not. This is because transfers between machines may not be necessary to perform consecutive operations.

The completion time for each job in the root node (NODE0) can be estimated as follows:

J1: $C_1^* = 20.5(\text{OP2}) + 19.5(\text{OP3}) + 1(\text{LOAD}) + 1(\text{UNLOAD}) = 42$.

J2: $C_2^* = 21 + 20.5 + 19.5 + 1 + 1 = 63$.

The evaluation function f^* at NODE0 is given by

$$f^*(\text{NODE0}) = \max\{C_1^*, C_2^*\} = 63 .$$

The status of all the jobs at the root node is represented as (JOB:READY0). The search tree generated is shown in Fig. 25.5 with the final solution path highlighted by the shaded nodes.

To illustrate the effectiveness of filtered beam search, we compare the solution to the example problem generated using filtered beam search with that generated by using breadth-first search and depth-first search. We do so by examining the number of possible paths (hence the number of nodes) in an exhaustive search tree (see Fig. 25.4); this gives us a worst-case (in the sense of involving most effort) scenario against which the other three search strategies are evaluated. A complete enumeration shows that there are 80 possible paths in our example that will lead to a goal node; each path corresponds to a feasible schedule.

Figure 25.4 shows the location of the two optimal paths (highlighted by the shaded nodes) with a makespan of 66 minutes, suppressing the 78

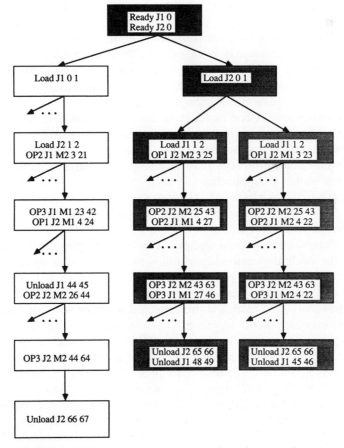

FIGURE 25.4. A segment of the exhaustive search tree.

TABLE 25.1. Number of Nodes Generated in Exhaustive Search

Level	Nodes Per Level	Branching Factor(b.f.)
0	1	2
1	2	2
2	4	5
3	20	10 of them have a b.f. of 2; the other 10 have a b.f. of 4.
4	60	20 of them have a b.f. of 2; the other 40 have a b.f. of 1.
5	80	40 of them have a b.f. of 1; the other 40 have a b.f. of 0.
6	40	
Total	207	

suboptimal ones. Only three paths are shown here; the rest are implied by the outgoing arrows at each node. The exhaustive search tree in Fig. 25.4 has seven levels. The number of outgoing arrows from each node indicates the number of possible successors sprouted from the node, or its branching factor (b.f.). Table 25.1 shows the total number of nodes generated in the exhaustive search.

We can use Fig. 25.4 to trace the search path of a breadth-first search, which expands all the successors of each node at each level of the search tree until it finds the goal node. Hence it resembles the exhaustive search up to the time when a goal node is found. In our case, the goal node is located at Level 5. The breadth-first search will examine all nodes up to Level 4 (i.e., 87 nodes) before it reaches one of the goal nodes at Level 5. So the total number of nodes needed to examine in this case is 88.

The depth-first search strategy expands from the root by generating only one offspring at a time until it reaches the goal node. Since there are 80 possible paths in our example that will lead to a goal node, there will be 80 possible depth-first search paths. The number of nodes generated will be either 6 (if any of the right-hand branch is sought) or at most 7 (if any of the left-hand branch is sought).

Using a filtered beam search strategy with a beam width of 2 and a filter width of 2, the node expansion algorithm is used to create only 2 successors at each node and out of all possible number of nodes at each level of the tree only 2 are chosen to expand and the rest are pruned. The value of the evaluation function (f^*), which is the expected makespan in our example, is used to decide which node to expand. The search tree generated by the filtered beam search is given in Fig. 25.5. The number next to each node is the f^*. The solution path is highlighted by the dark thick line.

As indicated in Table 25.2, only a total of 17 nodes are examined using filtered beam search. The filter width k limited the branching factor of each node to be 2. The beam width w ensures that only 2 nodes are expanded at each level. The number of nodes are limited to be at most 4($k \times w$ or 2×2) at each level of the search tree. Hence only 4 nodes are examined in Levels 3 and 4 as compared to 20 and 60, respectively, in the case of the exhaustive search.

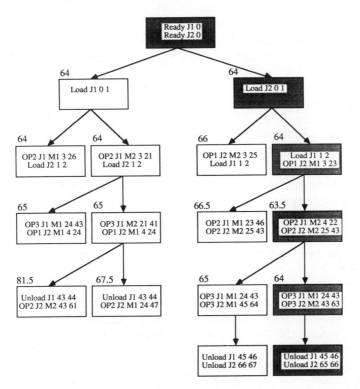

FIGURE 25.5. The filtered beam search tree (beam width = 2, filter width = 2).

Both filtered beam search and breadth-first search generate a schedule with a makespan of 66 minutes. But the number of nodes to be examined in breadth-first search is 88; whereas in the case of filtered beam search ($w = 2$, $k = 2$) it is only 17! Depth-first search involves the least effort (the number of nodes generated is 6, or at most 7). However, the a priori likelihood that depth-first search will generate an optimal path in this example is only 2 out of 80, or 2.5%. The likelihood of generating an optimal solution using depth-first search will get even worse for more

TABLE 25.2. Number of Nodes Generated in Filtered Beam Search ($w = 2$, $k = 2$)

Level	Nodes Possible*	Nodes Generated	Nodes Pruned
0	1	1	0
1	2	2	0
2	4	4	2
3	20	4	2
4	60	4	2
5	80	2	1
Total	207	17	7

Nodes Possible* = Number of nodes generated in the exhaustive search complex problems as the size of the potential search tree grows. Similarly, the computational advantages of filtered beam search over breadth-first search are even more striking as the size of the scheduling problem increases.

Computational experience with filtered beam search on FMS scheduling (3) indicates the following:

1. As beam width and/or filter width increases, the solution quality will improve at the expense of computational time and space requirements.
2. The improvement in performance diminishes as the beam/and or filter width increases.
3. The solution quality is better than depth-first search (even for beam width = 1).
4. The solution mechanisms is more cost efficient than breadth-first search.

It is important to realize that the use of heuristic information during the search process allows us to explore the most promising branches first; hence it is not surprising that increasing beam width w and filter width k beyond certain values (which our experimentation indicates are relatively small) does not significantly improve the quality of the solution generated. This key feature allowed us to generate good solutions without incurring significant computational overhead.

Best-First Search. Best-first search is one that always selects the most promising node as the next node to expand, irrespective of the level of the partial search graph in which it appears. The purpose of this procedure is to reduce the number of nodes expanded as compared to blind-search algorithms. Its effectiveness depends on the choice of f^*. A somewhat similar approach is adopted in hill climbing in which the most promising node is chosen from the most recent node. In best-first search, the most promising node is selected from all the nodes in the partially developed search graph (and hence the name "best-first"). The procedure can be summarized as follows:

1. Put the start node s on a list, called OPEN, of unexpanded nodes. Calculate $f^*(s)$ and associate its value with the node s.
2. If OPEN is empty, exit with failure; no solution exists.
3. Select from OPEN a node i at which f^* is minimum. If several nodes qualify, choose a goal node if there is one; otherwise, choose among them arbitrarily.
4. Remove node i from OPEN and place it on a list, called CLOSED, of expanded nodes.

5. If node i is a goal node, exit with success; a solution has been found.
6. Expand node i, creating nodes for all its successors. For every successor node j of i:
   (a) Calculate $f^*(j)$.
   (b) If j is neither in list OPEN nor in list CLOSED, then add it to OPEN, with its f^* value. Attach a pointer from j back to its predecessor i.
   (c) If j was already on either OPEN or CLOSED, compare the f^* value just computed for j with the value previously associated with the node. If the new value is lower, then
      (i) Substitute it for the old value.
      (ii) Point j back to i instead of to its previously found predecessor.
      (iii) If node j was on the CLOSED list, move it back to OPEN.
7. Go to step 2.

Best-first search is good with there is a natural measures of goal distance and a good path may look bad at shallow levels (16). In a sense, breadth-first search ($f^*(i) =$ depth of node i), uniform-cost search ($f^*(i) =$ cost of path from root to node i), and depth-first ($f^*(i) =$ negative of the depth of the node) can be viewed as special cases of the best-first search technique. The best-first search procedure can be extended to AND/OR graphs as well (14).

25.3.4. Optimal Search Strategies

The search techniques discussed so far may not always generate an optimal solution. A variety of search techniques have been proposed and used to obtain optimal solutions to design and manufacturing problems. However, they typically require substantial computational time and/or storage space for all but the smallest problems. Optimal search strategies are useful, though, as a benchmark against which other search techniques can be compared in terms of solution quality and computational and storage requirements.

A variety of optimal search strategies have been presented in the literature. For example, the *British Museum procedure* is an exhaustive search procedure that examines all possible solution paths and simply picks the best one; clearly, such an approach is effective only if the problem size is very small. Other widely used optimal methods include branch-and-bound and dynamic programming.

A*. The A* procedure is designed to find the least cost path from the start node to the goal node in a state space graph. The procedure is similar to the branch-and-bound procedure used widely in operations research (16). The essential idea in A* is to expand that node in the search graph which is likely to be on the shortest path from the start node to the goal node. Unlike the search strategies discussed in Section 25.3.3, A* attempts to find an optimal solution.

The key feature of A* is the definition of the evaluation function f^*. Associated with each node n is a cost, $f^*(n) = g(n) + h^*(n)$, where $g(n)$ is the actual cost of reaching node n from the start node and $h^*(n)$ is the estimated cost of reaching the goal node from node n. At each stage of the search process, A* expands the node with the least cost f^*. It is a well-known result (ref. 9, p. 233) that A* will always find the shortest path from the start node to the goal node provided $h^*(n)$ never overestimates the distance from node n to the goal node.

The A* procedure can be summarized as follows:

1. Put the start node s on a list, called OPEN, of unexpanded nodes. Calculate $f^*(s)$ and associate its value with the node s.
2. If OPEN is empty, exit with failure; no solution exists.
3. Select from OPEN a node i at which f^* is minimum. If several nodes qualify, choose a goal node if there is one; otherwise, choose among them arbitrarily.
4. Remove node i from OPEN and place it on a list, called CLOSED, of unexpanded nodes.
5. If node i is a goal node, exit with success; a solution has been found.
6. Expand node i, creating nodes for all its successors. For every successor node j of i:
   (a) If j is neither in list OPEN nor in list CLOSED, then estimate $h^*(j)$ and compute $f^*(j) = g(j) + h^*(j)$, where $g(j) = g(i) + c(i, j)$; $c(i, j) = $ length of the arc from node i to node j and $g(s) = 0$, where $s = $ start node.
   (b) If j is already on either OPEN or CLOSED, then direct its pointers along the path that gives the shortest $g(j)$.
   (c) If node j required pointer adjustment and was on the CLOSED list, move it back to OPEN.
7. Go to step 2.

In a sense, A* is a family of algorithms with a common structure. A specific instance of A* is obtained by specifying a particular form of h^*. If h^* is always zero, then A* reduces to the uninformed search strategy, the uniform-cost search. A variant of A* can be applied to AND/OR graphs as well.

25.3.5. Searching AND/OR Graphs

Problems represented by AND/OR graphs have the unique feature that each residual problem posed by some subset of candidates can be viewed as a conjunction of several subproblems that may be solved independently of each other, may require specialized treatments, and should therefore be represented as separate entities. This suggests that subproblems be treated as individual nodes in some graph, even though none of these subproblems alone constitutes a complete solution to the original problem. However, all the subproblems must be solved before the parent is considered solved.

A search problem to be solved by using the problem reduction approach requires the specification of a start node, a set of terminal nodes, and a set of operators for reducing goals to subgoals. The terminal nodes are primitive problems that are known to be solvable. The AND/OR search graph consists of two types of links: AND links and OR links. The complete solution to the problem represented by an AND/OR graph is represented by AND/OR subgraph, called a *solution subgraph*, which has the following properties:

1. It contains the start node.
2. All its terminal nodes are primitive problems, known to be solved.
3. If it contains an AND link, it must also contain all its sibling AND links.

The procedure uses backward reasoning from the initial goal unlike state space search algorithms, which use forward reasoning. At any stage of the search, a test could be conducted to find out if the explicated portion of the searched graph contains a solution graph. That test is best described as an attempt to *label* the start node *solved* (or *unsolvable*) through the use of the following *solve-labeling* rules. A node is said to be *solved* if one of the following conditions holds:

1. The node is in the set of primitive problems.
2. The node has AND nodes as successors and they are solved.
3. The node has OR nodes as successors and any of them is solved.

A node is said to be *unsolvable* if one of the following conditions is true:

1. The node has no successors and is not in the set of terminal nodes.
2. The node has AND nodes as successors and at least one of them is unsolvable.
3. The node has OR nodes as successors and all of them are unsolvable.

As an illustration of AND/OR graph search procedure, the following is an AND/OR beam search procedure:

1. Put the start node s on a list, OPEN, of unexpanded nodes.
2. From the search tree constructed so far, compute the b most promising solution trees $T_1, T_2, \ldots, T_b$; b is the branching factor.

3. For each tree T_k, $k = 1, \ldots, b$, select from OPEN the w best nodes for expansion (i.e., the nodes with the w least values of f^*; resolve ties arbitrarily); w is the beam width. Move these wb nodes from OPEN to a list, CLOSED, of expanded nodes. Remove these nodes from OPEN and place it in a list called CLOSED.

4. For each node n, $n = 1, 2, \ldots, wb$, do:

> If n is a terminal node,
> then

(a) Label node n solved.

(b) If the solution of n makes any of its ancestors solved, label these ancestors solved.

(c) If the start node is solved, exit with T as the solution tree.

(d) Remove from OPEN any node with a solved ancestor.

> else, if node n has no successors,
> > then

(a) Label node n unsolvable.

(b) If the unsolvability of n makes any of its ancestors unsolvable, label all such ancestors unsolvable as well.

(c) If the start node is labeled unsolvable, exit with failure.

(d) Remove from OPEN any node with an unsolvable ancestor.

> else, expand node n, generating all its immediate successors and, for each successor m representing a set of more than one subproblem, generating successors of m corresponding to the individual subproblems. Attach, to each newly generated node, a pointer back to its immediate predecessor and compute f^* for each newly generated node. Place all the new nodes that do not yet have descendants on OPEN. Finally, recompute $f^*(n)$ and f^* at each ancestor of n.

5. *Go to step* 2.

25.3.6. Recent Search Techniques

In Section 25.3.1, the basic hill climbing technique and its drawbacks were described. Some of the recent research has focused on techniques that draw on this simple method. Consider, for example, the iterated hill climbing method. Like the hill climbing method, it starts the search process and attempts to search along the path of steepest ascent until no uphill moves are possible. When it reaches a local optima, it does not stop like the basic hill climbing method; instead, it starts the search process from another starting point; this process continues until it exceeds a prespecified time limit. But this process can be computationally expensive for large problems, since the process evaluates all adjacent points to the current point before deciding which uphill move to apply. One way of avoiding the substantial computational overhead is to try alternatives only an uphill move is found without ensuring that better uphill moves are not available. This is illustrative of several search methods that attempt to circumvent local optima by using the strategy of sustained exploration. By sustained exploration, I refer to the strategy of not giving up when a local optima is reached but reinitiating the search process until some prespecified cutoff time is reached.

Two of the methods that have received substantial attention in recent years are simulated annealing and tabu search. Simulated annealing (11) is essentially a stochastic hill climbing method in which moves that worsen the solution are occasionally allowed in an attempt to reduce the chances of being stuck in a local optima. Simulated annealing has been applied to a variety of problems, such as the traveling salesman problem (10), the graph partitioning problem (8), and VLSI design (11). For a more extensive treatment of simulated annealing, see Otten and van Ginneken (13).

Tabu search is an open-ended metaheuristic specifically designed to circumvent local optimas (5, 6). On reaching a local optima, tabu search forbids undoing what has been done recently. A tabu list of forbidden attributes of moves is maintained. The algorithm systematically searches the solution space around it and makes the move that does not violate the tabu status attached to the attribute(s) of the move and that least worsens the solution. The worsening of the solution is called hill climbing and is carried out in the hope that the worsened solution will open up a new path and somewhere down the path, a better solution will be found. This is the mechanism used to circumvent the local optima. However, the tabu status attached to the attribute(s) may be overridden if the move satisfies the aspiration criteria. A commonly used aspiration criterion is that the tabu status is overridden if the move results in a solution that is better than the best found until now. At each move of the search process, the tabu list is maintained by removing from the list the oldest attribute added and adding to the list the attribute(s) of the current move (FIFO). The success of tabu search depends on the attributes made tabu, the size of the tabu list, and the aspiration criteria; these three aspects are closely related to one another. The tabu list size has been found to be closely related to the tabu attributes (7, 12). If the tabu list is very small, then the search process may fall back to one of the previously generated solutions and the process may be repeated. Cycling is said to have taken place in this case. If the tabu list is too large, then the list of available moves decreases, thereby decreasing the flexibility of the search. A mechanism to override the strong dependence of the search process on the tabu list size and the tabu attribute is to build into the search process a long-term memory function. The long-term memory function is most often used to obtain better results for large and complex problems and in the extremely naive case, the long-term memory function is merely a randomization process. After a fixed number of iterations, the long-term memory is invoked in the hope that the randomization will lead to a situation where the search process charters unfathomed solution space.

25.4. OTHER ISSUES

In the search techniques discussed so far, the emphasis has been on generating optimal solutions or "good" solutions without incurring significant computational time or storage space. From the point of view of intelligent systems, additional improvements in the quality of the solution generated can result by exploiting two features. First, the focus of intelligent search strategies has been on improving the solution by devising solution techniques that can also exploit the characteristics of specific application domains; this requires the representation and use of domain-specific information by the system. Our work on scheduling using filtered beam search (3, 4) is an instance where substantial improvements in solution quality have been obtained by effective representation and use of domain knowledge. Second, improvements in solution quality can be obtained by learning from experience in much the same way as humans do. The ability to learn from experience is a fundamental requirement of intelligence. In developing intelligent search strategies, learning can occur, for example, by (a) determining what strategies lead to successful (or unsuccessful) outcomes, (b) systematically generating an explanation for the success or failure of a particular set of strategies using background knowledge of the domain, and (c) using that higher-level control knowledge to generate superior solutions. Learning techniques have not been used widely to improve search techniques used in manufacturing and design. Artificial intelligence researchers have explored a variety of learning methods (e.g., similarity-based learning and explanation-based learning). However, they typically have been applied to problems that researchers and practitioners have typically not been interested in (e.g., toy problems such as the blocks world, diagnostic problems in expert systems, and natural language processing); their effectiveness in solving combinatorially complex problems is not immediately apparent.

At present, we are developing a learning mechanism to improve the performance of traveling salesman problems. Our approach to learning is based on the use of analytical methods to learn from experience but with a small number of observed examples. This approach is distinguished by its view of observed examples as instances of concepts in the context of background knowledge of the problem domain. Hence it seeks to generalize from a single example by analyzing why that example is an instance of a concept; the explanation identifies the relevant features of the example, which constitute sufficient conditions for describing that concept. The idea is to maintain the basic structure of the explanation while separating it from the unimportant details of the original specific example. The resulting knowledge cannot only be repeatedly used to improve solution quality and system performance, but it can be the basis of an important trait of an intelligent system: the ability to learn, in this case, from successful and unsuccessful search strategies and explain why some search strategies are effective. Such knowledge can be used as control knowledge to

- Improve the efficiency of the search process
- Improve the quality of the solution generated
- Direct the solution mechanism along paths it would not otherwise explore

It is important to note that the proposed approach is different from the more widely studied approach known as the empirical learning or similarity-based learning; this technique involves examining multiple examples of a concept in order to determine the features they have in common. The empirical approach assumes that an intelligent system can learn from examples without having much prior knowledge of the domain under study. The analytical approach, on the other hand, attempts to formulate a generalization after observing a single example but it requires that the system be provided with a great deal of domain knowledge at the outset.

The learning process, which will be developed and tested in the context of traveling salesman problems, consists of the following steps:

1. *Develop a domain theory.* Knowledge of the domain is key to the successful implementation of knowledge-based systems. Our earlier research has already used such knowledge in developing efficient algorithms. The emphasis here is to use it as background knowledge to support learning.

2. *Construct a complete explanation for a specific example by taking advantage of knowledge of the domain theory. Various mechanisms will be developed to do so.* For example, an explanation can be constructed by proving that an example is a member of a concept; this is equivalent to determining a logical proof that demonstrates how an example meets a set of sufficient conditions defining a particular concept. Another approach, particularly applicable to problem solving domains that require the use of one or more algorithms, would specify an explanation as a set of causally connected actions that demonstrate how a goal state is achieved.

3. *Remove branches of the explanation that are more specific than needed for the operationality of the resulting plan or proof.*

4. *Generalize the remaining explanation without invalidating its underlying structure.*

5. *Create a meta-rule that summarizes the resulting generalized explanation and index it so that it can be used to aid future problem solving.* Thus explanations can be reused to improve system performance.

To summarize, future efforts in developing intelligent search strategies will focus on two additional capabilities: sustained exploration and learning while searching. The first capability will attempt to limit the scope of the search by using knowledge about the solution space and the application domain so as to avoid spending computational effort on potential solution paths that are not likely to be promising. The second capability, on the other hand, attempts to avoid narrowing the search irrevocably so as not to eliminate the consideration of potential solution paths that are very promising, although it may not have appeared to be so initially.

REFERENCES

1. Barr, Avron, and Feigenbaum, Edward A. (1981). *The Handbook of Artificial Intelligence*, Vol. I, William Kaufmann, Los Altos, CA.
2. De, Suranjan (1988). A Knowledge-Based Approach to Scheduling in an FMS. In: *Approaches to Intelligent Decision Support*, edited by R. G. Jeroslow, *Annals of Operations Research*, J. G. Baltzer, Scientific Publishing, Amsterdam, Vol. 12, pp. 109–134.
3. De, S, and Lee, A. (1990). FMS Scheduling Using Filtered Beam Search, *Journal of Intelligent Manufacturing*, Vol. 1, No. 3, pp. 165–183.
4. De, S., and Lee, A. (1992). Flexible Assembly Scheduiing Using a Knowledge-Based Approach, *Expert Systems with Applications: An International Journal.* to be published.
5. Glover, Fred (1989). Tabu Search—Part I, *ORSA Journal of Computing*, Vol. 1, No. 3, pp. 190–206.
6. Glover, Fred (1990). Tabu Search—Part II, *ORSA Journal of Computing*, Vol. 2, No. 1, pp. 4–32.
7. Hertz, A., and de Werra, D. (1987). Using Tabu Search Techniques for Graph Coloring, *Computing*, Vol. 29, pp. 345–351.
8. Johnson, David S., Aragon, Cecilia R., McGeoch, Lyle A. and Scheveon, Catherine (1989). Optimization by Simulated Annealing: An Experimental Evaluation; Part I, Graph Partitioning, *Operations Research*, Vol. 37, No. 6, pp. 865–892.
9. Kanal, Laveen, and Kumar, Vipin (Eds.) (1988). *Search in Artificial Intelligence*, Springer-Verlag, Berlin.
10. Kirkpatrick, S. (1984). Optimization by Simulated Annealing: Quantitative Studies, *Journal of Statistical Physics*, Vol. 34, pp. 975–986.
11. Kirkpatrick, S., Gelatt, C. D. Jr., and Vecchi, M. P. (1983). Optimization by Simulated Annealing, *Science*, Vol. 220, pp. 671–680.
12. Malek, M., Guruswamy, M., Owens, H., and Pandya, M. (1989). Serial and Parallel Simulated Annealing and Tabu Search Techniques for the Traveling Salesman Problem, *Annals of Operations Research*, Vol. 21, pp. 59–84.
13. Otten, R. H. J. M., and van Ginneken, L. P. P. P. (1989). *The Annealing Algorithm*, Kluwer Academic Publishers, Norwell, MA.
14. Pearl, J. (1984). *Heuristics: Intelligent Search Strategies for Computer Problem Solving*, Addison-Wesley, Reading, MA.
15. Syslo, Maciej M., Deo, Narsingh, and Kowalik, Janusz S. (1983). *Discrete Optimization Algorithms: With Pascal Programs*, Prentice-Hall, Englewood Cliffs, NJ.
16. Winston, Patrick H. (1984). *Artificial Intelligence*, 2nd ed., Addison-Wesley, Reading, MA.

CHAPTER TWENTY SIX

Simulated Annealing and Tabu Search

WEN-CHYUAN CHIANG

College of Business Administration, University of Tulsa,
Tulsa, Oklahoma

GENARO J. GUTIERREZ, and PANAGIOTIS KOUVELIS

College of Business Administration, The University of Texas at Austin

26.1. INTRODUCTION

Many engineering design and manufacturing system design problems finally result in the solution of a combinational optimization problem. Solving a combinatorial optimization problem amounts to finding the "best" or "optimal" solution among a finite or a countably infinite number of alternative solutions (26a), which in the case of engineering or manufacturing system design will imply finding the optimal among a set of feasible alternatives. Examples of such engineering design problems can be found in the areas of VLSI design, communication and computer system design, and design of mechanical parts and mechanisms. Let us describe one such problem from the area of VLSI design. The general placement problem in VLSI design is to place a set of circuit modules on a chip such that a certain objective function is minimized. The ultimate goal is to minimize the total chip area occupied by the circuit modules and the interconnection between the modules (31, 32, 33). In the area of manufacturing system design, examples of combinatorial optimization application can be found in the area of flexible manufacturing system (FMS) design, layout design, material handling system, and warehousing and storage system design. Late in the chapter we elaborate on the solution of such a manufacturing system design problem. For reference on such problems see Kusiak (19), Kusiak (20), and Kouvelis (15).

Over the years it has been shown that many of the practical combinatorial

Intelligent Design and Manufacturing, Edited by Andrew Kusiak.
ISBN 0-471-53473-0 © 1992 John Wiley & Sons, Inc.

optimization problems, such as those encountered in engineering and manufacturing systems, belong to the class of NP-complete problems. A thorough overview of problems in this class can be found in Garey and Johnson (5). A direct consequence of the property of NP-completeness is that optimal solutions usually cannot be obtained in a reasonable amount of computation time. However, the large NP-complete problems frequently encountered in engineering or manufacturing system design still must be solved. The designer is limited to one pragmatic option: to look for quickly obtainable solutions at the risk of suboptimality. There are a large number of approximate (heuristic) algorithms tailored to the solution of specific combinatorial optimization problems. However, it is desirable for the engineering designer to have approximate algorithms applicable to a wide variety of combinatorial optimization problems encountered in the engineering and manufacturing practice. Such algorithms could become the necessary design support tools in the demanding era of concurrent engineering, where the interface between engineering and manufacturing design becomes a necessity. Recently, a class of such algorithms are the *intelligent heuristics*. Two of these intelligent heuristic approaches, the *simulated annealing* method and the *tabu search* approach, are the main subjects of this chapter.

The *simulated annealing (SA)* is a high-quality approximation algorithm, which is applicable to a wide array of engineering and manufacturing system design problems. In nature, it is a randomization algorithm and can be asymptotically viewed as an optimization procedure. In any practical implementation, however, it behaves as an approximation algorithm. The name of the algorithm derives from an analogy between solving optimization problems and simulating the annealing of a solid as proposed by Metropolis et al. (24). *Tabu search (TS)* is a higher level heuristic, or metastrategy, for solving combinatorial optimization problems. The technique is designed in a way that allows it to be superimposed on any procedure whose operation can be characterized as performing a sequence of moves which lead the procedure from one solution to another. Both the SA and TS have been designed in a way that they avoid getting trapped into inferior quality locally optimal solutions, which is a behavior exhibited by many local search heuristics.

In order to present the application of SA and TS methods for design problems in a constructive way, we have structured the chapter in the following manner. Both methods are applied on the same manufacturing system design problem, the *unidirectional flowpath design of automated guided vehicle system (AGVSs)*, which we introduce in detail in Section 26.2. A tutorial level exposition of the simulated annealing method is given in Section 26.3, and its application to our problem is presented in Section 26.4. Section 26.5 exposes the basics of the tabu search approach, and its application to our problem is given in Section 26.6. Section 26.7 compares the performance of the two methods on this specific design application, and Section 26.8 includes some concluding remarks.

26.2. THE UNIDIRECTIONAL FLOWPATH DESIGN PROBLEM FOR AUTOMATED GUIDED VEHICLE SYSTEMS (AGVSs)

Automated guided vehicle systems (AGVSs) have been of great interest to industry over the last two decades. The number of applications of these systems has increased to a point when AGVSs are considered to be a basic concept in material handling. Although initial applications of AGVSs were generally limited to warehouses, in recent years an increasing number of applications in manufacturing systems have been reported (18, 25, 29). AGVs play an important role in many low to medium volume manufacturing operations, including flexible manufacturing systems (FMSs), and in recent developments of flexible assembly systems.

Automated guided vehicles (AGVs) are electronically guided vehicles capable of undertaking material transport missions (i.e., loading, transportation, and unloading of materials) without human intervention. The vehicles are independently addressable and centrally controlled. An important part of an AGVS is an information system that handles the transfer of information between a central computer and peripheral computers (controllers) on-board the vehicles. The vehicles move along transportation networks appropriate to the specific application of the AGVS. An example of one such transportation network is given in Fig. 26.1. Guidance systems are used

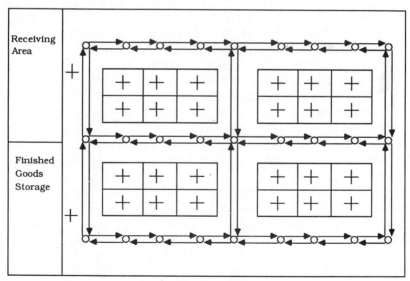

O	Nodes in the network representation (27)
→	Arcs in the network representation (60)
+	Workstations in the system (26)

FIGURE 26.1. Example of an AGVS layout and its network representation.

to determine the vehicle paths and to physically implement the desired transportation network. There are two types of guidance systems: those using off-board fixed paths and those using on-board software programmed paths. Wire guided (i.e., vehicles follow a wire buried in the floor) and painted line guided (i.e., vehicles follow fluorescent painted line by sensing the reflections of the path with the use of ultraviolet light sources attached to them) vehicle systems are examples of off-board fixed paths. On-board software programmed paths (in many cases referred to as "virtual flow paths") allow vehicle movements without physical guide paths. Finally, the vehicles have automatic battery charging systems, which dispatch them to a battery charger just before the power fails.

When designing an automatic guided vehicle transport system, there are several important issues to consider. The major design issues are vehicle configuration (the two basic types are unit load carrier and the tractor/ trailer combination), vehicle guidance method (wire guided, painted line guided, and virtual flow paths), choice of control systems (that involves setting vehicle dispatch rules, intersection control, and stopping criteria), and location of load/unload stations and of battery charging stations. Provided that a guide path is used for the movement of vehicles, the designer of the AGVS must also be concerned with the direction of flow on each segment of the path. The direction of traffic along the vehicle paths (in many cases referred to as "flowpaths") can be defined as either unidirectional or bidirectional. Unidirectional flow results when vehicles are restricted to travel only in one direction along a given segment of the flowpath. Bidirectional flowpaths allow vehicles to travel in both directions. Unidirectional flowpaths have been extensively implemented since they reduce the possibility of vehicle blocking, require fewer controls, and are more economical (6). Bidirectional flowpaths, on the other hand, reduce travel time of the vehicles at the expense of sophisticated control systems.

In this chapter we address the flowpath design issue and in particular we concentrate on the design of unidirectional AGVSs. To achieve an effective use of AGVs in a manufacturing application requires a well thought-out flowpath design, since it affects the total distance vehicles travel to accomplish the material transport tasks. We discuss approaches that result in unidirectional flowpaths that minimize the total distance traveled by loaded vehicles. Later we review relevant literature on the subject. We review only literature directly related to the issue of AGVS design. For a detailed coverage of operations research approaches to the broader issue of material handling system design refer to Maxwell (22), Matson and White (21), and Kusiak (18).

While a vast amount of engineering work appeared on the design and implementation of AGV hardware (2a, 12a, 14, 25), very little research has appeared on the optimization of AGVS operations. Furthermore, a majority of the papers written on AGVSs present empirical results based on simulation models (12, 26–28). Very few published studies with generally applicable results are available, and our discussion concentrates on them.

A design methodology with general applicability is presented by Maxwell and Muckstadt (23) for AGVSs with unit load carriers operating on unidirectional flowpaths. Kuhn (17) and Müller (25) present simple formulas for determining the optimal number of vehicles in an AGVS. Sharp and Liu (30) proposed an analytical method for configuring the network of a fixed path, closed loop MHS. Such a system could be implemented with AGVs.

Egbelu and Tanchoco (4) state some of the advantages of bidirectional flowpaths. In Gaskins and Tanchoco (6) the unidirectional flowpath design problem is stated. They formulate the problem as a nonlinear binary integer program. The complexity of the formulation makes it practically applicable only for small sized problems. Reasons supporting the use of unidirectional flowpaths can be found in Egbelu and Tanchoco (3, 4).

The motivating idea behind the use of intelligent heuristics, such as SA and TS, for the design of large sized unidirectional AGVSs is that we can achieve computational efficiency, which the previously presented methods lack, with a reasonable level of cost effectiveness of the resulting flowpath design. We spend the rest of the chapter on the presentation of the SA and TS methodologies and their application to the flowpath design problem.

26.3. THE SIMULATED ANNEALING METHOD FOR COMBINATORIAL OPTIMIZATION PROBLEMS

Simulated annealing (SA) is an algorithmic approach to solving combinatorial optimization problems. The relatively recent work of Kirkpatrick et al. (13) and Cerny (2) stimulated considerable interest in the method. The name of the algorithm derives from an analogy between solving optimization problems and simulating the annealing of solids as proposed by Metropolis et al. (24). For our purpose, simulated annealing can be viewed as an enhanced version of the iterative improvement method, in which an initial solution is repeatedly improved by making small local changes until no such changes yield a better solution. Simulated annealing randomizes the local search procedure and in some instances allows for solution changes that worsen the solution. That constitutes an attempt to reduce the probability of becoming stuck in a poor but locally optimal solution.

The SA algorithm operates as follows. Starting from an initial solution (let us call it σ) the algorithm generates at random a new solution σ' in the neighborhood of the original solution. Let $\text{Cost}(\sigma)$ and $\text{Cost}(\sigma')$ denote the objective function values of these two solutions, and $\Delta C = \text{Cost}(\sigma') - \text{Cost}(\sigma)$ the change in the value of the objective function as we move from solution σ to σ'. Let us say that we want to minimize the objective function.

If the change ΔC represents a reduction in the value of the objective function, then our move to the new solution σ' is accepted. If the change represents an increase in the objective function value, then the move is accepted with a specified probability. The acceptance probability function usually takes the form of $\exp(-\Delta C/T)$, where T is a control parameter. This implies that for larger values of T, moves with large positive ΔC can be accepted. As the value of T decreases, only moves with smaller positive ΔC have a reasonable probability to be accepted, and as the value of T approaches zero, no moves with positive ΔC will be accepted at all. The SA procedure is described in a pseudoalgorithmic way next.

A Generic Simulated Annealing Procedure

1. Get an initial configuration σ.
2. Get an initial temperature $T > 0$.
3. While the solution is not "frozen" do the following two steps:
    3.1. For $1 \leq i \leq L$ do
        3.1.1. Pick random neighbor σ' of σ.
        3.1.2. Let $\Delta C = \text{Cost}(\sigma') - \text{Cost}(\sigma)$.
        3.1.3. If $\Delta C \leq 0$ then set $\sigma = \sigma'$.
        3.1.4. If $\Delta C \geq 0$ the set $\sigma = \sigma'$ with probability $e^{-\Delta C/T}$.
    3.2. Set $T = r$ times T.
4. Return σ.

As can be seen from the above description, the SA procedure is more of an approach than a specific algorithm. To implement it we must make a variety of choices for the values of the parameters and the meaning of the undefined terms. The choices fall into two categories:

Problem Specific	Generic
Configuration	Initial temperature
Neighborhood of a configuration	Cooling ratio
Cost of a configuration	Epoch length
Initial configuration	Frozen system

We feel that all the above terms, and their importance in the SA procedure, can be clarified with the use of a specific example on an implementation of the SA procedure. Thus, in the next section, we describe in detail the meaning of all these terms and appropriate choice of parameters for a SA algorithm for the unidirectional flowpath design problem (UFDP).

Before we proceed to the example, we would like to mention that the simulated annealing algorithm converges with probability 1 to global optimality if certain conditions hold (1). One set of conditions for asymptotic convergence of SA specifies that an infinite number of solutions should be generated for each value of temperature T. Also, if $\{T_k\}$ is an infinite sequence of temperatures generated by the algorithm, then it should hold that $\lim_{k \to \infty} \{T_k\} = 0$. However, it is obvious that in actual implementations of the algorithm the asymptotic property of convergence to global optimality can only be approximated. That is, only a finite number of transitions will be generated at every temperature, and the condition $\{T_k\} \to 0$ for large k will hold only approximately. Due to the above approximations, SA is a heuristic procedure for difficult combinatorial problems will the UFDP in AGVSs.

26.4. SIMULATED ANNEALING HEURISTICS FOR THE UNIDIRECTIONAL FLOWPATH DESIGN PROBLEM OF AGVS

26.4.1. Selection of the Problem-Specific Parameters

In this section we discuss the application of the simulated annealing approach to the unidirectional flowpath design problem. The choices we have to make can be classified into two classes, namely, problem specific and generic. Below we fill in the details for the various problem-specific terms necessary for the development of a simulated annealing algorithm for the unidirectional flowpath design problem.

The *solution* is a flowpath design of our material handling network. The flowpath design in a "solution" is not restricted to be unidirectional. It could have bidirectional arcs. This implies that the SA algorithm examines also nonfeasible solutions. We use the term *feasible* solution to characterize those solutions that are legal solutions to the original problem (i.e., they are unidirectional flowpath designs). The neighborhood of a solution consists of those solutions that result from the initial one by reversing the direction of an existing directed arc in the current solution or by adding to the current solution a directed arc between two nodes.

Since the solution space is enlarged by including nonfeasible solutions, appropriate penalty terms have to be added to the cost to make a nonfeasible solution less attractive. Thus the cost C of a solution σ will be the total distance traveled by all material flows in the system, assuming that the flows travel the shortest possible distance path between their origin and destination for the given flowpath design, plus a penalty term for nonfeasible solutions. For a specific solution σ, let us denote by $D_\sigma = (d_{ij}^\sigma)$ the shortest path distance matrix between any two pairs of nodes for the specific solution σ. Then the total distance traveled by all material flows M_{ij} (where M_{ij} denotes the volume to be transported from station i to station j) in the solution σ is $\Sigma_{(i,j) \in A} M_{ij} d_{ij}^\sigma$. Then

$$C(\sigma) = \sum_{(i,j)\in A} M_{ij}d_{ij}^{\sigma} + \text{Penalty}(\sigma) .$$

There are different ways that we can develop a penalty function for a specific solution σ. The most interesting ones are the following:

1. Penalty $(\sigma) = \Delta > 0$, if solution σ is nonfeasible (i.e., it includes some bidirectional arcs), and zero otherwise. This penalty function definition is independent of how many bidirectional arcs are included in the flowpath design. Then the cost of solution σ is given by

$$C_1(\sigma) = \sum_{(i,j)\in A} M_{ij}d_{ij}^{\sigma} + \Delta \cdot \mathbf{1}\{\sigma \text{ nonfeasible}\}$$

(the function $\mathbf{1}\{\cdot\}$ is an indicator function), where A is the set of all possible arcs in our material handling network and (i, j) denotes an arc having direction from i to j.

2. Penalty $(\sigma) = \Sigma_{(i,j)\in BD(\sigma)} \delta_{ij}$, where $\delta_{ij} > 0$ and the set $BD(\sigma) = \{(i, j) \in A$: flow from $i \to j$ and from $j \to i$ is allowed in $\sigma\}$. This penalty function charges δ_{ij} if the flow between nodes i and j is bidirectional (which implies both arcs (i, j) and (j, i) are included in the flowpath design) and zero if it is not for all arcs (i, j). Then the cost of solution σ is given by

$$C_2(\sigma) = \sum_{(i,j)\in A} M_{ij}d_{ij}^{\sigma} + \sum_{(i,j)\in BD(\sigma)} \delta_{ij} .$$

3. Penalty $(\sigma) = \Delta \cdot \mathbf{1}\{\sigma \text{ nonfeasible}\} + \Sigma_{(i,j)\in BD(\sigma)} \delta_{ij}$. This definition combines the previous two penalty function definitions. The resulting cost for solution σ is

$$C_3(\sigma) = \sum_{(i,j)\in A} M_{ij}d_{ij}^{\sigma} + \sum_{(i,j)\in BD(\sigma)} \delta_{ij} + \Delta \cdot \mathbf{1}\{\sigma \text{ nonfeasible}\}$$

The choice of the cost function to be used in the implementation of the SA algorithm has a significant effect on the performance of the algorithm. It is quite easy to argue for the inferiority of the cost function $C_1(\sigma)$ for a larger Δ (for implementation purposes Δ has to be chosen an order of magnitude larger than the lower bound on the quantity $C_2(\sigma) = \Sigma_{(i,j)\in A} M_{ij}d_{ij}^{\sigma}$). A graphical aid in the presentation of our argument is Fig. 26.2. Let us assume, without loss of generality, that we start our SA algorithm from a feasible solution. For a period of time t_0 the algorithm will be evaluating feasible solutions. At time t_0 the accepted candidate solution in the neighborhood of our current solution is a nonfeasible one, which can happen with a positive probability due to the random nature of the SA algorithm. For this solution the penalty Δ will be charged. The algorithm

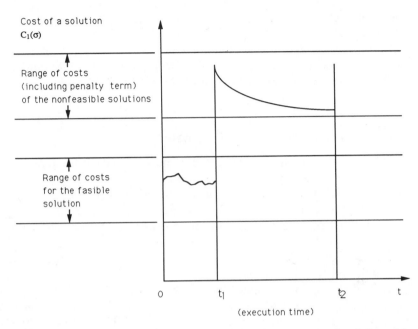

Cost of a solution
$C_1(\sigma)$

Range of costs
(including penalty term)
of the nonfeasible solutions

Range of costs
for the fasible
solution

0   t_1 t_2 t

(execution time)

FIGURE 26.2. A graphical explanation of the poor performance of SA with cost function C_1.

will proceed by searching the neighborhood of the current flowpath design with one bidirectional arc. Adding bidirectional arcs to that design will result, in general, in reduction of the function $C_1(\sigma)$ since the penalty term is independent of the number of arcs in the flowpath design. This search pattern will finally force SA to a nonfeasible local optimal with a large number of bidirectional arcs. According to the definition of the neighborhood of a solution, there is not going to exist any feasible solution in the neighborhood set of that nonfeasible local optimal, and as a consequence the SA will terminate at some time t_1 with a bidirectional flowpath design. That is an undesirable outcome for our purposes, and we suggest that the use of the cost function $C_1(\sigma)$ be avoided in implementations of the SA algorithm. This limits our choices to the cost functions $C_2(\sigma)$ and $C_3(\sigma)$. We cannot a priori make any theoretical argument for the relative superiority in performance of the SA with the cost function one of these. Thus we implemented SA algorithms for each one of these cost functions. The version of the SA algorithm with cost function $C_2(\sigma)$ is referred to as "SA with variable penalties (SA–VP)," and the version of the SA algorithm with cost function $C_3(\sigma)$ is referred to as "SA with fixed and variable penalties" or "SA with mixed penalties (SA–MP)." We discuss the relative performance of the two SA algorithms in Section 26.7.

For the SA algorithms, we used as an initial solution the flowpath design resulting from application of a simple heuristic presented in Kouvelis et al.

(16) (it is referred to as heuristic KGC-1). Starting from the initial solution, the SA generates at random a new solution σ' in the neighborhood of the original solution σ, and for which the change in the value of the cost function ΔC is quickly calculated. For the acceptance probability function in the case of ΔC we use the usual form in the literature (13a) of $\exp(-\Delta C/T)$, where t is a control parameter.

26.4.2. Selection of the Generic Parameters

There are four generic parameters that must be specified for implementation of the SA algorithm. A choice of these parameters is referred to as a cooling schedule. In general, the performance of SA algorithms is sensitive to the cooling schedules specified (11). Our experimental process of coming up with a satisfactory cooling schedule for our SA implementations is described next.

The selection of the "cooling" parameters has an impact on the quality of the solutions that the SA algorithm obtains. In general, a slower schedule, (i.e., one that starts from a higher temperature T, has a larger cooling ratio r, and has a larger temperature length L) will lead to better solutions. On the other hand, slower schedules tend to use more CPU time. Our objective is to select a set of cooling parameters that are expected to produce good, though not necessarily optimal, solutions in reasonable time. Our selection of the cooling parameters was based on extensive analysis of a base case problem, which consisted of a configuration with 20 nodes, 31 undirected arcs, and 10 different production routes. This problem is representative of the set of problems on which the resulting cooling schedule was tested. Default values were fixed for all parameters, and then one at a time all parameters were tested over a reasonable range. To obtain meaningful averages, each run was replicated 10 times with different streams of pseudo random numbers. The default values selected are as follows: the temperature length, L, was fixed at 15, the cooling ratio, r, was set at 0.95, and the initial temperature was set so that there is a 0.90 probability of accepting a cost increase of \$1000.

Temperature length, L, was tested at 1, 5, 10, 15, 20, 25, and 30. The averages, best, and worst cases are reported for models (SA–VP) and (SA–MP) in Figs. 26.3(a) and 26.3(b), respectively. We observed a marked improvement from 1 to 5, and a somewhat less dynamic improvement from 5 to 10. Performance for values of L larger than 10 is only marginally better. On the other hand, CPU times increase markedly for values of L larger than 10. For example, for $L = 10$, average CPU time is 0.93 minute on an IBM 3081, whereas for $L = 15$, average CPU time is 1.62 minutes. Thus for practical purposes, we decided to set $L = 10$ in the SA cooling schedule.

The *cooling ratio*, r, was tested at 0.7, 0.8, 0.9, 0.95, and 0.99. Averages, best, and worst cases are reported in Fig. 26.4 for model (SA–VP). Similar results were observed for (SA–MP). We can see that the graphs in Fig., 26.4

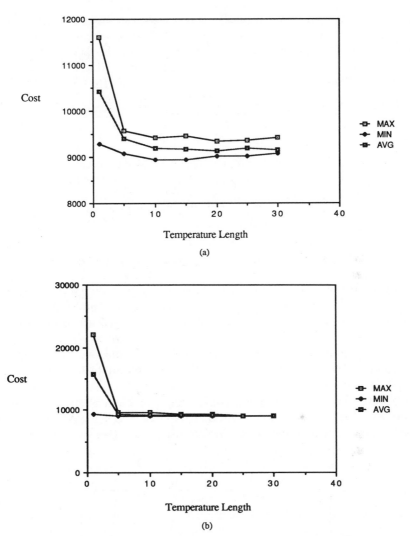

FIGURE 26.3. (a) Sensitivity of (SA–VP) to temperature length. (b) Sensitivity of (SA–MP) to temperature length.

are relatively flat. This apparent insensitivity of (SA–VP) can better be understood by observing that the three cooling parameters are not independent; thus if L and t are set at sufficiently high values, the results obtained by the algorithm are relatively insensitive to r. Although there appears to be no difference in performance for different values of r, we observed a reduction in standard deviation as r increases. For example, standard deviation of (SA–VP) went from 183.8 to 152.1 as r went from 0.9 to 0.95. Based on these observations, we decided to fix the cooling ratio r at its default value of 0.95.

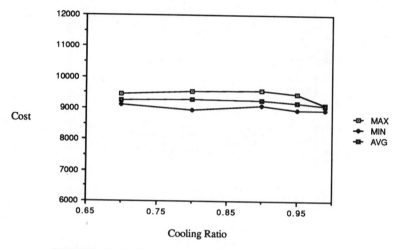

FIGURE 26.4. Sensitivity of (SA–VP) to cooling ratio.

Initial temperature, T, expressed as the initial probability of accepting a cost increase of $1,000, was tested at 0.1, 0.2, ... , 0.8, and 0.9. Averages, best, and worst cases for (SA–VP) are shown in Fig. 26.5. Similar results were observed for (SA–MP). We could not observe an increase in performance at higher initial temperatures, and CPU times were insensitive to initial temperature. Finally, T was fixed at its default of 0.9, this being a conservative choice.

In Section 26.7, we describe computational results obtained on a set of test problems of our two SA models, (SA–VP) and (SA–MP). For both models, we use a cooling schedule consisting of $T = 0.9$, $r = 0.95$, and $L = 10$.

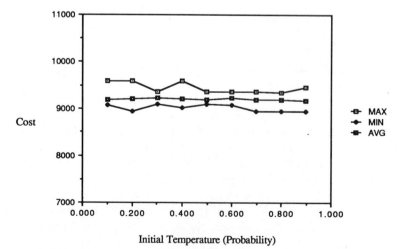

FIGURE 26.5. Sensitivity of (SA–VP) to initial temperature.

26.5. THE TABU SEARCH METHOD FOR COMBINATORIAL OPTIMIZATION PROBLEMS

Tabu search (TS) is a heuristic approach for solving combinatorial optimization problems. It is an adaptive procedure that can be superimposed on many other methods to prevent them from being trapped at locally optimal solutions. The method is presented in detail in Glover (8–10). In this section we summarize, from an application standpoint, the main features of the method. The method is presented in a diagrammatic way in Fig. 26.6.

The method starts with an initial current "solution." The term "solution" can have multiple meanings. It could be a feasible solution for the problem (the most usual case), a nonfeasible one, or even a partial solution (not all the variables of the problem have been assigned values). Using some local changes, we generate from the current solution a list of candidate solutions.

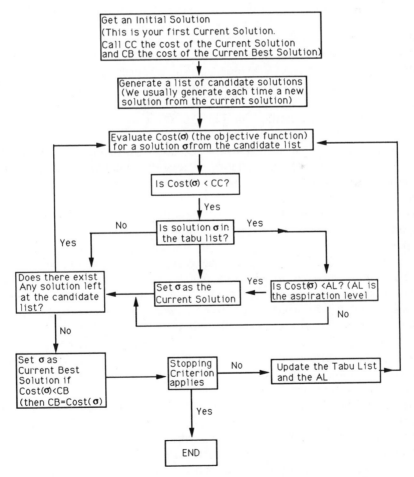

FIGURE 26.6. A simple form of a TS approach for cost minimization problems.

If the change results in a large number of candidate solutions, we might decide to restrict ourselves only to a subset of them. Then we have to evaluate the solutions in the candidate list. Let us assume for purposes of our presentation that we are dealing with a minimization problem and that Cost($\cdot$) is our objective function. Select the best solution σ from the candidate list, [i.e., the solution with minimum value of Cost(σ)]. If this selection is forbidden (i.e., tabu), we proceed to select the next best selection in the candidate list. Otherwise, σ is selected as the current solution.

The *tabu list* keeps track of a set of solutions that we would not want to examine (are "tabu" or forbidden). The purpose of this tabu list is to avoid cycling in the algorithm. For example, if the search hits a local minimum, the best move in the next iteration will result in an increase in cost. If the old solution is not excluded from the search, there is a good probability that the algorithm will return to the previous selection since it reduces the cost. Thus by making the move to the previous selection a tabu move, we avoid cycling the algorithm around the local minimum. The tabu status of a solution can be overridden with the use of certain *aspiration criteria*, as we shall see later. The tabu list and the aspiration criteria are the basic mechanisms with which TS avoids getting trapped into a local optimal solution. Let us say that our current solution belongs to the tabu list. Then the next step will be to see if it satisfies the aspiration criteria. In our simple exposition of the TS method in Fig. 26.6, we used a single aspiration criterion (that of an aspiration level AL, which is of a cost nature—it could be, for example, the cost of the current best solution for the problem so far). So if the cost of the current solution is less than the specified aspiration level, the tabu status of the solution is overridden and can still be admissible as the next current solution.

After going through the tabu and aspiration criteria checkpoints, the solution will be either admissible as a new current solution or it will be dropped from further consideration. In the latter case, we return to the candidate list of solutions. For the other case, we proceed to check the cost of the new current solution with that of the current best solution (the best solution that has been found so far). We update the current best solution if necessary, and we proceed to generate a new list of candidate solutions around the new current solution. The iterative process for that new list starts. The procedure goes on until one of the stopping criteria applies. The usual stopping criteria are that a total number of iterations since the start of the algorithm have elapsed or a certain number of iterations since the last current best solution was found has occurred.

Explaining the TS method at a rather abstract level is quite difficult. The application of the TS method for the solution of the unidirectional flowpath design problem for AGVSs, which is described in the next section, will help us to clarify some of the previously used terms (tabu list, aspiration level) and provide a more intuitive explanation for the method.

26.6. TABU SEARCH IMPLEMENTATION FOR THE UNIDIRECTIONAL FLOWPATH DESIGN OF AGVSs

In this section we describe two different implementations of the TS approach to the unidirectional flowpath design problem. Single examples of each implementation demonstrate the basic characteristics of the TS method and help to clear the understanding of the different terms used.

26.6.1. A Simple and Fast Implementation of TS (Tabu I)

The TS method starts with the initial solution flowpath design resulting from application of the same heuristic as the one used for the SA method. As for the SA method, the *solution* in our TS approach is a flowpath design of our material handling network, and it is not restricted to be unidirectional. It could include bidirectional arcs (i.e., we examine also nonfeasible solutions).

Let us call A the set of all potential arcs in our material handling network, and let $L(\sigma)$ denote the set of all arcs included in the current solution σ. The cost associated with a solution subsequently denoted as $C(\sigma)$ is calculated as

$$C(\sigma) = \sum_{(i,j)\in A} M_{ij}d_{ij}^{\sigma} + \text{Penalty}(\sigma) \,,$$

where for the penalty definition we use definition 3 of Section 4.1; that is

$$\text{Penalty}(\sigma) = \Delta \cdot \mathbf{1}\{\sigma \text{ nonfeasible}\} + \sum_{(i,j)\in BD(\sigma)} \delta_{ij} \,.$$

The procedure to generate the next solution σ' to enter the candidate list, using as input the current solution σ, operates in the following way (written in a pseudoalgorithmic format):

For all arcs $a \in A$ do the following:

1. If $a \in L(\sigma)$ consider excluding arc a from the flowpath design and call $\sigma'(a)$ the resulting solution [i.e., $L(\sigma'(a)) = L(\sigma)\backslash\{a\}$]. Evaluate $C(\sigma'(a))$.
2. If $a \notin L(\sigma)$ consider including a to the flowpath design, with $\sigma'(a)$ denoting the new solution [i.e., $L(\sigma'(a)) = L(\sigma) \cup \{a\}$].
3. Evaluate $C(\sigma'(a))$. Let $a^* = \arg\min\{C(\sigma'(a))\}$ (i.e., the arc whose inclusion or exclusion from the flowpath design results in the maximum benefit), and $C(\sigma'(a^*))$ the next solution in the candidate list.

The attribute of a solution that characterizes its tabu status is the arc whose inclusion or exclusion from the current solution led us to the next solution. The tabu list keeps track of a specified number (this is the tabu list size parameter τ) of arcs that were instrumental in generating the last τ solutions. So, using the above introduced notation of $\sigma'(a^*)$, we investigate its tabu status by checking if a^* is part of the tabu list. The tabu list is continuously updated. If there are already τ arcs in the tabu list, and we need to keep information for an additional one, we simply erase the first arc in the tabu list and store the new information. Each time that a new arc enters the tabu list, we also associate an aspiration level with it [referred to as $AL(a)$]. The aspiration level $AL(a) = C(\sigma'(a))$, where $\sigma'(a)$ is the solution at the point that a entered the tabu list. So the tabu status of an arc can be overridden only if the cost of the resulting solution is less than the above aspiration level.

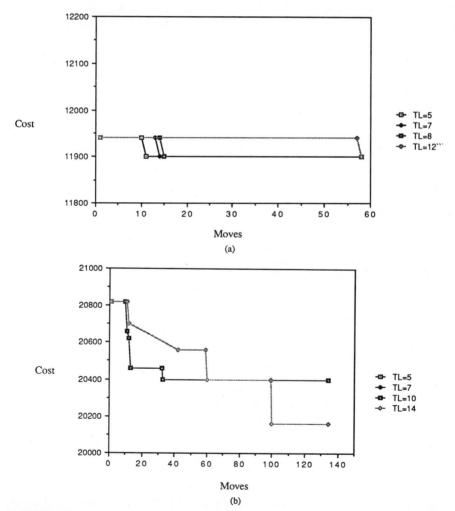

FIGURE 26.7. Sensitivity analysis of tabu size for 4×6 problems. (b) Sensitivity analysis of tabu size for 10×3 problems.

The important parameter of the TS implementation is the tabu list size τ. We have performed sensitivity analysis in Fig. 26.7. For the computational results of this method, which are reported in Section 26.7, we used $\tau = 7$. Our stopping criteria for the method are the usual ones, as mentioned in Section 26.5. The example that follows demonstrates the basic ideas behind the Tabu I implementation.

Example 1. In this example we use a simple material handling network for illustration purposes. The network has nine nodes and 24 potential arcs. The network is of grid structure (a rectangular grid), with any two adjacent nodes on the grid having distance of 1 (see Fig. 26.8a). There are nine different types of material flows that have to be accommodated in the network. The table below provides more detailed information on the flows. For each flow, the origin, the destination, and the volume are provided.

Flow	Origin	Destination	Volume
1	1	3	290
2	1	5	250
3	1	8	240
4	3	1	120
5	5	1	510
6	5	8	530
7	8	1	270
8	8	3	290
9	8	5	330

For our cost calculating we use $\delta_{ij} = \delta = 1{,}000 \; \forall i, j$, and $\Delta = 10{,}000$.

A heuristic solution to the problem is depicted in Fig. 26.8b. This solution serves as the initial current solution. Let us now set the procedure for generating the next candidate solution. We develop the following table:

Arc(a)	$C(\sigma'(a))$	Arc(a)	$C(\sigma'(a))$
(1, 2)	18,800	(2, 1)	$+\infty$
(1, 4)	$+\infty$	(4, 1)	18,840
(2, 3)	7,380	(3, 2)	$+\infty$
(2, 5)	19,380	(5, 2)	19,400
(3, 6)	19,380	(6, 3)	18,960
(4, 5)	$+\infty$	(5, 4)	19,380
(4, 7)	19,380	(7, 4)	19,700
(5, 6)	18,380	(6, 5)	19,380
(5, 8)	$+\infty$	(8, 5)	18,180
(6, 9)	19,380	(9, 6)	18,960
(7, 8)	19,380	(8, 7)	19,700
(8, 9)	18,960	(9, 8)	19,380

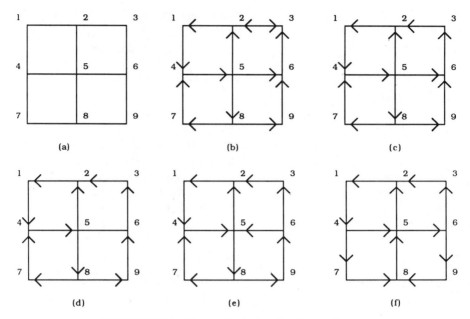

FIGURE 26.8. Flowpath design for Example 1.

Now let us explain some of these numbers. Let us consider arc(1, 2). The arc(1, 2) is not currently included in the flowpath design. If we include it in the flowpath design and name $\sigma'(1, 2)$ the resulting new flowpath design, its cost $\sigma'(1, 2) = 18{,}800$. Let us now consider the arc(1, 4). The arc(1, 4) is included in the current flowpath design. If we exclude this arc from the flowpath design and name $\sigma'(1, 4)$ the resulting new flowpath design, we have resulted in a nonfeasible solution. Material flows 1, 2, and 3 cannot be accommodated with that flowpath design. Any nonfeasible solution is assigned a very large cost in our computer implementation of the TS method (or $+\infty$ for theoretical purposes). Now let us look at arc(2, 3). It belongs to the current flowpath design. if arc(2, 3) is excluded from the flowpath design, then the resulting flowpath design $\sigma'(2, 3)$ will have $C(\sigma'(2, 3)) = 7{,}380$. Similar interpretations have all the other numbers in the above table. From the above table we can see that $a^* = (2, 3)$, which implies that the new candidate solution will have the arc(2, 3) excluded from the flowpath design. The new flowpath design is depicted in Fig. 26.8c.

At this point in time, the tabu list is empty. So the new candidate solution goes through the tabu status check and becomes the new current solution. The tabu list is now going to contain one arc—arc(2, 3)—and the aspiration level associated with that arc is $AL(2, 3) = 7{,}380$.

Using the flowpath design of Fig. 26.8c, we proceed to generate a new candidate solution. We develop the following table (the numbers have similar interpretation as above):

Arc(a)	$C(\sigma'(a))$	Arc(a)	$C(\sigma'(a))$
$(1,2)$	13,800	$(2,1)$	$+\infty$
$(1,4)$	$+\infty$	$(4,1)$	17,840
$(2,3)$	18,380	$(3,2)$	$+\infty$
$(2,5)$	18,380	$(5,2)$	8,400
$(3,6)$	18,380	$(6,3)$	$+\infty$
$(4,5)$	$+\infty$	$(5,4)$	18,380
$(4,7)$	18,380	$(7,4)$	8,700
$(5,6)$	7,960	$(6,5)$	18,380
$(5,8)$	$+\infty$	$(8,5)$	17,180
$(6,9)$	18,380	$(9,6)$	7,960
$(7,8)$	18,380	$(8,7)$	8,700
$(8,9)$	7,960	$(9,8)$	18,380

From the preceding table we can see that $a^* = (5,6)$ with $C(\sigma'(5,6)) =$ 7,960. Since $(5,6)$ was included in the previous flowpath design, this implies that $\sigma'(5,6)$ will exclude it. Arc$(5,6)$ is not in the tabu list. It therefore becomes the new current solution. However, the current best solution so far has a cost of 7,380 (the previous flowpath design). The tabu list increases to two arcs, with the second arc being $(5,6)$ and aspiration level associated with it AL$(5,6) = 7,380$.

Using the new current solution, which is depicted in Fig. 26.8d, we proceed to generate the next candidate solution. We develop the following table:

Arc(a)	$C(\sigma'(a))$	Arc(a)	$C(\sigma'(a))$
$(1,2)$	18,960	$(2,1)$	$+\infty$
$(1,4)$	$+\infty$	$(4,1)$	18,420
$(2,3)$	18,380	$(3,2)$	$+\infty$
$(2,5)$	18,960	$(5,2)$	10,000
$(3,6)$	18,960	$(6,3)$	$+\infty$
$(4,5)$	$+\infty$	$(5,4)$	18,960
$(4,7)$	18,960	$(7,4)$	9,280
$(5,6)$	7,380	$(6,5)$	7,960
$(5,8)$	$+\infty$	$(8,5)$	17,760
$(6,9)$	18,960	$(9,6)$	$+\infty$
$(7,8)$	18,960	$(8,7)$	9,280
$(8,9)$	$+\infty$	$(9,8)$	18,960

From the above table we can see that $a^* = (5,6)$ with $C(\sigma'(5,6)) =$ 7,380. Since $C(\sigma'(5,6))$ is not less than the aspiration level, the solution $\sigma'(5,6)$ cannot become a new current solution. We look at the second best solution in the above table. This is $a = (6,5)$ with $C(\sigma'(6,5)) = 7,960$. The $(6,5)$ was not part of the previous flowpath design, and consequently

$\sigma'(6, 5)$ will be a flowpath design that includes arc$(6, 5)$. Arc $(6, 5)$ is not in the tabu list. Consequently, it can becomes the new current solution. The tabu list increases to three arcs, with the third arc being $(6, 5)$ and aspiration level associated with it $AL(6, 5) = 7,960$. The new flowpath design is depicted in Fig. 26.8e.

The algorithm proceeds in a similar manner and after 20 iterations it reached a current best solution with cost 7,240 (it is depicted in Fig. 26.8f). In our implementation of TS for this example we used a tabu list of size 5.

26.6.2. A Different TS Implementation (Tabu II)

The Tabu II implementation exhibits many similarities with the Tabu I. The initial solution, the definition of a solution, the use of penalty terms, and the stopping criteria are the same as those of Tabu I. Also the function and creation of the tabu list and the associated aspiration criteria are the same. The difference between the two implementations is in the way the list of candidate solutions is generated. Tabu II uses a greedy procedure to generate the next solution.

The procedure to generate the next solution σ' to enter the candidate list, using as input the current solution σ, operates in the following way (written in pseudoalgorithmic format):

1. For all arcs $a \in A$ do the following:
   1.1. let $L(\sigma'(a)) = \cup_{a' \in TL} \{a'\}$, where TL is the tabu list.
   1.2. If $a \notin L(\sigma)$ and $a \notin TL$, then set $L(\sigma'(a)) = L(\sigma'(a)) \cup \{a\}$.
   1.3. For all material flows do the following:
       1.3.1. Calculate the shortest distance path for the flow to travel its designated route ignoring arcs $a \notin L(\sigma)$ and $a \in TL$.
       1.3.2. Include all arcs in the shortest path $L(\sigma'(a))$.
   1.4. Calculate $C(\sigma'(a))$.
2. Let $a^* = \arg \min \{C(\sigma'(a))\}$ and $\sigma'(a^*)$ be the next solution in the candidate list.

The important parameter of the Tabu II implementation is the tabu list size τ. We have performed sensitivity analysis in Fig. 26.8. For the computational results of this method, which are reported in Section 26.7, we used $\tau = 7$. The example that follows demonstrates the basic ideas behind the Tabu II implementation.

Example 2. In this example we use a bigger material handling network for illustration. The network has 20 nodes and 62 potential arcs. The network is of grid structure (a rectangular grid), with any two adjacent nodes on the grid having distance of 1 (see Fig. 26.9a). There are 10 different types of material flows that have to be accommodated in the network. The table below provides more detailed information on the flows. For each flow, the

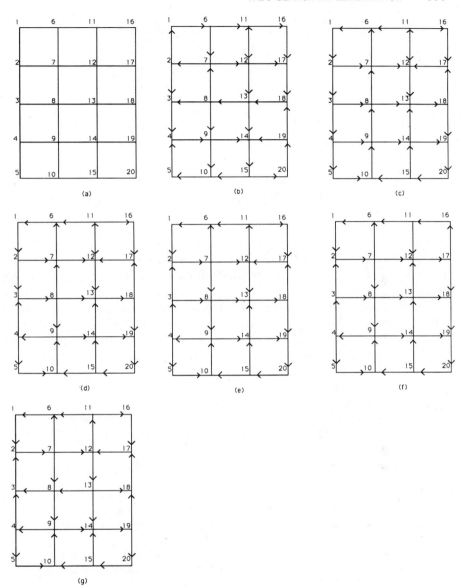

FIGURE 26.9. Flowpath design for Example 2.

stations it has to visit (routing of the flow) and the volume are provided. For our cost calculating we use $\delta_{ij} = \delta = 1{,}000 \ \forall i, j$, and $\Delta = 0$.

A heuristic solution to the problem is depicted in Fig. 26.9b. This solution serves as the initial current solution. To facilitate the presentation of our example, we develop an iteration table (see Table 26.1) in which the iteration sequence, current solution, current cost, best move (second best move, and other best moves), tabu status of an arc, and aspiration level are shown. Now let us describe in some more detail the first iteration.

Flow	Route	Volume
1	1 7 12 20 18 14 12	50
2	7 9 3 18 14	60
3	7 1 9 20 5 14	100
4	7 1 12 9 20 14	10
5	7 3 18 14	20
6	7 12 20 16 14	90
7	7 1 16 20 5 18 14	30
8	7 3 12 1 5 14	110
9	7 9 18 14	40
10	7 16 12 18 3 20 9 5 14	120

Iteration 1. We start from the initial solution generated by the heuristic (see Fig. 26.9b). We now have to generate the next current solution. For that we generate a list of candidate solutions. To do that we proceed as follows. We pick an arc, let us say arc(1, 6). We exclude all other arcs from the material handling network with the exception of those in the tabu list (at this point in time the tabu list is empty). Arc(1, 6) belongs to the current flowpath design, so we exclude it in the new one. Now, without imposing any other restrictions on the arc directions, we proceed to find the shortest distance

TABLE 26.1. Iteration Table for Example 2

	Iteration 1	Iteration 2	Iteration 3	Iteration 4
Current objective Function value (Z)	13,850	14,010	12,490	12,030
Best move	arc(1, 6) = 0 Z = 14,010	arc(3, 4) = 0 Z = 12,490	arc(2, 3) = 0 Z = 12,090	arc(12, 11) = 0 Z = 12,030
Second best move	arc(6, 11) = 0 Z = 14,010	arc(3, 8) = 0 Z = 13,070	arc(2, 7) = 0 Z = 12,210	arc(4, 9) = 0 Z = 12,030
Third best move	arc(3, 4) = 0 Z = 14,530	arc(2, 7) = 0 Z = 13,110	arc(4, 9) = 1 Z = 12,490	arc(5, 10) = 0 Z = 12,090
...	...	...	...	...
Current solution	Fig. 26.9c	Fig. 26.9d	Fig. 26.9e	Fig. 26.9f
Tabu state and (aspiration level)		arc(1, 6) = 1 (13,850)	arc(1, 6) = 1 (13,850) arc(3, 4) = 1 (12,490)	arc(1, 6) = 1 (13,850) arc(3, 4) = 1 (12,490) arc(2, 3) = 1 (12,090)

Note: arc(x, y) = 1 means arc(x, y) is included in the flowpath design
    arc(x, y) = 0 means arc(x, y) is excluded in the flowpath design

paths for the different flows. We start with flow 1 that has as its route $(1, 7, 12, 20, 18, 14, 12)$. We need to calculate the shortest distance paths between the pairs of nodes $(1, 7)$, $(7, 12)$, $(12, 20)$, $(20, 18)$, $(18, 14)$, $(18, 14)$, and $(14, 12)$. We need to do similar calculations for all subsequent flows. This finally will lead to the calculation of shortest distance paths between large sets of pair nodes. We summarize all these shortest path distance calculations in the table below:

(Origin, Destination)	Shortest Path
(1, 5)	(1, 2, 3, 4, 5)
(1, 7)	(1, 2, 7)
(1, 9)	(1, 2, 3, 4, 9)
(1, 12)	(1, 2, 7, 12)
(1, 16)	(1, 2, 7, 12, 11, 16)
(3, 12)	(3, 8, 7, 12)
(3, 18)	(3, 8, 13, 18)
(3, 20)	(3, 8, 13, 18, 19, 20)
(5, 14)	(5, 10, 9, 14)
(5, 18)	(5, 10, 9, 14, 13, 18)
(7, 1)	(7, 6, 1)
(7, 3)	(7, 6, 1, 2, 3)
(7, 9)	(7, 6, 1, 2, 3, 4, 9)
(7,12)	(7, 12)
(7, 16)	(7, 16, 11, 16)
(20, 9)	(20, 15, 10, 9)
(20, 16)	(20, 15, 10, 9, 8, 7, 12, 11, 16)
(9, 3)	(9, 8, 7, 6, 1, 2, 3)
(9, 5)	(9, 8, 7, 6, 1, 2, 3, 4, 5)
(9, 18)	(9, 8, 13, 18)
(9, 20)	(9, 14, 19, 20)
(12, 1)	(12, 11, 6, 1)
(12, 9)	(12, 11, 6, 1, 2, 3, 4, 9)
(12, 18)	(12, 13, 18)
(12, 20)	(12, 13, 18, 19, 20)
(14, 12)	(14, 13, 18, 17, 12)
(16, 12)	(16, 17, 12)
(16, 14)	(16, 17, 12, 13, 18, 19, 20, 15, 14)
(16, 20)	(16, 17, 12, 13, 18, 19, 20)
(18, 3)	(18, 17, 12, 11, 6, 1, 2, 3)
(18, 14)	(18, 19, 20, 15, 14)
(20, 5)	(20, 15, 10, 9, 8, 7, 6, 1, 2, 3, 4, 5)
(20, 14)	(20, 15, 14)
(20, 18)	(20, 15, 14, 13, 18)

All the arcs in the shortest distance paths are included in the flowpath design, which is referred as $\sigma'(1, 6)$ (it is depicted in Fig. 26.9c). The cost for the flowpath design [i.e., $C(\sigma'(1, 6))$] is calculated. This procedure is repeated for all arc(i, j). That generated our candidate list. The solution with the lowest cost is first examined in the candidate list. In this case $\phi s'(1, 6)$ happens to have the lowest cost $[C(\sigma'(1, 6)) = 14{,}010]$. Since no arcs are yet tabu, we can proceed and use $\sigma'(1, 6)$ as the next current solution. We also enter arc$(1, 6)$ in the tabu list and associate with it an aspiration level of 14,010.

Iteration 2. The best move in this iteration is to remove arc$(3, 4)$ from the flowpath design, which yields $C(\sigma'(3, 4)) = 12{,}490$. Since arc$(3, 4)$ is not in the tabu list, the solution is selected as the current solution. Arc$(3, 4)$ is put into the tabu list with aspiration level 12,490. The resulting flowpath design is depicted in Fig. 26.9d.

Iteration 3. The best move in this iteration is to remove arc$(2, 3)$ from the flowpath design, which yields $C(\sigma'(2, 3)) = 12{,}090$. Since arc$(2, 3)$ is not in the tabu list, the solution is selected as the current solution. Arc$(2, 3)$ is put into the tabu list with aspiration level 12,090. The resulting flowpath design is depicted in Fig. 26.9e.

Iteration 4. The best solution in this iteration is to exclude arc$(12, 11)$ from the flowpath design, which yields $C(\sigma'(12, 11)) = 12{,}030$. Since arc$(12, 11)$ is not in the tabu list, the solution is selected as the current solution. Arc$(12, 11)$ is put into the tabu list with aspiration level 12,030. The resulting flowpath design is depicted in Fig. 26.9f.

The algorithm proceeds is a similar manner and updates the tabu list in a similar manner to Example 1. After 36 iterations, it reached a current best solution with cost 11,350 (it is depicted in Fig. 26.9g). In our implementation we used a tabu list of size 7.

26.7. COMPUTATIONAL RESULTS

In this section we summarize the results of our computational experiments with 45 test problems of different sizes ranging from 9 nodes and 24 arcs to 30 nodes and 94 arcs.

We tested a simulated annealing algorithm, as well as the two tabu search algorithms described in the previous sections. A comparison of models (SA–VP) and (SA–MP) for the SA runs is summarized in Table 26.2. Although the differences in performance between the two models are small, we can observe that model (SA–VP) dominates model (SA–MP) across all three dimensions examined (CPU requirements, average relative error, and

TABLE 26.2. Relative Performance of SA Models

Model	Average CPU (seconds)[a]	Number of Times It Failed to Produce the Best Solution[b]	Average Relative Error[c](%)	Worst Case Error (%)
(SA–VP)	75.20	34	0.7	6.6
(SA–MP)	85.23	28	1.1	8.5

[a] On an IBM-3081 computer.
[b] There were 17 ties.
[c] This error is computed based on the best solution of the two models.

worst case error). For Tabu I we report the cost of the best solution obtained and CPU time after 10, 20, and 50 iterations (labeled TSI-10, TSI-20, and TSI-50, respectively in Table 26.3). For Tabu II we report the cost of best solution obtained and CPU time after 10, 20, 50, and 100 iterations (labeled TSII-10, TSII-20, TSII-50, and TSII-100, respectively).

In Table 26.3 we summarize the performance of the different algorithms. The fastest heuristic is TSI-10 with an average error of 5.05%, and a run time of 7.77 seconds. By roughly doubling the run time (14.88 CPU seconds) algorithm TSII-10 reduced the average error to 1.6%. We can observe from Table 26.3 that any other algorithm increases significantly the CPU time without a significant increase in performance. Moreover, when we compare TSII-10 with SA–VP, which is the best SA implementation (see Table 26.3), we can see that SA is dominated across all dimensions. Thus it seems that for this problem the tabu search methodology produces superior results.

TABLE 26.3. Relative Performance of Algorithms

Model	Average CPU (seconds)	Number of Times It Failed to Produce Uni-directional Layout	Average Relative Error[b](%)	Worst Case Error (%)
TSI-10	7.77	4	5.05	17.44
TSII-10	14.88	0	1.60	9.97
TSI-20	15.40	3	4.60	21.01
TSII-20	28.73	0	1.15	7.83
TSI-50	37.49	3	4.00	17.92
TSII-50	70.33	0	0.47	6.02
SA	94.19	0	2.71	13.58
TSII-100	134.90	0	0.32	6.02

[a] On an IBM-3081 computer.
[b] This error is computed relative to the best solution obtained for each problem (by any algorithm) counting only the cases in which each algorithm produced a unidirectional layout.

Although the quality of the solutions of both search algorithms improves with the number of iterations, the improvement beyond 10 iterations is marginal. This is due, in part, to the fact that the heuristic used to generate the starting points produced solutions of reasonably high quality.

26.8. CONCLUSION

Intelligent heuristics, like simulated annealing and tabu search, are important methodological tools for the design practitioner. Searching for an optimal (or very good) engineering or manufacturing system design usually results in solving a difficult combinatorial optimization problem. Both SA and TS are generic approaches that can be applied to any combinatorial optimization problem that the designer faces. Based on the extensive computational results reported in the literature, and as we demonstrated for our specific application (the UFDP), the designer should expect that a high quality heuristic solution to the problem will be obtained with reasonable computational requirements. The above approaches are flexible enough to allow designers to accommodate their experience and understanding of the problem when specifying the required algorithmic parameters. Sensitivity analysis on the various algorithmic parameters is needed for achieving the best possible performance of the above approaches. As the concurrent engineering approach forces designers to look for solutions to a set of design problems addressed simultaneously in a timely manner, we expect to see more and more applications of the intelligent heuristics to such problems.

REFERENCES

1. Aarts, E., and Korst, J. (1989). *Simulated Annealing and Boltzmann Machines: A Stochastic Approach to Combinatorial Optimization and Neural Computing*, Wiley, New York.

2. Cerny, V. (1985). A Thermodynamical Approach to the Traveling Salesman Problem: An Efficient Simulation Algorithm, *Journal of Optimization Theory Applications*. No. 45, pp. 41–51.

2a. Dahlstrom, K. (1981). Where to Use AGV Systems, Manual Forkliftz, Traditional Fixed Roller Conveyor Systems Respectively, *Proceedings of the 1st. Int. Conf. on AGVS*, Stratford-upon-Avon, pp. 173–182.

3. Egbelu, P. J., and Tanchoco, J. M. A. (1984). Characterization of Automatic Guided Vehicle Dispatching Rules, *International Journal of Production Research*, Vol. 22, No. 3, pp. 359–374.

4. Egbelu, P. J., and Tanchoco, J. M. A. (1986). Potential for Bi-Directional Guide-Path for Automated Guided Vehicle Based Systems, *International Journal of Production Research*, Vol. 24, No. 5, pp. 1075–1097.

5. Garey, M. R., and Johnson, D. S. (1979). *Computers and Intractability: A Guide to the Theory of NP-Completeness*, Freeman, San Francisco.

6. Gaskins, R. J., and Tanchoco, J. M. A. (1987). Flowpath Design for Automated Guided Vehicle Systems, *International Journal of Production Research*, Vol. 25, No. 5, pp. 667–676.

7. Gaskins, R. J., and Tanchoco, J. M. A. (1988). Flow Path Design for Automated Guided Vehicle Systems, *International Journal of Production Research*, Vol. 25, No. 5, pp. 667–676.

8. Glover, F. (1990). Tabu Search, Part 1, *ORSA Journal on Computing*, Vol. 1, No. 3, pp. 190–206.

9. Glover, F. (1990). Tabu Search, Part 2, *ORSA Journal on Computing*, Vol. 2, No. 1, pp. 4–32.

10. Glover, F. (1990). Tabu Search: A Tutorial, *Interfaces*, Vol. 20, No. 4, pp. 74–94.

11. Golden, B. L., and Skiscim, C. C. (1986). Using Simulated Annealing to Solve Routing and Location Problems, *Naval Research Logistics Quarterly*, No. 33, pp. 261–279.

12. Grobeschallau, W., and Heinzel, R. (1983). A New Planning Method for AGVS with Computer Graphics. In: *Proceedings of the 2nd. International Conference on AGVSs*, Stuttgart, Germany, edited by H. J. Warnecke, pp. 31–40.

12a. Kanewurf, X. (1979). Computer Controlled Guided Vehicles, *Proceedings of the 3rd. Int. Conf. on Automation in Warehousing*, Chicago, pp. 177–184.

13. Kirkpatrick, S., Gelatt, C. D., and Vecchi, M. P. (1983). Optimization by Simulated Annealing, *Science*, No. 220, pp. 671–680.

13a. Laarhoven, X., and Aarts, E. (1987). *Simulated Annealing: Theory and Applications*. D. Reidel Publishing Co., Dordrecht.

14. Koff, G. A., and Boldrin, B. (1985). Automated Guided Vehicles. In: *Materials Handling Handbook*, edited by R. A. Kulwiec, Wiley, New York, pp. 273–314.

15. Kouvelis, P. (1991). Design and Planning Problems in Flexible Manufacturing Systems: A Critical Review, *Journal of Intelligent Manufacturing*, in press.

16. Kouvelis, P., Gutierrez, G. J., and Chiang, W.-C. (1990). Heuristic Unidirectional Flowpath Design Approaches for Automated Guided Vehicle Systems, *International Journal of Production Research*, to be published.

17. Kuhn, A. (1983). Efficient Planning for AGVSs by Analytical Methods. In *Proceedings of the 2nd International Conference on AGVSs*, Stuttgart, Germany, edited by H. J. Warnecke, pp. 1–10.

18. Kusiak, A. (1985). Material Handling in Flexible Manufacturing Systems, *Material Flow* 2, pp. 79–95.

19. Kusiak, A. (Ed.) (1986). *Flexible Manufacturing Systems: Methods and Studies*, North-Holland, Amsterdam.

20. Kusiak, A. (1990). *Intelligent Manufacturing Systems*, Prentice-Hall, Englewood Cliffes, NJ.

21. Matson, J. O., and White, J. A. (1982). Operational Research and Material Handling, *European Journal of Operational Research*, pp. 309–318.

22. Maxwell, W. L. (1981). Solving Material Handling Design Problems with OR, *Industrial Engineering*, Vol. 13, No. 4, pp. 58–69.

23. Maxwell, W. L., and Muckstadt, J. A. (1982). Design of Automatic Guided Vehicle Systems, *IIE Transactions*, Vol. 14, No. 2, pp. 114–124.

24. Metropolis, N., Rosenbluth, A. W., Rosenbluth, M. N., Teller, A. H., and Teller, E. (1953). Equation of State Calculations by Fast Computing Machines, *Journal of Chemical Physics*, Vol. 21, No. 6, pp. 1087–1091.

25. Müller, T. (1983). *Automated Guided Vehicles*, IFS Publication, Bedford.

26. Newton, D. (1985). Simulation Model Helps Determine How Many Automated Guided Vehicles Are Needed, *Industrial Engineering*, Vol. 17, No. 2, pp. 68–78.

26a. Papadimitriou, X., and Steiglitz, X. (1982). *Combinatorial Optimization: Algorithms and Complexity*, Prentice Hall, Englewood Cliffs, NJ.

27. Robinson, R. L., and Tuan, B. E. (1982). Computer Aided Color Graphics for Vehicle Scheduling and Dispatching, *IEEE Proceedings of the International Conference on Cybernetics and Society*, Seattle, pp. 125–127.

28. Russell, R. S., and Tanchoco, J. M. A. (1984). An Evaluation of Vehicle Dispatching Rules and Their Effect on Shop Performance, *Material Flow*, Vol. 1, No. 4, pp. 271–280.

29. Schneider, F. (1981). Robomatic System for Dress-up of Engines and Transmission at OPEL, Germany. In: *Proceedings of the 1st International Conference on AGVS*, Stratford-upon-Avon, Edited by R. H. Hollier, pp. 199–212.

30. Sharp, G. P., and Liu, F. F. (1990). An Analytical Method for Configuring Fixed Path–Closed Loop Material Handling Systems, *International Journal of Production Research*, Vol. 28, No. 4, pp. 757–783.

31. Suaris, P. R., and Kedem, G. (1988). An Algorithm for Quadrisection and Its Application to Standard Cell Placement, *IEEE Transactions on Circuits and Systems*, Vol. 35, No. 3, pp. 294–303.

32. Wong, D. F., and Liu, C. L. (1986). A New Algorithm for Floorplan Design, *Proceedings 23rd ACM/IEEE Design Automation Conference*, Las Vegas, NV, pp. 101–107.

33. Wouters, C., Aarts, E., and Van Berkel, K. (1987). Optimization of Gate-Matrix Layouts Using Simulated Annealing, Report, Philips Research Laboratories, Eindhoven, The Netherlands.

CHAPTER TWENTY SEVEN

Approximate Reasoning in Manufacturing

H.-J. ZIMMERMANN

Institute for Operations Research, RWTH Aachen, Aachen, West Germany

27.1. COMPLEXITY, UNCERTAINTY, AND FUZZY SET THEORY

Production planning and control have been major areas of application for quantitative methods from operations research for a long time. Forecasting methods, linear programming, simulation, and other methods have been applied to the area of production planning, in which generally a more aggregated and medium-term view is predominant. In short-term, detailed production control, these analytic approaches are not very useful because the relationships that have to be modeled are, by far, too complex and stochastic. Therefore one either uses simple local priority rules for scheduling or one relies on the experience of practitioners, which seems very difficult to model mathematically. In recent years, methods of artificial intelligence have increasingly been used to model human experience quantitatively and have been combined with the computational efficiency of electronic data processing methodology. The resulting systems are generally called knowledge-based systems or expert systems (ESs).

Two of the major problems that have to be solved to obtain good ESs is to reduce complexity and to represent formally human knowledge, which is generally available only in linguistic terms and not in formal, mathematical forms. It almost always contains implicit or explicit uncertainties in various ways. Reduction of complexity and modeling of uncertainties can be considered as two of the main objectives of fuzzy set theory. Concerning complexity, Zadeh (23), the founder of fuzzy set theory, formulated his "principle of incompatibility" as follows:

Intelligent Design and Manufacturing, Edited by Andrew Kusiak.
ISBN 0-471-53473-0 © 1992 John Wiley & Sons, Inc.

As the complexity of a system increases, our ability to make precise and yet significant statements about its behaviour diminishes until a threshold is reached beyond which precision and significance (or relevance) become almost mutually exclusive characteristics.

With respect to uncertainty one might feel that probability theory or the Bayesian approach used frequently in ESs is sufficient to model it adequately. The following few examples might indicate that this is not always true. Considering the following statements, it becomes obvious that they are all uncertain in some way. They can, however, not all be modeled adequately using probability theory.

"*The probability of meeting the due date is .8.*" This is obviously a probabilistic statement that can easily be modeled by probability theory.

"*The chances of meeting the due date are good.*" Applying probability theory here becomes already more difficult.

"*We have good chances of not delaying the delivery too long.*" The formal formulation of this statement is even more tricky. This holds to an even higher degree with respect to the next statement, which could well be part of the human experience we want to include in a knowledge base.

"*Experience shows that raw material* A *is generally within the quality standards; considering, however, that it comes from supplier* B *this is rather unlikely.*"

Before the application of fuzzy set theory in manufacturing is described in more detail, it might be appropriate to present those parts of the basic theory that will be needed.

The first publications in fuzzy set theory by Zadeh (22a) and Goguen (3a, 3b) show the intention of the authors to generalize the classical notion of a set and a proposition (statement) to accommodate fuzziness in the sense of linguistic uncertainty. Zadeh (ref. 22a, p. 339) writes:

The notion of a fuzzy set provides a convenient point of departure for the construction of a conceptual framework which parallels in many respects the framework used in the case of ordinary sets, but is more general than the latter and, potentially, may prove to have a much wider scope of applicability, particularly in the fields of pattern classification and information processing. Essentially, such a framework provides a natural way of dealing with problems in which the source of imprecision is the absence of sharply defined criteria of class membership rather than the presence of random variables.

"Imprecision" here is meant in the sense of vagueness rather than the lack of knowledge about the value of a parameter as in tolerance analysis. Fuzzy set theory provides a strict mathematical framework (there is nothing fuzzy about fuzzy set theory!) in which vague conceptual phenomena exist.

A classical (crisp) set is normally defined as a collection of elements or objects $x \in X$, which can be finite, countable, or overcountable. Each single element can either belong to or not belong to a set A, $A \subset X$. In the former case, the statement "x belongs to A" is true, whereas in the latter case this statement is false.

Such a classical set can be described in different ways: One can either enumerate (list) the elements that belong to the set; describe the set analytically, for instance, by stating conditions for membership ($A = \{x \mid x \leq 5\}$); or define the member elements by using the characteristic function, in which 1 indicates membership and 0 nonmembership. For a fuzzy set, the characteristic function allows various degrees of membership for the elements of a given set.

Definition 1. If X is a collection of objects denoted generically by x, then a fuzzy set $\tilde{A}$ in X is a set of ordered pairs:

$$\tilde{A} = \{(x, \mu_{\tilde{A}}(x)) \mid x \in X\} .$$

$\mu_{\tilde{A}}(x)$ is called the membership function or grade of membership (also degree of compatibility or degree of truth) of x in $\tilde{A}$, which maps X to the membership space M. [When M contains only the two points 0 and 1, $\tilde{A}$ is nonfuzzy and $\mu_{\tilde{A}}(x)$ is identical to the characteristic function of a nonfuzzy set.] The range of the membership function is a subset of the nonnegative real numbers whose supremum is finite. Elements with a zero degree of membership are normally not listed.

Definition 2. A fuzzy number $\tilde{M}$ is a convex normalized fuzzy set $\tilde{M}$ of the real line $\mathbb{R}$ such that:

1. There exists exactly one $x_0 \in \mathbb{R}$ with $\mu_{\tilde{M}}(x_0) = 1$ (x_0 is called the mean value of $\tilde{M}$).
2. $\mu_{\tilde{M}}(x)$ is piecewise continuous.

Today this definition is very often modified. For the sake of computational efficiency and ease of data acquisition, trapezoidal membership functions are often used. Figure 27.1 shows such a fuzzy set, which could be called "approximately 5."

Definition 3. A linguistic variable is characterized by a quintuple $(x, T(x), U, G, \tilde{M})$ in which x is the name of the variable; $T(x)$ (or simply

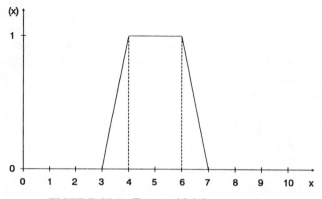

FIGURE 27.1. Trapezoidal fuzzy number.

T) denotes the term set of x, that is, the set of names of linguistic values of x, with each value being a fuzzy variable denoted generically by x and ranging over a universe of discourse U that is associated with the base variable u; G is a syntactic rule (which usually has the form of a grammar) for generating the name, X, of values of x; and M is a semantic rule for associating with each X its meaning, $\tilde{M}(X)$, which is a fuzzy subset of U. A particular X, that is, a name generated by G, is called a term. It should be noted that the base variable u can also be vector-valued (Fig. 27.2).

In order to use fuzzy sets the usual set-theoretic operations have to be defined. In this context it should be pointed out that the set-theoretic intersection corresponds to the logical "and," the set-theoretic union to the logical (inclusive) "or," and the set-theoretic complement to the logical "negation."

Zadeh (22a) defined the above-mentioned operations for fuzzy sets $\tilde{A}$, $\tilde{B}$ as follows:

Definition 4. The membership function $\mu_{\tilde{C}}(x)$ of the intersection $\tilde{C} = \tilde{A} \cap \tilde{B}$ is pointwise defined by

$$\mu_{\tilde{C}}(x) = \min\{\mu_{\tilde{A}}(x), \mu_{\tilde{B}}(x)\}, \qquad x \in X.$$

Definition 5. The membership function $\mu_{\tilde{D}}(x)$ of the union $\tilde{D} = \tilde{A} \cup \tilde{B}$ is pointwise defined by

$$\mu_{\tilde{D}}(x) = \max\{\mu_{\tilde{A}}(x), \mu_{\tilde{B}}(x)\}, \qquad x \in X.$$

Definition 6. The membership function of the complement of a fuzzy set $\tilde{A}$, $\mu_{\mathcal{C}\tilde{A}}(x)$ is defined by

$$\mu_{\mathcal{C}\tilde{A}}(x) = 1 - \mu_{\tilde{A}}(x), \qquad x \in X.$$

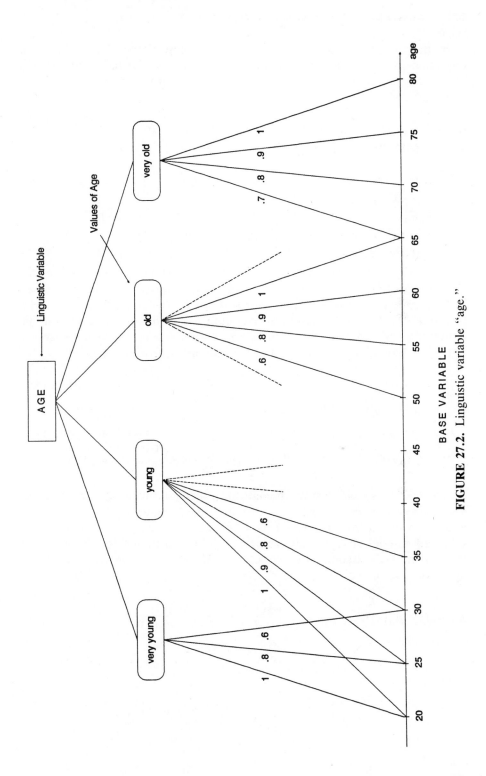

BASE VARIABLE

FIGURE 27.2. Linguistic variable "age."

705

In the meantime, a large number of alternative operators have been suggested in the literature. They all fall into one of the following classes: The intersection can be modeled by triangular norms or t-norms and the union by t-conorms or s-norms.

Definition 7. The t-norms are two-valued functions from $[0, 1] \times [0, 1]$ that satisfy the following conditions:

1. $t(0, 0) = 0$; $t(\mu_{\tilde{A}}(x), 1) = t(1, \mu_{\tilde{A}}(x)) = \mu_{\tilde{A}}(x)$, $x \in X$.
2. $t(\mu_{\tilde{A}}(x), \mu_{\tilde{B}}(x)) \leq t(\mu_{\tilde{C}}(X), \mu_{\tilde{D}}(x))$,
   if $\mu_{\tilde{A}}(x) \leq \mu_{\tilde{C}}(x)$ and $\mu_{\tilde{B}}(x) \leq \mu_{\tilde{D}}(x)$ (monotonicity).
3. $t(\mu_{\tilde{A}}(x), \mu_{\tilde{B}}(x)) = t(\mu_{\tilde{B}}(x), \mu_{\tilde{A}}(x))$ (commutativity).
4. $t(\mu_{\tilde{A}}(x), t(\mu_{\tilde{B}}(x), \mu_{\tilde{C}}(x))) = t(t(\mu_{\tilde{A}}(x), \mu_{\tilde{B}}(x)), \mu_{\tilde{C}}(x))$ (associativity).

The functions t define a general class of intersection operators for fuzzy sets. The operators belonging to this class of t-norms are, in particular, associative and therefore it is possible to compute the membership values for the intersection of more than two fuzzy sets by recursively applying a t-norm operator. Examples of t-norms and s-norms are, for instance:

$$t_1(\mu_{\tilde{A}}(x), \mu_{\tilde{B}}(x)) = \max\{0, \mu_{\tilde{A}}(x) + \mu_{\tilde{B}}(x) - 1\} \qquad \text{bounded difference},$$

$$s_1(\mu_{\tilde{A}}(x), \mu_{\tilde{B}}(c)) = \min\{1, \mu_{\tilde{A}}(x) + \mu_{\tilde{B}}(x)\} \qquad \text{bounded sum},$$

$$t_2(\mu_{\tilde{A}}(x), \mu_{\tilde{B}}(x)) = \mu_{\tilde{A}}(x) \cdot \mu_{\tilde{B}}(x) \qquad \text{algebraic product},$$

$$s_2(\mu_{\tilde{A}}(x), \mu_{\tilde{B}}(x)) = \mu_{\tilde{A}}(x) + \mu_{\tilde{B}}(x) - \mu_{\tilde{A}}(x) \cdot \mu_{\tilde{B}}(x) \qquad \text{algebraic sum}.$$

These operators have nice mathematical properties. According to empirical investigations, however, they do not model the "linguistic and" well, which is used in our day-to-day language (30). This is much better achieved by so-called compensatory or averaging operators. Examples for this type are the "symmetric" summation operator M, the "symmetric difference operator" N, or the "γ operator" G:

$$M(\mu_{\tilde{A}}(x), \mu_{\tilde{B}}(x)) = \frac{\mu_{\tilde{A}}(x) + \mu_{\tilde{B}}(x) - \mu_{\tilde{A}}(x) \cdot \mu_{\tilde{B}}(x)}{1 + \mu_{\tilde{A}}(x) + \mu_{\tilde{B}}(x) - 2\mu_{\tilde{A}}(x) \cdot \mu_{\tilde{B}}(x)},$$

$$N(\mu_{\tilde{A}}(x), \mu_{\tilde{B}}(x)) = \frac{\max\{\mu_{\tilde{A}}(x), \mu_{\tilde{B}}(x)\}}{1 + |\mu_{\tilde{A}}(x) - \mu_{\tilde{B}}(x)|},$$

$$G(x) = \left(\prod_{i=1}^{m} \mu_i(x) \right)^{(1-\gamma)} \left(1 - \prod_{i=1}^{m} (1 - \mu_i(x)) \right)^{\gamma},$$

$$x \in X, 0 \leq \gamma \leq 1.$$

27.2. EXPERT SYSTEMS, FUZZY LOGIC, APPROXIMATE REASONING, AND PLAUSIBLE REASONING

27.2.1. Expert Systems

It is assumed that the structure of expert systems is well known and does not need to be described here again. A few properties shall, however, be mentioned in order to define the type of expert systems that is meant here (ref. 28, p. 261):

1. The intended area of application is restricted in scope, ill structured, and uncertain. Therefore no well-defined algorithms are available nor are any efficient enough to solve the problem.
2. The system is—at least in part—based on expert knowledge that is either embodied in the system or can be obtained from an expert and can be analyzed, stored, and used by the system.
3. The system has some inference capability and can use the knowledge—in whatever way it is stored—to draw conclusions.
4. The interfaces to the user on one side and the expert on the other side should be such that the expert system can be used directly and that no "knowledge engineer" is needed between the system and the expert.
5. Since heuristic elements are contained in an expert system, no guarantee of optimality or correctness is provided. Therefore, in order to increase user acceptance, it is considered to be at least desirable that the system contains a "justification" or "explaining" module.

From the many kinds of knowledge representation only production rules are considered here. Knowledge and facts are then represented as follows:

$$\frac{\text{If } X \text{ then } Y \quad \text{(knowledge)}}{x \text{ is true} \qquad \text{(facts)}}$$

Therefore Y is true (conclusion).

The following assumptions are normally made in classical knowledge representation and inference methods:

1. X and Y are crisp statements of the type:
   The piece of furniture has four legs.
   Eve is 15 years old.
   Paul is the father of Eve.
   (*Description*)
2. The statements are deterministic in character; that is, they are certain, either true or false, and either fully supported or refuted by evidence.
   (*Qualification*)

3. If the statements include quantifiers, only two quantifiers are used: the existential quantifier ("There exists at least one") and the universal quantifier ("all"). (*Quantification*)

4. The X in the "facts" are fully identical with the X in the "knowledge statement." (*Matching*)

5. The if–then relationship is deterministic (true, reliable, correct, and certain). (*Inference*)

These are partly quite unrealistic assumptions with respect to human expert knowledge. They shall therefore be relaxed in the following.

27.2.2. Fuzzy Logic, Approximate Reasoning, and Plausible Reasoning

To understand fuzzy logic it is useful first to reconsider briefly the classical dual logic. Logical operators are normally defined by truth tables. If A and B are two statements that can be either true (truth value equals 1) or false (truth value equals zero), then the following truth table can be constructed:

A	B	$\wedge$	$\vee$	x	$\Rightarrow$	$\Leftrightarrow$	?
1	1	1	1	0	1	1	1
1	0	0	1	1	0	0	1
0	1	0	1	1	1	0	0
0	0	0	0	0	1	1	0

There are $2^{2^2} = 16$ truth tables (columns), each defining an operator. Assigning meanings (words) to these operators is not difficult for the first four or five columns: the first obviously characterizes the "and," the second the "inclusive or," the third the "exclusive or," and the fourth and fifth the implication and the equivalence. We shall have difficulties, however, in interpreting the remaining nine columns in terms of our language. If we have three truth values rather than two, this task of assigning meanings to truth tables becomes even more difficult.

So far, it has been assumed that each statement, A and B, could clearly be classified as true or false. If this is no longer true then additional truth values, such as "undecided" or "similar," can and have to be introduced, which leads to the many existing systems of multivalued logic. It is not difficult to see how the above-mentioned problems of two-valued logic in "calling" truth tables or operators increase as one move to multivalued logic. For only two statements and three possible truth values there are already $3^{3^2} = 729$ truth tables! The uniqueness of interpretation of truth tables, which is so convenient in boolean logic, disappears immediately because many truth tables in three-valued logic look very much alike.

The reasoning that corresponds to the use of production rules is the "modus ponens":

$$(A \wedge (A \Rightarrow B)) \Rightarrow B .$$

Here the five assumptions apply that were mentioned in Section 2.1. Fuzzy logic (FL), approximate reasoning (AR), and plausible reasoning (PR) relax these assumptions and try to design inference procedures that are much closer to human reasoning than dual logic.

In *fuzzy logic* the truth values are taken to be no longer 0 or 1 but terms of the linguistic variable "true," that is, "definitely false," "rather false," ... "almost true," "definitely true," and so on.

In *Approximate reasoning* in addition to the assumptions of FL the statements (A, B, etc.) no longer have to be crisp but can contain fuzzy expressions defined by fuzzy sets.

In *plausible reasoning* in addition to the assumption of AR the requested identity of the components of the facts with those of the rules is slightly relaxed. One could, for instance, proceed as follows: Let $\tilde{A}$, $\tilde{A}'$, $\tilde{B}$, $\tilde{B}'$ be fuzzy statements; then the generalized modulus ponens reads

> Premise: x is $\tilde{A}'$.
> Implication: If x is $\tilde{A}$ then y is $\tilde{B}$.
>
> ――――――――――――――――――――――――――――――
>
> Conclusion: y is $\tilde{B}'$.

For instance (15a),

> Premise: This tomato is very red.
> Implication: If a tomato is red then the tomato is ripe.
>
> ――――――――――――――――――――――――――――――
>
> Conclusion: This tomato is very ripe.

A conclusion is obviously reached by using the operators $\wedge$ (and) or $\vee$ (or) to combine the facts (antecedents) and the $\Rightarrow$ (implication). Very often the min and the max operators are used for $\wedge$ and $\vee$. In principle, however, all operators mentioned in Section 27.1 can and have been used. Very many implication operators have also been suggested and scrutinized (15a). Computationally, the implication operators are, however, often not very efficient. Therefore in the next section, a procedure is demonstrated that simplifies the computations and still leads to quite good results.

27.3. APPROXIMATE REASONING FOR MANUFACTURING CONTROL

The task to be solved is the scheduling of a flexible manufacturing system (FMS). It is assumed that the loading has already been performed, for instance, by fuzzy linear programming (6a).

The basic problem of release scheduling can be regarded as dispatching parts for a single capacity unit (the FMS) with several work stations: As long as there are working places unused and appropriate pallets with fixtures are available, new parts can be released into the FMS. Once the upper limit of parts has been reached, the remaining parts have to wait in a kind of waiting queue until one of the parts leaves the FMS. Then the decision, which part have to be released next, will be made according to the following procedure.

The decision of which part to release next mainly depends on date criteria of the parts under consideration, the impact of parts on machine utilization, and perhaps some kind of external priority. For the date criteria, we furthermore distinguish between the slack time of a part and the time the part has already waited for processing.

The impact on machine utilization can be twofold. First, one has to take care that the machines are used as uniformly as possible—thus trying to avoid bottlenecks. For this purpose a criterion of "uniformity of utilization" is defined. On the other hand, one wants to ensure a good utilization in the shift with reduced personnel, during which no parts can be fixed on pallets. On the contrary, parts can only be processed as long as they do not need any manual operation, be it for changing a pallet or in case of failure. This shall be taken into consideration by the concept "processing time until the next fixturing." The external priority can be given by the plant manager or some other responsible person.

The resulting hierarchy is shown in Fig. 27.3.

Example 1. Slack time is usually defined as the difference between the due date and the remaining operating time. In the case of the release scheduling for a FMS, three points have to be taken care of. First, the due dates are given for a given lot size. Thus the remaining operation time of the whole lot also has to be taken into consideration. Second, in a FMS there are normally several machines that are able to perform the same operation but that may take different times. Here we consider the best case: the fastest

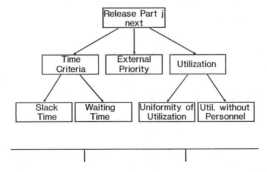

FIGURE 27.3. Criteria hierarchy for release scheduling.

machine for each remaining operation. Third, it has to be taken into account that several parts of the same lot may be processed within the FMS at the same time. For this purpose a scalar, a, is introduced, which corresponds to the average number of parts of the lot which are processed at the same time. The time when the decision is to be made by t_0 is denoted and one defines

$$s = a(\text{due date} - t_0) - \text{remaining operation time}$$

as indicator for the slack time of a lot. The slack time of a part as the second criterion can hardly be determined exactly in FMS because the processing time of the following operations depends on the choice of the appropriate machine. Hence the slack time is considered as a linguistic variable as defined in Section 27.1. As possible terms the primary terms are restricted to "sufficient," "short," and "critically short." The universe contains all possible values for the indicator, that is, in general all real numbers within a reasonable interval. Then the meaning of the terms can be defined by giving the degree of membership as a function of the above-defined indicator as base variable. Only piecewise linear membership functions, which can be defined by four parameters as shown in Fig. 27.4, are used.

Thus the suggested meaning of the linguistic variable "slack time of a lot" is given in Table 27.1. These parameters were obtained by extensive simulation studies for a specific structure of orders to be processed in a specific FMS.

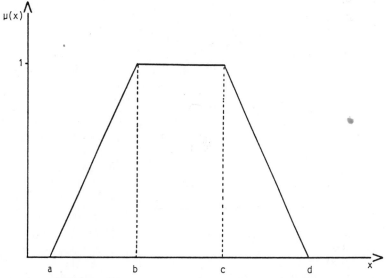

FIGURE 27.4. Piecewise linear membership function.

TABLE 27.1. Definition of the Linguistic Variable "Slack Time of a Lot"

Term	Parameter			
	a	b	c	d
Critically short	—	—	3.0	4.5
Short	3.0	4.5	6.0	8.0
Sufficient	6.0	8.0	—	—

Example 2. As a second example of how to define a linguistic variable the "processing time until the next fixturing" is considered. As base variable the empirical distribution function of the processing times until the part needs to be refixtured or is finished is chosen. Thus one can look at all parts with a *long* "processing time until the next fixturing" and not have to restrict the considerations to the part with the longest time or to a previously given number of parts with the longest processing time. Among those parts we are indifferent as to our decision, and thus we can distinguish them according to the other criteria of the hierarchy given in Fig. 27.3. As further terms, "short" and "medium" are allowed. Then the meaning of the term of the membership function can be defined by the parameters given in Table 27.2. As already mentioned previously, these parameters make sense only for the examined problem structure. They have to be redefined for other applications.

The term sets for the other criteria for release scheduling are given in Table 27.3.

TABLE 27.2. Definition of the Linguistic Variable "Processing Time Until the Next Fixturing"

Term	Parameter			
	a	b	c	d
Long	0.6	0.8	1.	1.
Medium	0.2	0.4	0.6	0.8
Short	0	0	0.2	0.4

TABLE 27.3. Term Sets of Linguistic Variables Used for Release Scheduling

Linguistic Variable	Terms
Waiting time	Long, medium, short
External priority	High, normal
Uniformity of utilization	High, medium, small

A small part of the criteria hierarchy as shown in Fig. 27.5 is now used to show how to derive the rule set of the hierarchically ordered criteria defined by linguistic variables.

1. IF   waiting time is long
   AND   slack time is critically short
   THEN date criteria are urgent   (1.0)

2. IF   waiting time is medium
   AND   slack time is critically short
   THEN date criteria are urgent   (0.8)

3. IF   waiting time is short
   AND   slack time is critically short
   THEN date criteria are urgent   (0.6)

4. IF   waiting time is long
   AND   slack time is short
   THEN date criteria are urgent   (0.5)

5. IF   waiting time is medium
   AND   slack time is short
   THEN date criteria are urgent   (0.2)

6. IF   waiting time is medium
   AND   slack time is short
   THEN date criteria are not urgent   (0.7)

Of course, a lot of additional rules might make sense. This depends on the specific application. The same dependence holds for the degree of sensibleness that is assigned to each rule and has been given in parentheses.

The principle for building such rules is very simple. In a first step one takes all possible combinations of terms for all linguistic variables in the assumption and combines them with all possible terms for the conclusion. In a second step one has to judge the resulting rules as to their sensibleness. This can be done by an expert (scheduler) and results in the "degrees of sensibleness" shown in parentheses for each rule above. With respect to computational efficiency, it is useful to define a threshold for the degree of

Date criteria

Waiting time   Slack time   **FIGURE 27.5.** Example hierarchy.

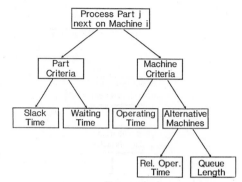

FIGURE 27.6. Criteria hierarchy for machine scheduling.

sensibleness so that one can drop less important rules. For the prototype that has been tested in simulation experiments, a set of 26 rules has been selected for release scheduling.

The machine scheduling procedure is very similar to dispatching when using priority rules. This means that no machine is allowed to wait if there is a part that can be processed on that machine. If there are several parts at a time waiting for a machine, then the method of approximate reasoning is used to choose a part from the waiting queue.

The hierarchy of decision criteria that are used is shown in Fig. 27.6.

The term sets for the linguistic variables are given in Table 27.4. Although some linguistic variables have the same labels or term sets as for the release scheduling, they are defined differently for machine scheduling by using different base variables and different membership functions.

Now the rule set for machine scheduling can be generated in a similar way as for the release scheduling. The inference is performed as follows: The antecedents, which are expressed in terms of linguistic variables, that is, as fuzzy sets, are first aggregated using the γ operator mentioned in Section 1. The "degree of sensibleness" is then interpreted as the degree to which the rule is valid. This is aggregated with the antecedent again using the γ operator with $\gamma = 1$ (no compensation!). Figure 27.7 sketches this procedure.

TABLE 27.4. Term Sets of the Linguistic Variables Used for Machine Scheduling

Linguistic Variable	Terms
Slack time	Critical, short, long enough
Waiting time	Long, medium, short
Operating time	Short, medium, long
Relative operating time	Short, medium, long
Queue length	Short, medium, long

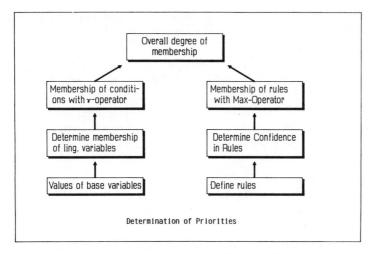

FIGURE 27.7. Principle of approximate reasoning for determination of priorities.

In order to evaluate the quality of the approaches described and the knowledge-based system for release and machine scheduling, a general simulation program for FMS was developed (5). It considers the following:

- Machines and other operating stations
- Tooling
- Storage capacities
- Transportation facilities
- Pallets and fixing units
- Workpieces

The data for the simulation runs were generated randomly but according to the structure of typical realized FMSs. This structure is mainly based on an observation of more than 150 FMSs installed all over the world (14).

The computations for the fuzzy LP that was used for loading took about 22 CPU-seconds for problems with 200 parts, 10 machines, and 9 time periods (with rolling time horizon) including the generation of all data. The problems contained about 500 variables and 800 restrictions. The density of the matrix was about 1.1%. For the solution of the LP problem we used APEX IV.

The computing time on the VAX was about 200 CPU-sections for 3200 operations. The time for a single decision is about 2×10^{-2} CPU-seconds on the VAX or 0.7 CPU-seconds on an IBM-AT.

The decisions of the knowledge-based system for *release scheduling* (KBR) have been compared with the following priority rule:

Release the part for which the number of the parts already release divided by the total lot size is minimal (*L*).

The knowledge-based system for *machine scheduling* (KBM) is compared with the often tested priority rule COVERT (C), which combines processing and slack time. Other priority rules have also been tested, but the complete series are only taken for the above-described priority rules, which seemed to be the best that could be found. This yields four combinations for release and machine scheduling, which have been evaluated according to:

- Mean in-process waiting time (*W*)
- Part of lots that have met their due dates (*D*)
- Mean machine utilization (*U*)

The aggregated results are given in Table 27.5. A comparison of KBR/ KBM with L/C shows that the knowledge-based system dominates the priority rules; that is, it is better according to each of the three objectives. It can also be observed that the combination L/KBM does not yield good results except for the machine utilization.

27.4. FUZZY CONTROL

Expert systems and fuzzy logic control (FLC) systems have certainly one thing in common: Both want to model human experience, human decision-making behavior. To quote some of the originators of FLC (9):

> The basic idea behind this approach was to incorporate the "experience" of a human process operator in the design of the controller. From a set of linguistic rules which describe the operator's control strategy a control algorithm is constructed where the words are defined as fuzzy sets. The main advantages of this approach seem to be the possibility of implementing "rule of the thumb" experience, intuition, heuristics, and the fact that it does not need a model of the process.

TABLE 27.5. Comparison of the Results for Scheduling

Scheduling		Release Machine *W* (min)	*D* (%)	*U* (%)
L	C	2923	61	79
L	KBM	2895	48	81
KBR	C	2859	96	77
KBR	KBM	2884	97	80

Certain complex industrial plants, for example, a cement kiln, can be controlled with better results by an experienced operator than by conventional automatic controllers. The control strategies employed by an operator can often be formulated as a number of rules that are simple to carry out manually but difficult to implement by using conventional algorithms. This difficulty arises because human beings use qualitative rather than quantitative terms when describing various decisions to be taken as a function of different states of the process. It is this qualitative or fuzzy nature of the human ways of making decisions that has encouraged control engineers to try to apply fuzzy logic to process control (17).

There are clear differences between expert systems and FLC:

1. The existing FLC systems originated in control engineering rather than in artificial intelligence.
2. FLC models are all rule-based systems.
3. In contrast to expert systems, FLC serves almost exclusively the control of (technological) production systems such as electrical power plants, kiln cement plants, and chemical plants; that is, their domains are even narrower than those of expert systems.
4. In general, the rules of FLC systems are not extracted from the human expert through the system but formulated explicitly by the FLC designer.
5. Finally, because of their purpose, their inputs are normally observations of technological systems and their outputs control statements.

The structure of a fuzzy logical controller is depicted in Fig. 27.8. The process of fuzzy control can roughly be described as shown in Fig. 27.9 (21). The essential design problems in FLC are the following:

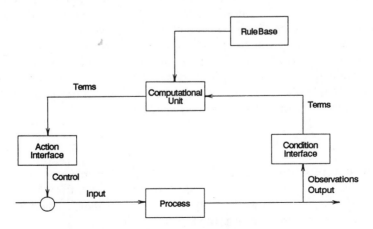

FIGURE 27.8. Fuzzy logic control system.

FIGURE 27.9. Functions of FLC.

1. Define input and control variables, that is, determine which states of the process are to be observed and which control actions are to be considered.
2. Define the condition interface; that is, fix the way is which observations of the process are expressed as fuzzy sets.
3. Design the rule base; that is, determine which rules are to be applied under which conditions.
4. Design the computational unit; that is, supply algorithms to perform fuzzy computations. Those will generally lead to fuzzy outputs.
5. Determine rules according to which fuzzy control statements can be transformed into crisp control actions.

Only one of the best known examples of the numerous applications of fuzzy control shall briefly be described here: the subway system of Sendai in Japan.

Since traffic conditions within Sendai's metropolitan area have become increasingly congested, the city planners commenced designing a subway system that would provide its citizens with the highest levels of comfort, safety, and efficiency. For these purposes, fuzzy control of the train operations seems appropriate to enable smooth acceleration, deceleration, and braking.

The operation strategies are divided into two functions: constant speed control (CSC) and train automatic stop control (TASC). CSC starts the train and keeps its speed below a specified limit. TASC decelerates the train speed in order to stop the train at a target position.

Conventional fuzzy control turned out to have disadvantages in this special application because it does not evaluate the results of selected control commands. To overcome this, it was decided to use predictive fuzzy control, where the control rules are as follows:

$$R_i: \quad \text{"If } (u \text{ is } C_1 \Rightarrow x \text{ is } A_i \text{ and } y \text{ is } B_i) \text{ then } u \text{ is } C_i\text{."}$$

In linguistic terms, this rule states: "If the state variable x is A_i and state variable y is B_i, when control action is selected for C_i at this time, then the control action C_i is issued as the output of the controller." x and y are state variables of the system and u is the control action.

The state variables are evaluations of the train system according to specific criteria and defined by fuzzy sets as follows (refs. 15, p. 64):

A. Safety Performance Indexes (S)
   1. Danger
   2. Safe
B. Comfort Performance Indexes (C)
   1. Good comfort (CG)
   2. Poor comfort (CP)
C. Energy Saving Performance Indexes (E)
   1. Energy-saved running (ES)
   2. Not energy-saved running (EN)
D. Traceability Performance Indexes (T)
   1. Good trace (TG)
   2. Accurate trace (TA)
   3. Low speed (TL)
E. Running Time Performance Indexes (R)
   1. In TASC zone (RT)
   2. Not in TASC zone (RF)
F. Stopgap Performance Indexes (G)
   1. Good stop (GG)
   2. Accurate stop (GA)

The exact definitions of the respective membership functions are omitted here (e.g., see ref. 22, p. 10). The possible control actions are:

1. PN = powering notch.
2. BN = braking notch.
3. DN = difference of notches.
4. N = control command notch (power notch/brake notch).

Based on these state variables and control actions, predictive fuzzy control rules are formulated. Examples of such rules are:

1. *CSC Rules*
   If (N *is* $0 \Rightarrow$ S is SS, C is CG and E is ES), then N is 0.
   If (DN is $0 \Rightarrow$ S is SS and T is TG), then DN is 0.
2. *TASC Rules*
   If (DN is $0 \Rightarrow$ R is RT and G is GG), the DN is 0.
   If (N is $1 \Rightarrow$ R is RT and C is CG), then N is 1.

In the ATO system, a control action is selected by evaluating all rules every 100 milliseconds. In simulation studies, the predictive fuzzy control ATO is compared with a conventional ATO according to the criteria "Riding Comfort," "Stopgap Accuracy," "Energy Consumption," and "Running Time" (ref. 22, p. 15).

1. *Riding comfort and stopgap accuracy.* The results of the simulation showed that the number of notch changes in the fuzzy ATO is about one-half, and the stopgap accuracy measured as the deviation of a given target is about one-third compared with those of the conventional ATO. This implies that the fuzzy ATO is able to control a train with good riding comfort and stopgap accuracy.

2. *Energy consumption and running time.* The investigations concerning these criteria showed a potential energy saving of over 10% and/or a shortened running time compared with the conventional ATO.

27.5. FUTURE DEVELOPMENTS

It has been shown how expert system technology and fuzzy control can be used to improve planning and control in manufacturing directly. These tools complement existing methods from operations research. They can also be used where other mathematical methods have failed so far.

In the meantime, fuzzy control has also been installed successfully in consumer goods, such as video cameras, vacuum cleaners, and car braking systems. Other industrial applications are robotics and container cranes. Most of these applications have been implemented in Japan. In the very recent past, however, shells for fuzzy control have also been developed in Europe and the United States. It is to be expected that in the very near future these techniques will conquer a much larger part of the market.

REFERENCES

1. Bolwijn, P. T., and Kumpe, T. (1986). Toward the Factory of the Future, *The McKinsey Quarterly*, Spring, pp. 40–49.
2. Buckley, J. J., Siler, W., and Tucker, D. (1986). A Fuzzy Expert System, *Fuzzy Sets and Systems*, Vol. 20, pp. 1–16.
3. Charles Draper Laboratories (1984). *Flexible Manufacturing Systems—Handbook*, Park Ridge, New York.
3a. Goguen, X. (1967). L-Fuzzy Sets, *JMAA*, Vol. 18, *pp.* 145–174.
3b. Goguen, X. (1969). The Logic of Inexact Concepts, *Synthese*, Vol. 19, pp. 325–373.
4. Hartley, J. (1984). *FMS at Work*, Bedford, United Kingdom.
5. Herling, Ch., and Hintz, G. W. (1986). *Ein Programm zur Simulation von flexiblen Fertigungssystemen*, Arbeits-bericht des Instituts für Wirtschaftswissenschaften Nr. 86/08, RWTH Aachen, Aachen.
6. Hintz, G. W. (1987). *Ein wissensbasiertes System zur Produktionsplanung und -steuerung für flexible Fertigungssysteme*, VDI-Verlag, Düsseldorf.
6a. Hintz, G. W., and Zimmermann, H.-J. (1989). A Method to Control Flexible Manufacturing Systems, *European Journal of Operational Research*, Vol. 41, pp. 321–334.

7. Kamp, A. W. (1978). *Ein Beitrag zur Ablaufplanung bei flexiblen Fertigungssystemen*, VDI-Verlag, Düsseldorf.

8. Kanet, J. J., and Adelsberger, H. H. (1987). Expert Systems in Production Scheduling, *European Journal of Operational Research*, Vol. 29, pp. 51–59.

9. Kickert, W. J. M., and Mamdani, E. H. (1978). Analysis of a Fuzzy Logic Controller, *Fuzzy Sets and Systems*, Vol. 1, pp. 29–24.

10. Kimemia, J. G. (1982). Hierarchical Control of Production in Flexible Manufacturing Systems, LIDS-TH-1251, Massachussetts Institute of Technology, Boston, MA.

11. Kusiak, A. (1986). Application of Operational Research Models and Techniques in Flexible Manufacturing Systems, *European Journal of Operational Research*, Vol. 24, pp. 336–345.

12. Kusiak, A. (Ed.) (1986). *Flexible Manufacturing Systems: Methods and Studies*, North-Holland, Amsterdam.

13. Kusiak, A., and Chen, M. (1988). Expert Systems for Planning and Scheduling Manufacturing Systems, *European Journal of Operational Research*, Vol. 34, pp. 113–130.

14. Mertins, K. (1985). Entwicklungsstand flexibler Fertigungssysteme—Linien-, Netz- und Zellenstrukturen, *ZwF*, Vol. 80, pp. 249–265.

15. Miyamoto, S., Yasunobo, S., and Ihara, H. (1987). Predictive Fuzzy Control and Its Application to Automatic Operation Systems. In: *Analysis of Fuzzy Information. Volume III: Applications in Engineering and Science*, edited by J. Bezdek, CRC Press, Boca Raton, FL, pp. 59–68.

15a. Mizumoto, M., and Zimmermann, H.-J. (1982). Comparison of Fuzzy Reasoning Methods, *Fuzzy Sets and Systems*, Vol. 8, pp. 253–283.

16. Mortimer, T. (Ed.) (1985). *Integrated Manufacture*, Bedford, United Kingdom.

17. Ostergaard, J. J. (1977). Fuzzy Logic Control of a Heat Exchanger Process. In: *Fuzzy Automata and Decision Processes*, edited by M. M. Gupta et al., Elsevier, Amsterdam, pp. 285–320.

18. Stecke, K. E. (1986). A Hierarchical Approach to Solving Machine Grouping and Loading Problems of Flexible Manufacturing Systems, *European Journal of Operational Research*, Vol. 24, pp. 369–378.

19. Stecke, K. E., and Talbot, F. B. (1985). Heuristics for Loading Flexible Manufacturing Systems. In: *Flexible Manufacturing*, edited by A. Raouf and S. I. Ahmad, North-Holland, Amsterdam pp. 75–83.

20. Stefik, M., et al. (1983). Basic Concepts for Building Expert Systems. in: *Building Expert Systems*, edited by F. Hayes-Roth et al., Addision-Wesley, Reading, MA, pp. 59–86.

21. Tong, R. M. (1984). A Retrospective View of Fuzzy Control Systems, *Fuzzy Sets and Systems*, Vol. 14, pp. 199–210.

22. Yasunobu, S., and Miyamoto, S. (1985). Automatic Train Operation System by Predictive Fuzzy Control. In: *Industrial Applications of Fuzzy Control*, edited by M. Sugeno, Elsevier Amsterdam, pp. 1–18.

22a. Zadeh, L. A. (1965). Fuzzy Sets, *Information and Control*, Vol. 8, pp. 338–353.

23. Zadeh, L. A. (1973). The Concept of Linguistic Variables and Its Application to

Approximate Reasoning, Memorandum No. ERL-M411, University of California, Berkeley.

24. Zadeh, L. A. (1977). A Theory of Approximate Reasoning, Memorandum No. UCB/ERL M77/58, University of California, Berkeley.

25. Zimmermann, H.-J. (1976). Description and Optimization of Fuzzy Systems, *International Journal of General Systems*, Vol. 2, pp. 209–215.

26. Zimmermann, H.-J. (1985). Applications of Fuzzy Set Theory to Mathematical Programming, *Information Sciences*, Vol. 36, pp. 28–58.

27. Zimmermann, H.-J. (1986). *Fuzzy Set Theory—and Its Applications*, Dordrecht, Lancaster.

28. Zimmermann, H.-J. (1987). *Fuzzy Sets, Decision Making and Expert Systems*, Kluwer Academic Publishers, Boston.

29. Zimmermann, H.-J. (1987). Fuzzy Sets in Expert Systems—Present Status and Future Developments. In: *Expert Systems in Production Engineering*, edited by G. Menges and N. Hövelmann, Springer, Berlin, pp. 47–59.

30. Zimmermann, H.-J., and Zysno, P. (1980). Latent Connectives in Human Decision Making, *Fuzzy Sets and Systems*, Vol. 4, pp. 37–51.

31. Zimmermann, H.-J., and Zysno, P. (1985). Quantifying Vagueness in Decision Models, *European Journal of Operational Research*, Vol. 22, pp. 148–158.

CHAPTER TWENTY EIGHT

Neural Networks in Acquisition of Manufacturing Knowledge

GERALD M. KNAPP and HSU-PIN (BEN) WANG

Department of Industrial Engineering, The University of Iowa, Iowa City, Iowa

1.1. INTRODUCTION

The usefulness of any knowledge-based system depends on the relevancy, consistency, organization, and completeness of the expert knowledge it contains. The problem of acquiring expert knowledge in a form usable by an expert system is known as knowledge acquisition. Knowledge acquisition has been identified as a major bottleneck to the implementation of knowledge-based system technology (4). The knowledge acquisition process is time consuming, costly, and error prone. Furthermore, the acquisition process ends with implementation of the knowledge-based system, and as a result, these systems cannot adapt to change. Another limitation is that most expert systems are unable to handle situations even slightly different from known prototype conditions. These severely limit expert system adaptivity, flexibility, and usefulness.

Much research has been conducted to automate knowledge acquisition using machine learning techniques (3, 8). In Knapp and Wang (6), machine learning techniques were applied to the problem of process planning. While the results were promising, the success of the approach was limited by a number of factors:

- The learning system time was highly dependent on problem size (number of features, attributes, and operations), which is common to many other machine learning methods.
- The system required special specification of a large number of heuristic "generalization rules." However, even this set did not produce all the

Intelligent Design and Manufacturing, Edited by Andrew Kusiak.
ISBN 0-471-53473-0 © 1992 John Wiley & Sons, Inc.

necessary generalizations needed for planning, the inadequacy of the explicit rule approach became particularly difficult to handle real-world problems.

- The learning system generated nonuseful inferences in addition to useful ones, because the learning process was not problem directed. The nonuseful inferences must then be stored, or sorted out and discarded. It is a computationally inefficient procedure.

A neuron has a number of branched dendrites and an axon, which are used to receive and pass information to other neurons. The neurons are connected with synapses to form a basic biocomputational system. Conceptually, the meaning of the connections is interpreted as the relations between the neurons. Figure 28.1 illustrates the basic architecture of neurons and connections. The number of connections among a network is so large that it provides the network with sophisticated capabilities such as logical derivation, objective perception in natural scenes, and so on.

The dendrites and axon are the channels for receiving and transmitting information. The synapses process limitations and stimuli. The reaction signal is produced and computed by the neurons. It is understood that the entire procedure of information processing and sharing is conducted in three steps:

1. Receiving information.
2. Processing information.
3. Responding to the information.

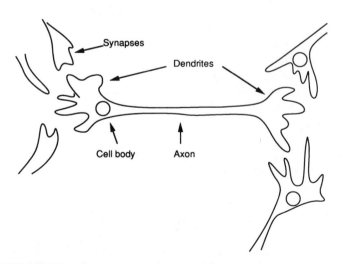

FIGURE 28.1. Basic architecture of neurons and connections.

The process keeps repeating until the network reaches a proper response toward the stimuli.

Neural networks have been widely used in pattern recognition and classification tasks such as vision and speech processing. Their advantages over traditional classification schemes and expert systems have been well documented (13):

- Parallel consideration of multiple constraints
- Capability for continued learning throughout the life of the system
- Graceful degradation of performance
- Ability to learn arbitrary mappings between input and output spaces.

Indeed, neural network computing shows great potential in performing complex data processing and data interpretation tasks. Neural networks are modeled after neurophysical structures of human brain cells and the connections among those cells. Such networks are characterized by exceptional classification and learning capabilities.

Neural networks differ from most other classes of AI tools in that the networks do not require clear-cut rules or knowledge to perform tasks. The magic of neural networks is the ability to make reasonable generalizations and perform reasonably on patterns that have never before been presented to the network. They learn problem-solving procedures by "characterizing" and "memorying" the special features associated with each training case and example, and "generalizing" the knowledge. Internally, this learning process is done by adjusting the weights tagged to the interconnections among those nodes of a network. The training (presentation of examples to the network) can be done in batches or individually in an incremental mode.

Neural networks are inspired by the biological systems in which large numbers of neurons, which individually function rather slowly, collectively perform tasks at amazing speeds that even the most advanced computer cannot match. These neurons are made of a number of simple processors, connected to one another by adjustable memory elements. Each connection is associated with a weight and the weight is adjusted by experience (training data).

Among the more interesting properties of neural networks is their ability to learn. Neural networks are not the only class of structures that learn. It is their learning ability coupled with the distributed processing inherent in neural network systems that distinguishes these systems from others.

Since neural networks learn, they are different from current AI expert systems in that these networks are more flexible and adaptive: they can be thought of as dynamic repositories of knowledge.

The structure of a network is loosely described in this section in order to provide a quick explanation of its problem-solving mechanism. A formal discussion on their structure is given later.

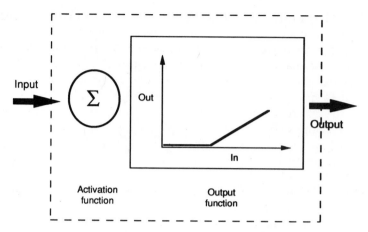

FIGURE 28.2. A neural network.

A neural network consists of a set of connected nodes and a set of signal processing rules as shown in Fig. 28.2. The nodes can be fully connected, sparsely connected, or feed-forward. Part of a fully connected network is shown in Fig. 28.3.

Learning rules define which network parameters (weights, thresholds, number of connections, etc.) change over time and in what way.

The node receives input from other nodes through weighted connections (links). The input node is activated by the total effect of the weighted signals (and some bias). The node's output is determined by processing the total sum of those weighted signals through a function, generally a sigmoid function. Output signals travel along other weighted links to connected nodes.

The simplest form of such neural networks is the two-layer associative network. As the name implies, there are only two layers of nodes in each associative network: input and output. The input patterns arriving at the input layer are mapped directly to a set of output patterns at the output

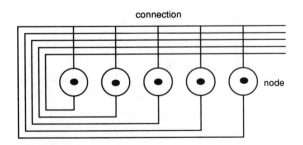

FIGURE 28.3. Part of a fully connected neural network.

layer. There are no hidden nodes; that is, there is no internal representation. These simple-structured networks have proved practically useful in a variety of applications. Obviously, the essential characteristics of such networks is that mapping is done from similar input patterns to simple output patterns.

Researchers have discovered that the procedures for neural networks to respond to the external stimulus are electrical reactions rather than chemical reactions. Over the years, researchers have tried to simulate a neural system by using physical devices, such as electrical circuits, resistors, and wires. The Hopfield neural network (HNN) is a prototype simulated neural system that possesses an extremely efficient computing ability (5).

A HNN has units such as neurons, synapses, dendrites, and an axon, which are made of electrical devices. The functions and implicit meanings of these units are exactly the same as they are in the bioneural system. The architecture of a HNN is shown in Fig. 28.4. The basic units are implemented as follows:

- Parallel input subsystem (dendrites, I_i, $i = 1, 2, 3, 4$)
- Parallel output subsystem (axon, V_i, $i = 1, 2, 3, 4$)
- Interconnectivity subsystem (synapses, the circuit in the schematic)

The neurons are constructed with electrical amplifiers in conjunction with the above-mentioned subsystems so as to simulate the basic computational features of the human neural system. The procedure for the HNN to process the external stimuli is similar to that of its biological counterpart. The functions of receiving, processing, and responding to the stimuli are translated into three major electrical functions performed by the HNN. These functions are:

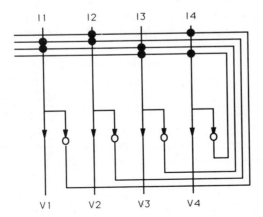

FIGURE 28.4. A schematic of a simplified four-neuron HNN.

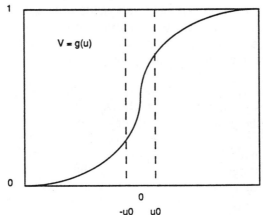

FIGURE 28.5. A typical neuron's sigmoid input–output relationship.

- Sigmoid function
- Computing function
- Updating function

The sigmoid function is a nonlinear increasing mathematical curve (see Fig. 28.5). This function describes how a neuron reacts to the external signal and generates its initial excitability value. The computing function, including all necessary information about its procedure for producing a reaction signal, is the major processor that processes the external information. The updating function is part of the computing function. The updating function prompts out the reaction signals toward the stimuli and passes these signals to the network to adjust the neurons' reactions. As such, the HNN is able to provide a proper response toward the external stimuli and settle into a stable state.

28.2. FORMAL DEFINITION OF NEURAL NETWORKS

Neural networks are a class of computing systems that use a highly parallel architecture to efficiently perform pattern recall, prediction, and classification tasks. These networks are loosely modeled after human networks of neurons in the brain and nervous systems. The networks consist of large numbers of simple processing units that communicate in parallel through weighted connections. The processing units are characterized by a state of activation, which is a function of the inputs to the units. Each unit calculates three functions:

1. Propagation function Net_i. Generally, the propagation function is calculated as the weighted sum of the outputs of all other units connected to unit i:

$$\text{Net}_i = \Sigma_j W_{ij} O_j , \tag{1}$$

where W_{ij} is the strength of the connection between units i and j, and O_j is the output level of unit j.

2. Activation function A_i. A function of Net_i and occasionally of time. Commonly used functions are the linear, logistic, and threshold functions.

3. Output function O_i. A function of A_i, O_i is often simply set equal to the activation, A_i.

The connections may be inhibitory or excitatory. Inhibitory connections tend to reduce the activation of units to which they are connected, while excitatory connections will tend to raise the activation.

Neural networks operate at two different levels of time: short-term response to environmental inputs (given by the state of activation of the units), and long-term changes in connection weights, which encode knowledge and change how the network reacts to its environment.

Many different neural network architectures have been developed. They differ in the types of propagation and activation functions used, how units are interconnected, and how learning is implemented. The type of paradigm used depends on the characteristics of the task to be performed. A major distinction among networks is whether the system will be used for recall (recognition), prediction, or classification. Recall systems are used for noise filtering and pattern completion (also called content addressable memories, or CAM). Examples are Anderson's Brain State in a Box (BSB) model (1) and the Hopfield network (5). Prediction systems can be used to estimate the behavior of complex systems. An example is the recent work of Rangwala and Dornfield (11) on prediction of machining parameters. Classification networks create (possibly arbitrary) mappings of input patterns into categories, represented by characteristic output patterns. Perceptron (12) is a classic example of this form of network.

28.2.1. Network Architecture

A three-layered perceptron architecture is used here for illustration purposes. The layers are organized into a feed-forward system, with each layer having full interconnection to the next layer, but no connections within a layer, nor feedback connections to the previous layer. The first layer is the input layer, whose units take on the activations equal to corresponding network input values. The second layer is referred to as a hidden layer, because its outputs are used internally and not considered as output of the network. The final layer is the output layer. The activations of output units are considered the response of the network.

The processing functions used in this illustration are as follows:

$$\text{Net}_i = \Sigma_j W_{ij} A_j + \phi , \tag{2}$$

$$A_i = O_i = 1/(1 - e^{\text{Net}_i}), \tag{3}$$

where ϕ is a unit bias (similar to a threshold). Short-term operation of the network is straightforward. The input layer unit activations are set equal to the corresponding elements of an input vector. These activations propagate to the hidden layer via weighted connections and are processed according to the functions above. The hidden layer outputs then propagate to the output layer and are again processed by the above functions. The activations of the output layer units form the network's response pattern.

We give a brief overview and analysis of the perceptron architecture to show how classification occurs within the network. Interested readers should refer to Rumelhart and McClelland (13) and Mirchandani and Cao (9) for further details. The analysis is given with respect to use of a threshold unit activation function, rather than the sigmoid function of Eq. (3). The basic operation of the network is similar for both functions.

A hidden or output unit utilizing a threshold function is either entirely deactivated or entirely activated, depending on the state of its inputs. Each unit is capable of deciding to which of two different classes its current inputs belong and may be perceived as forming a decision hyperplane through the n-dimensional input space. The orientation of this hyperplane depends on the value of the connection weights to the unit.

Each unit thus divides the input space into two regions. However, many more regions (and of much more complex shape) can be represented by considering the decisions of all hidden units simultaneously. Mirchandani and Cao (9) point out that the maximum number of regions M representable by H hidden nodes having n inputs each is given by

$$M = \sum_{k=0}^{n} \binom{H}{k}. \tag{4}$$

The actual number of representable regions will likely be lower, depending on the efficiency of the learning algorithm (note that if two units share the same hyperplane, or if three or more hyperplanes share a common intersection point, the number of representable classes will be reduced).

Each unit in a hidden layer may be viewed as classifying the input according to microfeatures. Additional hidden layers and the output layer then further classify the input according to higher level features composed of sets of microfeatures. For instance, a two-input unit perceptron with no hidden layers cannot learn the exclusive "OR" classification of its inputs, since this requires classifying two disjoint regions in the input space into the same class. The addition of a hidden layer results in the formation of a higher level feature that integrates the two disjoint regions into one, permitting correct classification.

The performance of a network utilizing a sigmoid unit activation function may be interpreted in a similar fashion. The sigmoid function, however, introduces "fuzziness" into the decision-making of a unit. The unit no longer divides the input space into two crisp regions, but rather assigns graded membership in one of the two regions. This fuzziness propagates through the network to the output. The goal of a classification system, however, is to make a decision regarding its inputs; it is therefore necessary to introduce a threshold decision at some point in the system. In our system, the output nodes use an output function different from the activation function:

$$O_i = 1 \quad \text{if } A_i > \varepsilon \text{ and } A_i = \max_j\{A_j\} \qquad (5)$$
$$= 0 \quad \text{otherwise}$$

where ε is a threshold value.

28.2.2. Back Propagation (BP)

The usefulness of the network comes from its ability to respond to input patterns in some orderly (desirable) fashion. For this to occur, it is necessary to train the network to respond correctly to a given input pattern.

Training, or knowledge acquisition, occurs by modifying the weights of the network to obtain the desired output. The most widely used learning mechanism for multilayered perceptrons is the back propagation (BP) algorithm (11).

The problem of finding the best set of weights to minimize error between the expected and actual response of the network can be considered a nonlinear optimization problem. The BP algorithm uses an iterative gradient descent heuristic approach to solving this problem.

The BP algorithm uses example input–output pairs to train the network. An input pattern is presented to the network, and the network unit activations are calculated on a forward pass through the network. The output unit activations represent the network's current response to the given input pattern. This output is then compared to a desired output for the given input pattern, and, assuming a logistic activation function, error terms are calculated for each output unit by the following operation:

$$\Delta_{oi} = \frac{\delta E}{\delta W_{ij}} = (T_i - A_{oi}) A_{hj} (1 - A_{hj}) , \qquad (6)$$

where E is the network error, T_i the desired activation of output unit i, A_{oi} the actual activation of output unit i, A_{hj} the actual activation of hidden unit j. The weights leading into the hidden nodes are then adjusted according to the following equation:

$$\Delta W_{ij} = -k\Delta_{oi} A_{hj} , \qquad (7)$$

where k is a small constant often referred to as the learning rate. The error terms are then propagated back to the hidden layer, where they are used to calculate the error terms for the units of the hidden layer as follows:

$$\Delta_{hi} = A_i(1 - A_i) \Sigma_k (\Delta_{ok} W_{jk}) . \qquad (8)$$

The weights from the input layer to the hidden layer are then adjusted as in Eq. (7).

Momentum terms may be added to Eq. (7) to increase the rate of convergence (11, 14). Such terms attempt to speed convergence by preventing the search from falling into shallow local minima during the search process. The strength of the error term will depend on the productivity of the current direction over a period of iterations and thus introduces a more "global" perspective to the search. In this study, a simple momentum strategy was used as follows:

- A momentum term was added to the weight update equation as follows:

$$\Delta W_{ij}(t + 1) = \eta \Sigma (\Delta_j O_i) + \alpha \Delta W_{ij}(t) , \qquad (9)$$

  where α is a momentum rate.
- If the total network error increases over a certain percentage, say 1%, from the previous iteration, the momentum rate α is immediately set to zero until the error is again reduced. This allows the search process to reorient itself if it gets off track. Once the error is again reduced, the momentum rate is reset to its original value.

During training, the total network error typically drops quickly during the initial iterations but easily becomes destabilized when using high learning rates. As the total network error converges toward zero, the rate of change in error gradually decreases, but the gradient descent search process can tolerate higher learning rates before destabilizing. In order to take advantage of this technique, a small acceleration factor ω was used to accelerate the learning rate from a small initial value (0.01) to some maximum value (0.75) over several thousand iterations. The acceleration occurs in a compound fashion, increasing the learning rate at each iteration according to the following:

$$\eta = \omega\eta , \qquad (10)$$

ensuring that large increases in the learning rate do not occur until the convergence process has stabilized. Use of the acceleration factor was highly effective, typically reducing convergence time by thousands of iterations. For this study, ω was set equal to 1.001. A more involved acceleration scheme, and description of its performance, can be found in Vogl et al. (14).

28.3. NEURAL NETWORK STRUCTURE AS A REPOSITORY OF PROCESS PLANNING KNOWLEDGE

Problem-solving tasks, such as process planning, may be considered pattern classification tasks. The process planner learns mappings between input patterns, consisting of the features and attributes of a part, and output patterns, consisting of sequences of machining operations to apply to these parts. Thus neural networks offer a promising solution for automating the learning of process planning knowledge.

28.3.1. Problem Representation

The process planning task may be represented by the transformation

$$F \otimes A \to C , \tag{11}$$

where F is a set of part features (symbols), A is a set of feature attributes, C is a set of feasible operation sequences, and $\to$ indicates a mapping function.

Process planners are interested in those features that are generated by some sequence of machining process. Typical features include holes (threaded or unthreaded), external cylinders (threaded or unthreaded), faces, slots, keyways, and gears. Each feature is associated with a set of attributes (or parameters) that define it from a manufacturing standpoint. These include dimensions, tolerances (both dimensional and geometric), and surface finish requirements.

Based on the particular values of a feature's attributes, the process planner can identify the sequence of operations necessary to produce the feature. Each sequence corresponds to a particular classification of the input pattern.

The transformation function $\to$ is developed by the process planner through experience. In the neural network, it is embedded in the network's connection weights through training.

Figure 28.6 demonstrates now inputs are physically presented to the network. Each known feature is associated with an input unit, which is highly active ($+1$) if the feature is present, and inactive (0) otherwise. Each known attribute is also associated with an input unit. The input unit takes on the value of the attribute, normalized to lie between 0 and 1.

The features composing the part being planned are presented to the network one at a time, along with their corresponding attributes. The network's response to the feature pattern represents selection of a machining operation to be applied to the feature. Every output unit corresponds to a particular machining operation. If the activation of the output unit is positive, it is interpreted as meaning that the selection of the corresponding machining operation is supported. A threshold mechanism selects the operation whose output unit has the highest positive activation above some

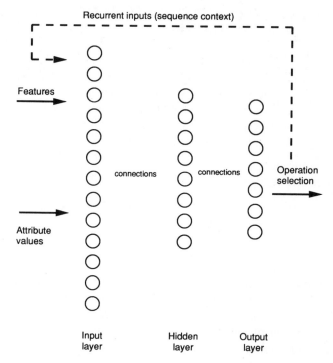

FIGURE 28.6. A neural network for process planning knowledge acquisition.

threshold [see Eq. (5)]. Such a mechanism can be implemented in a neural network via lateral inhibition in the output layer (7, 10).

Note that output units are assigned to single operations rather than sequences. To learn sequencing constraints, it is necessary to provide encoding of operation position within a sequence. Encoding may be accomplished directly by assigning one output unit to each possible sequence or, for each operation, to each possible position of that operation in a sequence. Neither of these encodings accurately reflects the approach of the expert process planner, however, and both require a large number of output units, many of which are infrequently used. An expert process planner builds a sequence of operations for each feature, with each operation being added depending on a specific, but not necessarily identical, set of attributes. In order to select operations singly, yet retain correct sequencing, the outputs of the network may be fed back as inputs (recurrence), thus serving as a context for a decision on the next operation in the sequence. Recurrence ends when a null output is obtained; that is, all output units have activations below the threshold value, signaling the sequence end. By using recurrence, output units are efficiently utilized, with little sacrifice in the final performance speed of the network.

28.3.2. Training

Training of neural networks may be distinguished by whether training patterns are presented incrementally or in batches. In incremental training, weights are updated after the presentation of each training input–output pair. An error measure, total pattern sum of squares error (PSS), is calculated for each iteration as

$$\text{PSS} = \Sigma \, o_{j=1}(T_j - A_j)^2 \,, \tag{12}$$

where o is the number of output units, and T_j and A_j are the training value and actual activation, respectively, for the output node. The training pattern is repeatedly presented until a tolerable error level is achieved, determined by

$$E \geqslant \text{TSS} \,, \tag{13}$$

where E is some error criterion.

In batch training, a set of patterns is presented to the network. The network error terms for each pattern are summed, and only at the end of presentation of all the patterns are the weights updated. Training stops only when the total sum of squares error over all patterns, TSS, falls below E, as given by

$$E \geqslant \text{TSS} = \Sigma p_{j=1} \, \text{PSS}_i \,, \tag{14}$$

where p is the number of patterns in the batch.

While the incremental approach more closely mimics the human learning experience, it tends to perform more poorly than the batch approach. The quality of system response becomes dependent on the order in which examples are presented, and also the extent to which error is minimized for each pattern. Minimization of error for a particular pattern may cause increased response error for another pattern (a form of forgetting). The batch approach minimizes response error over all example patterns, resulting in better overall performance. However, a complete set of examples covering all contingencies will rarely be available for a realistic application (consider the rapidly evolving product line of a manufacturing firm). Therefore some combination of batch and incremental modes might be useful.

28.3.3. Example

To demonstrate the neural network approach, a training set of example process plans was generated for spur gears of varying dimensions, tolerances, and surface finishes. Each gear consisted of five features: a hole, a keyway, two faces (identical in dimensions and tolerances), and the gear teeth.

TABLE 28.1. Example Features and Associated Attributes

Feature	Attributes
Hole	Depth, diameter, size tolerance, positional tolerance, circularity, straightness, surface finish
Face	Diameter, thickness (between opposing faces), size tolerance, parallelism, surface finish
Keyway	Width, depth, inset, size tolerance, positional tolerance, surface finish
Gear (involute)	Diametral pitch, error in action, surface finish

Each feature was associated with a set of attributes, defined in Table 28.1. For each example, the values for the attributes were randomly assigned within common manufacturing ranges for such features, subject to physical constraints (e.g., the inside hole could not have diameter greater

```
if (depth/diameter ratio)>=3 then
        if (diameter>2) then
                ''center drill''
                ''trepan''
        else
                ''gundrill''
        end if
else
        if (((diameter, 0.75) and
                ((size tolerance<=0.003) or
                (position tolerance<=0.005))))
        then
                ''center drill''
                ''twist drill''
        else
                ''twist drill''
        end if
if (straight<=0.001)
then
        ''ream''
end if
if (circularity<=0.001)
then
        ''counter-bore''
end if
if (surface finish<=16)
then
        ''hone''
end if
```

Figure 28.7. Some rules used to generate process plans for hole features.

⟨hole⟩
attributes:
  diameter=1.1011
  depth=8.2909
  size tolerance=+/-0.0105
  positional tolerance=+/-0.0117
  straightness tolerance=+/-0.0245
  circularity tolerance=+/-0.0033
  surface finish=32
process sequence:
  gundrill

⟨keyway⟩
attributes:
  width=0.2434
  depth=7.8141
  inset=0.8812
  size tolerance=+/-0.0188
  positional tolerance=+/-0.0053
  surface finish=16
process sequence:
  broach
  finish broach

⟨planes⟩
attributes:
  diameter=20.1368
  thickness=8.2909
  size tolerance=+/-0.0008
  parallelism tolerance=+/-0.0081
  surface finish=63
process sequence
  mill
  finish mill
  grind

⟨gear⟩
attributes:
  diametral pitch=5.6169
  error in action=+/-0.0026
  surface finish=8
process sequence:
  hob
  finish hob
  gear shaving

Figure 28.8. Examples of generated process plans.

than that of the faces). In order to accurately evaluate the performance of the network, the process plans were generated artificially using rules derived from manufacturing handbooks (1a, 2). These rules then constituted the domain transformation function → to be learned by the network. Figure 28.7 shows some rules used to generate the operation sequences for hole features. Note that these rules are nontrivial and, in fact, require evaluation of relationships between attributes (e.g., depth to diameter ratio). Figure 28.8 demonstrates some example process plans generated.

Table 28.2 defines the machining operations known to the network. A number of these operations, such as milling and honing, are used in the manufacture of more than one feature type.

A total of 19 attributes and 15 machining operations were defined for the four feature types (note that three of the attributes correspond to the same dimension—hole depth, face thickness, keyway depth—and can be considered a single attribute). The input layer consists of 38 units: 4 units corresponding to the four feature types, 19 units corresponding to the attributes, and 15 inputs corresponding to the recurrent feedback inputs. The output layer consists of 15 units, each corresponding to a particular machine operation.

Since the domain rules were known in advance, it was possible to determine an approximate lower bound on the number of hidden units needed using the guidelines set forth by Mirchandani and Cao (9). The number of regions required to be uniquely identified was found for each feature by taking the number of rule conditions (each condition corresponding to a hyperplane in the input space) times the maximum operation sequence length for that feature. The number of regions was then summed

TABLE 28.2. Known Machining Operations

Operation	Used for
Center drills	Hole
Trepan	Hole
Gundrill	Hole
Twist drill	Hole
Ream	Hole
Counterbore	Hole
Hone	Hole, keyway
Broach	Keyway
Finish broach	Keyway
Mill	Face, gear
Finish mill	Face, gear
Grind	Face, gear
Hob	Gear
Finish Hob	Gear
Gear shaving	Gear

over all four features, yielding a total of 120 regions. An approximate lower bound can then be found by determining an H satisfying the inequality:

$$M \leq \sum_{k=0}^{n} \binom{H}{k}$$

For M = 120 regions, the inequality was minimally satisfied for $H = 7$ nodes. The bound proved fairly accurate; 8 nodes were required during the simulations in order to achieve convergence during training.

The training set contained 30 examples. The network was trained in a number of different modes in order to characterize the performance differences between incremental and batch learning. The network was trained in four modes: a pure incremental mode, in which the network was trained on each pattern separately (requiring 30 training batches to learn all 30 examples); a combined approach, in which the network was trained in batches of 10 new examples (3 batches); a second combined approach in which the batch size was again 10, but 5 of the examples were randomly chosen from previously learned batches (6 batches), with the aim of reducing forgetting effects; and a pure batch mode, in which all 30 examples are presented simultaneously (1 batch).

In addition, the network was trained, in batch mode, with a 75-example training set in order to examine the effect of increasing training set size on performance.

For all simulations, the error criteria, E, was equal to 0.7, and the threshold, e, was also equal to 0.7. Training was performed on SUN SPARCstations, using a modified version of the BP network simulator developed by McClelland and Rumelhart (12a).

28.3.4. Results

After training, additional gear process plans were generated (randomly) for testing the network's trained performance. The trained network was tested separately with both the training and testing examples sets. For each training session, the following information was collected:

- Example set used (training or testing).
- Batch size.
- Overlap size.
- Number of iterations per batch (training only). For incremental training, the total number of iterations over all batches.
- Iterations per minute (training only). This information is included to give an idea of simulation speeds on a sequential computer and is not indicative of the speed of an actual neural network implementation.

- Percentage of essentially correct plans missing operations (COR/ MISS). The percentage of feasible plans that constituted a subset of the correct plan but were missing necessary finishing operations.
- Percentage of essentially correct plans with extra operations (COR/ EXTRA). The percentage of feasible plans that contain the correct plan as a subset but include unnecessary finishing operations.
- Percentage of incorrect but feasible plans (INC/FEAS). The percentage of plans generated by the network which are feasible for the particular feature but not applicable for the given feature instance (using a trepan operation instead of gundrill or twist drill for a small diameter hall).
- Percentage of infeasible plans (INFEAS). The percentage of plans generated that do not represent a possible plan, either due to inclusion of an incorrect operation (such as twist drill for a gear feature) or to improper operation ordering (finish mill before mill).
- Percentage of correct process plans (COR). The percentage of generated process plans containing no errors.

This information is summarized in Table 28.3. The percentages do not tally exactly to 100% because some plans had both extra and missing operations and thus were counted twice. The results indicate a consistent decrease in total convergence time required to learn a fixed number of examples with increasing batch size, suggesting that the batch approach is more efficient than the incremental approach from a training perspective. The number of training iterations per batch increases considerably, however, with increased batch size (note difference in convergence times between batch sizes 30 and 75). This can be attributed to the need for the system to consider more constraints simultaneously for larger batch size.

Because smaller batch sizes do not consider all constraints simultaneously, "forgetting" effects can occur: Learning new examples occurs at the expense of previously learned examples. That this has occurred can be evidenced by examining the performance of the network on the example set used to train it. Note that for the single batch case, performance was 100% correct for the training example set, whereas performance dropped to 49.2% for the strictly incremental approach (batch size 1). Increasing batch size yielded better performance, but the simulations required significantly more CPU time.

The forgetting effects can be negated to some extent by including previously learned examples in new examples batches as context. Note that with an overlap of 5 examples, and 5 new examples (a total batch size of 10), performance (total number of completely correct plans) improved over the case of batch size 10 with no overlap.

The majority of the planning errors obtained were not simply random errors but instead were inclusion and exclusion of finishing operations.

TABLE 28.3. Network Performance

Batch Size	Overlap Size	Example Set	Convergence (iter)	Speed (iter/min)	COR/MIS (%)	COR/EXT (%)	INC/FEA (%)	INFEAS (%)	COR (%)
1	0	Training	7,500	745	38.5	11.7	7.5	—	49.2
		Testing	—	—	47	9.1	1.4	0.8	41.6
10	0	Training	5,960	80	8.3	20.8	4.2	4.1	65
		Testing	—	—	19.1	25.8	5	5.8	46.7
5	5	Training	5,810	80	6.6	19.2	2.5	1.6	71.7
		Testing	—	—	15	21.7	3.3	5	61.7
30	0	Training	4,880	25	—	—	—	—	100
		Testing	—	—	7	18	2.5	1.6	73.5
75	0	Training	13,120	12	—	—	—	—	100
		Testing	—	—	—	10	1.6	0.8	83.3

Because examples are randomly drawn from the population (as is the case here), with each example representing a particular instance of the mapping function, the example set only provides an approximate model of the actual transformation function. Thus the quality of the internal model learned by the network is directly related to the sample size (number of examples). Note the increase in performance for the 75-example case over the 30-example case.

Another factor influencing error levels is the threshold, e. While the combined number of missing and extra operations did not significantly change for thresholds between 0 and 0.5, the threshold used [see Eq. (5)] had a significant effect on the ratio between the extra and missing operations. Raising the threshold caused an increase in missing operations and a decrease in extra operations, with a decrease in the threshold having the reverse effect.

Strictly infeasible plans were rare. In almost all cases, the infeasibility resulted from an inappropriate operation occurring at the end of a plan sequence, generally with much weaker activation than the correct operations within the plan. In all cases, the occurrence of the errors could be eliminated by raising the threshold. In addition, using larger batch sizes decreased the occurrence of these errors.

For a constant threshold, increasing the batch size led to an increase in both the use and misapplication of operations that were rare in the training example set, such as hone. This can be seen in the increase of COR/EXTRA for batch size 10 over batch size 1. However, with larger batch sizes, COR/EXTRA again began to decrease. For the small batch size, it appeared that the network was only able to learn those operations that were core to almost all process plans. As an example, the strictly incremental mode (batch size 1) never generated the operations gundrill, trepan, or hone, which occur infrequently in the training example set. Larger batch sizes generated these operations but, for batch size 10, frequently misapplied them.

An important issue in neural networks is training convergence rate, the average rate at which the network TSS approaches the error criterion E. The convergence rate is affected by a number of factors, including learning and momentum rates, and problem complexity (not necessarily a direct function of problem size, but rather a function of the resolution of classifications). In general, with all other factors held constant, increasing either the momentum or learning rates resulted in a decrease in convergence time. However, above a certain level, instability will occur, with the network eventually settling into a poor local error minima.

Another important issue for neural networks is size complexity (number of units and connections). For the given problem representation, the number of input units is directly equal to the number of features plus attributes and thus grows linearly with the size of the input space. Likewise, the number of output nodes grows linearly as the number of machining operations to be represented. The number of hidden nodes required is primarily a function of the problem complexity, in terms of the number of decision regions required. However, growth is slow, better than linear on the number of regions required. The slow growth in the number of hidden units protects against explosive growth in the number of connections required, bounded by $H \times \max(O, F \otimes A)$, where O is the number of machine operations, and H is the number of hidden units.

28.4. SUMMARY

A neural network approach for the automatic acquisition of process planning knowledge has been introduced. This approach overcomes the time complexity associated with earlier attempts using machine learning techniques. The example demonstrated here shows the potential of the approach for use on real-world problems. The neural network approach uses a single methodology for generating useful inferences, rather than using explicit generalization rules. Because the network only generates inferences as needed for a problem, there is no need to generate and store all possible inferences ahead of time. Future research will examine whether the network approach can also be applied to interfeature sequencing of operations and automatic recognition of machining features directly from CAD models. Another area to be examined includes robustness of the approach for human-generated examples, which, unlike the examples here, may contain noise in the form of cross-talk between example classifications. Cross-talk may occur as a result of human inconsistency or because the human planner is considering information unavailable to the network. In either case, automatic detection of such occurrences can improve the performance of the network.

REFERENCES

1. Anderson, J. (1977). Neural Networks with Cognitive Implications, in D. LaBerge & S. Samuels (Eds.), *Basic Processes in Reading Perception and Comprehension*, Lawrence Erlbaum Associates, Hillsdale, NJ, pp. 413–451.

1a. Chang, T. C., and Wysk, R. A. (1986). *An introduction to Automated Process Planning*, Prentice-Hall, Englewood Cliffs, NJ.

2. Dall, D. B. (Ed.) (1976). *SME Tool and Manufacturing Engineers Handbook*, McGraw-Hill, New York.

3. Dietterich, T. G., and Michalski, R. S. (1981). Inductive Learning of Structural Descriptions: Evaluation Criteria and Comparative Review of Selected Methods, *Artificial Intelligence*, Vol. 16, pp. 257–294.

4. Hayes-Roth, F., Waterman, D., and Lenat, D. B. (1983). *Building Expert Systems*, Addison-Wesley, Reading, MA.

5. Hopfield, J. J. (1982). Neural Networks and Physical Systems with Emergent Collective Computational Abilities, *Proceedings of the National Academy of Sciences*, Vol. 74, pp. 2554–2558.

6. Knapp, G. M., and Wang, H. P. (1989). Inductive Learning in Computer-Aided Process Planning, *Proceedings of the 1989 IIE Integrated Systems Conference & Society for Integrated Manufacturing Conference*, Atlanta, GA, November, pp. 398–403.

7. Kohonen, T. (1982). Self-Organized Formation of Tolologically Correct Feature Maps, *Biological Cybernetics*, Vol. 43, pp. 59–69.

8. Michalski, R. S., Carbonell, J. G., and Mitchell, T. M. (Eds.) (1986). *Machine Learning: An Artificial Intelligence Approach*, Volumes I and II, Tioga, Palo Alto, CA.

9. Mirchandani, G., and Cao, W. (1989). On Hidden Nodes for Neural Nets, *IEEE Transactions on Circuits and Systems*, Vol. 36, No. 5, pp. 661–664.

10. Pao, Y. H. (1989). *Adaptive Pattern Recognition and Neural Network*, Addison-Wesley, Reading, MA.

11. Rangwala, S. S., and Dornfield, D. A. (1989). Learning and Optimization of Machining Operations Using Computing Abilities of Neural Networks, *IEEE Transactions on Systems, Man, and Cybernetics*, Vol. 19, No. 2, pp. 299–314.

12. Rosenblatt, F. (1962). *Principles of Neurodynamics*, Spartan, New York.

12a. McClelland, J.L., and Rumelhart, D. E. (1986). *Parallel Distributed Processing*, Vol. 1, MIT Press, Cambridge, MA.

13. Rumelhart, D. E., and McClelland, J. L. (1988). *Explorations in Parallel Distributed Processing: A Handbook of Models, Programs, and Exercises*, MIT Press, Cambridge, MA.

14. Vogl, T. P., Mangis, J. K., Rigler, A. K., Zink, W. T., and Alkon, D. L. (1988). Accelerating the Convergence of the Back-Propagation Method, *Biological Cybernetics*, Vol. 59, pp. 257–263.